New Ideas in Low Dimensional Topology

K&E Series on Knots and Everything — Vol. 56

New Ideas in Low Dimensional Topology

Edited by

Louis H Kauffman

University of Illinois at Chicago, USA

V O Manturov

Bauman Moscow State Technical University, Russia

&

Chelyabinsk State University, Russia

World Scientific

NEW JERSEY • LONDON • SINGAPORE • BEIJING • SHANGHAI • HONG KONG • TAIPEI • CHENNAI

Published by

World Scientific Publishing Co. Pte. Ltd.
5 Toh Tuck Link, Singapore 596224
USA office: 27 Warren Street, Suite 401-402, Hackensack, NJ 07601
UK office: 57 Shelton Street, Covent Garden, London WC2H 9HE

Library of Congress Cataloging-in-Publication Data
New ideas in low dimensional topology / edited by L.H. Kauffman (University of Illinois at Chicago, USA), V.O. Manturov (Bauman Moscow State Technical University, Russia & Chelyabinsk State University, Russia).
pages cm. -- (Series on knots and everything ; vol. 56)
Includes bibliographical references.
ISBN 978-981-4630-61-0 (hardcover : alk. paper)
1. Low-dimensional topology. 2. Topological manifolds. I. Kauffman, Louis H., 1945–
II. Manturov, V. O. (Vasilii Olegovich)
QA612.14.N49 2015
514'.32--dc23

2014035528

British Library Cataloguing-in-Publication Data
A catalogue record for this book is available from the British Library.

Printed in Singapore by B & Jo Enterprise Pte Ltd

Introduction

This book consists in a selection of articles devoted to new ideas and developments in low dimensional topology. Low dimensions refer to dimensions three and four for the topology of manifolds and their submanifolds. Thus we have papers related to both manifolds and to knotted submanifolds of dimension one in three (classical knot theory) and two in four (surfaces in four dimensional spaces). Some of the work involves virtual knot theory where the knots are abstractions of classical knots but can be represented by knots embedded in surfaces. This leads both to new interactions with classical topology and to new interactions with essential combinatorics.

The first paper in this volume, by J. Scott Carter, is a pictorial introduction to knotted foams in four dimensional space, an analog of knotted trivalent graph embeddings in three dimensional space. The second paper, by J. Scott Carter and S. Kamada, is an introduction to the construction of manifolds in many dimensions via branched coverings. The third paper by R. Fenn is a description of some of the variations on knots that occur in virtual knot theory and its generalizations. The fourth paper, by S. Gukov and I. Saberi, is an introduction to the remarkable ideas in physics that are related to constructions of link homology. Link homology itself is a new subject in the study of invariants of knots and links.

In this approach, homology theories are associated with knots and links that categorify classical link invariants so that a graded Euler characteristic of the homology reproduces the classical invariant (e.g. the Alexander polynomial or the Jones polynomial). Such categorifications have their roots in certain physical ideas in the sense that they are related to Floer homology and its concept to use the Chern–Simons functional as a Morse function on the moduli space of connections on a three manifold. But the new relations to physics are subtle and involve delicate conjectures in string theory. The fifth paper is by A. Haydys and concerns the structure of Dirac operators in relation to the Seiberg–Witten equations that have been revolutionary in handling invariants of four manifolds. The sixth paper, by D. P. Ilyutko, V. O. Manturov and I. M. Nikonov is a study of graph links. Graph links are a generalization of knot theory that comes from studying knots and virtual knots in terms of their Gauss codes. It is a new and significant development in combinatorial topology. The seventh paper, by A. Juhász, is a concise and detailed survey of Heegaard–Floer homology. The eighth paper, by J. Juyumaya and S. Lambropoulou, is a description of their recent research on framed braids and Hecke algebras. The ninth paper, by L. H. Kauffman, is an introduction to new ideas in virtual knot cobordism, including a description of a generalization of the Lee homology and Rasmussen invariant to virtual knots and links due to H. Dye, A. Kaestner and L. H. Kauffman and based on work of V. O. Manturov. The tenth paper, by H. R. Morton, is a survey of classical and quantum methods for distinguishing mutant knots and links. Mutants have long been a test case for invariants, as many invariants are unable to distinguish them. The 11th paper, by J. Przytycki, is a study of homology theories generalizing cyclic homology that are related to algebraic structures in knot theory. The 12th paper, by D. Rolfsen, is a study of the ordering of knot groups, a consideration that has led to numerous good results in recent years. The 13th paper, by D. Ruberman and N. Saviliev, is a study obtaining Casson-type invariants from the Seiberg–Witten equations.

It should be clear to the reader that many if not all of the developments described in this volume are related to physics or motivated by physical considerations. We are looking forward to the further developments that will make these relationships between the pure mathematics of low dimensional topology and physical phenomena even more intimate.

Louis H. Kauffman
and
Vassily O. Manturov
September 2014

Contents

New Ideas in Low Dimensional Topology

Reidemeister/Roseman-Type Moves to Embedded Foams in 4-Dimensional Space

J. Scott Carter

Department of Mathematics,
University of South Alabama, Mobile, AL 36688, USA
carter@southalabama.edu

The dual to a tetrahedron consists of a single vertex at which four edges and six faces are incident. Along each edge, three faces converge. A 2-foam is a compact topological space such that each point has a neighborhood homeomorphic to a neighborhood of that complex. Knotted foams in 4-dimensional space are to knotted surfaces, as knotted trivalent graphs are to classical knots. The diagram of a knotted foam consists of a generic projection into 3-space with crossing information indicated via a broken surface. In this paper, a finite set of moves to foams are presented that are analogous to the Reidemeister-type moves for knotted graphs. These moves include the Roseman moves for knotted surfaces. Given a pair of diagrams of isotopic knotted foams, there is a finite sequence of moves taken from this set that, when applied to one diagram sequentially, produces the other diagram.

1. Introduction

Consider the spine of the tetrahedron that is obtained by embedding four copies of the topological space that is homeomorphic to the alpha-numeric character Y in each of the triangular faces of the tetrahedron and coning the result to the barycenter. This 2-dimensional space (Figs. 1 and 2), Y^2, has a single vertex, four edges, and six 2-dimensional faces. Three faces are incident to each edge, and

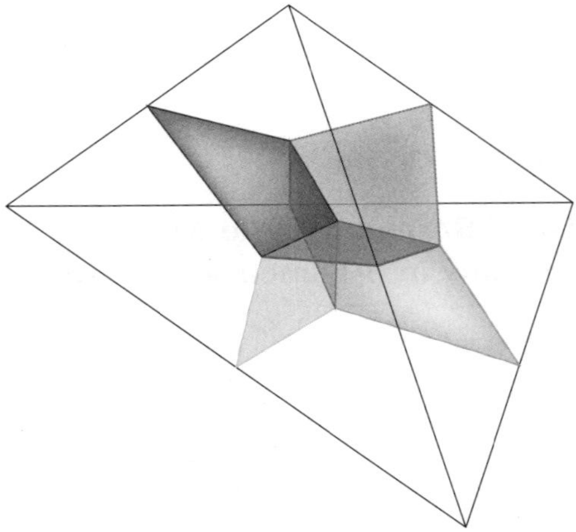

Fig. 1.

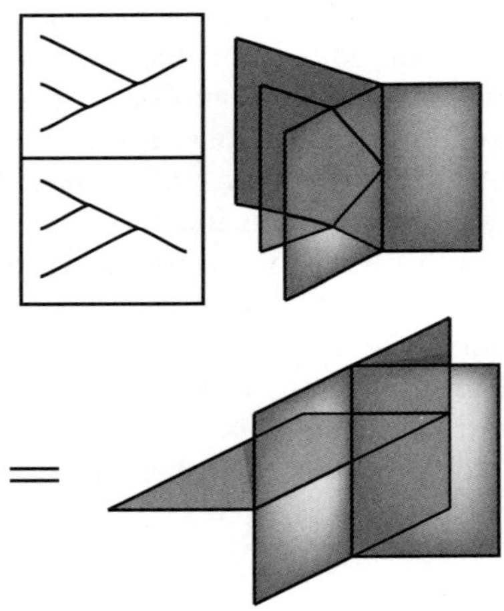

Fig. 2.

a neighborhood of a point in an open edge is homeomorphic to $\mathsf{Y} \times (-1, 1)$. A 2-*dimensional foam* (2-*foam*) is a compact topological space, F, such that any point has a neighborhood that is homeomorphic to a neighborhood of a point in Y^2. Thus a foam is stratified into isolated singular points, 1-dimensional edges at which three sheets meet, and 2-dimensional faces. The boundary of a foam is a trivalent graph. A *closed foam* has empty boundary. Analogous concepts exist in all dimensions. Just as a trivalent graph can be embedded and knotted in 3-space, a 2-foam can be embedded and knotted in 4-dimensional space.

The space Y^2 can be interpreted via a movie of the associativity rule when this is expressed in terms of binary trees. Foams are important since they include special spines of 3-dimensional manifolds [10, 11]. They are related to categorifications of the HOMFLYPT (FLYTHOMP) polynomial [7, 8, 16]. Knotted closed 2-foams and their higher dimensional generalizations can be used to represent 3-cycles in a homology theory of G-family of quandles and other more general algebraic structures.

The purpose of this paper is to present an analogue of the Reidemeister-type moves for knotted foams in 4-space. Just as knotted trivalent graphs (spacial graphs) contain classical knots and links as a subset, knotted foams include embedded surfaces in 4-dimensional space. Thus the moves that will be presented will include the Roseman moves [13]. Indeed, the proof that the given set of moves is sufficient to transform two diagrams of isotopic foams follows closely Roseman's original proof of the sufficiency of his set of seven moves.

To achieve the goals of presenting a set of moves to foams and demonstrating their sufficiency, the local pictures that are used to describe knottings of foams are given. These local crossings are obtained from the Reidemeister-type moves to graphs. The sufficiency of such moves are obtained by examining the generic critical points and transverse intersections of the self-intersection strata. Indeed, the description of the moves for foams are precisely an analysis of critical behavior and intersections between self-intersections or edges of the foam.

Theorem 1.1. *Let K_0 and K_1 be 2-foams without boundary embedded in 4-space with diagrams D_i for $i = 0, 1$. K_0 and K_1 are isotopic if and only if there is a sequence $D_{j/n}$ of diagrams for $j = 0, \ldots, n$ such that $D_{j/n}$ differs from $D_{(j-1)/n}$ by one of the Reidemeister/Roseman-type moves (see Fig. 3).*

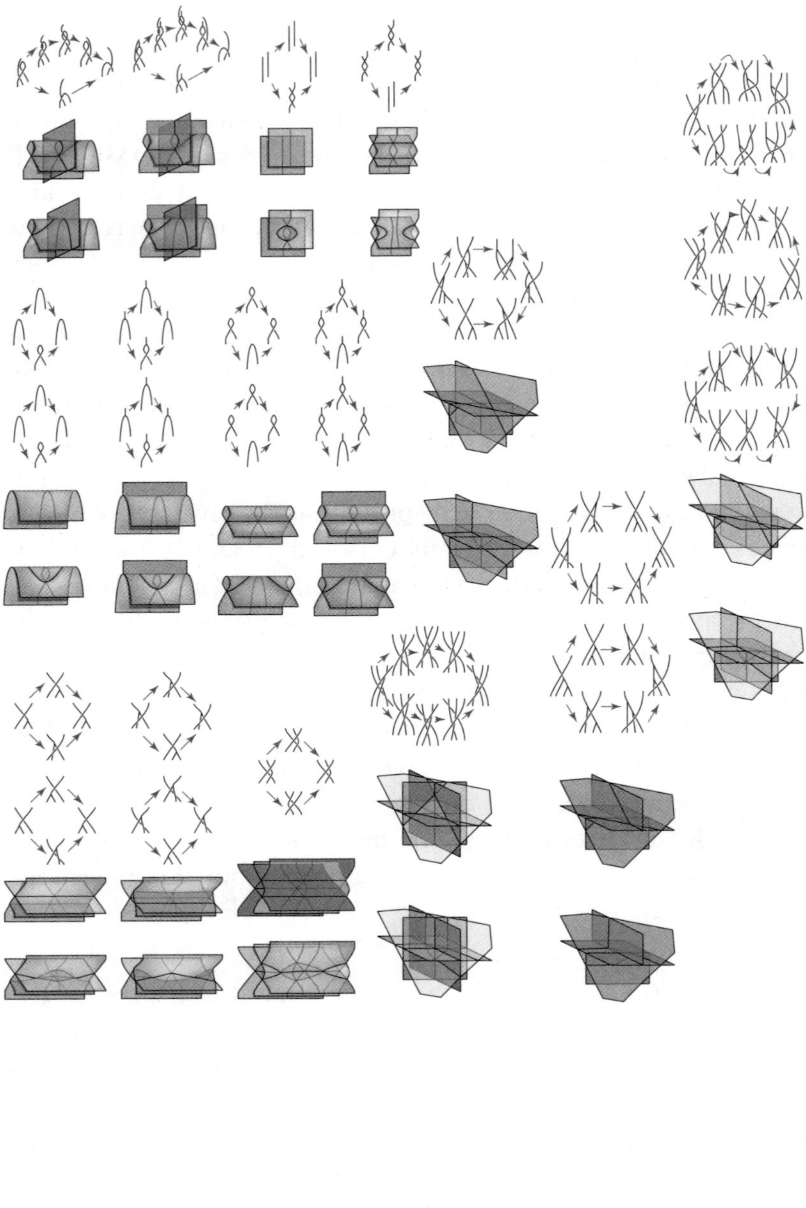

Fig. 3.

The movie parametrizations do not always agree with the projections as drawn, but they are topologically equivalent. In the rightmost column not all necessary crossings are indicated, but these are easy to guess given the remaining information. The proof of Theorem 1.1 will be presented in Sec. 6. First (Sec. 2), we discuss the Reidemeister moves for knotted trivalent graphs. Then (Sec. 3) we develop a short digression on the interactions among critical points, vertices and crossings. These critical interactions provide the atomic pieces used to construct foams. A second digression (Sec. 4) develops an idea first presented by Turaev [14] that indicates why only certain moves are needed as long as all possible type-II moves are present. Section 5 is a short section on critical points of foams and their edges. The idea of the proof of Theorem 1.1 is to analyze the possible critical points and intersections among the self-intersections and the singular strata. Section 7 examines two additional moves that change the topology of a foam, but for which a regular neighborhood is preserved. Section 8 points to unfinished endeavors.

2. Critical Points and Crossings — 1-Dimensional Case

Consider a trivalent graph that is embedded in 3-dimensional space. A generic projection onto the plane will have isolated transverse double points and points of no higher multiplicity. The points to which trivalent vertices project are not double points. If a height function is chosen in the plane, the graph may be assumed to have non-degenerate critical points that are either maxima or minima.

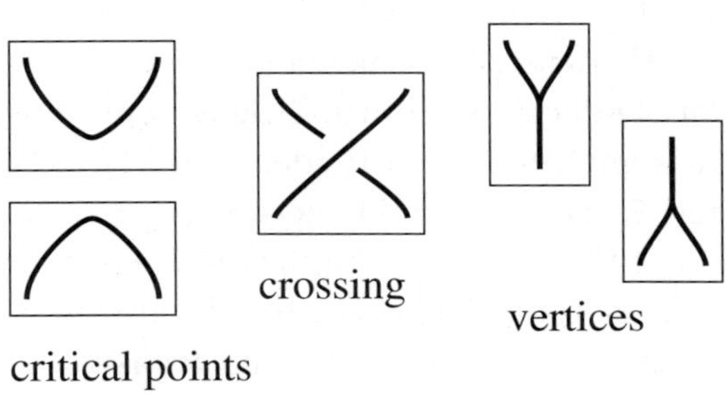

Fig. 4.

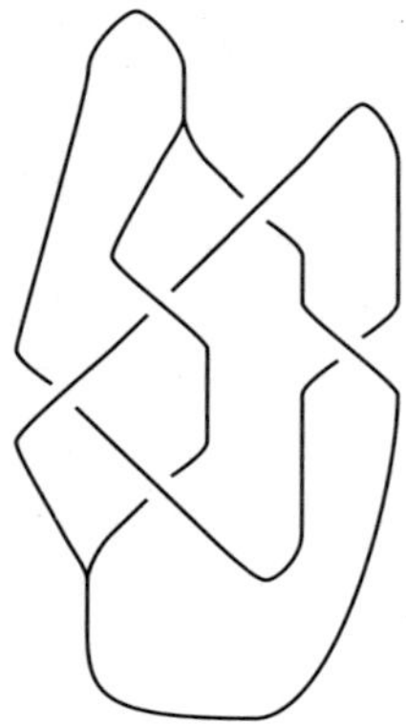

Fig. 5.

Furthermore, one can rearrange the graph in space so that the crossing points, vertices, and critical points all lie at distinct levels as indicated shown in Fig. 5.

The vertices of Y, the crossing points, and the critical points are 0-dimensional. To quantify the moves to trivalent graphs, we examine the transverse intersections and critical points of the corresponding sets in $\mathbb{R}^2 \times [0, 1]$ as an isotopy occurs. For example, a Reidemeister type-II move is a critical point of the 1-dimensional crossing set that is engendered as the projection of spacial graph moves in the plane. A Reidemeister type-III move is the transverse intersection between the trace of a crossing and the 2-dimensional sheet consisting of an arc of the diagram times the isotopy parameter. Similarly, there are two scenarios in which a vertex passes through a transverse arc: in one the vertex passes below the arc while in the other the vertex passes under the arc. A Reidemeister type-I move corresponds to a critical point on the double-decker set during the isotopy. The twisting of a trivalent vertex is analogous to a Reidemeister type-I move.

Figures 6 and 7 show the Reidemeister-type moves for trivalent spacial graphs and the corresponding critical events in the surfaces that represent the isotopies. The written discussion will now elaborate upon the nature of critical points and transverse intersections.

First consider a type-I move (indicated as an RI-move). A single crossing is involved. The crossing occurs along an arc of the knot

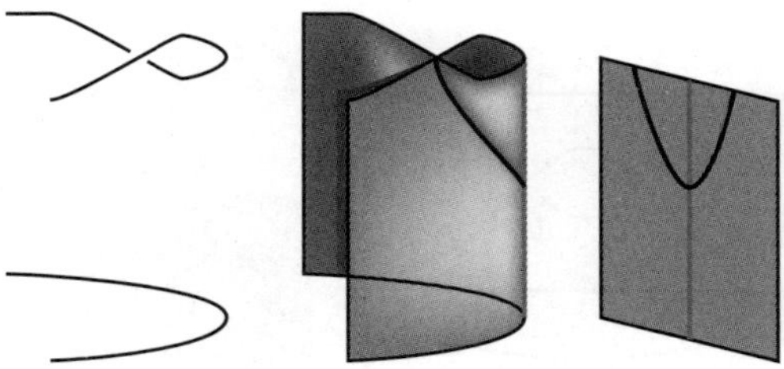

Fig. 6.

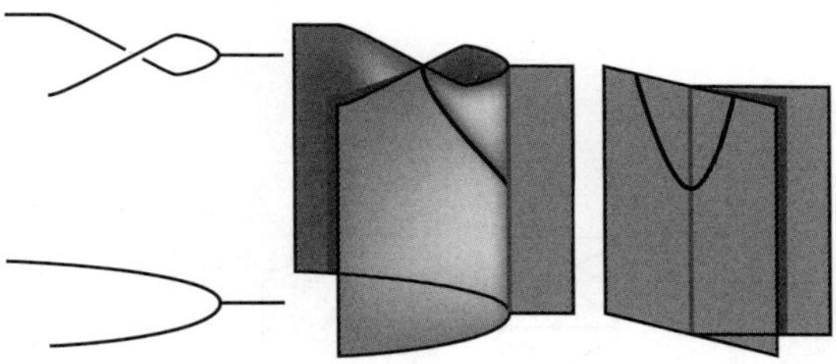

Fig. 7.

diagram. As the loop surrounded by the repeated crossing shrinks, the double-decker points on the arc converge to a simple critical point. This move, then, corresponds to a generic critical point of the double-decker set.

The *twisted vertex move* (Tw) (Fig. 7) also corresponds to a critical point of the 1-dimensional double decker set of the isotopy. We remark here that in the case of both the twisted vertex move and the type-I move, only one type of crossing is illustrated. Obviously, the move also holds with the opposite crossing.

The *type-II move* (RII) is shown in Fig. 8. It involves a pair of distinct double points that converge during the isotopy to a critical point on the double point set. Note that on the double-decker set there are a pair of critical points — one for each sheet involved in the crossing — and the critical levels coincide since they are equivariant with respect to the involution on the double-decker set.

The double points and the vertices of a trivalent graph are the 0-dimensional singularities. The critical points of the double point

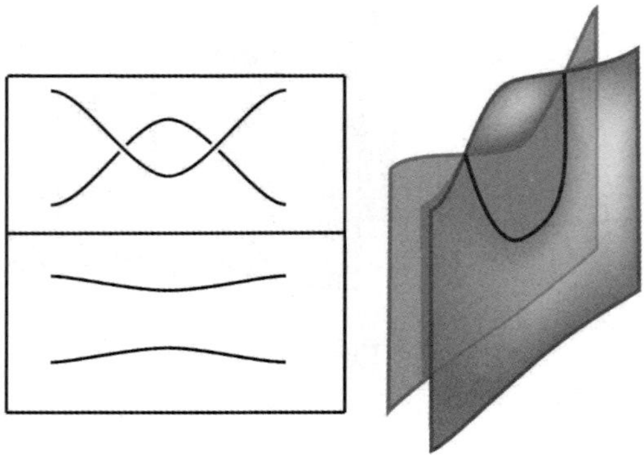

Fig. 8.

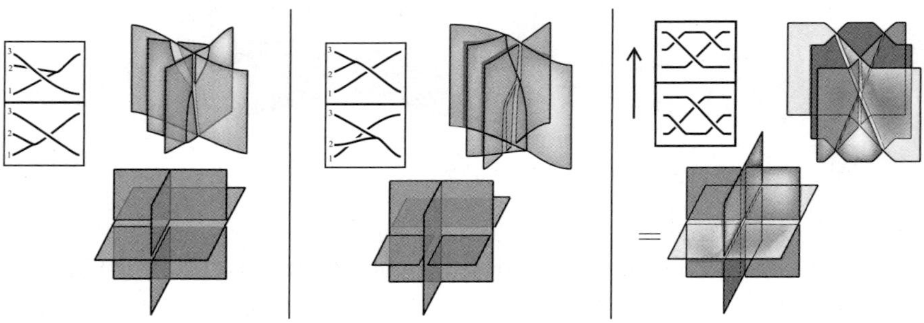

Fig. 9.

set, then correspond to the type-I and type-II Reidemeister moves, and the twisted vertex. Critical points for the set of vertices do not strictly correspond to moves for trivalent spacial graphs since they affect the topology of the graph. They will, however, represent aspects of the corresponding foam, and they will be discussed in Sec. 5.

The *type-III move* (RIII) and the moves in which a vertex passes over (IY) or under (YI) a transverse arc are all manifestations of a 1-dimensional set passing transversely through a 2-dimensional set in the 3-dimensional space of the isotopy direction times the plane of projection. The moves are depicted as broken surface diagrams in Fig. 9.

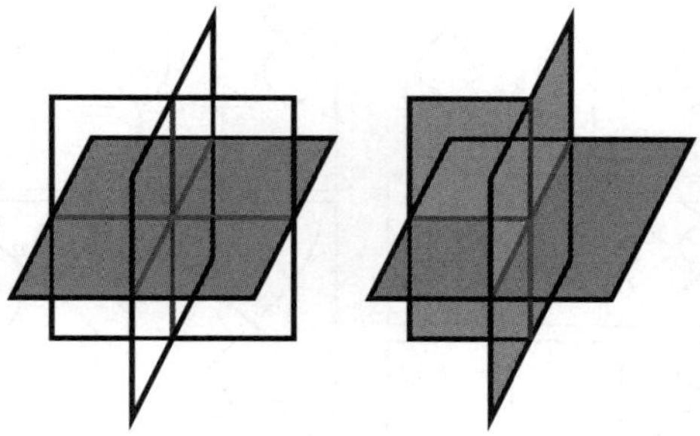

Fig. 10.

Figure 10 shows the projections of the type-III move and either move in which a vertex passes through a transverse arc are depicted. The 1-dimensional set is a vertical arc, and the 2-dimensional set is horizontal. In the case of the triple point, any one of the crossings between top/middle, top/bottom, or middle/bottom could serve as a vertical arc while either the bottom arc, middle arc, or the top arc (respectively) traces out the horizontal sheet. In the case of $\mathsf{Y} \times [0,1]$ intersecting the transverse sheet, the arc formed from the vertex intersects the transverse sheet which is either entirely above or entirely below the $\mathsf{Y} \times [0,1]$.

Next we complete the proof of the following classical result. See for example [1, 12].

Theorem 2.1. *For $i = 0, 1$, let $k_i : G \hookrightarrow \mathbb{R}^3$ be spacial embeddings of a trivalent graph G represented by diagrams D_i. Then $k_0(G)$ and $k_1(G)$ are isotopic embeddings if and only if there is a sequence of diagrams $D_{j/n}$ for $j = 0, \ldots, n$ such that $D_{j/n}$ differs from $D_{(j-1)/n}$ by a planar isotopy or an application of one of the moves indicated in Fig.* 11.

Proof. If the diagrams D_0 and D_1 differ by a finite sequence of these moves, then they are (clearly) isotopic.

Suppose that the two diagrams are isotopic. The 0-dimensional sets for the diagram D_0 are the crossing points and vertices. During an isotopy, these trace out 1-dimensional sets. Using the isotopy

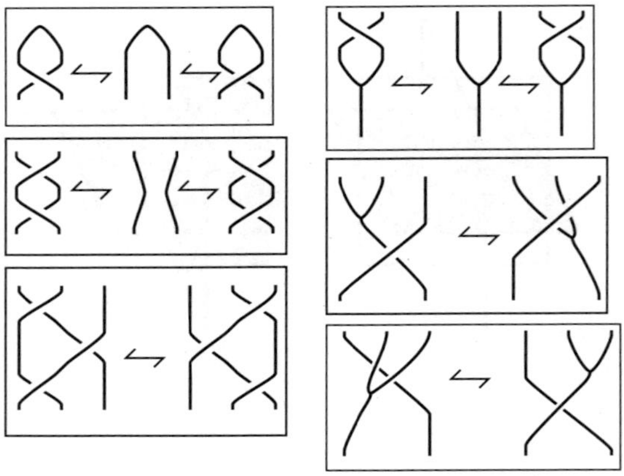

Fig. 11.

direction as a Morse function for this 1-dimensional set, we quantify the critical points. As mentioned above, the critical points of the vertex set alter the topology of the underlying graph. So we only consider the critical points of crossing set. These correspond to type-I and type-II Reidemeister moves. The transverse intersections between an arc of crossing points (or edge of $\mathsf{Y} \times [0,1]$) and the 2-dimensional sheets formed from arcs of the diagram times the isotopy parameter correspond to the type-III moves or the vertex of a Y crossing over or under a transverse sheet (the YI or IY moves).

Thus a given isotopy may be modified so that the critical points and transverse intersections occur at differing times. Moreover, by compactness, we may assume that there are only finitely many such critical points or intersections. This completes the proof. □

We remind the reader that the move considered in the second row, second column of Fig. 11 is called *the YI-move*. The move in the third row, second column is called *the IY-move*. The move in the first row of the second column is called *twisted vertex move*.

3. Critical Points, Crossings, and Vertices

In many circumstances, it is necessary to impose a height function upon the plane into which a knotted trivalent graph is projected.

For example, in developing categorical/algebraic interpretations of the graph, a height function is chosen, the trivalent vertex represents a multiplication or comultiplication operator, and the generic moves that reflect the changes in height are axioms in the algebraic setting. In particular, multiplication and comultiplication can be defined in terms of each other in the Frobenius algebra setting. This coincidence between algebraic structures occurs as a result of generic perturbations of the height functions. The moves indicated to the right together with the exchange of distant critical points are sufficient to readjust the height function of any embedded trivalent graph.

Each of these moves can be interpreted as a local picture of the projection of a knotted 2-foam in 4-space into the plane. In this case, a foam has generic folds that either end in cusps or end along the edge of a foam. Thus a fold ends in a cusp, a double point curve can pass over a fold, an edge passes over a fold, or a fold is created at the trivalent edge. In this paper, we will not study the interactions of such moves, nor will we attempt to incorporate them into a movie-move theorem such as that given in [3] or [6] which contains a heuristic analysis for immersed surfaces.

Instead, the interpretations are presented here since they facilitate the illustration of knotted foams. Specifically, we can define a diagram of a knotted foam as a projection with crossing information indicated along the double curves. When an author desires to illustrate a foam, he or she can make a detailed movie in which each still in the movie has a height function, and slow the action down until successive stills differ by simple moves. The moves are birth, death, zippers (which are all illustrated in Fig. 16), type-I, type-II, type-III, IY, YI, associators, and the commutation of distant critical points). Then the author can use the local projections to draw the pieces of a knotted foam. The drawing is, of course, in the plane of the paper, and keeping track of the folds helps indicate many of the details of the foam.

It is an interesting aspect of the geometry that the corresponding moves with minima in place of maxima follow from the moves given here. In particular the zig-zig moves (at the top of

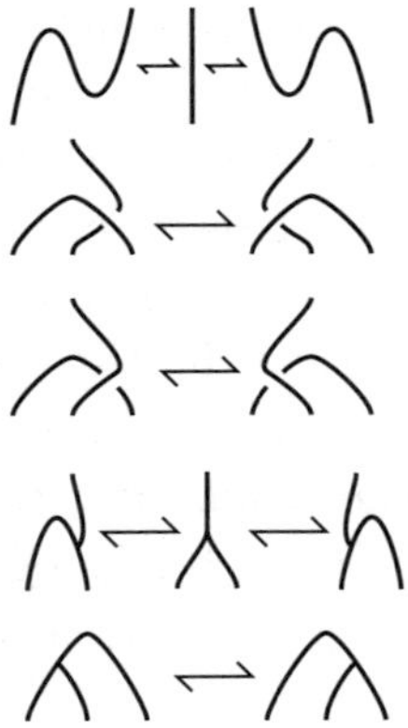

Fig. 12.

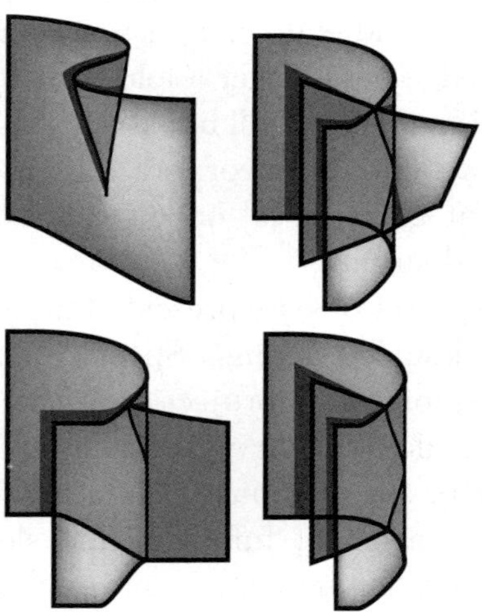

Fig. 13.

Fig. 12) and the commutation of distant critical points allow the "upside-down" versions to hold. While this seems to be well-known among experts, and indeed it is a standard exercise, a proof is provided in Fig. 14.

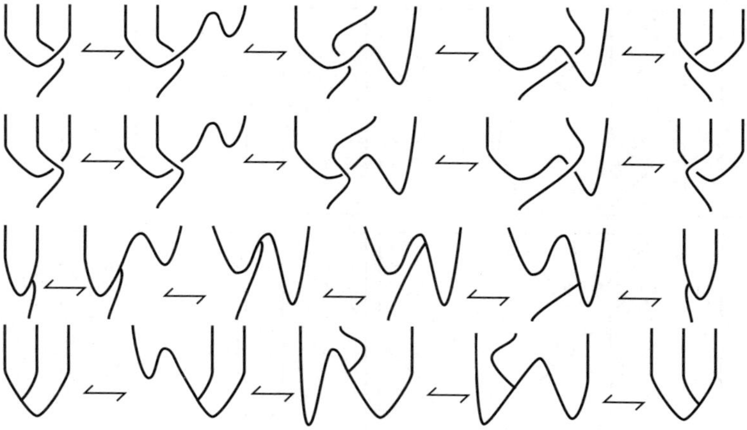

Fig. 14.

4. Turaev's Trick

In this section, we continue to work in the un-oriented category. For a type-III move there are six possible initial crossings for which the move is valid. Each of these follows from the type-III move that is given and the two possible type-II moves given. Similarly, the YI and IY moves are stated with specific crossing information given. Using Turaev's idea, we indicate that the moves with alternate crossing information given follow from the given moves and the type-II moves given. It is not difficult to generalize these proofs to the situations in which arcs are oriented. In the case of the Reidemeister type-III moves, the situation can be thought of very algebraically. Starting from the braid relation $aba = bab$, we prove: $\overline{a}ba = \overline{a}bab\overline{b} = \overline{a}aba\overline{b} = ba\overline{b}$. Similarly, $ab\overline{a} = \overline{b}bab\overline{a} = \overline{b}aba\overline{a} = \overline{b}ab$. The remaining three identities follow similarly.

5. Critical Points for Foams

Let $k : F \hookrightarrow \mathbb{R}^4$ denote an embedding of a closed foam F into 4-space. Consider a generic projection $p : \mathbb{R}^4 \to \mathbb{R}^3$. Then after

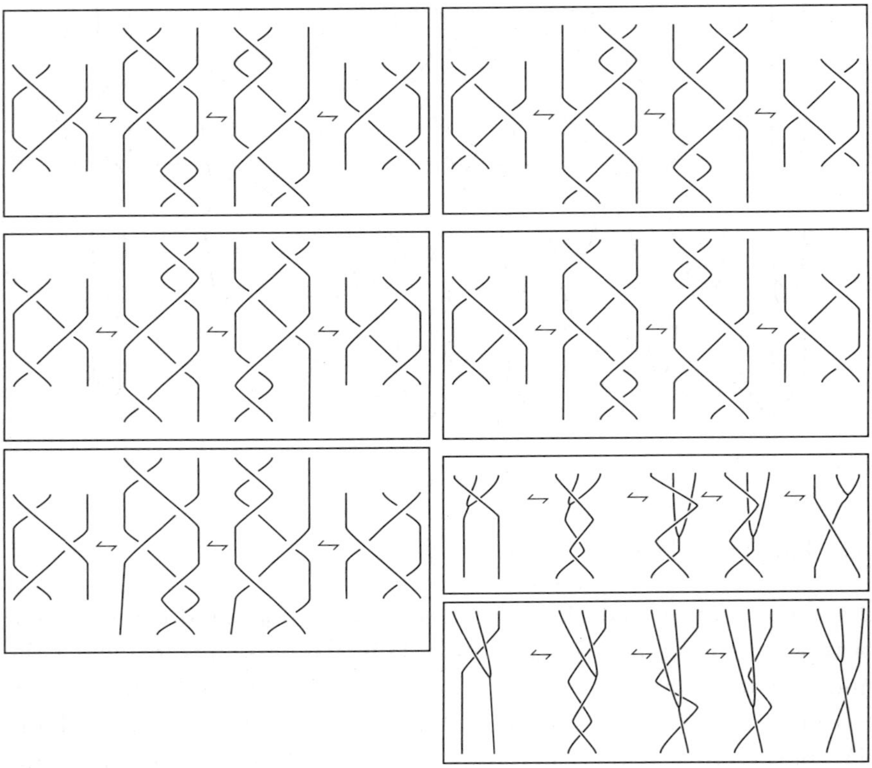

Fig. 15.

a small perturbation, if necessary, a height function on the image $p(k(F))$ can be chosen so that the critical points of the foam, its edge set, and the double point set are all at distinct levels. Moreover, the generic triple points, the intersections of transverse sheets with the edge set, the vertices are all at distinct levels, and these levels are distinct from the critical points of the previous sentence. The critical points are births or deaths of simple closed curves or saddle points as indicated in Fig. 16. The critical points for the edge set are also illustrated here; these are called *zipper moves*.

The remaining local pictures are the time-elapsed versions of the type-I, II and III Reidemeister moves, the YI and IY moves, the associator move (which is the neighborhood of a vertex in a foam),

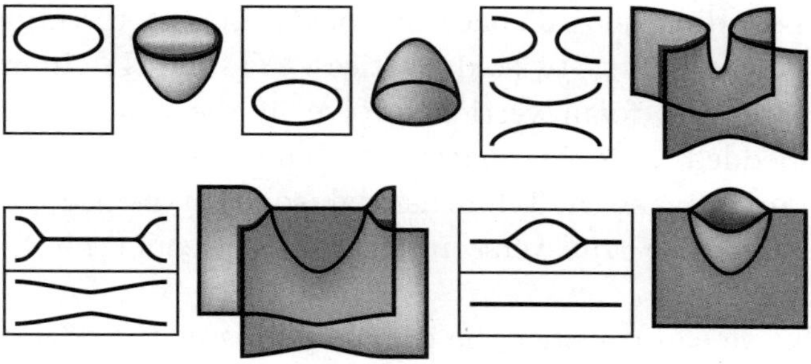

Fig. 16.

and a twisted vertex move (Tw). As mentioned above, in drawing (projecting to the plane), the edges and double curves can pass over the folds which correspond to the traces of the critical points in the stills.

The analysis of this section allows us to describe a knotted foam by means of a movie when necessary.

6. Critical and Intersection Behaviors for Isotopies of Foams

We turn now to proving the main result, Theorem 1.1. The sketch of the proof goes as follows. Branch points, twisted vertices, triple points, and the intersections between a transverse sheet and the edges of the foam are 0-dimensional and hence isolated. The non-degenerate critical points in the isotopy directions correspond to each of the moves RI, Tw, RIII, YI, and IY being invertible. For the RI, Tw, YI, and IY moves, there are two types of invertibilty: elliptic and hyperbolic, that depend on the structure nearby. The critical points of the double point set correspond to the Roseman bubble and saddle moves for the double point set. These correspond to the elliptic and hyperbolic confluence of the double point set or the two types of invertibility of the the RII-move.

The next set of moves occur when, in the isotopy direction, a transverse sheet intersects any one of the 0-dimensional sets. That is, an embedded sheet becomes an embedded 3-dimensional solid in the 4-dimensional spacetime of the isotopy direction. Meanwhile, branch points, twisted vertices, YI and IY intersections, triple points, and foam vertices evolve in the time direction and yield embedded arcs in spacetime. The transverse intersection between an arc and a 3-dimensional solid in 4-space is an isolated point. These transverse intersections account for five of the remaining moves.

Finally, we can consider the double points or the edges of foams evolving in spacetime to become 2-dimensional sets. The transverse intersection between two surfaces in 4-space consists of isolated points. These transverse intersections can also be used to reinterpret some of the previous moves; more importantly, they include a move below that is called the YY-move. The rest of the proof consists of recognizing that all of the possible critical points and transverse intersections have been identified.

Four moves to foams are illustrated in Fig. 17 indicate the critical points of the branch points and twist vertices. The movie versions indicate the two possible liftings into 4-dimensional space. The branch points and twisted vertices are 0-dimensional and therefore are isolated. In the isotopy direction, these points form 1-dimensional sets whose non-degenerate critical points correspond to the moves indicated. At the top of the illustration, an *elliptic confluence of branch points* is indicated. Analogously, and immediately below is an *elliptic confluence of twisted vertices.* The next two moves are the *hyperbolic confluence of branch points or of twisted vertices*, respectively. In the movie parametrizations, we consider the source and target graphs to appear as the top or bottom boundary of either foam. The path around the left indicates the foam on the left and that on the right indicates the foam on the right. It is easy to imagine a singular graph that encapsulates the singularity that occurs as the branch points or twisted vertices converge.

In a similar fashion, the triple points of the projection of a knotted foam are 0-dimensional and isolated. In the isotopy direction, a non-degenerate critical point corresponds to the annihilation/creation of a pair triple points. Only one of six possible lifts is indicated in the movie version here. However, all six possible liftings are needed in

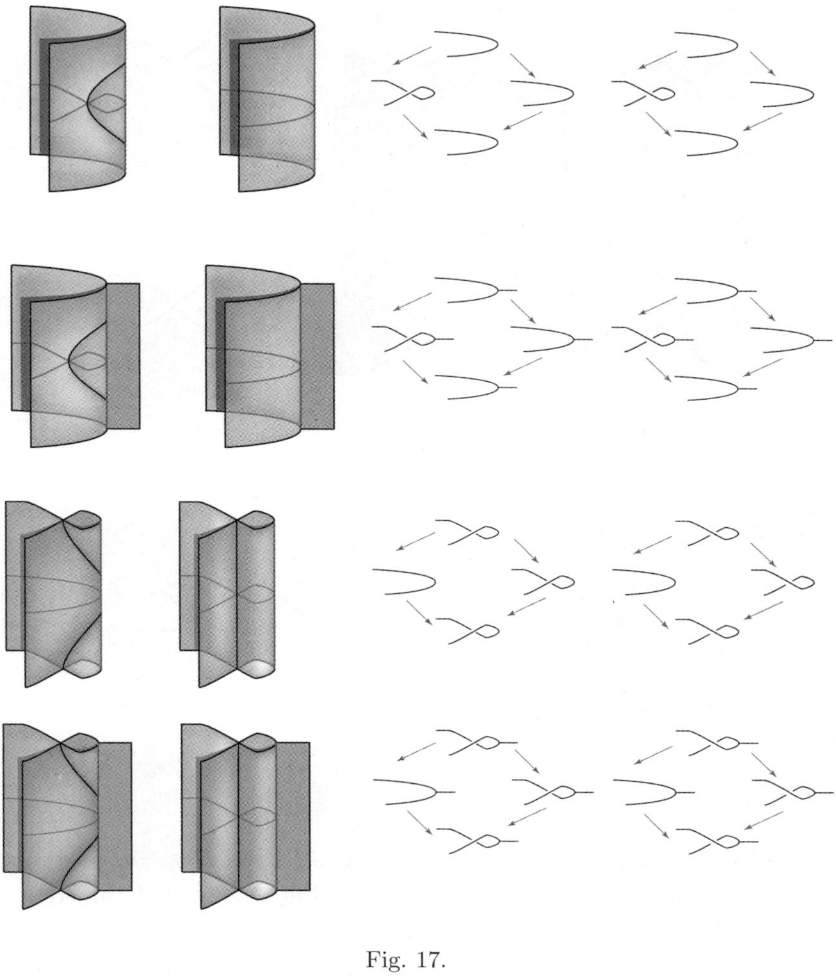

Fig. 17.

order to facilitate the higher dimensional analogue of Tureav's trick. The move illustrated here is called *type-III type-III-inverse move.*

The move depicted in Fig. 19 is a *YI-bubble move* or a *IY-bubble move.* The crossing points between an edge and a transverse sheet are 0-dimensional. A critical point in the isotopy direction creates or annihilates such a pair. There are two possible sets of crossings for such a move. These correspond to the YI or the IY moves as illustrated to the right of the projected surfaces.

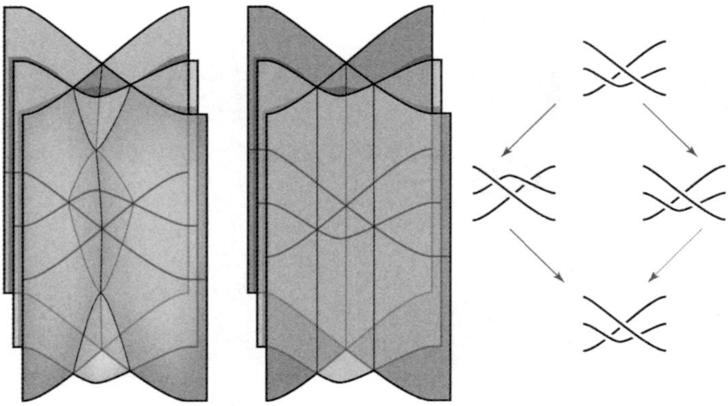

Fig. 18.

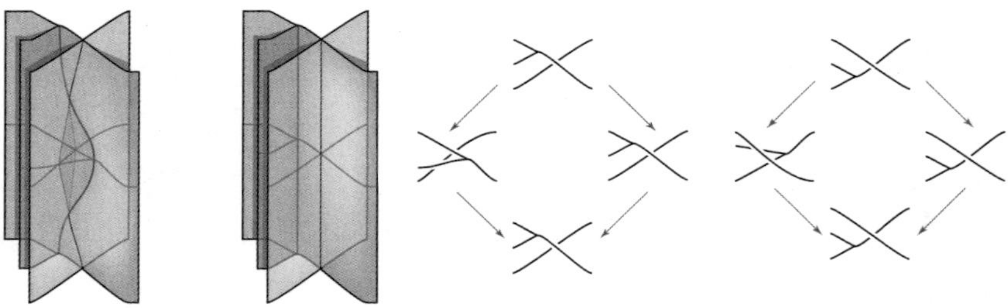

Fig. 19.

The move depicted in Fig. 20 is a *YI-saddle move* or an *IY-saddle move*. The analysis of the crossing points and the edge of the foam follows exactly as in the preceding paragraph. On the left-hand side of the move, there are a pair of 0-dimensional multiple points that are formed as the transverse intersection between an edge of the foam and an embedded sheet. A critical point in the isotopy direction annihilates or creates such a pair of crossing points. At the critical point, the edge of the foam is tangent to the embedded sheet. There are two lifts into 4-space of this move and these are represented by the movie moves to the right of the illustrations of the foams.

The moves illustrated in Fig. 21 are the *type-II bubble* or *type-II saddle moves*. These correspond to optimal or saddle critical points of the 1-dimensional double point set of the foam. Both are found

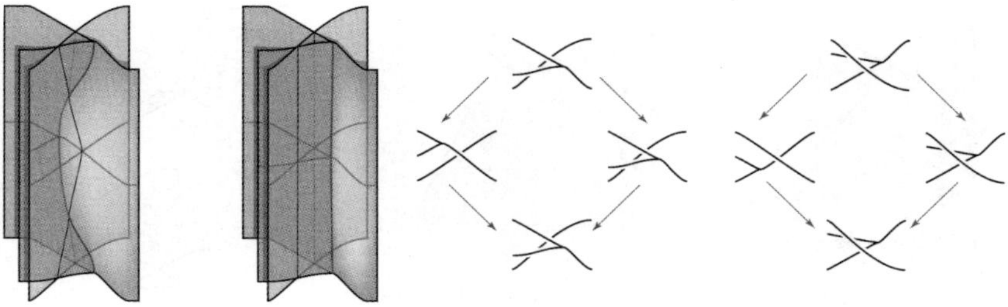

Fig. 20.

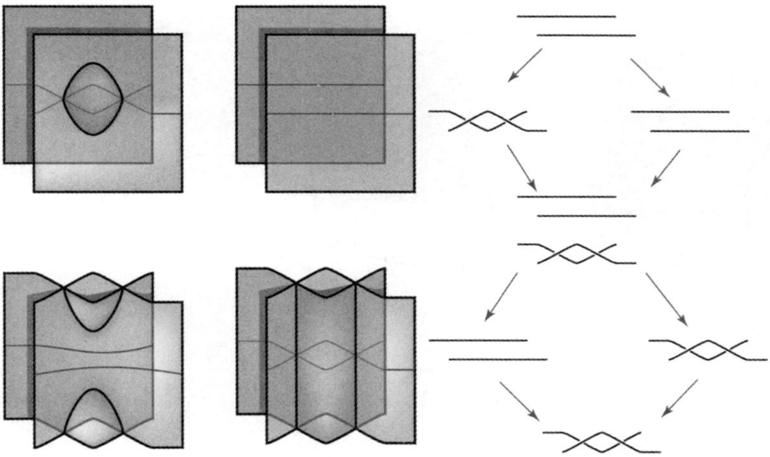

Fig. 21.

in Roseman's list of moves to knotted surfaces since neither involves an edge of a foam. In the isotopy direction the double point set is 2-dimensional, and these moves represent surface critical points. In the movie parametrizations only one of the two possible crossings are illustrated.

Figure 22 shows *pushing a branch point* (*or a twisted vertex*) *through a transverse sheet.* The transverse intersection between a 3-dimensional solid and a 1-dimensional arc in 4-space is an isolated point. The branch point or twist vertex evolves in spacetime to be an arc and the nearby embedded sheet evolves to be 3-dimensional. As a result, on the right-hand side of either move there is a triple

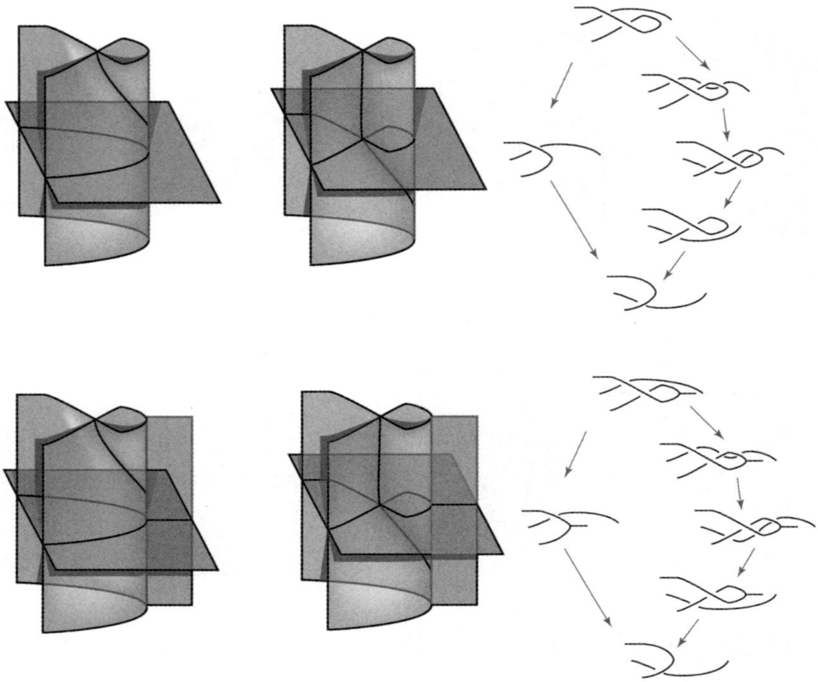

Fig. 22.

point. The movie parametrizations do not indicate both possible liftings. Furthermore neither coincides with a strict interpretation as horizontal cross-sections of the figures: it is more convenient to draw the transverse sheet horizontally. Nevertheless, there is a sequence of slices that gives the movie versions indicated.

The three movie moves that are indicated in Fig. 23 are called from left to right, *the IYI-move*, *the IIY-move*, and the *YII-move.* When these three moves are combined with all possible type-III type-III-inverse moves, then any foam of the form $\mathsf{Y} \times [0,1]$ can pass over, under, or through a pair of transverse sheets. To make that statement more precise, let us consider the YII-move that is depicted upon the right of Fig. 23. The $\mathsf{Y} \times [0,1]$ is completely below the two transverse sheets. Another possibility is that it lies above the two transverse sheets. Or it could be above the second sheet and below the

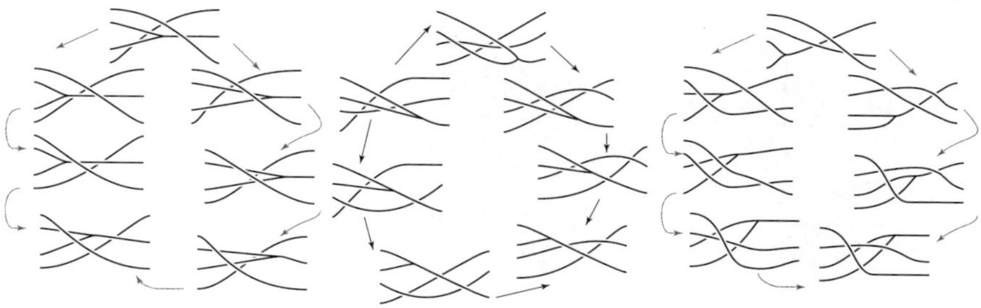

Fig. 23.

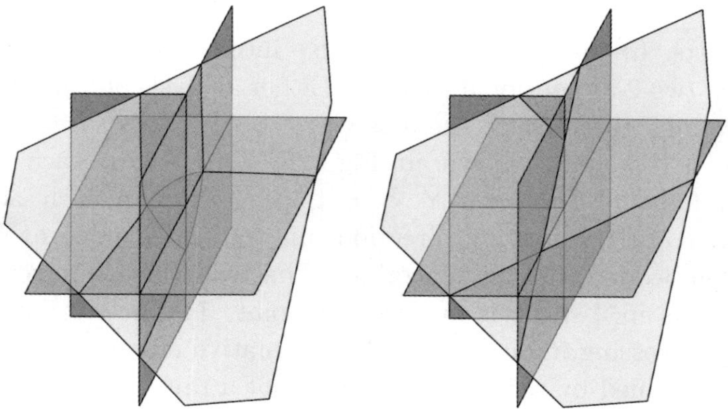

Fig. 24.

third sheet. In any of these cases the movie move can be performed. Similar situations hold for the IYI and the IIY-moves. However, in the presence of a sufficient class of type-III type-III inverse moves, these alternative crossings follow from those that are indicated here.

Each of the YII, IYI, and IIY moves projects in spacetime to the same configuration. Figure 24 shows a foam of the form $\mathsf{Y} \times [0,1]$ intersecting a pair of intersecting embedded disks. We can consider the intersection of an edge of $\mathsf{Y} \times [0,1]$ with one of these sheets as an isolated vertex that can pass through the remaining sheet. (In spacetime, this is a 1-dimensional sheet — crossing $\times$ interval — intersecting a 3-dimensional sheet). Alternatively, we may consider the edge of $\mathsf{Y} \times [0,1]$ moving through the double line of the remaining

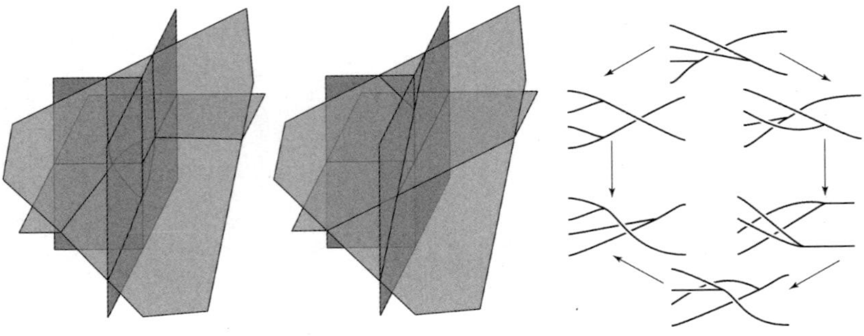

Fig. 25.

two sheets. In this case the transverse intersection in spacetime is between two 2-dimensional surfaces. Either analysis of the transverse intersections gives rise to the move.

The YY-move is shown in Fig. 25. Two edges, each coming from a foam of the form $\mathsf{Y} \times [0,1]$, pass through each other in spacetime. In the isotopy direction, an embedded edge evolves to be an embedded disk in spacetime. The transverse intersection of two 2-disks in 4-space is an isolated point. The movie illustration indicates crossing information. The alternative crossing information can be obtained by reflection. There is not a precise match between the movie parametrization and the surfaces as projected here, but it is not difficult to rearrange the surfaces by isotopy in 3-space to make the figures match.

The YYI-move and the *IYY-move* are shown in Figs. 26 and 27. A vertex at the juncture of the foam Y^2 passes through a transverse sheet. In spacetime the vertex evolves into a 1-dimensional set, and as before, the transverse embedded sheet becomes 3-dimensional. The two movie moves that are illustrated include crossing information. Either the transverse sheet is completely above or completely below the foam Y^2.

The tetrahedral move, or quadruple point move is indicated in Fig. 27 with one possible collection of crossing information. The moves that involve different crossing information follow from a higher dimensional analogue of Turaev's trick and the various type-III type-III-inverse moves (see [5] for an implementation). One of the triple

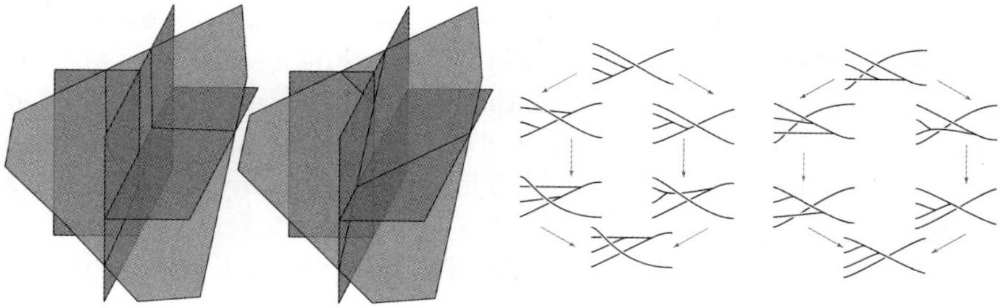

Fig. 26.

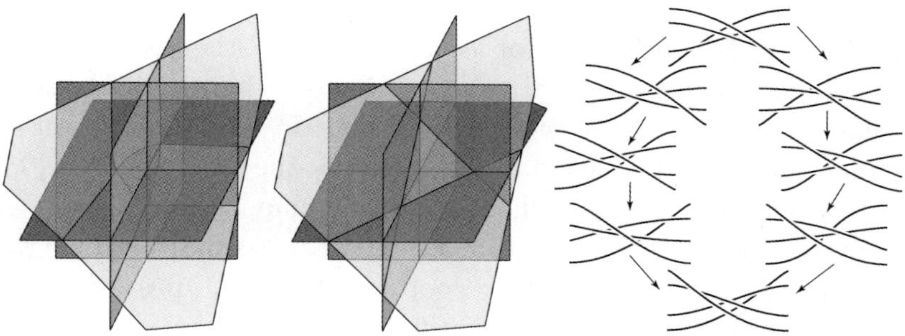

Fig. 27.

points (chosen among four possible) passes through a transverse sheet. The triple point evolves in spacetime to form a 1-dimensional arc. The transverse sheet evolves as a 3-dimensional solid. The transverse intersection in spacetime is an isolated point. This point is also the intersection of two double point arcs. There are three ways to parse the double point set into a pair of disjoint arcs. These are counted as pairs of two-element subsets of $\{1, 2, 3, 4\}$; specifically, $\{\{1 \cap 2\}, \{3 \cap 4\}\}$, $\{\{1 \cap 3\}, \{2 \cap 4\}\}$, or $\{\{1 \cap 4\}, \{2 \cap 3\}\}$ where the numbers indicate labels on each of the surfaces that intersect on either side of the move.

In the 11 paragraphs that precede this paragraph, each of the potential Roseman-type moves has been described as a critical point of a multiple point stratum or the transverse intersection between intersection strata or embedded sheets. We observe that these descriptions are exhaustive. The 0-dimensional sets are the branch points, twist vertices, triple points, or intersections between edges of the foam and a transverse sheet. The 1-dimensional sets are the double point arcs. Clearly, the 2-dimensional sets are the embedded sheets in the foam. Every possible critical point or transverse intersection has been accounted. Thus these are the codimension 1 singularities.

Now let us complete the proof of Theorem 1.1. First, if two diagrams differ by any one of the moves depicted, then they are isotopic. Next suppose that two diagrams represent isotopic embeddings. We have a map $\mathcal{K}: F \times [0,1] \to \mathbb{R}^4$ such that the restrictions $\mathcal{K}|_{F\times\{i\}} = K_i$ to the ends represent the given knottings of the foam F. Moreover, for each $t \in [0,1]$, the foam $F \times \{t\}$ is embedded. The isotopy $\mathcal{K}$ can be adjusted slightly, if necessary, so that the composition $p_2 \circ \mathcal{K}$ has generic singularities on the singular sets of 3-dimensional foam that is the product $F \times [0,1]$. By compactness, there will be finitely many singularities. The isotopy can be perturbed further, if necessary, so that each singular point lies at a different time coordinate. The types of generic singularities that need to be quantified are critical points for the 1- and 2-dimensional strata of the isotopy and transverse intersections between strata.

The catalogue of 0-dimensional singular points on a given knotted foam are as follows:

(0.1) branch points or twist vertices that result from an RI or Tw move to a knotted trivalent graph,
(0.2) triple points or the intersection between an edge of $\mathsf{Y} \times [0,1]$ and an embedded sheet, or
(0.3) vertices of the foam.

The 1-dimensional singular points of a knotted foam are as follows:

(1.1) the double point arcs that are caused by the transverse intersection between a pair of sheets of the foam, or
(1.2) the edge set of a foam.

Any face of the foam F is 2-dimensional and therefore non-singular.

Under the assumption that the singularities of the isotopy are generic and isolated, we examine each singular situation.

First we consider the critical points of 1-dimensional sets. The singular points described in items (0.1), (0.2), and (0.3) evolve in spacetime to 1-dimensional sets. During the isotopy, the topology of the foam remains constant. Therefore, there are no critical points for the vertex set listed as item (0.3). (However, in the following section, we discuss such topological changes.) The critical points for branch points and twist vertices (0.2) are the elliptic and hyperbolic confluences of each. These are accounted for in the first illustration of this section. The critical points for the RIII, YI, or IY moves correspond to the type-III type-III inverse moves, the YI or IY bubble and saddle moves.

In examining the critical points of the 2-dimensional sets in the isotopy, we first observe (in a manner similar to that above) that the edge set of the foam F remains unchanged during the isotopy. Thus we only have to consider the critical points of the double point set. These correspond to the type-II saddle and bubble moves.

The remaining singularities to be considered are those that are caused by the transverse intersections between two strata. In 4-space, we have intersections between a 3-dimensional stratum and a 1-dimensional stratum (3-1), or between a pair of 2-dimensional strata (2-2). To track the (3-1) intersections, we consider arcs formed in spacetime by the vertices listed in items (0.1) through (0.3) and examine their transverse intersections with an embedded sheet. The resulting intersections are as follows:

- (3.1.B) a branch point passing through a transverse sheet,
- (3.1.Tw) a twisted vertex passing through a transverse sheet,
- (3.1.V) a vertex passing through a transverse sheet as in the YYI move,
- (3.1.YI) the intersection between an edge and a transverse sheet passing through a third sheet (moves YII, IYI and IIY),
- (3.1.III) the intersection between a triple point and a fourth transverse sheet (quadruple point move),
- (2.2.YY) the intersection between two edges of the foam (move YY),

(2.2.YD) the intersection between an edge of the foam and a double curve (moves YII, IYI and IIY), and

(2.2.DD) the intersection between two double point arcs (the quadruple point move).

We observe that there are coincidences between (3.1.III) and (2.2.DD), as well as coincidences between (3.1.YI) and (2.2.YD).

Finally, we point out that the assumptions on general position are warranted. First, the isotopy can be fixed within a neighborhood of its 1-dimensional strata (the trace of the vertices, branch points, twist points, triple points, and edge intersections with transverse sheets) so that the critical points of these sets are non-degenerate. Next, without affecting the 1-dimensional strata, the double point set can be adjusted to have non-degenerate critical points. Any adjustment to the double point set can be extended to a neighborhood of the double point set within the foam. Intersections between a pair of double point arcs, a pair of edges or a double point arc can also be made to be generic and transverse as can intersections between vertices and sheets. There perturbations are then extended over the remainder of the $F \times [0, 1]$. This completes the proof.

7. Moves that Change the Topology of the Underlying Foam

It is important to remark that not only are embedded trivalent graphs studied for their own sake, but a given knotted handlebody in 3-space deformation retracts to an embedded trivalent graph. The graph, however, is not unique. Two graphs that "carry" such a knotted handlebody differ by the so-called IH-move. Up to equivalence, the IH-move is given via the movie parametrization of the basic foam Y^2. The theory of knotted handlebodies embedded in 3-space is equivalent to the theory of knotted trivalent graphs modulo the IH-move.

A similar situation holds in 4-space. We can include among the Roseman moves two additional moves that are indicated below. The first of these is the invertibility of the IH-move. In the theory of

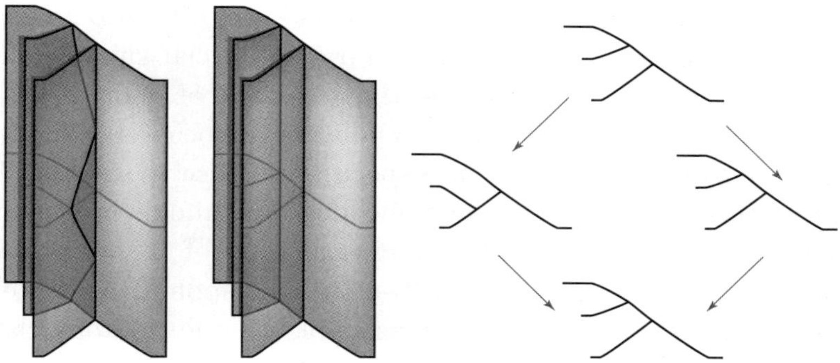

Fig. 28.

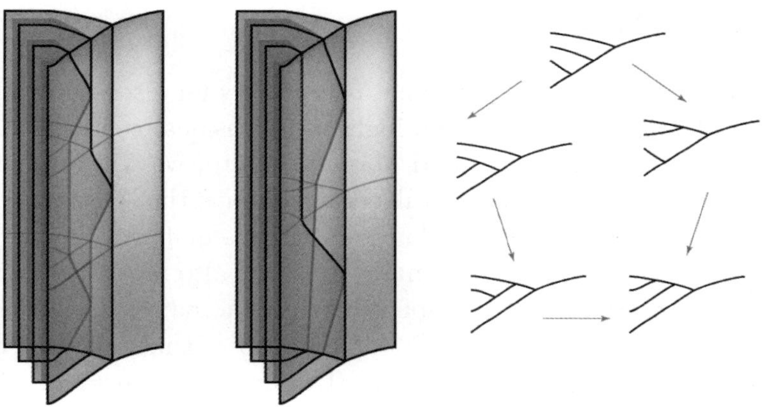

Fig. 29.

special spines for 3-manifolds it is sometimes called the *lune move* or the *orthogonality condition*. The reason for the latter name comes from the Tureav–Viro [15] invariants, the neighborhood of a vertex of a foam is colored by representations of $U_q(sl_2)$, and the move corresponds to the orthogonality condition for the $6j$-symbol. See also [4]. Observe that if a foam is embedded in 3-space, then regular neighborhood of the foam is invariant under this move. So similarly, a regular neighborhood in 4-space of such a foam is also invariant since it can be obtained from the neighborhood in 3-space by the cartesian product with an open interval.

The remaining move is the (3,2)-*move* or *Elliott–Beidenharn* move. Again when considered as a move to special spines, following [10, 11], the move preserves the topology of the underlying 3-manifold. It is easy to see that a regular neighborhood of the foam is preserved under the move in 3-space and hence in 4-space. The resulting theory is the theory of embedded 4-manifolds in 4-space.

The lune move corresponds to a critical point of the vertex set in deformation of a foam, and the $(3, 2)$-move corresponds to a vertex in a 3-dimensional foam. Thus it is reasonable to study isotopy classes of knotted foams modulo these additional two relations.

8. Future Work

This paper is a technical piece that is necessary for a serious study of knotted foams and their 3 (and higher)-dimensional generalizations. In work with Atsushi Ishii and Masahico Saito, we will establish a cohomology theory for certain algebraic systems that is sufficient to define nontrivial invariants of knotted 2-foams in 4-space.

The inclusion of the penultimate section is also meant to indicate the initial stages in the study of 3-dimensional foams. In particular, one can construct movies of 3-dimensional foams embedded in 5-space by including the Roseman/Reidemeister moves of Theorem 1.1, the orthogonality and Eilliott–Beidenharn moves, critical points of surfaces and critical points of the edge sets. Thus the critical points of surfaces correspond to 0, 1, 2, and 3-handles that are attached to the solid sheets of 3-foams. Moreover, the edge set of a 3-foam is a 2-foam. Important moves to 3-foams are easy to establish. A full Roseman-type theorem is unknown to the author at this time, but determining one should be routine and geometrically tedious.

Finally, it is worth mentioning that there is an underlying categorical motivation here that is related to the tangle hypothesis of Baez and Dolan [2]. Here we are considering the interaction between a braiding and a Frobenius structure as well as the identities among relations of these. The precise location of knotted foams in the Baez-Dolan table is an interesting taxonomic problem.

Acknowledgments

This paper was studied with the support of the Ministry of Education Science and Technology (MEST) and the Korean Federation of Science and Technology Societies (KOFST). I owe much of the approach here to a conversation that I had with Osamu Saeki. This paper is part of an on-going project with Atsushi Ishii and Masahico Saito. In addition, I have had many valuable conversations with Seiichi Kamada, Shin Satoh, and the faculty and students at the TAPU workshops and seminars.

References

[1] J. W. Alexander and G. B. Briggs, On types of knotted curves, *Ann. Math.* **28** (1926/27) 562–586.

[2] J. C. Baez and J. Dolan, Higher-dimensional algebra. III; n-categories and the algebra of opetopes, *Adv. Math.* **135** (1998) 145–206.

[3] J. S. Carter, *An Excursion in Diagrammatic Algebra: Turning a Sphere from Red to Blue* (World Scientific, 2012).

[4] J. S. Carter, D. E. Flath and M. Saito, *The Classical and Quantum 6j-symbols* (Princeton Univ. Press, 1995).

[5] J. S. Carter, D. Jelsovsky, L. Langford, S. Kamada and M. Saito, Quandle cohomology and state-sum invariants of knotted curves and surfaces, *Trans. Amer. Math. Soc.* **355** (2003) 3947–3989.

[6] J. S. Carter, J. Rieger and M. Saito, A combinatorial descriptions of knotted surfaces and their isotopies, *Adv. Math* **127** (1997) 1–51.

[7] M. Khovanov, $sl(3)$ link homology, *Algebr. Geom. Topol.* **4** (2004) 1045–1081.

[8] M. Mackaay and P. Vaz, The universal $sl3$-link homology, *Algebr. Geom. Topol.* **7** (2007) 1135–1169.

[9] S. V. Matveev, Transformations of special spines, and the Zeeman conjecture (in Russian), *Izv. Akad. Nauk SSSR Ser. Mat.* **51** (1987) 1104–1116; 1119; translation in *Math. USSR-Izv.* **31** (1988) 423–434.

[10] S. V. Matveev, Distributive groupoids in knot theory (in Russian), *Mat. Sb. (N.S.)* **119** (1982) 78–88; 160.

[11] R. Piergallini, Standard moves for standard polyhedra and spines, in *Third National Conference on Topology* (Italian) (Trieste, 1986), *Rend. Circ. Mat. Palermo* (2) Suppl. No. **18** (1988) 391–414.

[12] K. Reidemeister, *Knot Theory*, Translation of the original (1932) book *Knotentheorie* (BSC Associates, Moscow, Idaho, 1983).

[13] D. Roseman, Reidemeister-type moves of surfaces in four-dimensional space, in *Knot Thoery*, Banach Center Publications, Vol. 42, pp. 347–380, Institute of Mathematics, Polish Academy of Sciences, Warszawa 1998.

[14] V. Turaev, The Yang–Baxter equation and invariants of links, *Invent. Math.* **92** (1988) 527–553.
[15] V. Turaev and O. Viro, State sum invariants of 3-manifolds and quantum $6J$-symbols, *Topology* **31** (1992) 865–902.
[16] P. Vaz, The diagrammatic Soergel category and sl(2) and sl(3) foams, *Int. J. Math. Math. Sci.* (2010), Art. ID 612360.

How to Fold a Manifold

J. Scott Carter* and Seiichi Kamada†

*Department of Mathematics,
University of South Alabama, Mobile, AL 36688, USA
carter@southalabama.edu

†Department of Mathematics, Osaka City University,
3-3-138 Sugimoto, Sumiyoshi-ku, Osaka 558-8585, Japan
skamada@sci.osaka-cu.ac.jp

1. Introduction

The English word "manifold" evokes images of an object that is layered or folded. Of course, the mathematical definition is quite different where the emphasis is upon "many" such as the multiplicity of descriptions afforded by coordinate charts or indeed the need to define many coordinate systems — one for each point in the space. Nevertheless, there are structures in which the colloquial meaning and the mathematical meaning coincide. Specifically, in the case of a covering space or a branched cover, we imagine that the manifold is stacked in layers above the base space. According to a classical theorem of Alexander [1], every compact connected orientable manifold can be expressed as an irregular simple branched covering of the sphere with branch loci consisting of a codimension 2 subcomplex. In dimensions two through four, we assume that the branch locus is a finite collection of points, a knot or link, or a knotted or linked surface [5, 8, 14]. In this paper, we demonstrate methods of folding these coverings. We prove the following results.

Theorem 1.1. *Let $k = -1, 0, 1$, or 2. Let $f : M^{k+2} \longrightarrow S^{k+2}$ be a 2-fold branched cover of S^{k+2} branched along a closed oriented (possibly disconnected) smooth submanifold $L^k \hookrightarrow S^{k+2}$. Then there is an embedding $\tilde{f} : M^{k+2} \hookrightarrow S^{k+2} \times D^2$ such that $p \circ \tilde{f} = f$, where $p : S^{k+2} \times D^2 \longrightarrow S^{k+2}$ is the projection onto the first factor.*

Such an embedding, $\tilde{f}$, will be called a *folded embedding* or a *embedded folding.* We will say that the cover is *folded.* In case of n-fold irregular simple branched coverings $f : M^{k+2} \longrightarrow S^{k+2}$, there are topological obstructions to achieving an embedding. Still in many cases, we can construct an immersion $\tilde{f} : M^{k+2} \looparrowright S^{k+2} \times D^2$ such that the composition $p \circ \tilde{f} = f$. In this case, we call $\tilde{f}$ an *immersed folding* or a *folded immersion.*

Theorem 1.2. *Let $k = -1, 0$ or 1. Let $f : M^{k+2} \longrightarrow S^{k+2}$ be a 3-fold simple branched cover of S^{k+2} branched along a closed oriented (possibly disconnected) smooth submanifold $L^k \hookrightarrow S^{k+2}$. Then there is an immersion $\tilde{f} : M^{k+2} \looparrowright S^{k+2} \times D^2$ such that the restriction of the projection onto the first factor is the covering map f. Thus the lift $\tilde{f}$ is an immersed folding.*

Often our techniques work for branched coverings of degree greater than 3 and in case $k = 2$. In this paper, we concentrate on the lower degree branched covers. We postpone the proof of Theorem 1.2 in case $k = 2$ for a subsequent paper. We also give examples of foldings of 3- and 4-dimensional spheres that are constructed from our techniques. The advantages to our constructions are that the embeddings and immersions are given by explicit descriptions for which invariants such as fundamental group or Fox colorings can easily be computed.

To illustrate the problems associated with constructing embedded foldings, consider the knot 7_4 which is given as the plat closure of the braid word $\sigma_2^3 \sigma_1^{-1} \sigma_2^3$. Since the knot has determinant 15, it is 3-colorable. Since it is a 2-bridge knot, the 3-fold branched cover of S^3 along the knot can be constructed with three 0-handles, two 1-handles, two 2-handles and three 3-handles. The union of the 0- and 1-handles is a 3-ball as is the other side of the decomposition.

So the covering space is also S^3. We construct, quite explicitly, an immersed folding. Yet, there is a simple closed curve of double points for this folding. We demonstrate why this knot does not bound an embedded folding. The result *does not* say that the 3-sphere cannot be embedded, but instead says that an embedding does not project canonically upon the standard sphere as a covering.

In the case of the 3-fold branched covering of S^4 branched along the 2-twist-spun trefoil, we have a folded immersion in $S^4 \times D^2$, but not a folded embedding. This example will be presented elsewhere.

Our main results are interesting in the light of theorems of Alexander [1], Hilden [5] and Montesinos [14] in dimension 3 and Iori and Piergallini [8] in dimension 4.

Theorem 1.3. (J.W. Alexander [1]) *For any closed oriented and connected m-manifold M^m, there exists a simple branched covering $f : M^m \to S^m$ for some degree.*

Theorem 1.4. (H. M. Hilden [5], J. M. Montesinos [14]) *For any closed oriented and connected 3-manifold M^3, there exists a 3-fold simple branched covering $f : M^3 \to S^3$ such that the branch set L is a link (or a knot).*

The following is a conjecture due to Montesinos.

Conjecture 1.5. *For any closed oriented and connected 4-manifold M^4, there exists a 4-fold simple branched covering $f : M^4 \to S^4$ such that L is an embedded surface in S^4.*

Some partial answers to this conjecture are known as follows.

Theorem 1.6. (R. Piergallini [15]) *For any closed oriented and connected 4-manifold M^4, there exists a 4-fold simple branched covering $f : M^4 \to S^4$ such that L is an immersed surface in S^4.*

Theorem 1.7. (M. Iori and R. Piergallini [8]) *For any closed oriented and connected 4-manifold M^4, there exists a 5-fold simple branched covering $f : M^4 \to S^4$ such that L is an embedded surface in S^4.*

It is known that every closed 3-manifold can be embedded in $\mathbb{R}^5$, and that the obstructions to embedding a closed orientable 4-manifold in $\mathbb{R}^6$ are given by the Pontryagin class and the signature of the 4-manifold. So in the 4-dimensional case, the surface of self-intersections will be related to one of these characteristic classes.

Complex projective space $\pm\mathbb{C}P^2$ can be obtained as a 2-fold branch cover of S^4 branched along the standardly embedded projective planes of normal Euler class ± 2 [13]. However, our results for 2-fold branched covers are dependent upon the existence of a Seifert manifold for 2-fold branched covers. The normal Euler classes prohibit the existence of such a Seifert solid. It is possible that an alternative chart movie description can handle this case, but as of this writing we do not have an alternative method for this.

Our principal technique is to generalize the notion of a chart to a chart surface (or curtain) in dimension 3. In dimension 4, we develop a 3-dimensional analogue called an interwoven solid. We recall from [10] that a *chart* is a labeled finite graph in the plane that has three types of vertices: 1-valent black vertices, 4-valent crossings, and 6-valent white vertices. The labels upon the edges incident at crossings and 6-valent vertices are required to satisfy additional conditions that we discuss below (Sec. 3.1).

A *curtain* is an immersed labeled compact surface with boundary in 3-space. The boundary is embedded as a knot or link which may pass through the sheets of the curtains. The singularities of the curtains are non-generic intersections among three sheets and transverse intersections between two sheets. The curtain can be put into general position with respect to a height function defined on 3-space. In this case, the critical points are quantified as changes to charts. These changes are as follows:

1. introducing or removing a single labeled edge that joins two black vertices but does not intersect another edge of the chart;
2. an application of any of the *chart moves* that are depicted in Fig. 1.

A knotted or linked surface in 4-dimensional space can be given via a movie description, for example. In such a movie a sequence

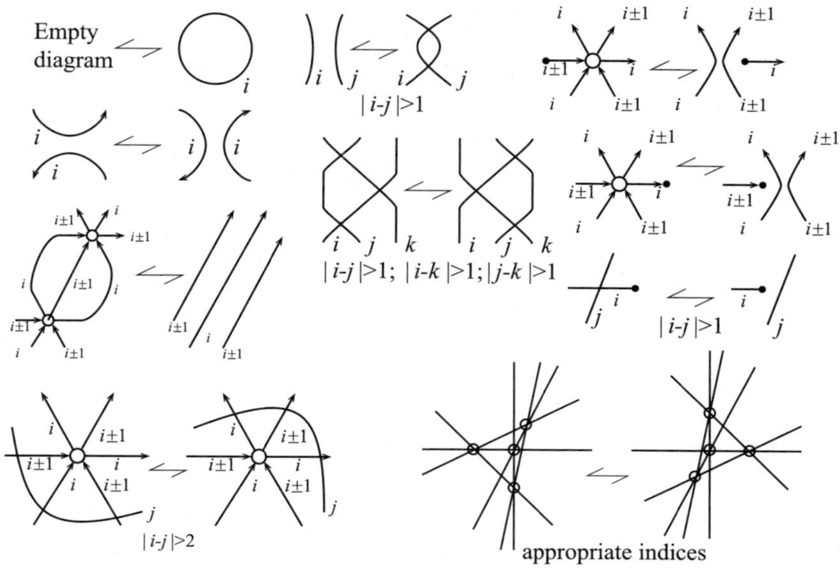

Fig. 1.

of knot and link diagrams is given in such a way that successive cross-sections differ by a critical point (birth or death of a simple closed curve or a saddle point) or by one of the Reidemeister moves that are indicated in Fig. 2.

In such a movie, we explicitly choose a height function for each of the stills and keep this height function consistent throughout the movie. Alternatively, an oriented surface knot or link can be put into surface braid form (via an oriented chart), and the movie description is determined by a critical point analysis of the corresponding braid chart. We will have cause to use both descriptions.

In order to produce a folded immersion of the branched covering of S^4 branched along the surface link, we need a tertiary chart-like structure: that of an interwoven solid.

Interwoven solids are to curtains as curtains are to charts. Thus an *interwoven solid* is an immersed 3-dimensional manifold in 4-space whose boundary is a knotted or linked surface. The boundary may pass through the solid, and the immersion is not in general position. An interwoven solid can be described as a sequence of

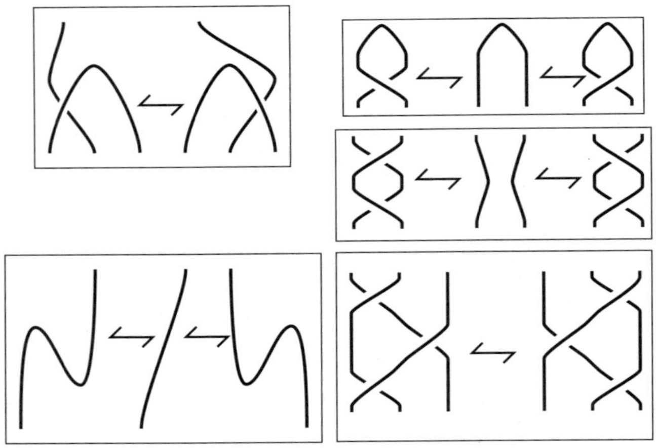

Fig. 2.

curtains (with nodes) where successive curtains differ by generic critical points or specific curtain moves. Here we will not list all the possible curtain moves, but leave that taxonomy for our future work in this direction. We give three specific examples of interwoven solids. First, the Seifert solid that a knotted or linked surface bounds in 4-space is an interwoven solid. Second, consider a genus n surface that is embedded in 3-space. An interwoven solid is the bounded 3-manifold that such a surface bounds. Third, our last example consists of the spun trefoil presented in movie form. The interwoven solid that yields its 3-fold branched cover is indicated as a sequence of curtains. Successive curtains in this case differ from each other by replacing a chart move and its inverse with a product of charts or *vice versa*.

So for each still in the movie (or braid movie) of the knotted surface, we construct a curtain. In the case of the 2-fold branched cover, the curtain is a Seifert surface. Each critical event for the knot movie induces a critical event between the curtains. Each will be described explicitly. At the end of the movie, it may be necessary to unravel the curtains. In our examples for the 3-fold branch covers of classical knots, we need to use chart moves to achieve the unraveling.

Here we will introduce some of the curtain movie moves that are necessary to achieve the unraveling of the interwoven solid. However, a full list is not necessary to articulate since we can, at the end of the movie, isolate the curtains into an immersed closed surface that is contained in a ball neighborhood in a 3-dimensional slice of 4-space. This surface can be eliminated since it is null-homologous in 3-space.

Here is an outline of the paper. We begin with a basic review of the permutation and braid groups. Then we describe simple branched covers of 2-dimensional surfaces. We extend the ideas of the second author to give a combinatorial description of permutation and braid charts. In particular, we demonstrate that vertices in charts correspond to specific critical events for surfaces in 3- and 4-dimensional space. We then use our ideas to develop embeddings of 2-fold branched covers of the 3- and 4-sphere branched over codimension 2 embedded submanifolds. Many examples are given with quite a bit of detail rendered. Three examples of embedded and immersed foldings of the 3-fold branched covers of the 3- and 4-sphere are given. The paper closes with a proof of Theorem 1.2 in case $k = 1$.

2. Some Basic Definitions

Throughout this work, we will be using the permutation and braid groups. The permutation group Σ_n is the set of bijective maps on $\{1, 2, \ldots, n\}$. It has a presentation:

$$\begin{aligned}\Sigma_n = \langle \tau_1, \tau_2, \ldots, \tau_{n-1} \ :\ & \tau_i\tau_j = \tau_j\tau_i && \text{if } |i-j| > 1;\\ & \tau_i\tau_{i+1}\tau_i = \tau_{i+1}\tau_i\tau_{i+1} && \text{if } i = 1, \ldots, n-2;\\ & \tau_i^2 = 1 && \text{if } i = 1, \ldots, n-1\rangle.\end{aligned}$$

The permutation group is a quotient of the braid group B_n which has the presentation:

$$\begin{aligned}B_n = \langle \sigma_1, \sigma_2, \ldots, \sigma_{n-1} \ :\ & \sigma_i\sigma_j = \sigma_j\sigma_i && \text{if } |i-j| > 1;\\ & \sigma_i\sigma_{i+1}\sigma_i = \sigma_{i+1}\sigma_i\sigma_{i+1} && \text{if } i = 1, \ldots, n-2\rangle.\end{aligned}$$

We will be considering knotted and linked subsets of the $(k+2)$-*dimensional sphere* for $k = -1, 0, 1, 2$ which is defined to be the set of unit vectors in $\mathbb{R}^{k+3}$. Specifically,

$$S^{k+2} = \left\{ (x_1, x_2, \ldots, x_{k+3}) \in \mathbb{R}^{k+3} : \sum_{j=1}^{k+3} x_j^2 = 1 \right\}.$$

This is the boundary of the unit $(k+3)$-disk

$$D^{k+3} = \left\{ (x_1, x_2, \ldots, x_{k+3}) \in \mathbb{R}^{k+3} : \sum_{j=1}^{k+3} x_j^2 \leq 1 \right\}.$$

We will frequently consider a fixed embedding of a $(k+2)$-dimensional disk in the sphere S^{k+2}. Thus many constructions occur in the disk and extend trivially outward to the sphere. We include $\mathbb{R}^{k+3}$ in $\mathbb{R}^{k+4}$ in a standard fashion as $\{(x_1, x_2, \ldots, x_{k+3}, 0) : x_j \in \mathbb{R}\}$. Furthermore, it will often be convenient to decompose the disk as a Cartesian product of lower dimensional disks. For example, $D^2 \simeq D^1 \times D^1 \simeq I \times I$, where I is the unit interval $I = [0, 1]$.

A *simple branched cover of* S^{k+2} *of degree* n is a compact orientable manifold M^{k+2} together with a surjective map $f : M^{k+2} \to S^{k+2}$ such that each point in S^{k+2} is of one of two types.

1. A *regular point* is a point that has a $(k+2)$-disk neighborhood N such that $f^{-1}(N)$ is homeomorphic to the disjoint union of exactly n copies of N.
2. A (*simple*) *branch point* is a point that has a neighborhood N such that $f^{-1}(N)$ consists of $n-1$ disk neighborhoods $N_1, \ldots, N_{n-1}$, the map $f|_{N_j}$ is a homeomorphism for $j = 1, \ldots, n-2$, and $f|_{N_{n-1}} : N_{n-1} \to N$ is a two-to-one branched covering map. In this case, there is a coordinate system $(x_1, \ldots, x_{k+2})$ of this component such that the restriction of f is given by $(x_1, x_2, x_3, \ldots, x_{k+2}) \mapsto (x_1^2 - x_2^2, 2x_1x_2, x_3, \ldots, x_{r+2})$.

A branched cover of degree n is also called an *n-fold branched cover*. Throughout, we will only deal with simple branched coverings, and so we will speak colloquially of branched coverings. We sometimes work in the PL-category.

3. 2-Dimensional Simple Branched Coverings

Let $f : M^2 \to S^2$ be an n-fold simple branched cover with branch set L, and let $\underline{f} : M^2 \setminus f^{-1}(L) \to S^2 \setminus L$ be the associated covering map; that is $\underline{f}$ is the restriction of f to the complement of the branch set.

Take a base point $*$ of $S^2 \setminus L$ to consider the fundamental group $\pi_1(S^2 \setminus L, *)$. The preimage $f^{-1}(*)$ of the base point $*$ consists of n points of M^2. Then we have a *monodromy* $\rho : \pi_1(S^2 \setminus L, *) \to \Sigma_n$, where the symmetric group Σ_n on letters $\{1, 2, \ldots, n\}$ is identified with the symmetric group on $f^{-1}(*)$. (A monodromy ρ depends on the identification between $\{1, 2, \ldots, n\}$ and $f^{-1}(*)$.) The covering map $\underline{f}$ is determined by the monodromy.

By the Riemann–Hurwitz formula, L consists of an even number of points. The idea of the monodromy is illustrated in the central figure of Fig. 3. The concept of a (permutation) chart was summarized in the introduction and is discussed in detail below.

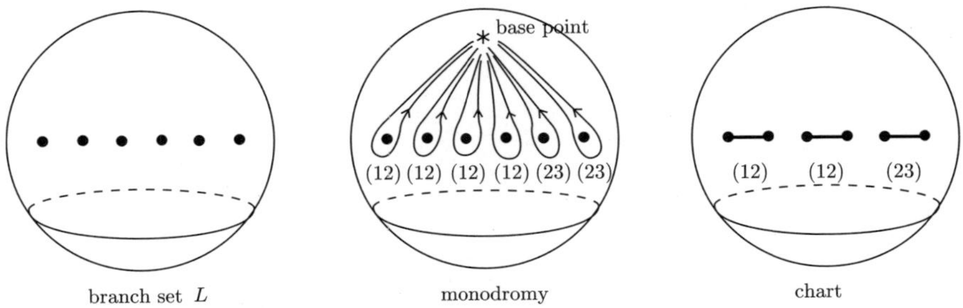

Fig. 3.

When a monodromy is described by a chart, it is easy to construct M^2. We explain it by using an example. Let Γ be the chart depicted on the right of Fig. 4. Consider three copies of S^2 labeled by 1, 2, and 3, say S^2_1, S^2_2 and S^2_3, respectively. On the copy S^2_1, draw the edges with label (12) of Γ, on the copy S^2_2, draw the edges with label (12) of Γ and those with label (23), and on the copy S^2_3, draw the edges with label (23). Cut the three 2-spheres along these edges, to obtain three compact surfaces, say M_1, M_2 and M_3, as indicated in Fig. 4.

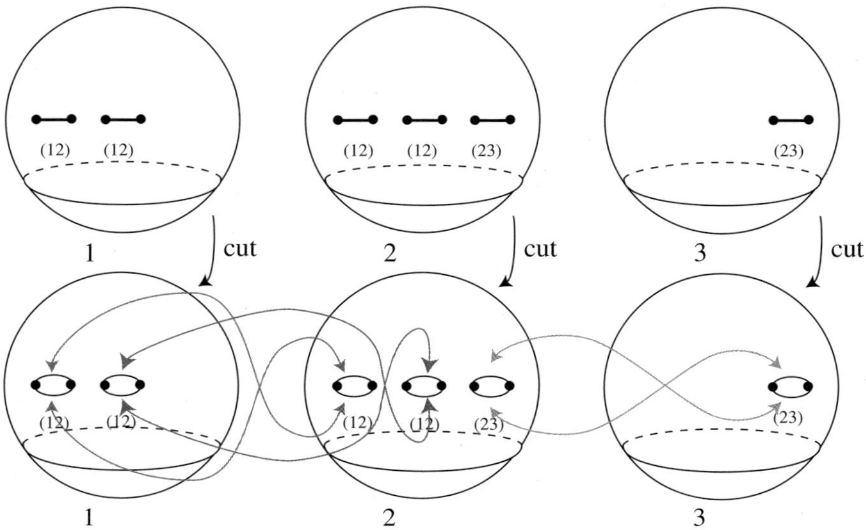

Fig. 4.

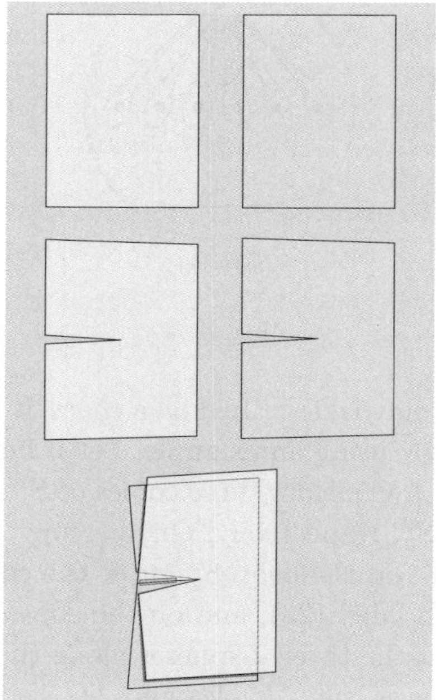

Fig. 5.

The surface M^2 is obtained from the union $M_1 \cup M_2 \cup M_3$ by identifying the boundary as follows: Let e be an edge with label (12) on S_1^2, and let e_+ and e_- be the copies of e in ∂M_1. Let e' be the corresponding edge on S_2^2, and let e'_+ and e'_- be the corresponding copies in ∂M_2. Then we identify e_+ with e'_-, and identify e_- with e'_+, respectively. All boundary edges of $M_1 \cup M_2 \cup M_3$ are identified in this fashion, and we have a closed surface. This is the desired M^2.

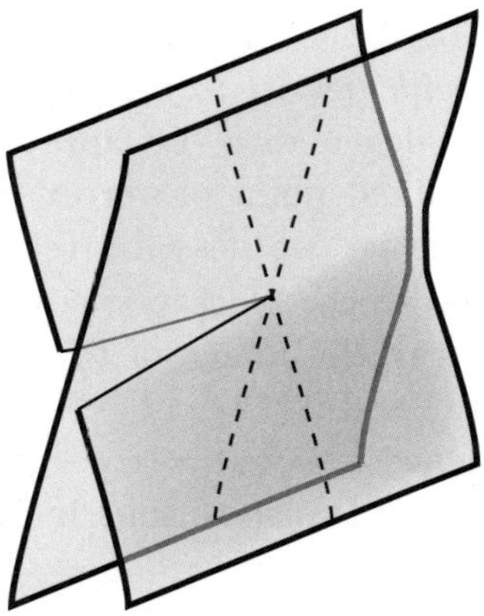

Fig. 6.

In a neighborhood of the branch points we can cut and paste (tape) as indicated with two sheets of paper in Fig. 5. The local picture of a branch point as a broken surface diagram is indicated in Fig. 6.

3.1. *Permutation and braid charts*

A *(permutation) chart of degree n*, (or an Σ_n*-chart*), is a labeled finite graph embedded in the 2-disk $D^2 \simeq I \times I$ which has three types of vertices. The labels written on the edges are taken from the set $\{1, 2, \ldots, n-1\}$ — these will correspond to the generators $\tau_1, \ldots, \tau_{n-1}$ of the permutation group Σ_n. The vertices are of the following type:

1. a *black vertex* is a mono-valent whose incident edge may have any label taken from $\{1, 2, \ldots, n-1\}$;

2. a *crossing* is 4-valent, and the labels on the incident edges are given in cyclic order i, j, i, j where $|i - j| > 1$;
3. a *white vertex* is 6-valent, and its incident edges have labels in cyclic order given by $i, i+1, i, i+1, i, i+1$ for some $i = 1, 2, \ldots, n-2$.

Necessarily, a chart has an even number of black vertices.

Let $\Gamma \subset D^2$ denote a chart with m black vertices $\{b_1, \ldots, b_m\}$. We identify D^2 with $[0,1] \times [0,1]$ and consider the projection $p_1 : [0,1] \times [0,1] \to [0,1]$ onto the first factor. We will assume that Γ is in *general position with respect to* p_1. That is, the critical points of the arcs are all non-degenerate (C^2-approximated by quadratic functions), and each critical point or vertex projects to a different time value. For convenience, we describe vertices as critical points of the chart. We may order the black vertices from left-to-right with respect to p_1. In this way the image of b_j under the projection p_1 onto the first factor is t_j and $0 < t_1 < t_2 < \cdots < t_m < 1$.

By the correspondence $i \leftrightarrow \tau_i = (i, i+1) \in \Sigma_n$, the labels of a chart are assumed to be transpositions in Σ_n. For a chart Γ, we consider a monodromy

$$\rho_\Gamma : \pi_1(D^2 \setminus L) \to \Sigma_n, \quad [\ell] \mapsto [\text{intersection word of } \ell \text{ w.r.t. } \Gamma],$$

where L ($=L_\Gamma$) is the set of black vertices $\{b_1, \ldots, b_m\}$. An intersection word is a sequence of elements of $\{1, \ldots, n-1\}$, which is regarded as an element of Σ_n by the correspondence $i \leftrightarrow \tau_i = (i\ i+1) \in \Sigma_n$.

If an edge terminates at a black vertex, then this becomes a simple branch point as above. If an edge terminates at a crossing, the ith and $(i+1)$th sheets are cut and re-glued while the jth and the $(j+1)$th sheets are cut and re-glued. The identifications occur in disparate sheets and the cutting can proceed. In the case that an edge terminates at a white vertex, the reassembly is indicated in Fig. 7. The resulting configuration resembles the intersection of the three coordinate planes in space. The resulting surface $M^2(\Gamma)$ with boundary is called the *n-fold irregular simple branched cover induced from the chart* Γ. The surface $M^2(\Gamma)$ extends to a simple branched cover of the 2-sphere, by trivially extending the cover over the complement of the disk in S^2.

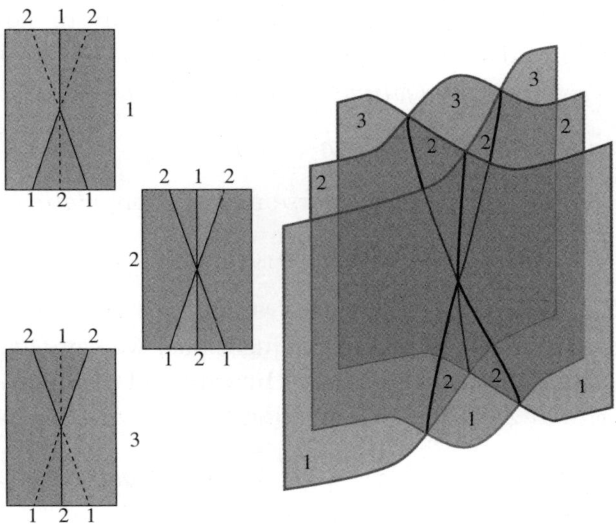

Fig. 7.

The classification of simple branched coverings was studied by J. Lüroth [12], A. Clebsch [4] and A. Hurwitz [7]. The classification theorem is stated as follows.

Theorem 3.1. *Let* $f : M^2 \to S^2$ *and* $f' : M^{2'} \to S^2$ *be* n*-fold simple branched coverings with branch sets* L *and* L'*, respectively. We assume that* M^2 *and* $M^{2'}$ *are connected. Then* f *and* f' *are equivalent if and only if* $\#L = \#L'$.

Hurwitz [7] studied branched coverings by using a system of monodromies of meridian elements of the branch set, called a *Hurwitz system*, and studied when two systems present the same (up to equivalence) branched coverings.

A Hurwitz system depends on a system of generating set of $\pi_1(S^2 \setminus L, *)$. For the generating system depicted in the middle of the first illustration of this section, the Hurwitz system is

$$\alpha = ((12), (12), (12), (12), (23), (23)).$$

Besides a choice of a generating system, a Hurwitz system depends on the identification of $\{1, 2, \dots, n\}$ and the fiber $f^{-1}(*)$.

Two Hurwitz systems present the same (up to equivalence) monodromy if and only if they are related by a finite sequence of *Hurwitz moves* and *conjugations*. The *Hurwitz moves* are

$$(a_1, \ldots, a_k, a_{k+1}, \ldots, a_m) \mapsto (a_1, \ldots, a_{k+1}, a_{k+1}^{-1} a_k a_{k+1}, \ldots, a_m)$$

for $k = 1, \ldots, n-1$ and their inverse moves. *Conjugations* are

$$(a_1, \ldots, a_m) \mapsto (g^{-1} a_1 g, \ldots, g^{-1} a_m g)$$

for $g \in \Sigma_n$. When two Hurwitz systems are related by a finite sequence of Hurwitz moves and conjugations, we say that they are *HC-equivalent.* (H and C stand for Hurwitz and conjugation.)

Due to Hurwitz [7], the classification theorem is stated as follows.

Theorem 3.2. *Let $f : M^2 \to S^2$ be an n-fold simple branched covering. Assume that M^2 is connected. Any Hurwitz system of f is HC-equivalent to*

$$((12), \ldots, (12), (13), (13), (14), (14), \ldots, (1, n), (1, n)).$$

Theorem 3.3. *Let $f : M^2 \to S^2$ be an n-fold simple branched covering, and ρ_f a monodromy of f. There exists a chart Γ such that $\rho_\Gamma = \rho_f$. (We call Γ a chart description of f or ρ_f.)*

Proof. For a detailed proof, see [10]. □

3.2. *Constructing a branched covering from a chart*

The surface $M^2(\Gamma)$ can be mapped into 3-dimensional space in general position by using the chart to create a *permutation movie* that we turn to describe now. Recall that the chart is in general position with respect to the projection p_1 onto the horizontal edge of the square. Each critical point of an edge is non-degenerate, and the critical points and vertices occur at distinct times. Let $t_0 = 0 < t_1 < t_2 < \cdots < t_{s-1} < 1 = t_s$ denote the critical values, and consider the intersection of Γ with the vertical line $L_i = p_1^{-1}((t_i + t_{i+1})/2)$.

Now read the intersection sequence of Γ with the lines L_i for $i = 1, \ldots, s$ in order. We read the sequence from top to bottom and write

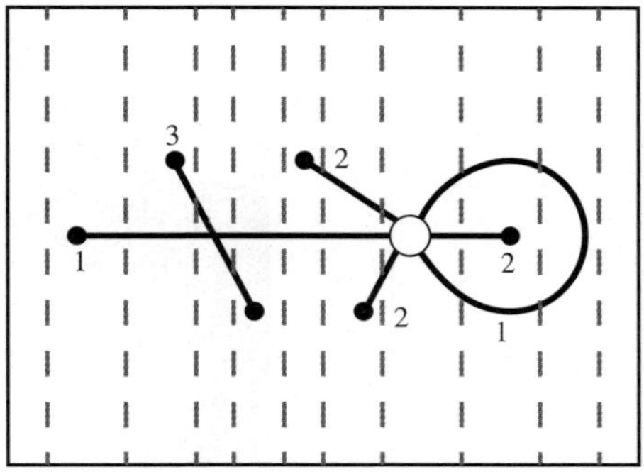

Fig. 8.

the sequence from left to right. For example, for the chart in Fig. 8, the sequence reads: $\emptyset; (1); (3,1); (1,3);\ (1);\ (2,1);\ (2,1,2);\ (1,2,1);$ $(1,1);\ \emptyset$. We rewrite this as a sequence of words in the permutation group Σ_n (here $n = 3$) as follows: $1;\ \tau_1;\ \tau_3\tau_1;\ \tau_1\tau_3;\ \tau_1;\ \tau_2\tau_1;\ \tau_2\tau_1\tau_2;$ $\tau_1\tau_2\tau_1;\ \tau_1\tau_1; 1$.

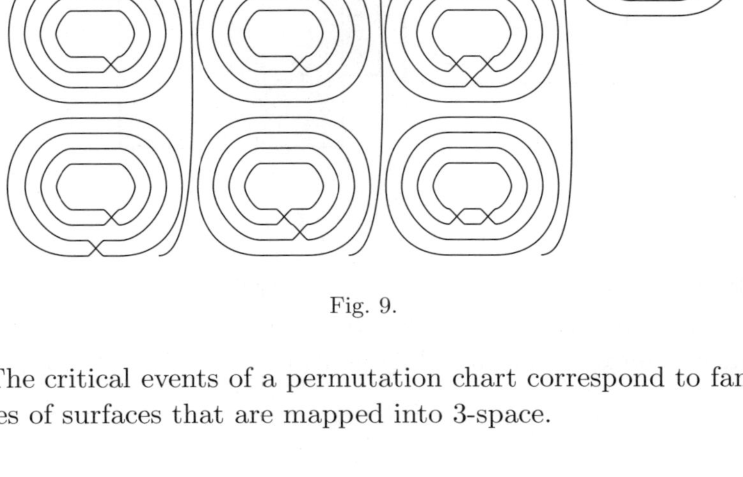

Fig. 9.

The critical events of a permutation chart correspond to familiar pieces of surfaces that are mapped into 3-space.

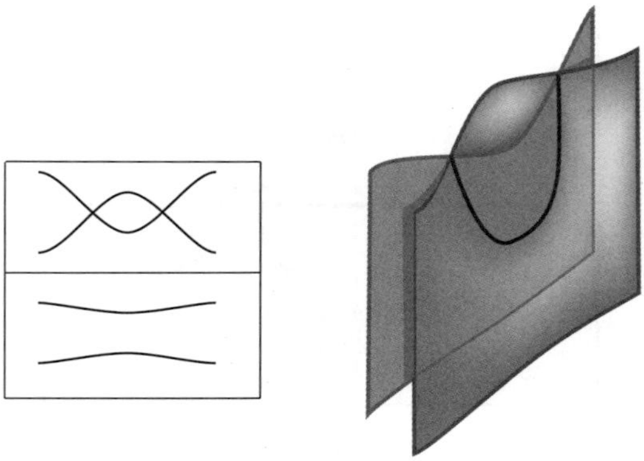

Fig. 10.

1. Type II moves. Critical events of the form $\subset^k_k$ correspond to the movie in which the creation of a pair of canceling generators $1 \Rightarrow \tau_k \tau_k$ occurs in a type II birth fashion. Critical events of the form ${}^k_k \supset$ correspond to the annihilation of the same pair of generators in a type II death fashion: $\tau_k \tau_k \Rightarrow 1$. See Fig. 10.

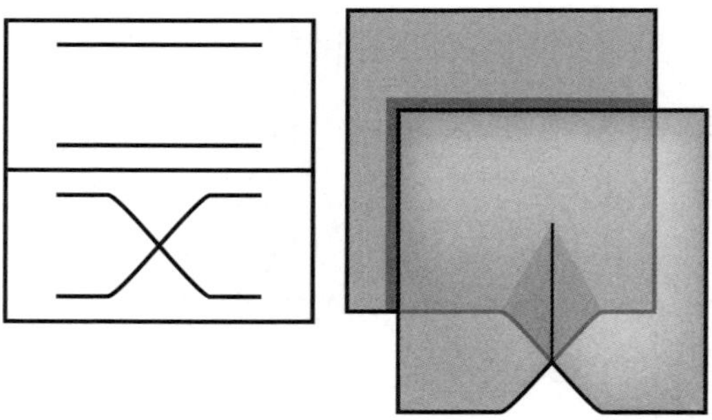

Fig. 11.

2. Black vertices. A black vertex corresponds to inserting or removing a generator from a word. In the local picture of the surface, a branch point occurs at a saddle point in the surface. See Fig. 11.

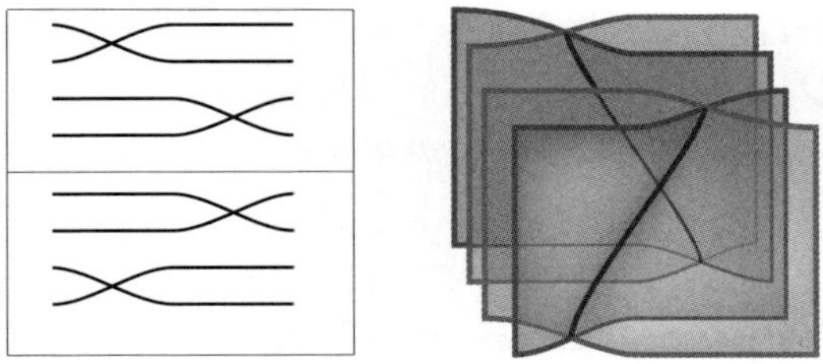

Fig. 12.

3. Crossing exchange. A crossing indicates the interchange of distant crossings. See Fig. 12.

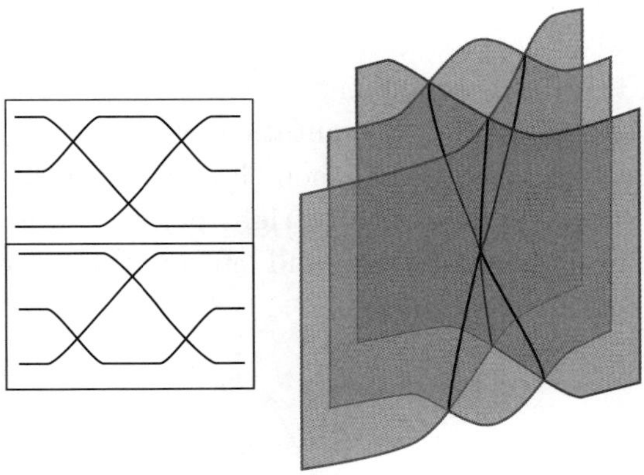

Fig. 13.

4. White vertices. A white vertex indicates a triple point of the surface induced by the relation $\tau_i\tau_{i+1}\tau_i = \tau_{i+1}\tau_i\tau_{i+1}$. See Fig. 13.

We have the following construction.

Proposition 3.4. *Let Γ denote a permutation chart of degree n in the disk D^2. Let $M^2(\Gamma)$ denote the n-fold irregular branched cover of D^2 induced by Γ. Then there is a general position map $\tilde{f} : M^2(\Gamma) \to D^2 \times I$ such that the projection onto D^2 induces the branched covering map.*

Let $F : \hat{M}^2(\Gamma) \to S^2$ denote the extension of the branch cover $M^2(\Gamma)$ over D^2 to the sphere S^2. Then there is a general position map $\tilde{F} : \hat{M}^2(\Gamma) \to S^2 \times [0,1]$ such that $p \circ \tilde{F} = F$ where p is the projection onto the first factor.

3.3. *Lifting a permutation movie to a braid movie*

A *braid chart of degree* n is a permutation chart in which orientations on the edges have been chosen so that white vertices and crossings are of the form indicated in the figure. Thus there are three incoming and three outgoing edges that alternate i, $i \pm 1$, i, $i \pm 1$, i, $i \pm 1$ in cyclic order. Similarly, at a crossing the edges with labels i and the edges with labels j are oriented consistently ($|i - j| > 1$) as indicate in Fig. 14. In this case, we say that there is a *flow though the white vertex* or *the crossing.* If a permutation chart can be consistently oriented to create a braid chart, then the permutation movie can be lifted to a braid movie. Specifically, right pointing edges with label i correspond to braid generators σ_i, and left pointing edges with label i correspond to σ_i^{-1}.

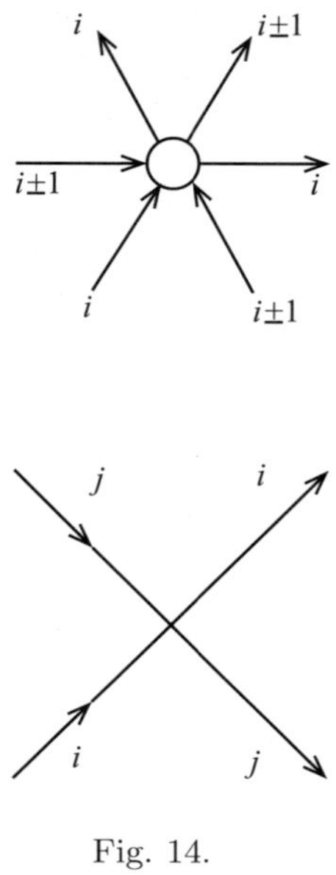

Fig. 14.

Let Γ_n denote a degree n permutation chart. If there is a consistent orientation on the edges of Γ_n, then the opposite orientation will also be consistent; we let $\Gamma_n^{\pm}$ denote the resulting braid charts. As before, let $M^2(\Gamma)$ denote the n-fold irregular simple branched cover of D^2 that is associated to the chart $\Gamma = \Gamma_n$, and let $\hat{M}(\Gamma)$ denote its extension to the 2-sphere.

Theorem 3.5. *If $\Gamma_n^{\pm}$ is a braid chart, then there is an embedding, $\tilde{F} : \hat{M}(\Gamma) \to S^2 \times [0,1] \times [0,1]$ of the n-fold irregular simple branched cover $F : \hat{M}(\Gamma) \to S^2$ such that the composition $p \circ \tilde{F}$ agrees with the covering map F where $p : S^2 \times [0,1] \times [0,1] \to S^2$ is the projection onto the first factor. In this way the covering F has a folded embedding.*

Proof. The lifting of each permutation to a braid induces an embedding of the surface $\hat{M}$ into $S^2 \times [0,1] \times [0,1]$. The projection to the first two factors $S^2 \times [0,1]$ induces the generic map of Proposition 3.4. □

3.4. *A partial lifting*

It is possible that a given chart cannot be oriented consistently so that all white vertices have a flow. A *semi-oriented chart* is a permutation chart Γ^* that includes a fourth type of vertex which is bivalent, and the chart is oriented such that (1) each bivalent vertex is either a source or a sink and (2) each white vertex and each crossing has a flow. At a *source* vertex the two emanating edges have the same label and point away from the vertex. At a *sink* the edges point towards the vertex.

Lemma 3.6. *Any permutation chart can be semi-oriented.*

Proof. Locally orient the edges so that there is a flow through all white vertices and all crossings. If there is an edge whose endpoints are either white vertices or crossings such that the local orientations do not match, then introduce a source or sink. Do the same for all such edges. □

Lemma 3.7. *A semi-oriented chart $\Gamma^* = \Gamma_n^*$ induces an immersion $\tilde{F} : \hat{M}^2(\Gamma^*) \looparrowright S^2 \times [0,1] \times [0,1]$ of the associated irregular simple branched covering $F : \hat{M}^2(\Gamma^*) \to S^2$ such that $p \circ \tilde{F} = F$ where p is the projection onto the first factor. In this way the covering F has a folded immersion.*

Note that this result gives Theorem 1.2 for $k = 0$. Moreover, we may assume that the chart is standard. Clearly, a standard chart can be oriented. So by Theorem 3.5, we have a stronger result for surfaces.

Theorem 3.8. *The n-fold simple irregular branched cover of S^2 branched over a finite set (even number) of points has an embedded folding.*

4. Two-Fold Branched Coverings of S^3 Branched Along a Knot

Our construction of the folding of the two-fold branched cover of S^3 branched along a knot or link will give rise to a Heegaard diagram for this 3-manifold in an interesting and natural fashion. In the subsequent sections, we also see handle decompositions for both 3- and 4-dimensional branched coverings. So for the readers convenience, we review the basic handle terminology.

4.1. *Review of handles*

Let m be a fixed positive integer. Here we are mostly interested in $m = 0, 1, 2, 3$ or 4. Let $j \in \{0, 1, \ldots, m\}$. A *j-handle* in an m-manifold N is a subset homeomorphic to an m-disk, but decomposed as the Cartesian product $D^j \times D^{m-j}$. The subset $S^{j-1} \times D^{m-j}$ of the boundary is called the *attaching region* or *A-region.* The sphere $S^{j-1} \times \{0\}$ is called the *attaching sphere* (*A-sphere*). The A-sphere is the boundary of the *core disk* $D^j \times \{0\}$. The *belt region* (*B-region*) is the subset $D^j \times S^{m-j-1}$ of the boundary. The *belt sphere* (*B-sphere*) is $\{0\} \times S^{m-j-1}$. It is the boundary of the *co-core disk* $\{0\} \times D^{m-j}$. See Fig. 15.

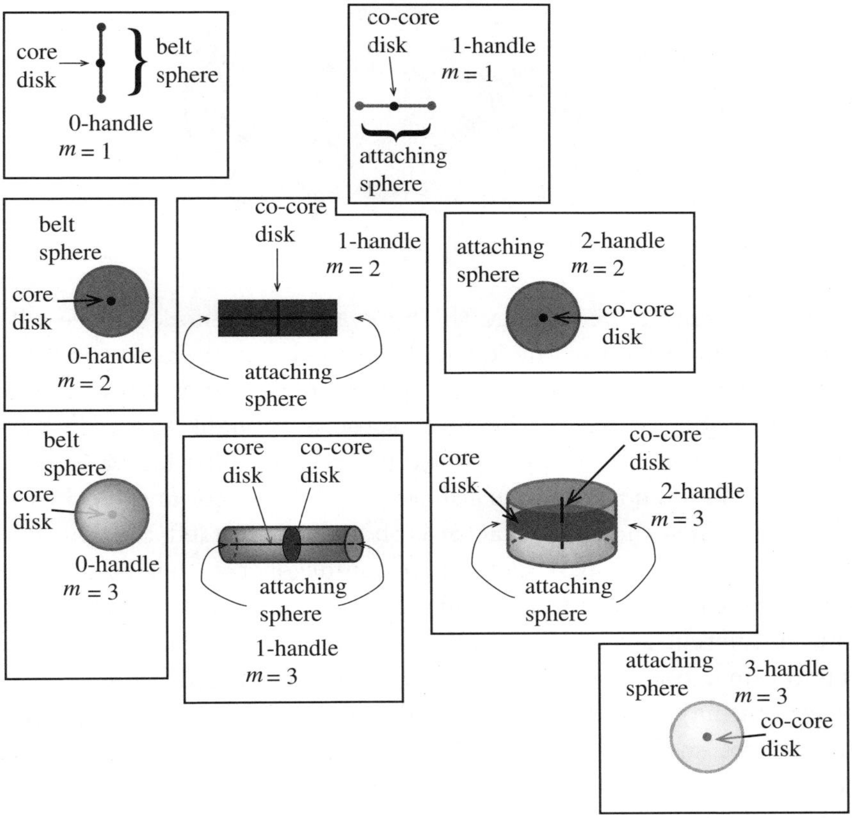

Fig. 15.

4.2. *The Seifert surface as a 2-dimensional chart*

There is a well-known construction for the 2-fold branched cover of S^3 branched along a knot or link. First, we choose an orientable Seifert surface for the knot or link. Then we take two copies of S^3, cut each along the Seifert surface, and glue the positive side of one copy to the negative side of the other. Our construction uses the Seifert surface as a 2-dimensional chart. Specifically, we choose a height function on S^3 that restricts to a non-degenerate height function on the Seifert surface of the knot or link.

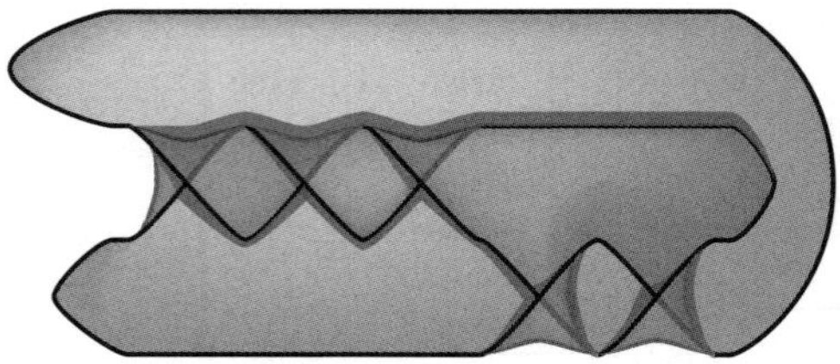

Fig. 16.

By convention we arrange the knot and the Seifert surface with the initial minimum to the left and the final maximum to the right. The knot may be taken to lie within a 3-disk that is parametrized as $[0,1] \times [0,1] \times [0,1]$. The first factor is thought to run left-to-right, the second from bottom-to-top, and the third factor is back-to-front. We often rotate each still 90° in clockwise (or anti-clockwise) direction. Then the second factor is thought to run left-to-right. The projection onto the first factor is a non-degenerate height function for the Seifert surface. In particular each critical point for the surface is at a different level, each crossing is at a different level, the surface has optimal points at the optimal points of the link (it may have other optima), and it may have saddle points. We cut the link and the Seifert surface between critical levels where the crossings are considered to be critical. Since Seifert surfaces are constructed by twisted bands, we will also arrange that the twists are achieved by particular critical motions near the crossings. The illustrations, Fig. 16 and Fig. 17, indicate these ideas for the knot 5_2.

For a Seifert surface $\mathcal{F} \subset [0,1] \times [0,1] \times [0,1]$ and for each $s \in [0,1]$, let $\Gamma_s = \Gamma_s(\mathcal{F}) \subset [0,1] \times [0,1]$ denote the cross-section of $\mathcal{F}$ at s, i.e.

$$\mathcal{F} \cap \{s\} \times [0,1] \times [0,1] = \{s\} \times \Gamma_s.$$

For a non-critical level s, the cross-section Γ_s is a permutation chart of degree 2. An orientation of the link induces an orientation of the Seifert surface, and an orientation for the cross-sectional charts. The conventions are indicated in Fig. 17, when we rotate the stills 90°: an arc pointing towards the right is positive and the corresponding black vertex is a sink; a left-pointing arc is negative and the corresponding black vertex is a source.

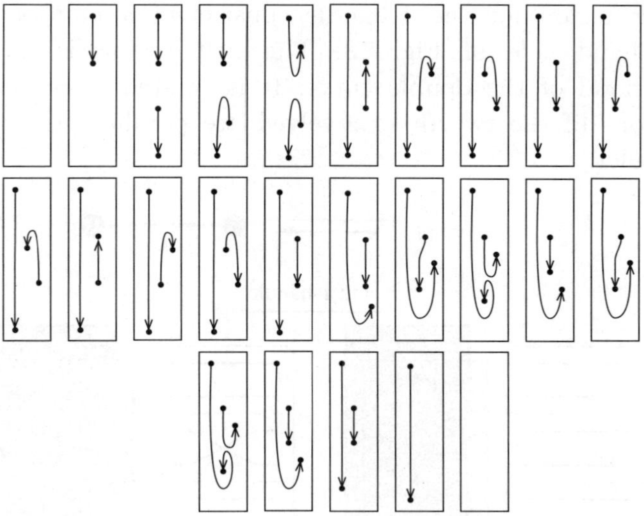

Fig. 17.

For each non-critical cross-section, the oriented cross-sectional chart Γ_s is a braid chart of degree 2 and it describes an embedding $F_s \hookrightarrow S^2 \times D^2$ of a closed oriented surface F_s as a surface braid (cf. [10]). We will call $F_s \subset S^2 \times D^2$ the *cross-sectional surface* or the *cross-sectional surface braid* at s.

The rest of the construction is to describe the changes to surface braids induced by critical events for the Seifert surface. Thus we will have an embedded folded solid over the cube $[0,1] \times [0,1] \times [0,1]$, and we will extend this to the trivial cover of the complementary disk in S^3.

4.3. *Illustrations of the critical events*

Here is a list of the critical events.

1. $\stackrel{1-H}{\Longrightarrow}$ or $\stackrel{2-H}{\Longrightarrow}$ — *handle attachment.* A local optimal point of knot diagram occurs. The move from left-to-right in the illustration (Fig. 18) represents a 1-handle being attached between the cross-sectional surfaces. The moves from right-to-left represents a 2-handle being attached between the cross-sectional surfaces. The attaching region for the 1-handle consists of a pair of disks on the left-hand side of Fig. 18. The belt region is an annular neighborhood of the double curve. It is twisted since the normal orientations of the two disks involved are parallel (as opposed to anti-parallel).

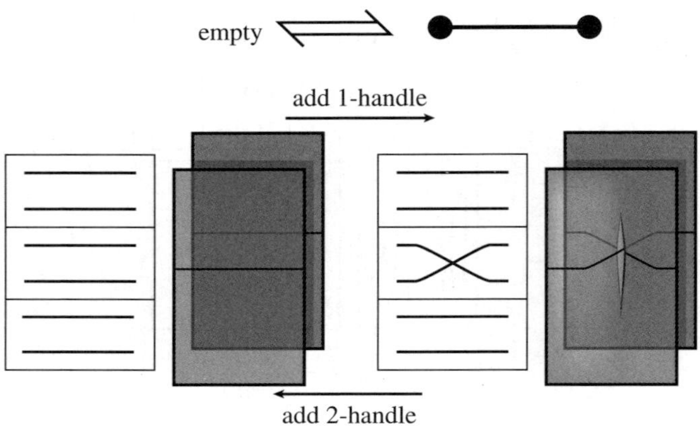

Fig. 18.

2. $\stackrel{IIb}{\Longrightarrow}$ — *a type II bubble move.* An optimum occurs in the interior of the Seifert surface. In this case the successive cross-sections differ by the inclusion of a simple closed curve. The cross-sectional surfaces are isotopic as surface braids since they differ by the chart move that is indicated in Fig. 19.

3. $\stackrel{IIs}{\Longrightarrow}$ — *A type II saddle move.* A saddle occurs in the interior of the Seifert surface. In this case, the relation $\tau_1 \cdot \tau_1 \rightarrow 1$ is followed by the relation $1 \rightarrow \tau_1 \cdot \tau_1$ in the permutation representation, and the sequence $\tau_1 \cdot \tau_1 \rightarrow 1 \rightarrow \tau_1 \cdot \tau_1$ is replaced by the identity sequence $\tau_1 \cdot \tau_1 \rightarrow \tau_1 \cdot \tau_1 \rightarrow \tau_1 \cdot \tau_1$ (or vice versa). In the braid case, the sequence $\sigma_1^{\pm 1} \cdot \sigma_1^{\mp 1} \rightarrow 1 \rightarrow \sigma_1^{\pm 1} \cdot \sigma_1^{\mp 1}$ is replaced by the identity sequence $\sigma_1^{\pm 1} \cdot \sigma_1^{\mp 1} \rightarrow \sigma_1^{\pm 1} \cdot \sigma_1^{\mp 1}$. The cross-sectional surfaces are isotopic as surface braids since they differ by the chart move that is indicated in Fig. 20.

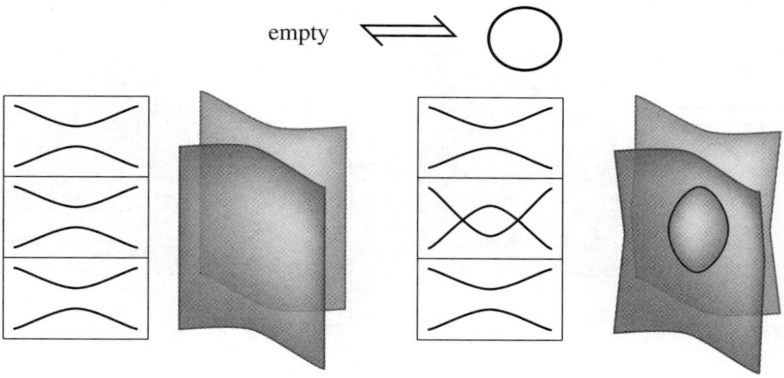

Fig. 19.

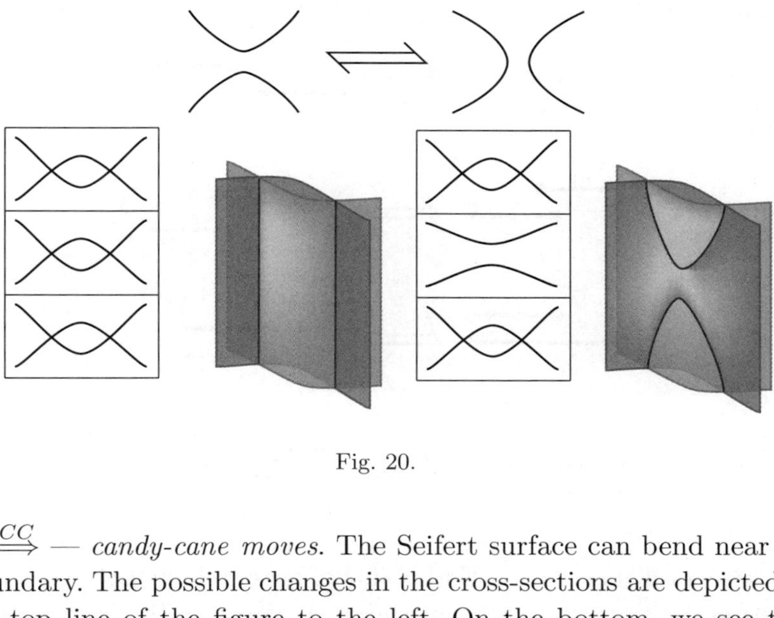

Fig. 20.

4. $\overset{CC}{\Longrightarrow}$ — *candy-cane moves.* The Seifert surface can bend near the boundary. The possible changes in the cross-sections are depicted on the top line of the figure to the left. On the bottom, we see that a type-II move followed by a branch point can be replaced by a branch point in the opposite direction. The illustration indicates the phenomenon in the permutation case, but the braid case is also easy to understand. In this case the crossing information is consistent.

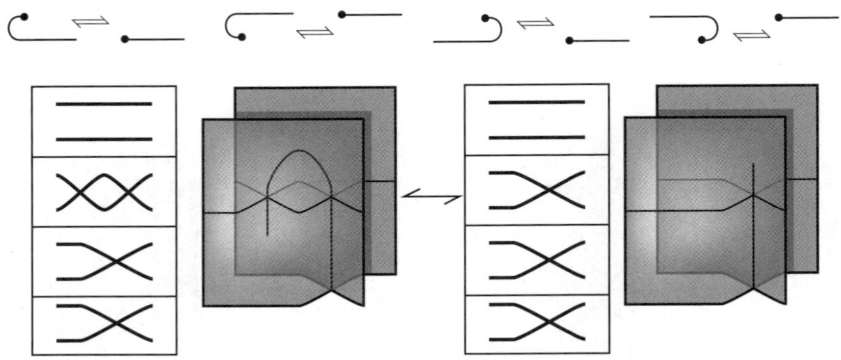

Fig. 21.

The cross-sectional surfaces are isotopic as surface braids, although the branch points corresponding the black vertices move slightly.

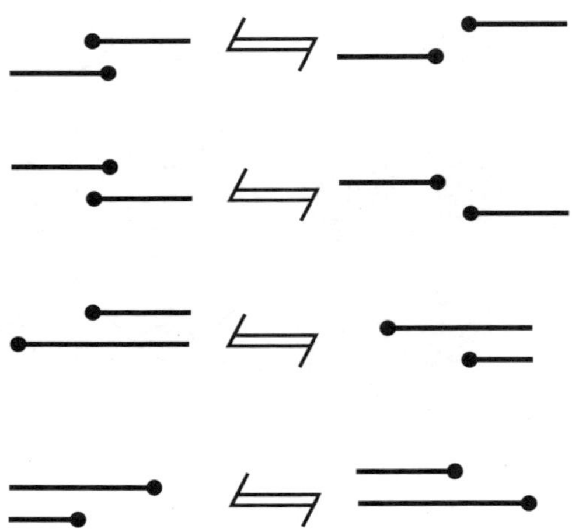

5. $\overset{X}{\Longrightarrow}$ — *exchange moves.* End points of arcs exchange places. These moves occur near a crossing in the diagram. The illustration given indicates the variety of ways in which these exchanges can take place. Within the surfaces constructed from the cross-sections, the situation is quite straightforward. The branch points and the double point arcs that terminate at the branch points are free to move within the surfaces as long as the arcs of double points do not intersect.

The cross-sectional surfaces are isotopic as surface braids because the branch points corresponding the black vertices move slightly.

6. $\stackrel{Z}{\Longrightarrow}$ — *type II zig-zag moves.* Cusps occur on the Seifert surface. One such cusp is indicated in Fig. 22. The cross-sectional surfaces are isotopic as surface braids because underlying charts are topologically equivalent.

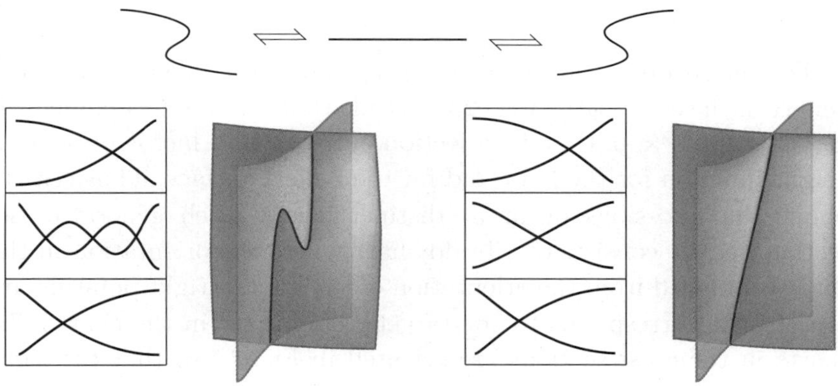

Fig. 22.

Lemma 4.1. *A given segment in a braid chart can be split into two via the moves above. Thus splitting an edge into two edges is also caused by adding a disjoint segment. In either case, a handle is attached to the underlying 3-manifold:*

Proof.

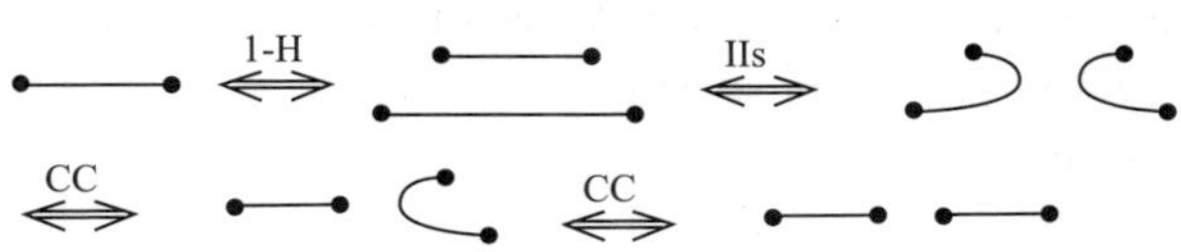

□

4.4. *Example*

In this section, we give an explicit description of the 2-fold branched cover of S^3 branched along the knot 5_2.

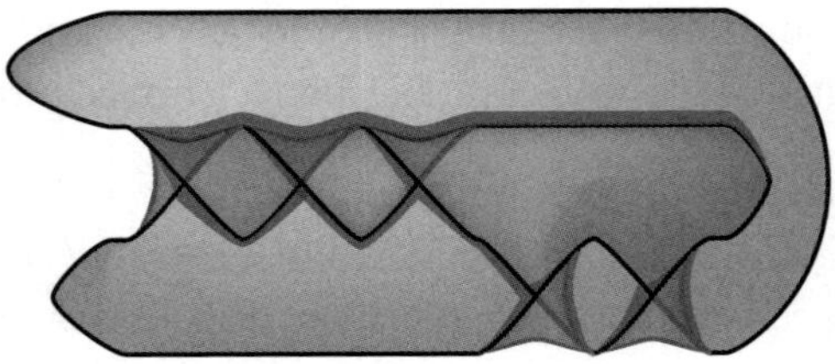

Fig. 23.

For the reader's convenience, we reproduce the illustration of the Seifert surface for 5_2 in Fig. 23. Recall, that the knot is embedded in $[0,1] \times [0,1] \times [0,1]$ and projection onto the first factor provides a height function for the knot and for the Seifert surface. Thus critical points and crossings occur at distinct levels. Each generic cross-section $p_1^{-1}(t_i)$ consists of a braid chart where the orientation in the chart is induced from the orientation of the knot: a right pointing arc in the knot corresponds to the target of an arrow in the chart. The charts in the cross-sections are oriented in Fig. 17 so that the right edges correspond to the portion of the knot diagram that is closest to the observer.

We read each cross-sectional chart as a sequence of braid words. The sequence always starts and ends at the empty word. A horizontal line intersecting the chart at a non-critical level intersects the arcs of a chart. If the arc is up pointing, the braid generator is positive; if down-pointing, then a negative generator is encountered. The critical points of the chart $\cap$ and $\cup$ correspond to type II moves. The black vertices correspond to branch points — the introduction or deletion of a braid generator or its inverse.

A critical event between charts is one of the changes in charts catalogued in the previous section.

Thus we have

- for each cross-section of a chart a braid word;
- for the set of cross-sections a "paragraph of braid words;"
- for the collection of charts an "essay" that consists of paragraphs of braid words;
- methods of getting from one word to another;

- methods of getting from one paragraph to the next;
- some standard introductory words and phrases that lead to the paragraphs and correspond to empty charts.

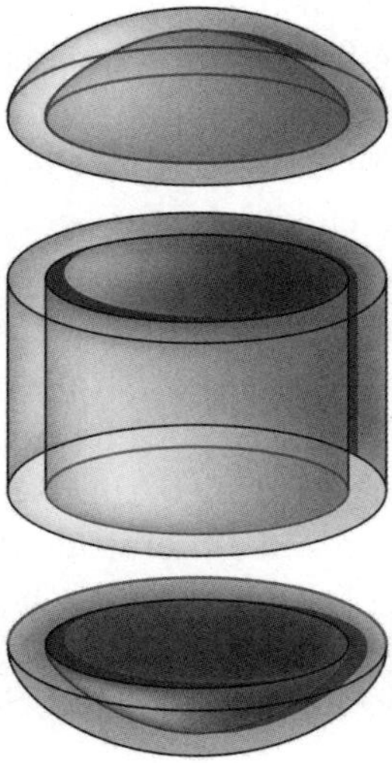

Fig. 24.

Before developing the braid word essay associated to 5_2, we discuss the introductory words and paragraphs. Each braid word represents a two-fold cover of a circle. The representation is achieved by taking the braid closure of each word. In particular, the braid closure of the empty word corresponds to two nested circles in the plane. Since the chart is embedded as a 2-disk in the 2-sphere, the unknotted, nested circles at the top and the bottom sequentially bound disks that cover the polar caps of the 2-spheres. The closure of an empty chart is depicted in Fig. 24. More generally, the pair of nested caps at the top of the diagram and the pair of nest bowls at the bottom appear at the top and bottom of any particular chart.

Thus the empty charts that appear in the cross-sections to the immediate left and right of the knot with Seifert surface represent a pair of nested, embedded spheres. These spheres, in turn, bound a pair embedded 3-balls (at each end of the chart sequence) that trivially cover the left and right ends of S^3. By turning the picture of the embedded knot on its side, we can imagine the 3-sphere as a northern polar cap which is a 3-ball, a temperate zone which is homeomorphic to $S^2 \times [0, 1]$ and in which the knot lies, and a southern polar cap which is another 3-ball.

Returning to the specifics of the sequence of charts for the Seifert surface of the knot 5_2, we have the following. We start with a pair of nested 2-spheres that were created as in the preceding paragraph and represented as an empty chart. At the end of each of the sequences below, we have a pair of nested 2-spheres that successively bound a pair of 3-balls. The notation $\mathrm{b}_j^{\pm}$ indicates that a crossing is added or subtracted via a black vertex at the jth position in the word. Thus a segment is flanked by b_j^{+} and b_{j+1}^{-}. The notation $\mathrm{II}_j^{\pm}$ indicates that a type II move has been performed that either adds or subtracts a pair of oppositely signed braid generators with the first insertion or deletion occurring at the jth position of the next word.

The braid essay associated to the Seifert surface of the knot 5_2:

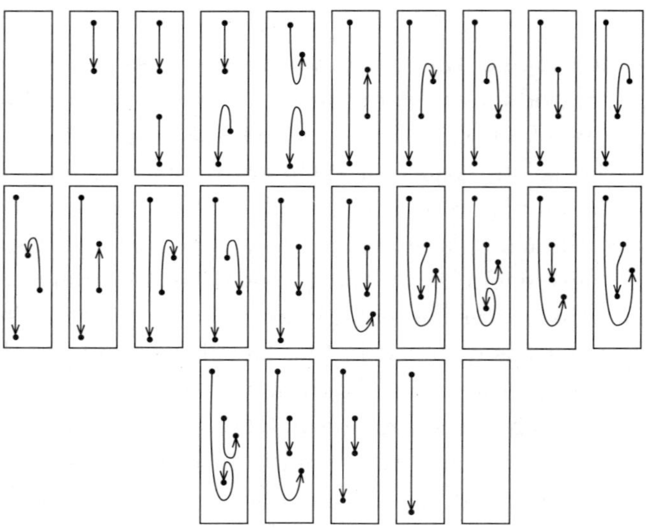

$$[(\emptyset)] \overset{1-H}{\Longrightarrow}$$

$$[(\emptyset) \xrightarrow{\mathrm{b}_0^+} (\sigma_1^{-1}) \xrightarrow{\mathrm{b}_1^-} (\emptyset)] \overset{1-H}{\Longrightarrow}$$

$$[(\emptyset) \xrightarrow{\mathrm{b}_0^+} (\sigma_1^{-1}) \xrightarrow{\mathrm{b}_1^-} (\emptyset) \xrightarrow{\mathrm{b}_0^+} (\sigma_1^{-1}) \xrightarrow{\mathrm{b}_1^-} (\emptyset)] \overset{CC}{\Longrightarrow}$$

$$[(\emptyset) \xrightarrow{\mathrm{b}_0^+} (\sigma_1^{-1}) \xrightarrow{\mathrm{b}_1^-} (\emptyset) \xrightarrow{\mathrm{II}_1^+} (\sigma_1^{-1}\sigma_1) \xrightarrow{\mathrm{b}_2^-} (\sigma_1^{-1}) \xrightarrow{\mathrm{b}_1^-} (\emptyset)] \overset{CC}{\Longrightarrow}$$

$$[(\emptyset) \xrightarrow{\mathrm{b}_0^+} (\sigma_1^{-1}) \xrightarrow{\mathrm{b}_1^+} (\sigma_1^{-1}\sigma_1) \xrightarrow{\mathrm{II}_1^-} (\emptyset) \xrightarrow{\mathrm{II}_1^+} (\sigma_1^{-1}\sigma_1) \xrightarrow{\mathrm{b}_2^-} (\sigma_1^{-1}) \xrightarrow{\mathrm{b}_1^-} (\emptyset)] \overset{IIs}{\Longrightarrow}$$

$$[(\emptyset) \xrightarrow{\mathrm{b}_0^+} (\sigma_1^{-1}) \xrightarrow{\mathrm{b}_1^+} (\sigma_1^{-1}\sigma_1) \xrightarrow{\mathrm{b}_2^-} (\sigma_1^{-1}) \xrightarrow{\mathrm{b}_1^-} (\emptyset)] \overset{CC}{\Longrightarrow}$$

$$[(\emptyset) \xrightarrow{\mathrm{b}_0^+} (\sigma_1^{-1}) \xrightarrow{\mathrm{II}_2^+} (\sigma_1^{-1}\sigma_1\sigma_1^{-1}) \xrightarrow{\mathrm{b}_3^-} (\sigma_1^{-1}\sigma_1) \xrightarrow{\mathrm{b}_2^-} (\sigma_1^{-1}) \xrightarrow{\mathrm{b}_1^-} (\emptyset)] \overset{X}{\Longrightarrow}$$

$$[(\emptyset) \xrightarrow{\mathrm{b}_0^+} (\sigma_1^{-1}) \xrightarrow{\mathrm{II}_2^+} (\sigma_1^{-1}\sigma_1\sigma_1^{-1}) \xrightarrow{\mathrm{b}_2^-} (\sigma_1^{-1}\sigma_1^{-1}) \xrightarrow{\mathrm{b}_2^-} (\sigma_1^{-1}) \xrightarrow{\mathrm{b}_1^-} (\emptyset)] \overset{CC}{\Longrightarrow}$$

$$[(\emptyset) \xrightarrow{\mathrm{b}_0^+} (\sigma_1^{-1}) \xrightarrow{\mathrm{b}_1^+} (\sigma_1^{-1}\sigma_1^{-1}) \xrightarrow{\mathrm{b}_2^-} (\sigma_1^{-1}) \xrightarrow{\mathrm{b}_1^-} (\emptyset)] \overset{CC}{\Longrightarrow}$$

$$[(\emptyset) \xrightarrow{\mathrm{b}_0^+} (\sigma_1^{-1}) \xrightarrow{\mathrm{II}_2^+} (\sigma_1^{-1}\sigma_1^{-1}\sigma_1) \xrightarrow{\mathrm{b}_3^-} (\sigma_1^{-1}\sigma_1^{-1}) \xrightarrow{\mathrm{b}_2^-} (\sigma_1^{-1}) \xrightarrow{\mathrm{b}_1^-} (\emptyset)] \overset{X}{\Longrightarrow}$$

$$[(\emptyset) \xrightarrow{\mathrm{b}_0^+} (\sigma_1^{-1}) \xrightarrow{\mathrm{II}_2^+} (\sigma_1^{-1}\sigma_1^{-1}\sigma_1) \xrightarrow{\mathrm{b}_2^-} (\sigma_1^{-1}\sigma_1) \xrightarrow{\mathrm{b}_2^-} (\sigma_1^{-1}) \xrightarrow{\mathrm{b}_1^-} (\emptyset)] \overset{CC}{\Longrightarrow}$$

$$[(\emptyset) \xrightarrow{\mathrm{b}_0^+} (\sigma_1^{-1}) \xrightarrow{\mathrm{b}_1^+} (\sigma_1^{-1}\sigma_1) \xrightarrow{\mathrm{b}_2^-} (\sigma_1^{-1}) \xrightarrow{\mathrm{b}_1^-} (\emptyset)] \overset{CC}{\Longrightarrow}$$

$$[(\emptyset) \xrightarrow{\mathrm{b}_0^+} (\sigma_1^{-1}) \xrightarrow{\mathrm{II}_2^+} (\sigma_1^{-1}\sigma_1\sigma_1^{-1}) \xrightarrow{\mathrm{b}_3^-} (\sigma_1^{-1}\sigma_1) \xrightarrow{\mathrm{b}_2^-} (\sigma_1^{-1}) \xrightarrow{\mathrm{b}_1^-} (\emptyset)] \overset{X}{\Longrightarrow}$$

$$[(\emptyset) \xrightarrow{\mathrm{b}_0^+} (\sigma_1^{-1}) \xrightarrow{\mathrm{II}_2^+} (\sigma_1^{-1}\sigma_1\sigma_1^{-1}) \xrightarrow{\mathrm{b}_2^-} (\sigma_1^{-1}\sigma_1^{-1}) \xrightarrow{\mathrm{b}_2^-} (\sigma_1^{-1}) \xrightarrow{\mathrm{b}_1^-} (\emptyset)] \overset{CC}{\Longrightarrow}$$

$$[(\emptyset) \xrightarrow{\mathrm{b}_0^+} (\sigma_1^{-1}) \xrightarrow{\mathrm{b}_1^+} (\sigma_1^{-1}\sigma_1^{-1}) \xrightarrow{\mathrm{b}_2^-} (\sigma_1^{-1}) \xrightarrow{\mathrm{b}_1^-} (\emptyset)] \overset{CC}{\Longrightarrow}$$

$$[(\emptyset) \xrightarrow{\mathrm{b}_0^+} (\sigma_1^{-1}) \xrightarrow{\mathrm{b}_1^+} (\sigma_1^{-1}\sigma_1^{-1}) \xrightarrow{\mathrm{b}_2^-} (\sigma_1^{-1}) \xrightarrow{\mathrm{b}_1^+} (\sigma_1^{-1}\sigma_1) \xrightarrow{\mathrm{II}_1^-} (\emptyset)] \overset{X}{\Longrightarrow}$$

$$[(\emptyset) \xrightarrow{\mathrm{b}_0^+} (\sigma_1^{-1}) \xrightarrow{\mathrm{b}_1^+} (\sigma_1^{-1}\sigma_1^{-1}) \xrightarrow{\mathrm{b}_2^+} (\sigma_1^{-1}\sigma_1^{-1}\sigma_1) \xrightarrow{\mathrm{b}_2^-} (\sigma_1^{-1}\sigma_1) \xrightarrow{\mathrm{II}_1^-} (\emptyset)] \overset{IIs}{\Longrightarrow}$$

$$[(\emptyset) \xrightarrow{\mathrm{b}_0^+} (\sigma_1^{-1}) \xrightarrow{\mathrm{b}_1^+} (\sigma_1^{-1}\sigma_1^{-1}) \xrightarrow{\mathrm{b}_2^+} (\sigma_1^{-1}\sigma_1^{-1}\sigma_1) \xrightarrow{\mathrm{II}_2^-} (\sigma_1^{-1}) \xrightarrow{\mathrm{II}_2^+} (\sigma_1^{-1}\sigma_1^{-1}\sigma_1)$$

$$\xrightarrow{\mathrm{b}_2^-} (\sigma_1^{-1}\sigma_1) \xrightarrow{\mathrm{II}_1^-} (\emptyset)] \overset{CC^2}{\Longrightarrow}$$

$$[(\emptyset) \overset{b_0^+}{\to} (\sigma_1^{-1}) \overset{b_1^+}{\to} (\sigma_1^{-1}\sigma_1^{-1}) \overset{b_2^-}{\to} (\sigma_1^{-1}) \overset{b_1^+}{\to} (\sigma_1^{-1}\sigma_1) \overset{II_1^-}{\to} (\emptyset)] \overset{X}{\Longrightarrow}$$

$$[(\emptyset) \overset{b_0^+}{\to} (\sigma_1^{-1}) \overset{b_1^+}{\to} (\sigma_1^{-1}\sigma_1^{-1}) \overset{b_2^+}{\to} (\sigma_1^{-1}\sigma_1^{-1}\sigma_1) \overset{b_2^-}{\to} (\sigma_1^{-1}\sigma_1) \overset{II_1^-}{\to} (\emptyset)] \overset{IIs}{\Longrightarrow}$$

$$[(\emptyset) \overset{b_0^+}{\to} (\sigma_1^{-1}) \overset{b_1^+}{\to} (\sigma_1^{-1}\sigma_1^{-1}) \overset{b_2^+}{\to} (\sigma_1^{-1}\sigma_1^{-1}\sigma_1) \overset{II_2^-}{\to} (\sigma_1^{-1}) \overset{II_2^+}{\to} (\sigma_1^{-1}\sigma_1^{-1}\sigma_1)$$
$$\overset{b_2^-}{\to} (\sigma_1^{-1}\sigma_1) \overset{II_1^-}{\to} (\emptyset)] \overset{CC^2}{\Longrightarrow}$$

$$[(\emptyset) \overset{b_0^+}{\to} (\sigma_1^{-1}) \overset{b_1^+}{\to} (\sigma_1^{-1}\sigma_1^{-1}) \overset{b_2^-}{\to} (\sigma_1^{-1}) \overset{b_1^+}{\to} (\sigma_1^{-1}\sigma_1) \overset{II_1^-}{\to} (\emptyset)] \overset{CC}{\Longrightarrow}$$

$$[(\emptyset) \overset{b_0^+}{\to} (\sigma_1^{-1}) \overset{b_1^+}{\to} (\sigma_1^{-1}\sigma_1^{-1}) \overset{b_2^-}{\to} (\sigma_1^{-1}) \overset{b_1^-}{\to} (\emptyset)] \overset{2-H}{\Longrightarrow}$$

$$[(\emptyset) \overset{b_0^+}{\to} (\sigma_1^{-1}) \overset{b_1^-}{\to} (\emptyset)] \overset{2-H}{\Longrightarrow}$$

$$[(\emptyset)].$$

Remark 4.2. The regions between the attachment of the last 1-handle and the first 2-handle represent homeomorphisms of the torus. One can easily trace meridional and longitudinal classes through the homeomorphism and determine the lens space that is the 2-fold branched cover of the knot 5_2. In more generality, it is straightforward to determine the lens space structure of the 2-fold branched cover of any 2-bridge knot.

In even more generality, when we choose a braid representation of a knot, there is a standard Seifert surface associated to such a representative. A Heegaard splitting of the 2-fold branched cover is determined since the optima of the diagram on the left of the diagram (recall we are arranging the knots horizontally) each corresponds to a 1-handle attachment while the optima on the right correspond to 2-handles. Attaching regions for 1-handles encircle the arcs in the braid chart. The braid induces a homeomorphism of the Heegaard surface. And so a representation of the braid group in the mapping class group is determined.

Remark 4.3. The Seifert surface need not be orientable in order to fold the 2-fold branched cover. We can use unoriented charts, and map each covering surface in $\mathbb{R}^3$. The union will be a 3-manifold mapped into $\mathbb{R}^4$. Furthermore, we can cut a non-orientable

Seifert surface along a simple closed curve and obtain an orientable surface bounded by a link one component of which is the original knot. We use the orientation reversing loop to define a sequence of semi-oriented charts. Below the sequence of semi-oriented charts associated to the three-half twisted Möbius band whose boundary is the trefoil is indicated.

Immediately following this remark, the braid essay that is associated is given. The crossing changes are indicated as $\sigma_1^{\pm} \stackrel{\chi_j^-}{\to} \sigma_1^{\mp}$. The birth and death of such a curve are indicated by $\stackrel{\Xi_\pm}{\Longrightarrow}$.

The braid essay associated to the three-half twisted Möbius band:

$[(\emptyset)] \stackrel{1-H}{\Longrightarrow}$

$[(\emptyset) \stackrel{\mathrm{b}_0^+}{\to} (\sigma_1) \stackrel{\mathrm{b}_1^-}{\to} (\emptyset)] \stackrel{\Xi_+}{\Longrightarrow}$

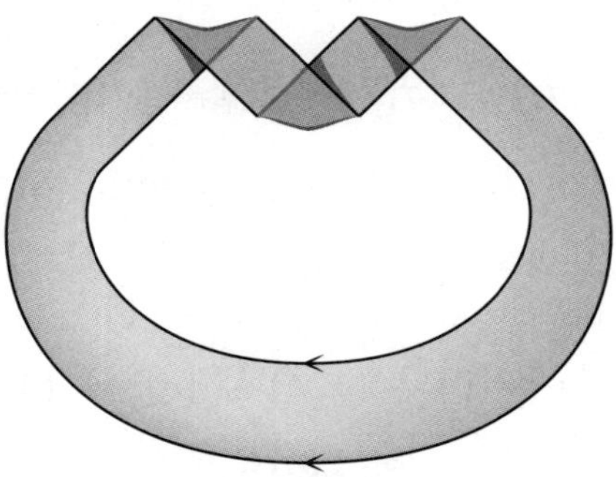

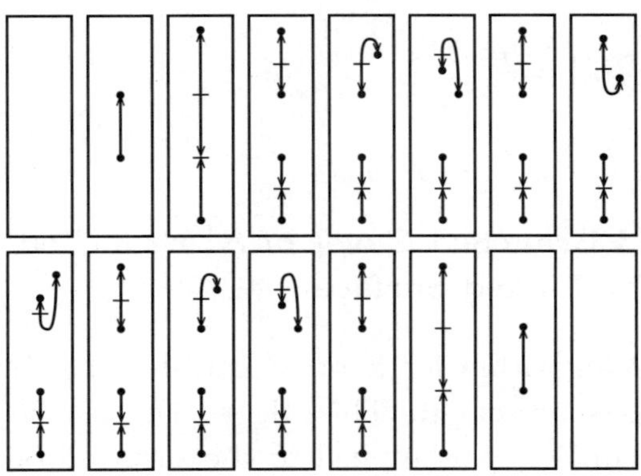

$$[(\emptyset) \xrightarrow{\mathrm{b}_0^+} (\sigma_1) \xrightarrow{\chi_1^-} (\sigma_1^{-1}) \xrightarrow{\chi_1^+} (\sigma_1) \xrightarrow{\mathrm{b}_1^-} (\emptyset)] \stackrel{1-H}{\Longrightarrow}$$

$$[(\emptyset) \xrightarrow{\mathrm{b}_0^+} (\sigma_1) \xrightarrow{\chi_1^-} (\sigma_1^{-1}) \xrightarrow{\mathrm{b}_1^-} (\emptyset) \xrightarrow{\mathrm{b}_0^+} (\sigma_1^{-1}) \xrightarrow{\chi_1^+} (\sigma_1) \xrightarrow{\mathrm{b}_1^-} (\emptyset)] \stackrel{CC}{\Longrightarrow}$$

$$[(\emptyset) \xrightarrow{\mathrm{II}_1^+} (\sigma_1\sigma_1^{-1}) \xrightarrow{\mathrm{b}_2^-} (\sigma_1) \xrightarrow{\chi_1^-} (\sigma_1^{-1}) \xrightarrow{\mathrm{b}_1^-} (\emptyset) \xrightarrow{\mathrm{b}_0^+} (\sigma_1^{-1}) \xrightarrow{\chi_1^+} (\sigma_1) \xrightarrow{\mathrm{b}_1^-} (\emptyset)] \stackrel{X}{\Longrightarrow}$$

$$[(\emptyset) \xrightarrow{\mathrm{II}_1^+} (\sigma_1\sigma_1^{-1}) \xrightarrow{\chi_1^-} (\sigma_1^{-1}\sigma_1^{-1}) \xrightarrow{\mathrm{b}_1^-} (\sigma_1^{-1}) \xrightarrow{\mathrm{b}_1^-} (\emptyset) \xrightarrow{\mathrm{b}_0^+} (\sigma_1^{-1}) \xrightarrow{\chi_1^+} (\sigma_1)$$
$$\xrightarrow{\mathrm{b}_1^-} (\emptyset)] \stackrel{CC}{\Longrightarrow}$$

$$[(\emptyset) \xrightarrow{\mathrm{b}_0^+} (\sigma_1) \xrightarrow{\chi_1^-} (\sigma_1^{-1}) \xrightarrow{\mathrm{b}_1^-} (\emptyset) \xrightarrow{\mathrm{b}_0^+} (\sigma_1^{-1}) \xrightarrow{\chi_1^+} (\sigma_1) \xrightarrow{\mathrm{b}_1^-} (\emptyset)] \stackrel{CC}{\Longrightarrow}$$

$$[(\emptyset) \xrightarrow{\mathrm{b}_0^+} (\sigma_1) \xrightarrow{\chi_1^-} (\sigma_1^{-1}) \xrightarrow{\mathrm{b}_1^+} (\sigma_1^{-1}\sigma_1) \xrightarrow{\mathrm{II}_1^-} (\emptyset) \xrightarrow{\mathrm{b}_0^+} (\sigma_1^{-1}) \xrightarrow{\chi_1^+} (\sigma_1) \xrightarrow{\mathrm{b}_1^-} (\emptyset)] \stackrel{X}{\Longrightarrow}$$

$$[(\emptyset) \xrightarrow{\mathrm{b}_0^+} (\sigma_1) \xrightarrow{\mathrm{b}_0^+} (\sigma_1\sigma_1) \xrightarrow{\chi_1^-} (\sigma_1^{-1}\sigma_1) \xrightarrow{\mathrm{II}_1^-} (\emptyset) \xrightarrow{\mathrm{b}_0^+} (\sigma_1^{-1}) \xrightarrow{\chi_1^+} (\sigma_1) \xrightarrow{\mathrm{b}_1^-} (\emptyset)] \stackrel{CC}{\Longrightarrow}$$

$$[(\emptyset) \xrightarrow{\mathrm{b}_0^+} (\sigma_1) \xrightarrow{\chi_1^-} (\sigma_1^{-1}) \xrightarrow{\mathrm{b}_1^-} (\emptyset) \xrightarrow{\mathrm{b}_0^+} (\sigma_1^{-1}) \xrightarrow{\chi_1^+} (\sigma_1) \xrightarrow{\mathrm{b}_1^-} (\emptyset)] \stackrel{CC}{\Longrightarrow}$$

$$[(\emptyset) \xrightarrow{\mathrm{II}_1^+} (\sigma_1\sigma_1^{-1}) \xrightarrow{\mathrm{b}_2^-} (\sigma_1) \xrightarrow{\chi_1^-} (\sigma_1^{-1}) \xrightarrow{\mathrm{b}_1^-} (\emptyset) \xrightarrow{\mathrm{b}_0^+} (\sigma_1^{-1}) \xrightarrow{\chi_1^+} (\sigma_1) \xrightarrow{\mathrm{b}_1^-} (\emptyset)] \stackrel{X}{\Longrightarrow}$$

$$[(\emptyset) \xrightarrow{\mathrm{II}_1^+} (\sigma_1\sigma_1^{-1}) \xrightarrow{\chi_1^-} (\sigma_1^{-1}\sigma_1^{-1}) \xrightarrow{\mathrm{b}_1^-} (\sigma_1^{-1}) \xrightarrow{\mathrm{b}_1^-} (\emptyset) \xrightarrow{\mathrm{b}_0^+} (\sigma_1^{-1}) \xrightarrow{\chi_1^+} (\sigma_1)$$
$$\xrightarrow{\mathrm{b}_1^-} (\emptyset)] \stackrel{CC}{\Longrightarrow}$$

$$[(\emptyset) \xrightarrow{\mathrm{b}_0^+} (\sigma_1) \xrightarrow{\chi_1^-} (\sigma_1^{-1}) \xrightarrow{\mathrm{b}_1^-} (\emptyset) \xrightarrow{\mathrm{b}_0^+} (\sigma_1^{-1}) \xrightarrow{\chi_1^+} (\sigma_1) \xrightarrow{\mathrm{b}_1^-} (\emptyset)] \stackrel{2-H}{\Longrightarrow}$$

$$[(\emptyset) \xrightarrow{\mathrm{b}_0^+} (\sigma_1) \xrightarrow{\chi_1^-} (\sigma_1^{-1}) \xrightarrow{\chi_1^+} (\sigma_1) \xrightarrow{\mathrm{b}_1^-} (\emptyset)] \stackrel{\Xi_-}{\Longrightarrow}$$

$$[(\emptyset) \xrightarrow{\mathrm{b}_0^+} (\sigma_1) \xrightarrow{\mathrm{b}_1^-} (\emptyset)] \stackrel{2-H}{\Longrightarrow}$$

$$[(\emptyset)] \stackrel{1-H}{\Longrightarrow}$$

5. The 2-fold Branched Cover of S^4 Branched Along a Knotted or Linked Surface

In [2] a Seifert algorithm for knotted surfaces that project without branch points is presented. This algorithm was adjusted by the second author in [9] to be applied to the surface braid case. In our respective books, [3] or [10] these algorithms are described in detail. The 2-fold branched cover of S^4 branched along an orientable knotted or linked surface can be constructed from two copies of S^4 both cut along the Seifert solid, and gluing the positive side of one copy to the negative side of the other.

The braid chart can be used to give a movie description of the branch locus. In such a movie, or in a standard movie description of the knotted surface, the Seifert solid is cut into 2-dimensional slices that are Seifert surfaces for the classical cross-sectional links. Thus for any such cross-section, the previous section provided an embedding of the 2-fold branched cover of that cross-section. These are, then, connected by either handle attachments or isotopies that are induced by the Reidemeister moves.

Specifically, the birth of a simple closed curve corresponds to a 1-handle attached between the successive 3-dimensional 2-fold branched covers. A 1-handle attached between successive movie stills corresponds to a 2-handle attached between successive 3-dimensional 2-fold branched covers.

Following the discussion of the 3-fold branched covers of classical knots, an example of 3-fold branched cover of the 4-sphere branched along the spun trefoil will be given. These examples are more complicated than the 2-fold branched coverings.

In Fig. 25 we indicate that the 2-fold branched cover of S^4 branched along an unknotted sphere is S^4, and we indicate an embedded folding. Meanwhile, the 2-fold branched cover of S^4 branched along an unknotted torus is $S^2 \times S^2$ and also has an embedded folding.

To summarize the constructions, the Seifert surface in dimension 3 creates a 2-dimensional chart or curtain. By placing this in general position with respect to a height direction, we can cut between critical levels and connect the resulting sequence of 2-dimensional charts via well-understood critical phenomena that correspond to chart moves. In this way, we can construct embedded foldings for the 2-fold branched cover of S^3 branched along a knot or link. Furthermore, we can use non-orientable Seifert surfaces to construct immersed foldings. In 4-space, we can use the Seifert algorithm

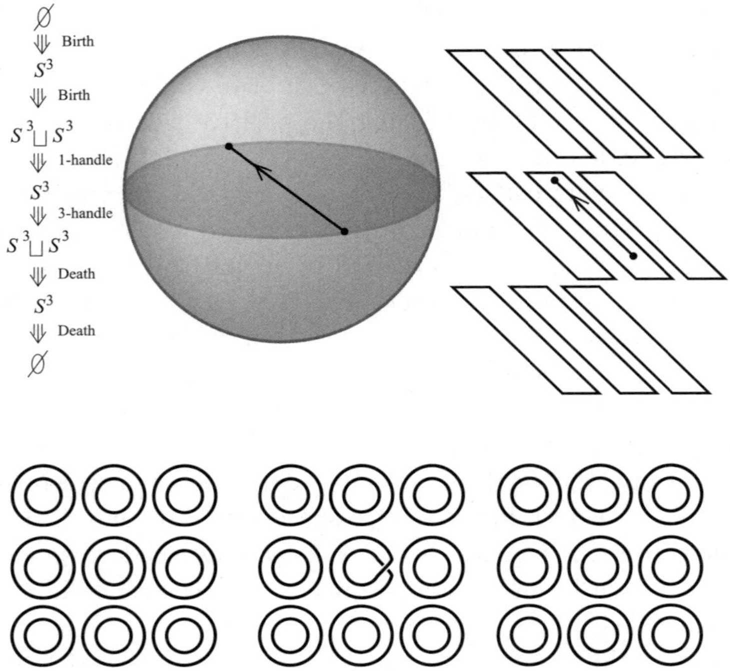

Fig. 25.

for knotted surfaces to construct a Seifert solid. This solid can be expressed in movie form or via the surface braid picture of the knotted orientable surface. The critical events and Reidemeister moves have well-understood effects upon the cross-sectional Seifert surfaces. A folding of the 2-fold branched cover of S^4 branched along a given surface knot or link is constructed, then, via interconnecting the corresponding covers of S^3. Thus we have proven Theorem 1.1.

6. Three-fold Simple Branched Covers

In this section, we construct embeddings and immersions of 3-fold branch covers of the 3- and 4-sphere branched over the trefoil, the knot 7_4, and the knotted sphere that is the spin of the trefoil. We begin with the trefoil as indicated Fig. 26. The trefoil is 3-colorable, and we can lift this coloring to a representation to the 3-string braid

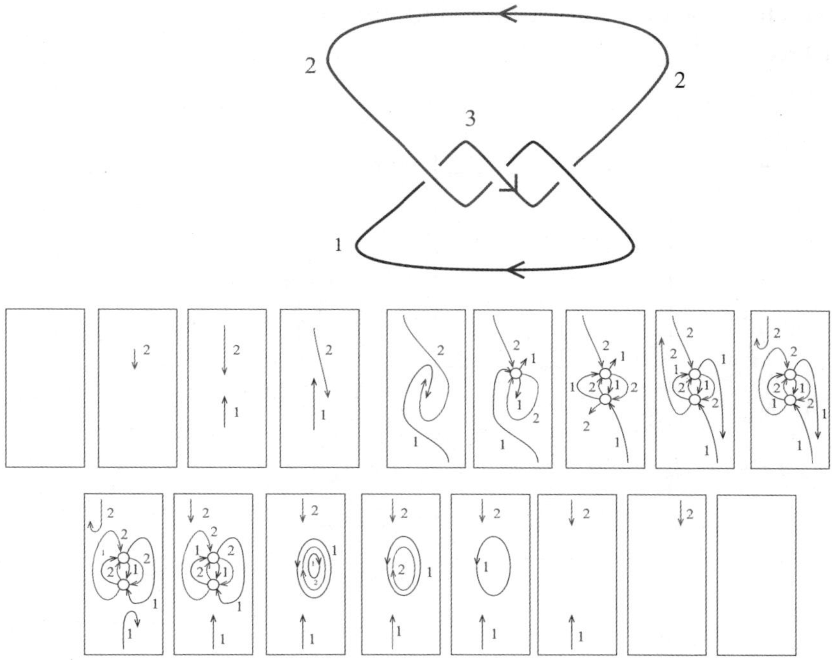

Fig. 26.

group (also the fundamental group of the knot itself) by assigning the braid generator σ_1 to the (blue) label 1, assigning the braid generator σ_2 to the (red) label 2, and assigning $\sigma_2^{-1}\sigma_1\sigma_2$ to the (green) label 3. From this coloring, a sequence of braid charts is constructed. The endpoints of the arcs in the charts correspond to black vertices, and as the charts are stacked, these end points trace the outline of the knot. Successive braid charts differ by a planar isotopy, by one of the chart moves, or by the addition or subtraction of a simple arc. As before, the addition or subtraction of an arc corresponds to attaching a handle between successive sheets in the covering. The sequence of charts can be interpolated in 3-space to create a (non-generically) immersed surface in 3-space that we call a *curtain*. To construct the 3-fold branch cover, we take three copies of S^3, cut each along the appropriately labeled sheets of the curtains, and re-glue. The embedding in this case is achieved by interpreting each chart as a braided surface and interpolating between successive surfaces.

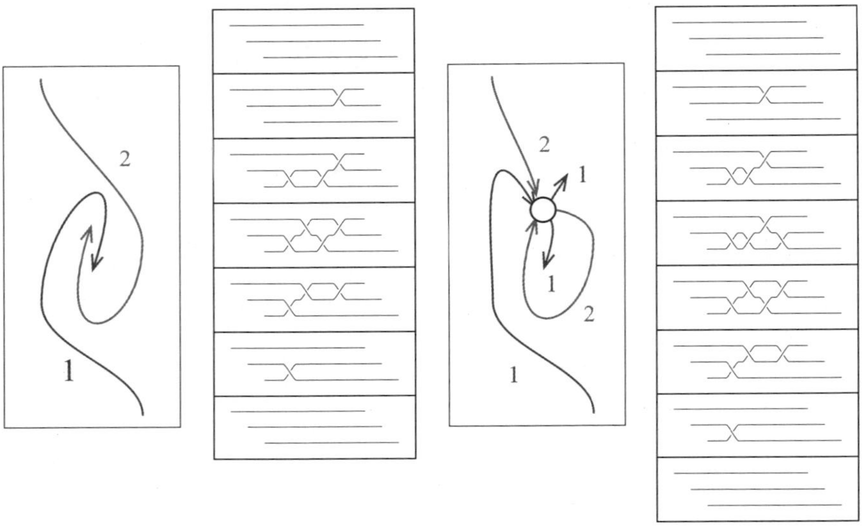

Fig. 27.

For two cross-sections of the curtains, we have indicated the resulting braid movies depicted in Fig. 27. The difference between these two movies is a C-III move that corresponds to pushing a branch point through a transverse sheet. At the level of surface braids, this is a manifestation of one of the Roseman moves. The complete sequence of such surface isotopies and handle additions describes the branched cover explicitly embedded in 5-space.

The corresponding immersed folding of the 3-fold branched cover of S^3 branched along the knot 7_4 is illustrated in Fig. 28. As before, the (blue) label 1 indicates the braid generator σ_1 and the (red) label 2 indicates σ_2. Here, however, nodes are introduced in the cross-sectional charts. Such a node indicates that a braid generator is switched with its inverse. In a cross-sectional surface braid, there is a self-transverse double point at such a node. In the case at hand there is a simple closed curve of such double points for the immersed folding of branched cover. It is known [11] that there is no surjective

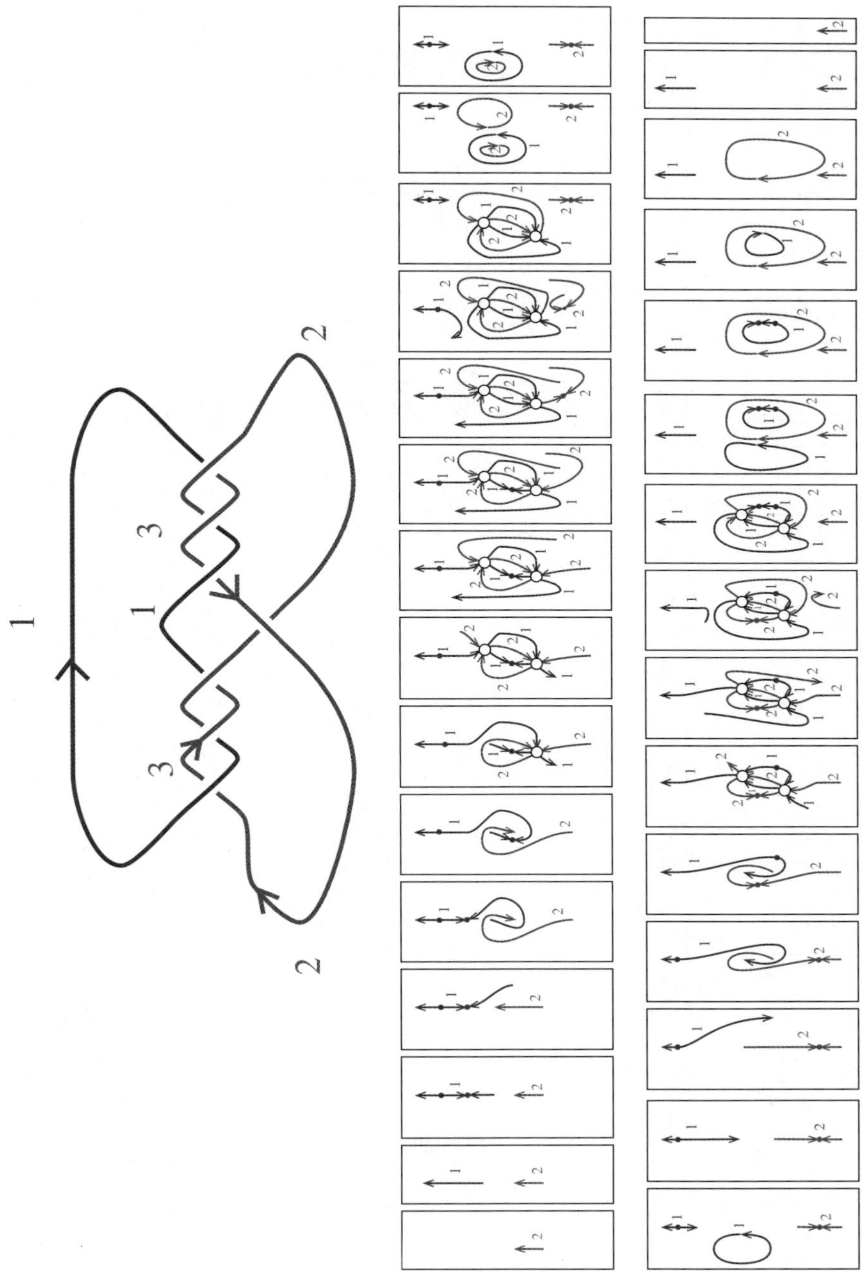

Fig. 28.

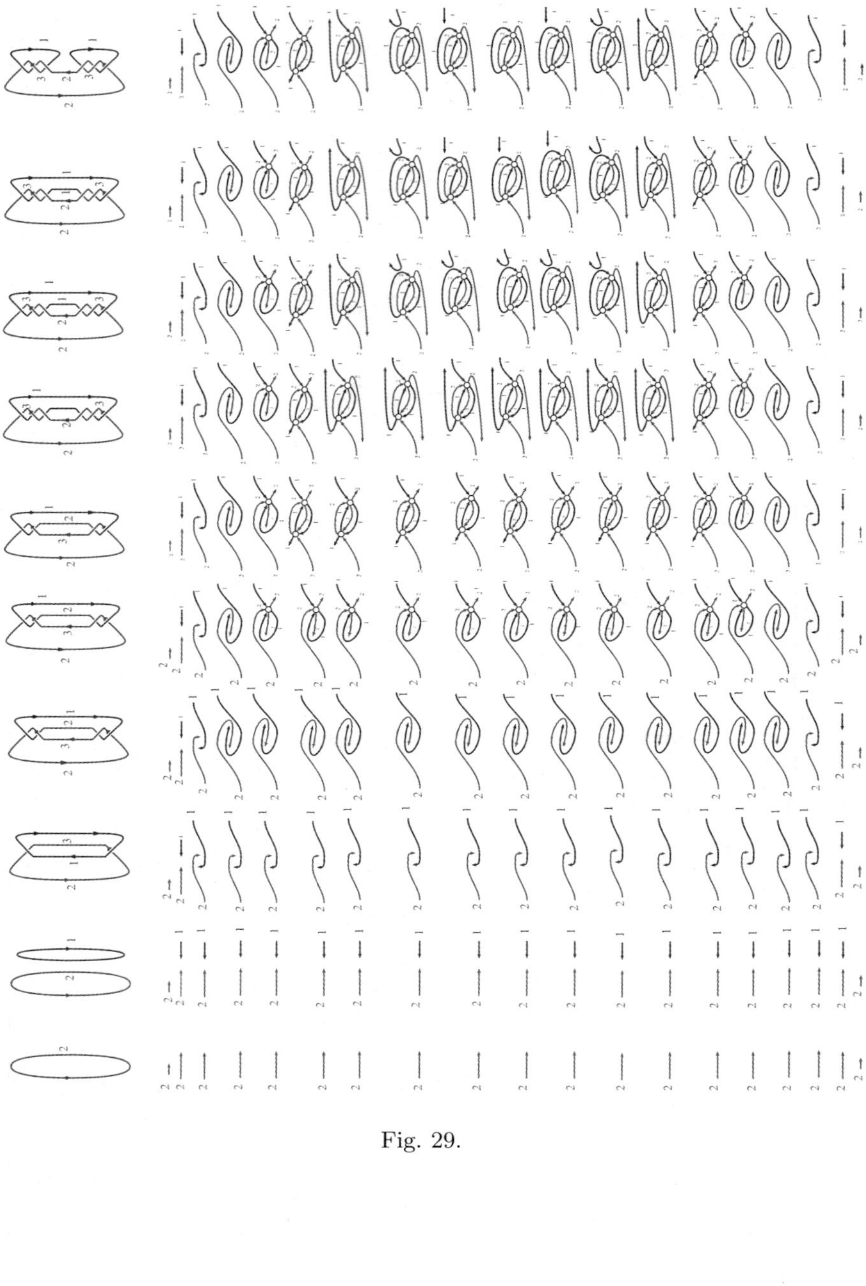

Fig. 29.

homomorphism between the fundamental groups of the complement of 7_4 and the complement of the trefoil in which the peripheral structure is preserved. Such a homomorphism is necessary for there to be an embedded folding. See also [6].

Our last example for this paper is a construction of an embedded folding of the 3-fold branch cover of S^4 branched along the spun trefoil. A movie for the spun-trefoil is palindromic with the square knot as the central step. We illustrate a curtain for each nontrivial cross-section. Each such curtain determines an embedding of the 3-fold branched cover of S^3 branched along the corresponding knot or link (in fact trivial link until the middle level). Two curtains in successive cross-sections differ by easy to understand moves to curtains. For example, moving from the empty curtain to the curtain that is a red disk (a disk labeled 2) corresponds to attaching a 1-handle between the second and third copies of S^3. The next few steps all correspond to performing a move and then undoing the same move. Thus the curtain move is to replace the identity with a pair of a move and its inverse. When a saddle is attached to the cross-sectional knot, a critical point of index 2 is attached between the curtains.

6.1. *Proof of Theorem 1.2 when* $k = 1$

Suppose that a knot or link K is given that is 3-colorable. Assume that the knot is given in braid form with braid index m. By convention, we write the knot as the closure of the braid β with the closing strings on the right of the braid, and orient the braid downward.

The coloring induces a *color vector* at the top of the braid. This is the sequence $(c_1, \ldots, c_m)$ of colors at the top-left of the braid with $c_j \in \{1, 2, 3\}$. Each braid generator or its inverse induces a transformation of this vector. In a few sentences, we will describe these transformations in terms of the curtains. First, we use the color vector to describe the structure of the curtains before the braiding occurs.

We may recolor the knot so that the first element, c_1, in the color vector is 1. By convention we indicate this element by the color blue in our illustrations. If $c_j = 2$, then we color this red. The color 1 corresponds to the transposition $(1, 2)$ while 2 corresponds to $(2, 3)$. The remaining color 3 corresponds to $(1, 3)$. Observe that in the permutation group $(1, 3) = (2, 3)^{-1}(1, 2)(2, 3)$.

In anticipation of a lifting of the cross-sectional chart of a curtain to the braid group, we construct a specific chart that reflects the color vector immediately before braiding occurs. There is a sequence of horizontal arcs oriented as right-pointing arrows. Using a standard coordinate system, the top most arc points from $(-m, m)$ towards (m, m). The arc immediately below this points from $(-m+1, m-1)$ towards $(m-1, m-1)$. Continue shortening the next arc by two units and move it one unit below the prior arc. Thus the lowest arc points from $(-1, 1)$ to $(1, 1)$. This top arc is labeled 1 (colored blue). Similarly, if $c_j = 1$, then the arrow from $(-m-1+j, m+1-j)$ to $(m+1-j, m+1-j)$ is labeled 1 (colored blue). If $c_j = 2$, then the arrow from $(-m-1+j, m+1-j)$ to $(m+1-j, m+1-j)$ is labeled 2 (colored red). Finally, if $c_j = 3$, then the arrow from $(-m-1+j, m+1-j)$ to $(m+1-j, m+1-j)$ is colored 1 but it is encircled by an anti-clockwise oriented oval that is colored 2.

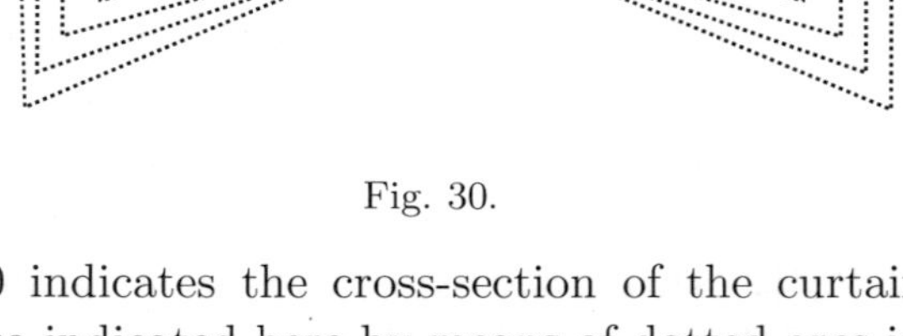

Fig. 30.

Figure 30 indicates the cross-section of the curtain before the braiding. Also indicated here by means of dotted arcs is an Hurwitz arc system that connects the end-points of the arrows to the base point which we take to be the origin. Moving up from this level, at the jth level the jth arc vanishes. If it is encircled by an oval, then this oval vanishes between the jth and $(j+1)$st level. Thus before the knot appears, the curtain is empty and represents three nested

2-spheres. After the first arc is born, the first and second spheres are connected by a 1-handle. As each maximal point of the knot (in braid form) is passed, another 1-handle is attached to the surface above. The ovals correspond to a type II, type II bubble move being performed so that a handle can be attached between the first and third sheets. The Hurwitz arcs on the left side of Fig. 30 record the color vector $(c_1, \ldots, c_m)$.

Figure 31 indicates the effect of the standard braid generator s_j and its inverse to the curtains. The grey bands indicate that other arcs may cross the Hurwitz arc that connects to the black vertex. The result of the action of the braid generator or its inverse upon the color vector is determined by reading the intersection sequence with the corresponding Hurwitz arcs.

At the bottom of the braid β, the curtain still consists of k embedded arrows with some of them encircled, but the topography is potentially complicated. However, the color vector has returned to its original state $(c_1, \ldots, c_m)$. We proceed to simplify the curtain by applying chart moves, or, if necessary, by introducing nodes. Here is how to proceed.

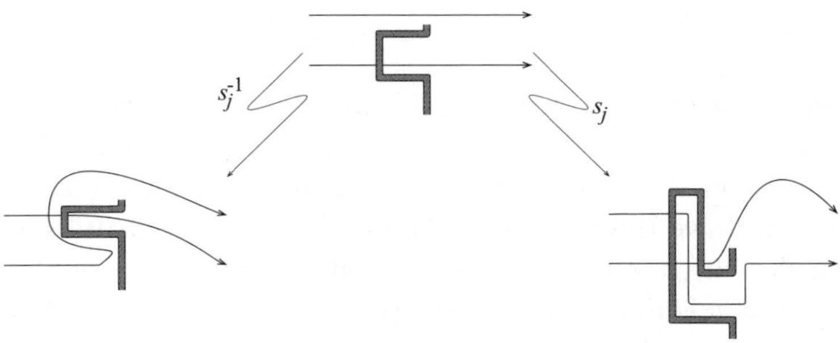

Fig. 31.

Consider the Hurwitz arc α_1 that appears on the top left. It intersects the arcs and ovals of the chart in a sequence w_1. In this way the word $w_1 c_1 w_1^{-1}$ represents the element $c_1 = (1, 2)$ in the permutation group. Consequently, the intersection sequence w_1 represents a trivial word or $(1, 2)$ in the permutation group Σ_3. The word w_1

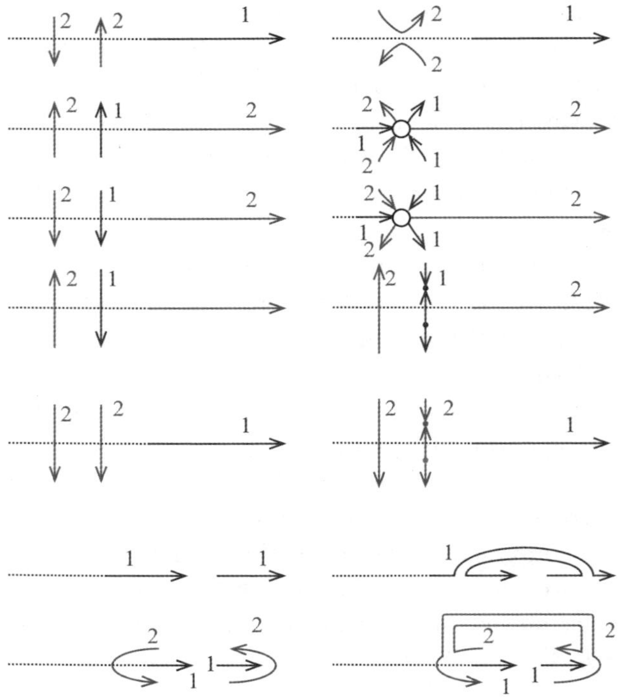

Fig. 32.

also represents a word in the braid group since the orientations of the oval and the original arrows can be used to determine a direction of intersection with the Hurwitz arc. That is to say, at this point each chart above the bottom of the braid β is oriented. We continue to consider the oriented case. If w_1 contains a syllable of the form vv^{-1}, where v is red or blue, then we may eliminate this syllable by performing a saddle move and using a segment of the Hurwitz arc α_1 as the core of the saddle. Such moves can be performed successively. Similarly, if w_1 ends in (red, blue) or in (red^{-1}, blue^{-1}), then a C-III move can be used to pull the black vertex at the end of the top arc to the left; doing so reduces the length of w_1.

Continuing in this way, we gradually shorten w_1. However, w_1 may represent a trivial word in the permutation group or $(1, 2)$, but may not represent a trivial word or σ_1 in the braid group. In that case, there is a syllable of the form vv in the word w, and we introduce a pair of nodes on one of the segments and on either side of the Hurwitz arc α_1. Doing so, changes the local orientation and the intersection sequence becomes $v^{-1}v$, so that a saddle cancellation can occur. Thus we simplify the intersection word until it is empty or σ_1. If the intersection word is σ_1, then by a saddle move, we may remove the intersection. At this stage, another saddle band can be attached, so that the points $(-m, m)$ and (m, m) are joined by an embedded arc that does not intersect the rest of the chart.

We proceed to successively simplify the intersection words between the Hurwitz arcs α_2 through α_m so that they are each trivial. When each intersection sequence is empty, the left end of any arc or oval can be attached to the right end by means of saddle bands. We obtain a copy of the original chart immediately before the braiding and a disjoint closed chart that has nodes upon it. The closed chart can be eliminated by C-I moves, by moving nodes through white vertices (if necessary) and by canceling pairs of nodes. The copy of the original chart is trivialized by mimicking the moves that occur before the braid β. This completes the proof.

Future Work

In a subsequent work, we will demonstrate how to construct immersed foldings of the 3-fold branched covers of S^4 branched along a knotted or linked surface. We are also examining simple branched covers of higher degree. In particular, we are interested in 5-fold branch covers of S^4 in the light of Iori and Piergallini's theorem. We have constructed an immersion of the 3-fold branched cover of S^4 branched along the 2-twist-spun trefoil. These ideas lead naturally to a study of higher dimensional knotting via a braid theory. In particular, diagrammatic methods can be extended by using charts, curtains, and additional colorings.

Acknowledgments

The genesis of this paper was a conversation that J.S.C. had with John Etnyre at Georgia Tech a few years ago. During the Sixth East Asia Knot School in Hiroshima (January 2011), J.S.C. and S.K. improved upon the original construction. At that time J.S.C. had support from NSF grant #0603926. J.S.C.'s visit to Hiroshima was generously supported by grant JSPS KAKENHI number 19204002 issued to Makota Matsumoto. This paper began when J.S.C. was visiting Kyungpook National University. His visit was supported by the Ministry of Education Science and Technology (MEST) and the Korean Federation of Science and Technology Societies (KOFST). S.K. is being supported by JSPS KAKENHI Grant Nos. 21340015 and 23654027. As part of that grant, he was able to visit Kyungpook, continue this conversation with J.S.C., and the authors developed the final steps in the process. We would also like to thank Makoto Sakuma and Daniel Silver for helpful bibliographical remarks.

References

[1] J. W. Alexander, Note on Riemann spaces, *Bull. Amer. Math. Soc.* **26** (1920) 370–372.
[2] J. S. Carter and M. Saito, A Seifert algorithm for knotted surfaces, *Topology* **36** (1996) 179–201.
[3] J. S. Carter and M. Saito, *Knotted Surfaces and Their Diagrams*, Mathematical Surveys and Monographs, Vol. 55. Amer. Math. Soc., Providence, RI, 1998.
[4] A. Clebsch, Zur Theorie der Riemann'schen Fläche, *Math. Ann.* **6** (1973) 216–230.
[5] H. M. Hilden, Three-fold branched coverings of S^3, *Amer. J. Math.* **98** (1976) 989–997.
[6] K. Horie, T. Kitano, M. Matsumoto and M. Suzuki, A partial order on the set of prime knots with up to 11 crossings, *J. Knot Theory Ramifications* **20** (2011) 275–303.
[7] A. Hurwitz, Ueber Riemann'sche Flächen mit gegebenen Verzweigungspunkten, *Math. Ann.* **39** (1891) 1–60.
[8] M. Iori and R. Piergallini, 4-manifolds as covers of the 4-sphere branched over non-singular surfaces, *Geom. Topol.* **6** (2002) 393–401.
[9] S. Kamada, Seifert circles for surface braids, RIMS Kokyuroku (RIMS, Kyoto) **813** (1992) 144–154.

[10] S. Kamada, *Braid and Knot Theory in Dimension Four*, Mathematical Surveys and Monographs, Vol. 95, Amer. Math. Soc., Providence, RI, 2002.

[11] T. Kitano and M. Suzuki, A partial order in the knot table, *Experiment. Math.* **14** (2005) 385–390.

[12] J. Lüroth, Note über Verzweigungsschnitte und Querschnitte in einer Riemann'schen Fläche, *Math. Ann.* **4** (1871) 181–184.

[13] W. S. Massey, The quotient space of one complex projective plane under conjugation is a 4-sphere, *Geom. Dedicata* **2** (1973) 371–374.

[14] J. M. Montesinos, Three-manifolds as 3-fold branched covers of S^3, *Quart. J. Math. Oxford Ser.* (2) **27** (1976) 85–94.

[15] R. Piergallini, *Four-manifolds as 4-fold branched covers of* S^4, *Topology* **34** (1995) 497–508.

Generalised Biquandles for Generalised Knot Theories

Roger Fenn
School of Mathematical Sciences, University of Sussex,
Falmer, Brighton, BN1 9RH, England
rogerf@sussex.ac.uk

This paper looks at generalised knot theories and defines the generalised biquandles which are most appropriate for the theories. I have tried to indicate how a general theory of knots could be integrated into one meta-theory. I have then shown how a theory of biracks and biquandles can be invented to associate with these theories and provide invariants.

1. Introduction

This paper is the culmination of many years work by the author. There are two notions: the first of which is a *generalised notion of knots* and the second is a *generalised notion of biquandle.* A generalised knot theory is defined by *moves* on a diagram and might be termed a *combinatorial knot theory.* The notion of a biquandle, being an algebraic labelling of the arcs, is generalised so that the relations satisfied by the biquandles are consistent with the moves of the theory. These biquandles can then be used to define invariants for the particular knot or link defined by the diagram. The invariants are usually polynomials or homology classes of certain associated spaces.

Some examples are given but more examples, computer generated, will be given in a later paper. This paper might be considered as the theoretical forerunner of the later papers.

The generalised knot theories considered in this paper are the following:

1.1. *Classical knots and links*

These hardly need an introduction and their invariants are legion. The application of biquandles in the form of the fundamental quandles has proved to be a complete invariant, see [8, 17, 21]. However the introduction of the Δ-move, [16], loosens the theory so that, for example, all knots of one component are now trivial.

Homotopy link theory is obtained from the classical theory by allowing self intersections of components, see [5].

1.2. *Virtual and welded knots and links*

In the theory of virtual knots and links, [18], the application of biquandles, even linear ones, has proved very effective, see [3, 6, 7]. The idea of a biquandle as an invariant of virtual knots was considered in [15]. There, the definition was linked to the idea of a *switch map*. In this paper the definition is rotated though a right angle to a *sideways map*. This simplifies some of the formulas but of course does not change the invariants.

The theory of welded knots is obtained by allowing the first forbidden move, [12, 24]. The applications of biquandle theory here still need work and this will be the subject of a future paper.

1.3. *Miscellaneous knot theories*

A few papers have been written on *doodles*, see for example [5, 13, 19]. The same is true of *singular knot theory*, [2, 11]. However, both are ripe for exploitation by generalised biquandles.

2. Definitions

2.1. *Generalised combinatorial knot theories*

Objects in a generalised knot theory can be defined combinatorially in terms of a *diagram*, K. The underlying graph, $|K|$, of a diagram is the image of one or more immersed oriented circles in general position in the plane with additional structure. This image, sometimes called a *shadow*, is an oriented 4-valent planar graph in which the vertices correspond to crossing double points of the immersion.

The *components* of the diagram are the images of the individual immersed circles. Generally the objects with one component are called *knots* and otherwise *links*. Unless emphasis is needed, all objects will be called knots.

The edges of the graph are called *arcs*. The arcs are oriented by the immersed circles and because of general position, this orientation crosses over at the vertices.

To get a diagram, each vertex is decorated by a *crossing type*. Classically, the decoration was indicated by a break to indicate an under and over crossing. Examples of other types of crossing decoration will be given below. Each crossing decoration has a positive and negative version which might not be distinct.

Fig. 1. Positive and negative classical crossings.

Some more crossing types are illustrated in Figs. 2 and 3.

Fig. 2. Doodle/flat and virtual/weld crossings.

Doodle, flat, virtual and weld crossings are involutive, that is their positive and negative versions are equal.

Fig. 3. Positive and negative singular crossings.

If C_i denotes a general crossing type, then we will indicate this with an i near a positive crossing and an $\bar{\imath}$ near a negative crossing. If the sign of a C_i crossing is irrelevant, we indicate with i', i'' etc.

Fig. 4. Positive and negative crossings of type C_i.

To sum up: a diagram, K, consists of its underlying graph, $|K|$, and a crossing type attached to its vertices.

2.2. *Converting diagrams into knots*

Two diagrams are said to be *equivalent* and hence represent the same object if they are related by a series of *moves* allowed by the theory. The moves generate an equivalence relation on the collection of diagrams. A *knot* is the equivalence class of a diagram. A knot with more than one component is sometimes called a *link*. There are various possible types of move depending on the decoration type at crossings. A move may be *allowable or forbidden* according to the theory.

The possible moves fall into five classes, $\mathbf{R_0}$–$\mathbf{R_4}$ listed below.

Moves $\mathbf{R_0}$–$\mathbf{R_3}$ are generalisations of the famous *Reidemeister* moves. So we will call $\mathbf{R_0}$–$\mathbf{R_4}$ *generalised* Reidemeister moves. Sometimes we will shorten this to **GR** *moves.*

$\mathbf{R_0}$**:** This is just a topological equivalence of the original immersion in the plane.

R₁: This move introduces or deletes a curl or monogon with nothing inside. The new or deleted crossing may be positive or negative.

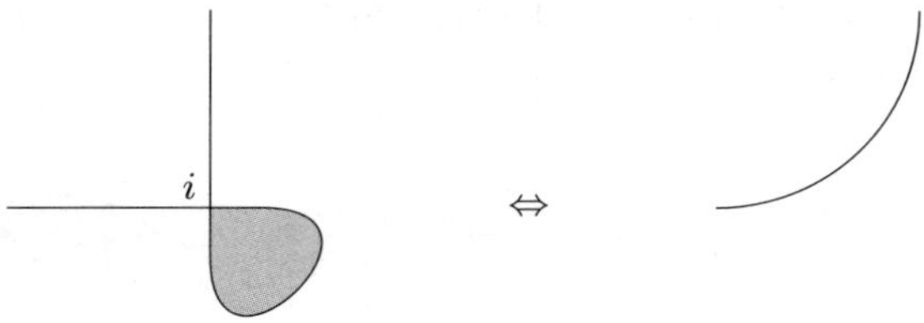

Fig. 5. Introducing or deleting a monogon in an $\mathbf{R_1}$ move.

In this case and subsequently, a shaded n-gon means an interior disjoint from the rest of the diagram.

$\mathbf{R_2}$**:** Here two parallel opposite crossings are introduced or deleted and the bigon defined between them is also introduced or deleted.

Fig. 6. Introducing or deleting a bigon in an $\mathbf{R_2}$ move.

We will make a blanket assumption that an $\mathbf{R_2}$ move is always possible and allowed.

$\mathbf{R_3}$**:** A crossing type C_i is said to *dominate* a crossing type C_j if two consecutive crossings of type C_i appear to pass a crossing of type C_j. This kind of crossing is called an $\mathbf{R_3}$ type.

Fig. 7. Crossing C_i dominating C_j in an $\mathbf{R_3}$ move.

Note that the order of the C_i crossing types is reversed along the moving arc.

The notion of one crossing type dominating another was first learnt by the author during a lecture by Lou Kauffman.

$\mathbf{R_4}$: Types C_i and C_j are said to *commute* if parallel copies of C_i and C_j are allowed to interchange.

Fig. 8. C_i commuting with C_j in an $\mathbf{R_4}$ move.

Types C_i and C_j are said to *anti-commute* if parallel copies of C_i and C_j are allowed to interchange and the sign of one crossing changes.

Fig. 9. C_i anti-commuting with C_j in an $\mathbf{R_4}$ move.

2.3. *Generalised Reidemeister moves and orientation*

Here we look at how different orientations of the crossings and arcs of the diagram affect the GR moves.

Orienting $\mathbf{R_0}$: A *mirror* transformation is an orientation reversing homeomorphism of the plane which changes the sign of the crossing types. The image of a diagram after a mirror transformation is called its *mirror* diagram.

If we reverse the orientation of some or all of the circles which are immersed in the plane, then we call the composition the *reverse* of the diagram.

If all the signs of the crossings are changed we call the result the *inverse* diagram.

A diagram which is equivalent to its mirror image is called *amphicheiral.* A diagram, equivalent to its reverse, is called *reversible.* A diagram, equivalent to its inverse, is called *invertible.*

For example, a simple loop in the plane is not amphicheiral and has a clockwise or anti-clockwise orientation depending on whether its tangent is right or left turning. A number 8 is reversible.

For classical knots a mirrored diagram is equivalent to an inverted diagram. This need not be the case for more general knots.

Orienting $\mathbf{R_1}$: There are four cases according to the orientation of the monogon and the sign of the crossing.

Orienting $\mathbf{R_2}$: Again there are four cases. If the bigon is oriented, it could turn clockwise or anti-clockwise. Otherwise the bigon has a flow either from a negative crossing to a positive one or the opposite.

Orienting $\mathbf{R_3}$: This is complicated and we need to make further definitions and impose conditions. Consider the moving path in an $\mathbf{R_3}$ move joining two crossings of the same type. As indicated earlier we will always assume that if the first crossing encountered when moving from left to right is positive (negative) then the second encountered after the move is also positive (negative) and conversely.

We now look at the orientations of the arcs and try and reduce the number of possibilities. We can rotate the diagram so that the moving arc appears to travel from left to right. There are now three possibilities:

1. The triangle formed by the three arcs is oriented. We then say that the move is *oriented*. Note that if one of the triangles is oriented clockwise, then after the move the triangle is oriented anti-clockwise and conversely.
2. All three arcs appear to move from left to right, then we say that the move is *braid-like*.
3. If neither of the above, then the move is *mixed*.

We will prove that we need only consider braid-like moves but first we need some definitions.

We let $R_3(u, v; \epsilon, \eta; \xi)$ denote the most general $\mathbf{R_3}$ move illustrated in Fig. 10.

Fig. 10. The $R_3(u, v; \epsilon, \eta; \xi)$ move.

The orientations of the "fixed" lines whose intersection defines the C_j crossing are given by u, v. There are four possibilities. The orientations of the C_i crossings defined by the moving arc are given by ϵ, η. Again there are four possibilities. The orientation of the C_j crossing is given by ξ and has two possibilities.

This makes up to 32 diagrams to consider. But we will reduce this to eight, using a trick invented by Turaev.

Lemma 2.1. *The following general* $\mathbf{R_3}$ *moves are equivalent;*

$$R_3(u, v; \epsilon, \eta; \xi) \cong R_3(v, \bar{u}; \eta, \bar{\epsilon}; \xi) \cong R_3(\bar{u}, \bar{v}; \bar{\epsilon}, \bar{\eta}; \xi) \cong R_3(\bar{v}, u; \bar{\eta}, \epsilon; \xi).$$

Proof. The proof is indicated by Figs. 11–13. In each case an $\mathbf{R_3}$ is about to be performed but first a bigon is created with an $\mathbf{R_2}$ move. Then an $\mathbf{R_3}$ move of a different kind is performed. Followed by an $\mathbf{R_2}$ which eliminates a bigon. The result is the completion of the original $\mathbf{R_3}$ move. The process is reversable.

Fig. 11. $R_3(u, v; \epsilon, \eta; \xi) \cong R_3(v, \bar{u}; \eta, \bar{\epsilon}; \xi)$.

Fig. 12. $R_3(v, \bar{u}; \eta, \bar{\epsilon}; \xi) \cong R_3(\bar{u}, \bar{v}; \bar{\epsilon}, \bar{\eta}; \xi)$.

Fig. 13. $R_3(\bar{u}, \bar{v}; \bar{\epsilon}, \bar{\eta}; \xi) \cong R_3(\bar{v}, u; \bar{\eta}, \epsilon; \xi)$.

□

We can conclude from the above lemma that the orientations of the straight lines, u and v, can be ignored so we might as well assume that the move is braid-like. That is, all the arcs are oriented from left to right. Let this move be designated $R_3(\epsilon, \eta; \xi)$.

Theorem 2.2. *As* $\mathbf{R_2}$ *moves are always allowable, we need only consider at most eight oriented* $\mathbf{R_3}$ *moves,* $R_3(\epsilon, \eta; \xi)$ *where* $\epsilon, \eta, \xi = \pm 1$.

Orienting $\mathbf{R_4}$: Suppose we are dealing with a commutation. There are three cases. If the bigon is oriented, it could be oriented clockwise or anti-clockwise. Otherwise, the arcs are parallel.

For an anti-commutation there are possibly six cases to consider: the ones above and mirror images.

2.4. *Examples with one type of crossing*

If there is only one type of crossing involved in an $\mathbf{R_3}$ move, then there are at most two allowable possibilities which need to be considered as we shall now show.

Lemma 2.3. *If there is only one type of crossing involved, then* (i) *an* $R_3(\epsilon, \eta; \xi)$ *is allowed if and only if an* $R_3(\bar{\epsilon}, \xi; \eta)$ *is allowed and* (ii) *an* $R_3(\epsilon, \eta; \xi)$ *is allowed if and only if an* $R_3(\xi, \bar{\eta}; \epsilon)$ *is allowed.*

Proof. Imagine that we would like to do the inverse of an $R_3(\bar{\epsilon}, \xi; \eta)$ and we add a cancelling $\bar{\epsilon}, \epsilon$ on the left to arrive at Fig. 14. Note that our blanket asumption allows this.

The move we would like to do is over the unshaded trigon.

Now assume that an $R_3(\epsilon, \eta; \xi)$ is allowed on the shaded trigon. If we do this and cancel the resulting cancelling $\epsilon, \bar{\epsilon}$ on the right, we find that we have performed an inverse of an $R_3(\bar{\epsilon}, \xi; \eta)$.

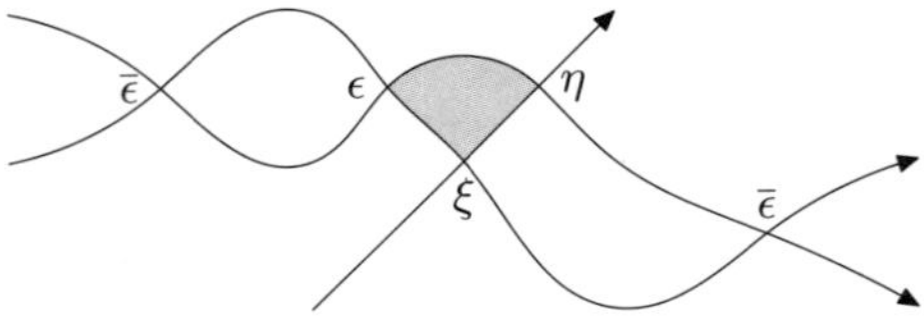

Fig. 14.

Similar arguments to the above prove the converse and yields (ii). □

Theorem 2.4. *If only one type of crossing is involved in an* $\mathbf{R_3}$ *move, then the possible allowable* $\mathbf{R_3}$ *moves can be reduced to* $R_3(+,+;+)$ *and* $R_3(+,+;-)$.

Proof. By (i) from the above lemma we can always assume that the first entry, ϵ, in $R_3(\epsilon,\eta;\xi)$ is positive. This leaves the four possibilities, $R_3(+,+;+)$, $R_3(+,-;+)$, $R_3(+,-;-)$ and $R_3(+,+;-)$. By (ii) $R_3(+,+;+)$ is equivalent to $R_3(+,-;+)$. By a combination of (i), (ii) and (i), $R_3(+,-;+)$ is equivalent to $R_3(+,-;-)$. □

Classical knots and links

There is only one type of classical crossing. This comes in a positive and negative version.

The move $\mathbf{R_1}$ is always allowed unless the knot is framed. Geometrically, framing can be interpreted as a knot with components made of ribbons homeomorphic to $S^1 \times I$.

In classical theory we need only consider one allowed move, $R_3(+,+;+)$, illustrated below.

Fig. 15. $R_3(+,+;+)$.

The allowed move corresponds to a topological situation. There is a hierarchy among the three arcs: one is on top, the moving one is in between and one is at the bottom.

However, if we consider the "forbidden" move, $R_3(+,+;-)$, then there is no such hierarchy.

Fig. 16. $R_3(+,+;-)$ the forbidden Δ move.

If all two moves are allowed, (the forbidden moves is called a Δ-move), then we get a theory discussed in [23] for example.

Flats and doodles

If a flat crossing has $\mathbf{R_1}$ and is its own inverse (involutive) but does not dominate itself, then we have the theory of doodles. See [13, 19].

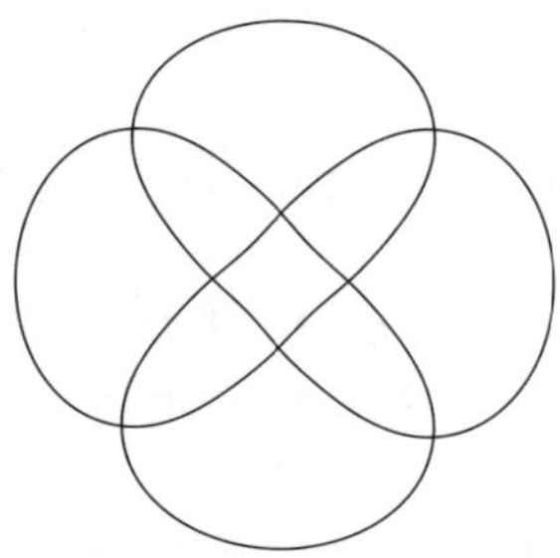

Fig. 17. The poppy: a nontrivial doodle with one component.

If $\mathbf{R_3}$ is allowed and it does dominate itself then the theory collapses, unless $\mathbf{R_1}$ is forbidden when each component is classified by its winding number.

Examples with Two Types of Crossing

Virtual, weld, free and singular knot theories

A virtual crossing has $\mathbf{R_1}$, is involutive and dominates everything including itself. Teamed with classical crossings this gets us the usual theory of virtual knots and links, [18].

Fig. 18. A virtual arc dominating a classical crossing.

Normally a classical crossing cannot dominate a virtual, but if we allow $R_3(-,+;v)$, where an arc appears to cross *over* a virtual, we get welded knot theory, [12]. This move is called the first forbidden move of virtual knot theory or the *weld move.*

Fig. 19. The weld move.

If we allow $R_3(+,-;v)$, the second forbidden move of virtual knot theory, where an arc appears to cross *under* a virtual, then the theory collapses, see [24].

The *virtualisation* move of Kauffman is equivalent to an $\mathbf{R_4}$ move where a virtual crossing anti-commutes with a classical crossing.

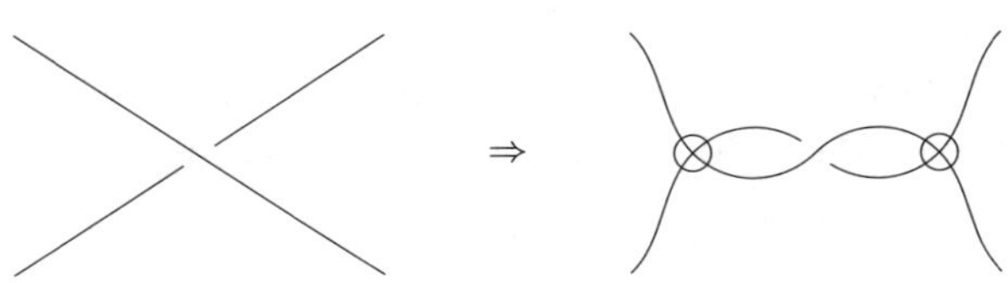

Fig. 20. The virtualisation move.

Virtualisation, in a neighbourhood of the crossing, changes the undercrossing to an overcrossing at the expense of introducing two new virtual crossings.

It can easily be seen that under a virtualisation the fundamental quandle quotiented out by the condition $a^{bb} = a$ is invariant.

Under this move a nontrivial classical knot can be made into a nontrivial virtual knot with trivial Jones' polynomial, [18]. It is unknown if this leads to a nontrivial classical knot with trivial Jones' polynomial.

Free knot theory is obtained by allowing the GR-moves full rein. So a free knot is represented by a virtual knot diagram in which the over and under arcs of a classical crossing are not distinguished and the virtualisation move is allowed, see [25].

Singular knot theory

If the virtual crossing dominates all, then, roughly speaking, the singular crossing dominates nothing, even itself. What this means precisely is that a singular knot diagram has classical and singular crossings. All moves of the form $R_3(s_1, s_2; s_3)$ are forbidden and dominance of classical crossings is defined by allowing $R_3(+, -; s)$ and $R_3(-, +; s)$ but forbidding $R_3(+, +; s)$ and $R_3(-, -; s)$.

The various dominances are summed up in the following diagrams where the positive singular crossing could just as easily be replaced by a negative one.

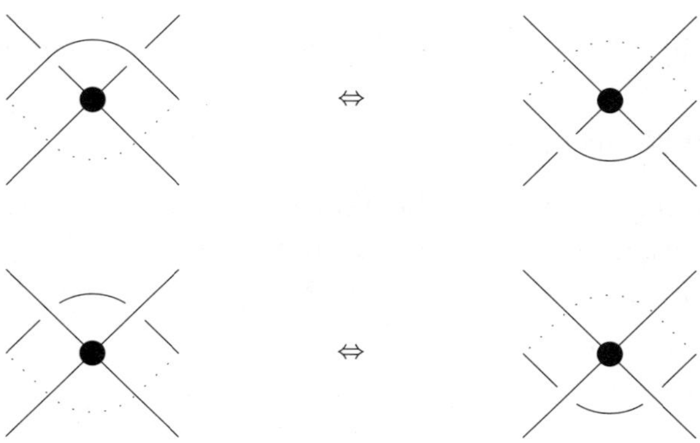

Fig. 21. Classical crossings dominating a singular crossing.

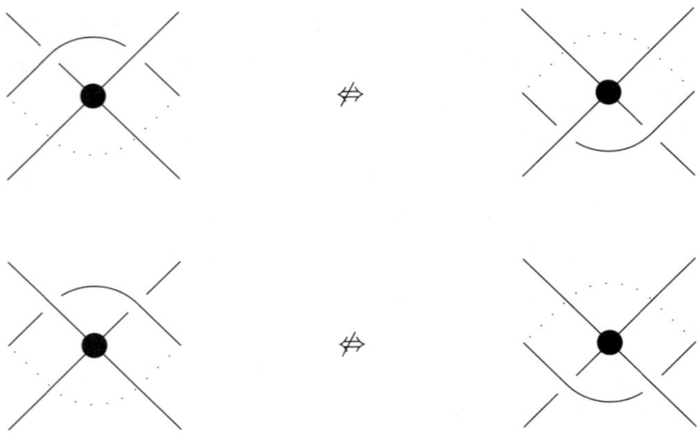

Fig. 22. Forbidden singular moves.

So as Theorem 2.2 predicts; there are eight possible R_3 moves, four of which are allowed and four of which are forbidden.

If we think of the singular crossing as joined by a blob of glue then the allowed and forbidden moves make geometric sense as does the R_4 move where a singular crossing commutes with a classical crossing.

Fig. 23.

Geometrically, the effect is to turn the singular crossing over, see [2].

Flattened knot theory

All the above theories can be replaced by flat versions in which classical crossings are replaced by flat ones. Thus for example we have flat virtual theory. However, I have never seen flat singular theory.

Examples with Many Types of Crossing

Homotopy links

Let D be a classical diagram of a link with n components. Then we call a crossing of type C_{ij} if the ith component crosses over the jth component. If $i \neq j$, then C_{ij} has a distinct inverse. Otherwise C_{ii} is involutive. The crossings satisfy the usual $\mathbf{R_3}$ moves for classical crossings.

As C_{ii} is involutive, over and under arcs are indistiguishable. It follows by standard arguments that individual components can be unknotted. The invariants of homotopy links are linking numbers and higher order Milnor numbers, [5, 22]. For a recent paper on these invariants, see [20]. For an aplication of quandles, see [16].

3. Algebraic Invariants (Generalised Biquandles)

In this section we will look at how invariants can be obtained by labelling the arcs of a diagram and noting how these change under the GR moves.

Let X be a set of *labels*. The labels will be denoted by lower case roman letters, $a, b, c, \ldots, x, y, z$. The set X may be finite with no structure, a topological space or a set with some algebraic structure such as a group, an algebra, a module over a ring, etc.

Let $F : X^2 \to X^2$ be a map. We assume that F respects any structure on X. Then F defines two binary operations by the rule, $F(a,b) = (b_a, a^b)$. For each $a \in X$ there are maps $f^a, f_a \colon X \to X$ defined by $f^a(x) = x^a$ and $f_a(x) = x_a$. The operation $(a,b) \to a^b$ is called the *up* operation and the operation $(a,b) \to a_b$ is called the *down* operation.

Both the notations a^b, b_a and f^a, f_a are useful and we will employ them throughout the paper.

Definition. The map $F : X^2 \to X^2$ is called a *sideways* map if the following three conditions hold:

1. $F : X^2 \to X^2$ is a bijection,
2. $f^a : X \to X$ is a bijection for all $a \in X$ and
3. $f_a : X \to X$ is a bijection for all $a \in X$.

We write $(f^a)^{-1}(x) = x^{a^{-1}}$ and $(f_a)^{-1}(x) = x_{a^{-1}}$. So, with bracketing on the right ignored, $x^{aa^{-1}} = x^{a^{-1}a} = x$ and $x_{aa^{-1}} = x_{a^{-1}a} = x$ for all $a, x \in X$.

The three conditions are independent as the following example shows.

Let $X = \{0, 1\}$, let $f_0 = f^0$ be the $0, 1$ interchange (01) and let $f_1 = f^1$ be the identity. Then $F(0, 0) = F(1, 1) = (1, 1)$ and so F is not a bijection.

On the other hand, if $f^0 = (01)$, f^1 is the identity and f_1 is the constant map with value 1 and f_0 is the constant map with value 0, then F is a bijection on X^2.

However the following is easy to prove.

Theorem 3.1. *Suppose f^a is a bijection for all $a \in X$ and f_b is the identity for all $b \in X$. Then the sideways map F is a bijection.*

Note that this is the situation corresponding to a quandle or rack, [8].

It follows from the conditions on a sideways map that in an expression such as b^a or b_a the element b, called the *base*, is well-defined. Let $S : X^2 \to X^2$ satisfy $S(b_a, a) = (a^b, b)$. The map S is called the *switch* associated to the sideways map, F.

Theorem 3.2. *The switch, S, associated to a sideways map F is well-defined by $S(x, y) = (f^z(y), z)$ where $z = f_y^{-1}(x)$. Moreover, S is a bijection on X^2.*

Proof. Notice that the expression (b_a, a) is uniquely defined because f_a is a bijection and (a^b, b) is uniquely defined because f^b is a bijection. If we put $(x, y) = (b_a, a)$, then $a = y$ and $b_a = x$ so $b = x_{a^{-1}} = z$, say.

The switch is a bijection because it has a clearly defined inverse given by $S^{-1}(a^b, b) = (b_a, a)$. □

Note the fact that it is a bijection depends only on the fact that $f^a, f_a \colon X \to X$ are bijections for all $a \in X$ and not on the fact that $F : X^2 \to X^2$ is a bijection. In an earlier paper on the subject, [15], the definitions and importance of the switch and sideways map are interchanged. The only reason for the change here is that the identities satisfied by the up and down operations by reason of axioms, to be defined later, are simplified.

The switch also defines binary operations and suppose $S(x, y) = (y \wedge x, x \vee y)$. Then these operations are related to the previously defined operations by $x \vee y = x_{y^{-1}}$ and $y \wedge x = y^{x_{y^{-1}}}$

Dually, the inverse, S^{-1}, is the switch associated to the sideways map, F^*, where $F^*(a, b) = (b^a, a_b)$.

As for the sideways map, the switch also defines functions $s^a \colon X \to X$ and $s_a : X \to X$ given by

$$S(a, b) = (s^a(b), s_b(a)).$$

The functions f and s are symmetrically related as follows:

$$f_a = s_a^{-1}, \; s_a = f_a^{-1}$$

and

$$f^b(a) = s^{s_a^{-1}(b)}(a), \; s^b(a) = f^{f_a^{-1}(b)}(a).$$

Clearly s_a is a bijection since f_a is a bijection. However, the author has been unable to prove a similar result for s^a and suspects that on occasions it may not be a bijection. It seems that one of the differences between the switch and the sideways approach is whether the associated maps of X are bijections.

Added in Proof. Andy Bartholomew has shown that s^a is indeed a bijection.

We now label the arcs of the diagram with elements of the labelling set X. The labels at a crossing are constrained by the corresponding sideways map and corresponding switch. For example consider a general crossing in the figure below such that in a

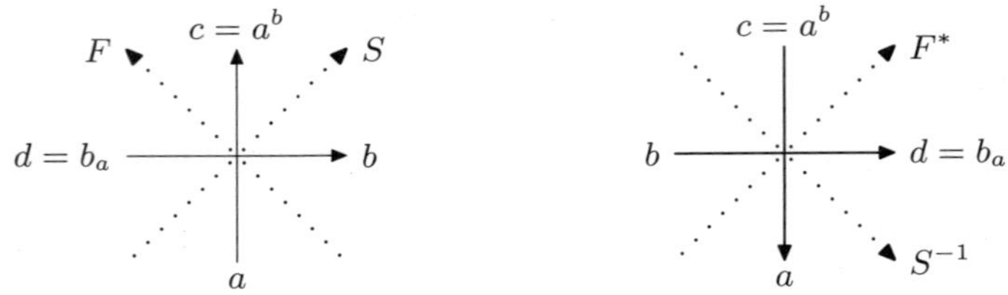

Fig. 24. Labelling the arcs around crossings $+ab$ and $-ab$.

positive crossing the incoming arcs are labelled d, a and the outgoing arcs are labelled b, c. For a negative crossing this is reversed.

Adjacent labels are determined by the opposite labels and the maps F and S. If the crossing is positive and a, b are given, then F determines $c = a^b$ and $d = b_a$. Similarly if d, a are given then S determines the outgoing labels $b = d_{a^{-1}}$ and $c = a^b$. Likewise d, c determine a, b by F^{-1} and c, b determine the incoming labels d, a by S^{-1}. We label the positive crossing $+(ab)_i$ if the crossing is of type C_i. If the crossing type is understood, then we just label the crossing $+ab$.

If the crossing type is negative, then the incoming labels replace a with b and d with c. The outgoing labels replace c with d and b with a. Otherwise the same equations hold. Now we label the crossing $-ab$.

To summarise: the arcs are labelled by single elements of the labelling set X and the crossings by signed pairs $\pm(ab)_i = \pm ab$ of the labelling set. The difference between the labelling of positive and negative crossings is that the up and down operations are interchanged.

4. Relations on the Labelling Due to the GR Moves

We now look at the above moves in more detail and see how they affect the labellings and the sideways and switch maps. It is important to realise that if a move is allowed, then the corresponding relation must be satisfied but, just as important for an effective invariant, if a move is forbidden then the corresponding relation must fail.

$\mathbf{R_1}$: Consider the move shown in the following figure before it is untwisted. There are two cases where the vertex is positive. On the left of the picture the monogon is oriented clockwise. The boundary of the monogon created by the twist is labelled by a and b so for consistency $a = b$ and the crossing is labelled aa.

Fig. 25. Labelling aa.

On the right of the figure the monogon is oriented anti-clockwise. The boundary of the monogon created by the twist is labelled by a^b and b_a so for consistency $a^b = b_a$.

In order for the labelling to be carried over after the untwisting we need the condition $a^a = a_a$ which is also the necessary condition on the right because now $a = b$.

Now consider the situation where the crossing is negative.

Fig. 26. Labelling $-aa$.

Looking at the labels on the arcs we can conclude the same deductions as in the positive case. This can be summed up by the equality $a^a = a_a$ for all a. We call this the *biquandle* condition. This is equivalent to the condition that the sideways map F preserves the diagonal in X^2.

Conversely suppose a monogon is created from a labelled diagram. Suppose the label on the incoming and outgoing arcs is x. We must find a label a for the monogon. If the biquandle condition holds, then the sideways map F which is a bijection on X^2 is also a bijection when restricted to the diagonal. So there is an $a \in X$ such that $F(a, a) = (a^a, a_a) = (x, x)$. By the biquandle condition, $a^a = a_a$.

R₂: This either creates or deletes a pair of double points of opposite sign. The possible orientations of the just created or about to be deleted bigon and associated labellings are shown in the following figures, (i), (ii), (iii) and (iv). In (iii) and (iv) the bigons are oriented clockwise and anti-clockwise respectively.

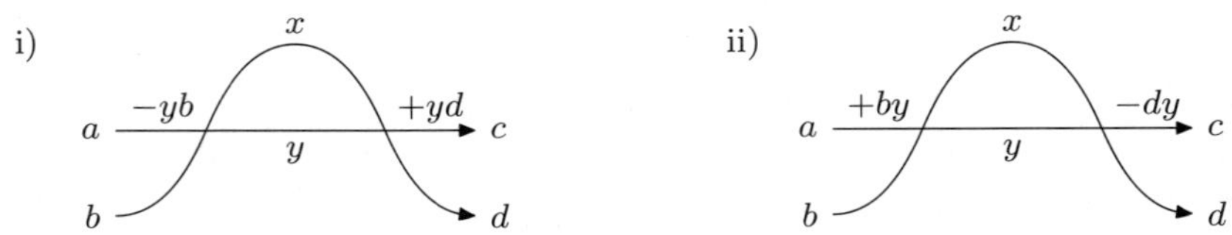

Fig. 27.

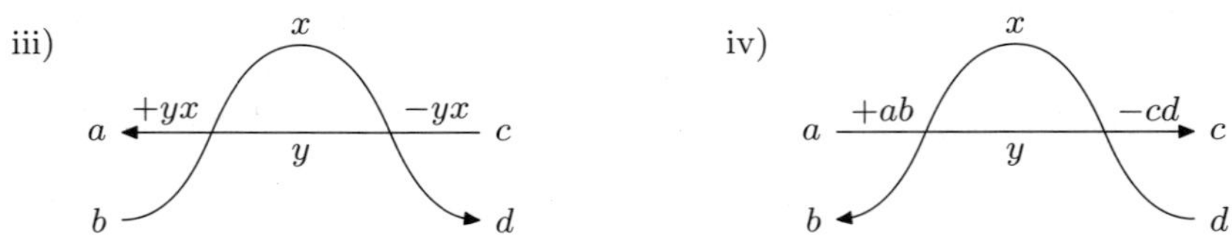

Fig. 28.

If the bigon has just been born then the label pairs a, c and b, d have equal members and we must find unique labels x, y for the bigon.

If the bigon is about to disappear then the labels x, y are given and we must show as a consequence that the label pairs a, c and b, d have equal members.

For case (i), using the crossing $+yd$ we have $c = y^d$ and $x = d_y$. Using the crossing $-yb$ we have $a = y^b$ and $x = b_y$. If the bigon is about to disappear, then $d = x_{y^{-1}} = b$ and $c = y^d = y^b = a$. If the bigon has just been created, then $y = a^{b^{-1}}$ and $x = b_{a^{b^{-1}}}$ define the labels.

For case (ii), a similar argument to (i) works. Using the crossing $+by$, we have $a = y_b$ and $x = b^y$. Using the crossing $-dy$ we have $c = y_d$ and $x = d^y$. If the bigon is about to disappear, then $d = x^{y^{-1}} = b$ and $c = y_d = y_b = a$. If the bigon has just been created, then $y = a_{b^{-1}}$ and $x = b^{a_{b^{-1}}}$ define the labels.

For case (iii), if the bigon is about to disappear then we have $F^{-1}(y, x) = (b, a) = (d, c)$ which is possible as F is invertible. If the bigon has just been created, then $(y, x) = F(b, a)$.

For case (iv), a similar argument to (iii) works. If the bigon is about to disappear, then we have $F(x, y) = (a, b) = (c, d)$. If the bigon has just been created, $(x, y) = F^{-1}(a, b)$ using the invertibility of F.

Involutive crossings

If a crossing is the same as its inverse, then this imposes the relations

$$S^2 = \text{id} \quad \text{and} \quad F = F^*$$

on the switch and the sideways map. Looking at the details it is easy to see the following.

Theorem 4.1. *If the crossing is involutive then the up and down operations satisfy $a^b = a_b$ for all a, b. That is, the up and down operations are the same.*

Let us call the conclusion of the above theorem, the *involution* condition. Note that the involution condition implies the biquandle condition.

R₃: Given that the second move is always possible the analysis of the implications of the third move is easier since we have reduced the consideration to braid like moves. This means for the $R_3(\epsilon, \eta; \xi)$ move the relation needed or not needed for the switch S of C_i and the switch T of C_j is the set-theoretic *Yang, Baxter* equation

$$(S^\epsilon \times 1)(1 \times T^\xi)(S^\eta \times 1) = (1 \times S^\eta)(T^\xi \times 1)(1 \times S^\epsilon).$$

For example, see [6].

Examples with one type of crossing

Now $S = T$ and as we have seen earlier, the relevant relations are

$$(S \times 1)(1 \times S)(S \times 1) = (1 \times S)(S \times 1))(1 \times S), \qquad \text{(i)}$$

$$(S \times 1)(1 \times S^{-1})(S \times 1) = (1 \times S)(S^{-1} \times 1))(1 \times S). \qquad \text{(ii)}$$

So for doodles, none of these should be valid. For classical knot theory, (i) must be true and (ii) fail. For Δ-theory, both must be true.

The condition (i) implies the following relations

$$a^{bc_b} = a^{cb^c}, \quad c_{ba^b} = c_{ab_a}, \quad {b_a}^{c_a} = {b^c}_{a^c}$$

as the following diagram makes clear.

The crossings in the above diagram appear to lie on the surface of a cube. The implications of this will be clear when we consider homology later.

To illustrate these ideas consider an invariant for doodle theory. To simplify matters suppose that R is a commutative associative integral domain. Let the actions be linear, so that the labelling set is an R-module and $a^b = \lambda a + \mu b$ where $\lambda, \mu \in R$. As actions are invertible, λ must be a unit and since the crossing is involutive, the down action $b_a = \lambda b + \mu a$ is the same.

However, doodles do not satisfy $\mathbf{R_3}$. So for a useful invariant neither must the labellings. In [7] it is noted that to satisfy $\mathbf{R_3}$ using the above actions, $\mu = 0$. So if $\mu \neq 0$, then we get a useful invariant of doodles, say a polynomial in λ, μ, as defined in [6].

$\mathbf{R_4}$**:** If S_1 and S_2 are the corresponding switches and F_1 and F_2 are the corresponding sideways maps, then for the fourth GR they either commute or anti-commute in pairs. So $S_1S_2 = S_2S_1$ and $F_1F_2 = F_2F_1$, or alternatively $S_1S_2 = S_2^{-1}S_1$ and $F_1F_2 = F_2^{-1}F_1$.

5. Homology

Associated to any knot theory and labelling set X is a topological space built of cubes. The dimension of this space is at least 2 but could be infinite. The 1-cells are defined by the labels and the 2-cells by pairs of labels, ab. The 3-cells if they exist are defined by triples and so on.

All the 1-cells are cycles and there is a fundamental group with generators the labels and relations $ab^{-1}(a^b)^{-1}b_a$.

As for homology the boundary of the square ab is given by

$$\partial(ab) = a - b + b_a - a^b.$$

For geometric insight see the face of the cubes in Fig. 29.

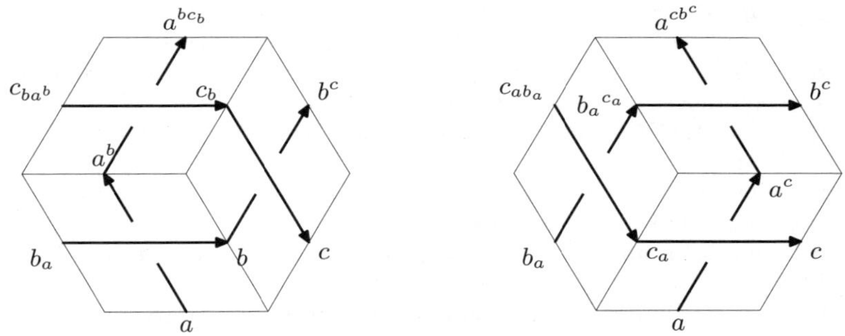

Fig. 29.

Any labelled knot in the theory defines a 2-cycle by adding up the labels of the crossings with signs.

The 3-cells are dependent on the existence or not of $\mathbf{R_3}$ moves, see Fig. 29. So for a labelled doodle, the 2-cycle defines the homology class exactly.

The homology in the classical and virtual case is a well-travelled path. See [4, 9, 10] for the case of quandle and rack homology where there are cells of arbitary dimension. The biquandle and birack homology is more mysterious and is worthy of further study.

As indicated in the Introduction a further paper will consider homology in greater detail and computer-generated examples will be given.

References

[1] A. Bartholomew and R. Fenn, Biquandles of small size and some invariants of virtual and welded knots, *J. Knot Theory Ramifications* **20** (2011) 943–954.

[2] J. S. Birman, New points of view in knot theory, *Bull. Amer. Math. Soc. (N.S.)* **28** (1993) 253–287.

[3] S. Budden and R. Fenn, Quaternion algebras and invariants of virtual knots and links II: The hyperbolic case, *J. Knot Theory Ramifications* **17** (2008) 305.

[4] J. Scott Carter, M. Elhamdadi, M. Graa and M. Saito, Co-cycle knot invariants from quandle modules and generalized quandle homology, *Osaka J. Math.* **42** (2005) 499–541.

[5] R. Fenn, *Techniques of Geometric Topology*, Cambridge Univ. Press, 1983.

[6] R. Fenn, Quaternion algebras and invariants of virtual knots and links I: The elliptic case, *J. Knot Theory Ramifications* **17** (2008) 279.

[7] R. Fenn, Tackling the trefoil, *J. Knot Theory Ramifications* **21** (2012) 1240004.

[8] R. Fenn and C. Rourke, Racks and links in codimension 2, *J. Knot Theory Ramifications* **1** (1992) 343–406.

[9] R. Fenn, C. Rourke and B. Sanderson, The rack space, in *Trans. Amer. Math. Soc.* **359** (2007) 701–740.

[10] R. Fenn, C. Rourke and B. Sanderson, An introduction to species and the rack space, in *Topics in knot theory* (Erzurum, 1992), NATO Adv. Sci. Inst. Ser. C Math. Phys. Sci., Vol. 399, pp. 33–55, Kluwer Acad. Publ., Dordrecht, 1993.

[11] R. Fenn, E. Keyman and C. Rourke, The singular braid monoid embeds in a group, *J. Knot Theory Ramifications* **7** (1998) 881–892.

[12] R. Fenn, R. Rimnyi and C. Rourke, The braid-permutation group, *Topology* **36** (1997) 123–135.

[13] R. Fenn and P. Taylor, Introducing doodles, in *Topology of Low-Dimensional Manifolds*, Proc. Second Sussex Conf., Chelwood Gate, 1977, Lecture Notes in Math., Vol. 722, pp. 37–43. Springer, Berlin, 1979.

[14] R. Fenn and V. Turaev, Weyl algebras and knots, *J. Geom. Phys.* **57** (2007) 1313–1324.

[15] R. Fenn, M. Jordan-Santana and L. Kauffman, Biquandles and virtual links, in *Topology Appl.* **145** (2004) 157–175.

[16] A. Inoue, Quasi-triviality of quandles for link-homotopy, *J. Knot Theory Ramifications* **22** (2013) 1350026.

[17] D. Joyce, A classifying invariant of knots, the knot quandle, *J. Pure Appl. Algebra* **23** (1982) 37–65.

[18] L. H. Kauffman, Virtual knot theory, *Eur. J. Combinatorics* **20** (1999) 663–690.

[19] M. Khovanov, Doodle groups, *Trans. Amer. Math. Soc.* **349** (1997) 2297–2315.

[20] R. Koytcheff, The Milnor triple-linking number of string links by cut-and-paste topology, arXiv:1209.6381.
[21] S. Matveev, Distributive Groupoids, *Math. USSR Sb.* **47** (1984) 73–83.
[22] J. Milnor, Link groups, *Ann. Math. (2)* **59** (1954) 177–195.
[23] H. Murakami and Y. Nakanishi, On a certain move generating link-homology, *Math. Ann* **284** (1989) 75–89.
[24] S. Nelson, Unknotting virtual knots with Gauss diagram forbidden moves, *J. Knot Theory Ramifications* **10** (2001) 931.
[25] V. O. Manturov and D. P. Ilyutko, *Virtual Knots, The State of the Art*, Series on Knots and Everything, Vol. 51, World Scientific, 2012.

Lectures on Knot Homology and Quantum Curves

Sergei Gukov[*] and Ingmar Saberi[†]
California Institute of Technology
[*] *gukov@theory.caltech.edu*
[†] *isaberi@caltech.edu*

Besides offering a friendly introduction to knot homologies and quantum curves, the goal of these lectures is to review some of the concrete predictions that follow from the physical interpretation of knot homologies. In particular, it allows one to answer questions like *Is there a direct relation between Khovanov homology and the A-polynomial of a knot?* which would not have been asked otherwise. We will explain that the answer to this question is "yes" and introduce a certain deformation of the planar algebraic curve defined by the zero locus of the A-polynomial. This novel deformation leads to a categorified version of the Generalized Volume Conjecture that completely describes the "color behavior" of the colored $\mathfrak{sl}(2)$ knot homology and, eventually, to a similar conjecture for the colored HOMFLY homology. Furthermore, this deformation is strong enough to distinguish mutants, and its most interesting properties include relations to knot contact homology and knot Floer homology.

0. Introduction

An alternative title of these lecture notes could be "Categorification and Quantization." However, these lectures will by no means serve as a complete introduction to the two topics of *quantization* and *categorification*. Each of these words represents not so much a single idea as a broad tool, program, or theme in physics and mathematics, and both are areas of active research and are still not fully understood. One could easily give a full one-year course on each topic separately.

Rather, the goal of these lectures is to serve as an appetizer: to give a glimpse of the ideas behind quantization and categorification, by focusing on very concrete examples and giving a working knowledge of how these ideas are manifested in simple cases. It is our hope that the resulting discussion will remain accessible and clear while shedding at least some light on these complex ideas, and that the interest of the reader will be piqued.

Imagine the category of finite-dimensional vector spaces and linear maps. To each object in this category is naturally associated a number, the dimension of that vector space. Replacing some collection of vector spaces with a collection of numbers in this way can be thought of as a *decategorification*: by remembering only the dimension of each space, we keep some information, but lose all knowledge about (for instance) morphisms between spaces. In this sense, decategorification forgets about geometry.

Categorification can be thought of as the opposite procedure. Given some piece of information (an invariant of a topological space, for instance), one asks whether it arises in some natural way as a "decategorification": a piece of data extracted out of a more geometrical or categorical invariant, which may carry more information and thus be a finer and more powerful tool. An answer in the affirmative to this question is a categorification of that invariant.

Perhaps the most familiar example of categorification at work is the reinterpretation of the Euler characteristic as the alternating sum of ranks of homology groups,

$$\chi(M) = \sum_{k\geq 0}(-1)^k \operatorname{rank} H_k(M). \tag{0.1}$$

In light of this formula, the homology of a manifold M can be seen as a categorification of its Euler characteristic: a more sophisticated and richly structured bearer of information, from which the Euler characteristic can be distilled in some natural way. Moreover, homology theories are a far more powerful tool than the Euler characteristic alone for the study and classification of manifolds and topological spaces. This shows that categorification can be of practical interest: by trying to categorify invariants, we can hope to construct *stronger* invariants.

While the idea of categorification is rooted in pure mathematics, it finds a natural home in the realm of topological quantum field theory (TQFT), as will be discussed in Sec. 4. For this, however, we first need to understand what "quantum" means by explaining the quantization program, which originated squarely within physics. Its basic problem is the study of the transition between classical and quantum mechanics. The classical and quantum pictures of a physical system make use of entirely different and seemingly unconnected mathematical formalisms. In classical mechanics, the space of possible states of the system is a symplectic manifold, and observable quantities are smooth functions on this manifold. The quantum mechanical state space, on the other hand, is described by a Hilbert space $\mathscr{H}$, and observables are elements of a noncommutative algebra of operators acting on $\mathscr{H}$. Quantization of a system is the construction of the quantum picture of that system from a classical description, as is done in a standard quantum mechanics course for systems such as the harmonic oscillator and the hydrogen atom. Therefore, in some sense, quantization allows one to interpret quantum mechanics as "modern symplectic geometry." We will give a more complete introduction to this idea in Sec. 3.

One main application of the ideas of quantization and categorification is to representation theory, where categorification, or "geometrization," leads naturally to the study of geometric representation theory [1]. Another area of mathematics where these programs bear much fruit is low-dimensional topology, which indeed is often called "quantum" topology. This is the arena in which we will study the implications of quantization and categorification, primarily for the reason that it allows for many concrete and explicit examples and computations. Specifically, almost all of our discussion will take place in the context of knot theory. The reader should not, however, be deceived into thinking of our aims as those of knot theorists! We do not discuss quantization and categorification for the sake of their applications to knot theory; rather, we discuss knot theory because it provides a window through which we can try and understand quantization and categorification.

1. Why Knot Homology?

A *knot* is a smooth embedding of a circle S^1 as a submanifold of S^3:

$$k : S^1 \hookrightarrow S^3, \quad K := \operatorname{im} k. \tag{1.1}$$

See e.g. Figs. 2 and 3 for some simple examples. Likewise, a *link* is defined as an embedding of several copies of S^1. Two knots are equivalent if the two embeddings k and k' can be smoothly deformed into one another through a family of embeddings, i.e. without self-intersections at any time. One should think of moving the knot around in the ambient space without breaking the string of which it is made.

In studying a knot, one usually depicts it using a planar *knot diagram*: this should be thought of as a projection of the knot from $\mathbb{R}^3 = S^3 \backslash \{\text{pt.}\}$, in which it lives, to some plane $\mathbb{R}^2 \subset \mathbb{R}^3$. Thus, a knot diagram is the image of an immersion of S^1 in $\mathbb{R}^2$, having only double points as singularities, and with extra data indicating which strand passes over and which under at each crossing. Examples of knot diagrams can be seen in Figs. 2–4.

It should be clear that there is no unique diagram representing a given knot. We could obtain very different-looking pictures, depending on the exact embedding in $\mathbb{R}^3$ and on the choice of plane to which we project. Two knot diagrams should of course be seen as equivalent if they depict equivalent knots, but this equivalence could be nontrivial and difficult to see. The situation is made a little more tractable by a theorem of Reidemeister, which states that two knot diagrams are equivalent if and only if they can be transformed into one another by a sequence of three simple transformations. These basic transformations are the *Reidemeister moves*, which are depicted in Fig. 1, and show replacements that can be made in any portion of a knot diagram to give an equivalent diagram.

Finding a sequence of Reidemeister moves that changes one given knot diagram into another, or showing that no such sequence exists,

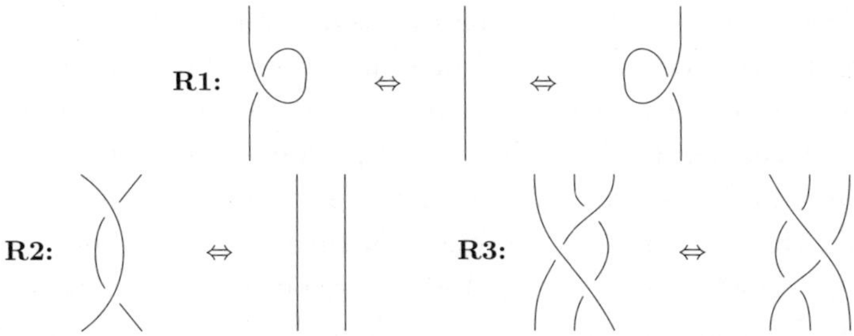

Fig. 1. The three Reidemeister moves, which generate all equivalences between knot diagrams.

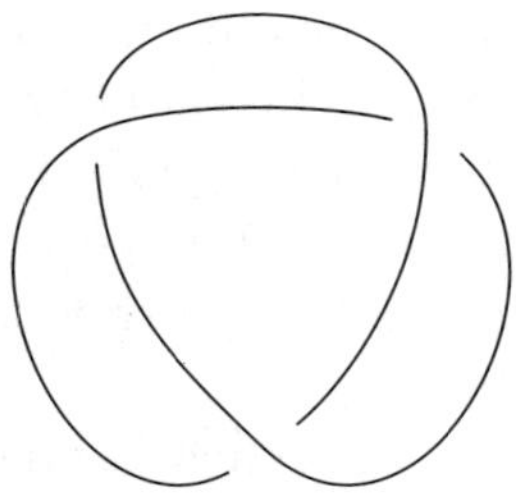

Fig. 2. The trefoil knot 3_1.

is still an *ad hoc* and usually intractable problem. As such, in attempting to classify knots, more clever methods are important. One of the most basic tools in this trade is a *knot invariant*: some mathematical object that can be associated to a knot, that is always identical for equivalent knots. In this way, one can definitively say that two knots are distinct if they possess different invariants. The converse, however, is not true; certain invariants may fail to distinguish between knots that are in fact different. Therefore, the arsenal of a knot theorist should contain a good supply of different invariants. Moreover, one would like invariants to be as "powerful" as possible; this just means that they should capture nontrivial information about the knot. Obviously, assigning the number 0 to every knot gives an invariant, albeit an extremely poor one!

Since one usually confronts a knot in the form of one of its representative knot diagrams, it is often desirable to have an invariant that can be efficiently computed from a knot diagram. Showing that some such quantity associated to a diagram is actually an invariant of knots requires demonstrating that it takes the same values on all equivalent diagrams representing the same knot. Reidemeister's theorem makes this easy to check: to show that we have defined a knot invariant, we need only check its invariance under the three moves in Fig. 1.[a]

Given the goal of constructing knot invariants, it may be possible to do so most easily by including some extra structure to be used in the construction. That is, one can imagine starting not simply with a knot, but with a knot "decorated" with additional information: for instance, a choice of a Lie algebra $\mathfrak{g} = \mathrm{Lie}(G)$ and a representation R of $\mathfrak{g}$. It turns out that this additional input data from representation theory does in fact allow one to construct various invariants (numbers, vector spaces, and so on), collectively referred to as *quantum group invariants.* A large part of these lectures will consist, in essence, of a highly unorthodox introduction to these quantum group invariants.

The unorthodoxy of our approach is illustrated by the fact that we fail completely to address a natural question: what on earth do (for instance) the quantum $\mathfrak{sl}(N)$ invariants have to do with $\mathfrak{sl}(N)$? Representation theory is almost entirely absent from our discussion; we opt instead to look at an alternative description of the invariants, using a concrete combinatorial definition in terms of so-called *skein relations.* A more full and traditional introduction to the subject would include much more group theory, and show the construction of the quantum group invariants in a way that makes the role of the additional input data $\mathfrak{g}$ and R apparent [2, 3]. That construction involves assigning a so-called "quantum R-matrix" to each crossing in a knot diagram in some manner, and then taking a trace around the knot in the direction of its orientation. The connection to

[a]Nonetheless, since a knot is intrinsically an object of three-dimensional topology that can be imagined without any use of diagrams, it might be hoped that one could give an obviously three-dimensional construction of various invariants that does not require a choice of a two-dimensional projection. As we discuss later in these notes, Witten's physical interpretation of the Jones polynomial in [2] does exactly this.

representation theory is made manifest; the resulting invariants, however, are the same.

Example 1.1. Suppose that we take an oriented knot together with the Lie algebra $\mathfrak{g} = \mathfrak{sl}(N)$ and its fundamental N-dimensional representation. With this special choice of extra data, one constructs the quantum $\mathfrak{sl}(N)$ invariant, denoted $P_N(K;q)$. Although it makes the connection to representation theory totally obscure, one can compute $P_N(K;q)$ directly from the knot diagram using the following skein relation:

$$q^N P_N(\nwarrow\!\!\!\!\nearrow) - q^{-N} P_N(\nearrow\!\!\!\!\nwarrow) = (q - q^{-1}) P_N(\uparrow\uparrow). \qquad (1.2)$$

(Note that we will sometimes write $P_N(K)$ for the polynomial P_N associated to the knot or link K, suppressing the variable q; no confusion should arise.) For now, one can think of q as a formal variable. The subdiagrams shown in (1.2) should be thought of as depicting a neighborhood of one particular crossing in a planar diagram of an oriented knot; to apply the relation, one replaces the chosen crossing with each of the three shown partial diagrams, leaving the rest of the diagram unchanged.

To apply this linear relation, one also needs to fix a normalization, which can be done by specifying P_N for the unknot. Here, unfortunately, several natural choices exist. For now, we will choose

$$P_N(\bigcirc) = \frac{q^N - q^{-N}}{q - q^{-1}} = \underbrace{q^{-(N-1)} + q^{-(N-3)} + \cdots + q^{N-1}}_{N \text{ terms}}. \qquad (1.3)$$

This choice gives the so-called *unnormalized* $\mathfrak{sl}(N)$ polynomial. Notice that, given any choice of $P_N(\bigcirc)$ with integer coefficients, the form of the skein relation implies that $P_N(q) \in \mathbb{Z}[q, q^{-1}]$ for every knot.

Notice further that, with the normalization (1.3), we have

$$P_N(\bigcirc) \xrightarrow[q \to 1]{} N, \qquad (1.4)$$

which is the dimension of the representation R with which we decorated the knot, the fundamental of $\mathfrak{sl}(N)$. We remark that this leads to a natural generalization of the notion of dimension, the so-called *quantum dimension* $\dim_q(R)$ of a representation R, which arises from the quantum group invariant constructed from R evaluated on the unknot.

Equipped with the above rules, let us now try to compute $P_N(q)$ for some simple links. Consider the *Hopf link*, consisting of two interlocked circles:

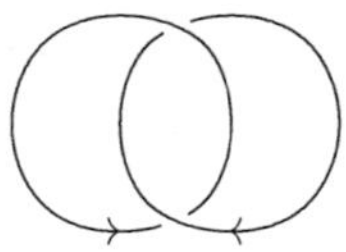

Applying the skein relation to the upper of the two crossings, we obtain:

$$q^N P_N\Big[\underbrace{\ \ }_{\text{Hopf link}}\Big] - q^{-N} P_N\Big[\underbrace{\ \ }_{\text{two unknots}}\Big] = (q - q^{-1}) P_N\Big[\underbrace{\ \ }_{\text{one unknot}}\Big]. \tag{1.5}$$

This illustrates a general feature of the skein relation, which occurs for knots as well as links: In applying the relation to break down any knot diagram into simpler diagrams, one will in fact generally need to evaluate P_N for links rather than just for knots, since application of the relation (1.2) may produce links with more than one component. This means that the normalization (1.3) is not quite sufficient; we will need to specify P_N on k unlinked copies of the unknot, for $k \geq 1$.

As such, the last of our combinatorial rules for computing $P_N(q)$ concerns its behavior under disjoint union:

$$P_N(\bigcirc \sqcup K) = P_N(\bigcirc) \cdot P_N(K), \tag{1.6}$$

where K is any knot or link. Here, the disjoint union should be such that K and the additional unknot are not linked with one another.

Caution: The discerning reader will notice that our final rule (1.6) is not linear, while the others are, and so is not respected under rescaling of $P_N(q)$. Therefore, if a different choice of normalization is made, it will *not* remain true that $P_N(k \text{ unknots}) = [P_N(\bigcirc)]^k$. The nice behavior (1.6) is particular to our choice of normalization (1.3). This can be expressed by saying that, in making a different normalization, one must remember to normalize only one copy of the unknot.

To complete the calculation we began above, let us specialize to the case $N = 2$. Then we have

$$P_2(\bigcirc) = q^{-1} + q$$

$\Rightarrow P_2($  $) = (q^{-1} + q)^2 = q^{-2} + 2 + q^2.$ (1.7)

Applying the skein relation (1.5) then gives

$q^2 P_2($  $) = q^{-2}(q^{-2} + 2 + q^2) + (q - q^{-1})(q + q^{-1})$

$$= q^{-4} + q^{-2} + 1 + q^2, \tag{1.8}$$

so that

$P_2($  $) = q^{-6} + q^{-4} + q^{-2} + 1.$ (1.9)

We are now ready to compute the $\mathfrak{sl}(N)$ invariant for any link.

From the form of the rules that defines this invariant, it is apparent that dependence on the parameter N enters the knot polynomial only by way of the combination of variables q^N. As such, we can define the new variable $a := q^N$, in terms of which our defining relations become

$aP_{a,q}($  $) - a^{-1}P_{a,q}($  $) = (q - q^{-1})P_{a,q}($ 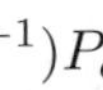 $),$ (1.10)

$$P_{a,q}(\bigcirc) = \frac{a - a^{-1}}{q - q^{-1}}. \tag{1.11}$$

Together with the disjoint union property, these rules associate to each oriented link K a new invariant $P_{a,q}(K)$ in the variables a and q, called the (unnormalized) *HOMFLY-PT polynomial* of the link [4]. This is something of a misnomer, since with the normalization (1.11) the HOMFLY-PT invariant will in general be a rational expression rather than a polynomial. We have traded the two variables q, N for q and a.

For various special choices of the variables a and q, the HOMFLY-PT polynomial reduces to other familiar polynomial knot invariants:

- $a = q^N$, of course, returns the quantum $\mathfrak{sl}(N)$ invariant $P_N(q)$.

- With the particular choice $a = q^2$ $(N = 2)$, the HOMFLY-PT polynomial becomes the classical *Jones polynomial* $J(L; q) \equiv P_2(q)$,

$$J(K; q) = P_{a=q^2,q}(K). \tag{1.12}$$

  Discovered in 1984 [6], the Jones polynomial is one of the best-known polynomial knot invariants, and can be regarded as the "father" of quantum group invariants; it is associated to the Lie algebra $\mathfrak{sl}(2)$ and its fundamental two-dimensional representation.
- $a = 1$ returns the *Alexander polynomial* $\Delta(K; q)$, another classical knot invariant. This shows that the HOMFLY-PT polynomial generalizes the $\mathfrak{sl}(N)$ invariant, in some way: the evaluation $a = 1$ makes sense, even though taking $N = 0$ is somewhat obscure from the standpoint of representation theory.

Now, the attentive reader will point out a problem: if we try and compute the Alexander polynomial, we immediately run into the problem that (1.11) requires $P_{1,q}(\bigcirc) = 0$. The invariant thus appears to be zero for every link! However, this does not mean that the Alexander polynomial is trivial. Remember that, since the skein relations are linear, we have the freedom to rescale invariants by any multiplicative constant. We have simply made a choice that corresponds, for the particular value $a = 1$, to multiplying everything by zero.

This motivates the introduction of another convention: the so-called *normalized* HOMFLY-PT polynomial is defined by performing a rescaling such that

$$P_{a,q}(\bigcirc) = 1. \tag{1.13}$$

This choice is natural on topological grounds, since it associates 1 to the unknot independent of how the additional input data, or "decoration," is chosen. (By contrast, the *unnormalized* HOMFLY-PT polynomial assigns the value 1 to the empty knot diagram.) Taking $a = 1$ in the *normalized* HOMFLY-PT polynomial returns a nontrivial invariant, the Alexander polynomial.

Exercise 1.2. Compute the normalized and unnormalized HOMFLY-PT polynomials for the trefoil knot $K = 3_1$ (Fig. 2). Note that one of these will actually turn out to be polynomial!

Having done this, specialize to the case $a = q^2$ to obtain the normalized and unnormalized Jones polynomials for the trefoil. Then specialize to the case $a = q$. Something nice should occur! Identify what happens and explain why this is the case.

Solution. Applying the skein relation for the HOMFLY-PT polynomial to one crossing of the trefoil knot gives

$aP_{a,q}(3_1) - a^{-1}P_{a,q}($  $) = (q - q^{-1})P_{a,q}($  $)$.

Then, applying the relation again to the Hopf link (as in the above example) gives

$aP_{a,q}($  $) - a^{-1}P_{a,q}($  $) = (q - q^{-1})P_{a,q}($  $)$.

Therefore, for the unnormalized HOMFLY-PT polynomial,

$P(3_1) = a^{-2}P($  $) + a^{-2}(q - q^{-1})\left[a^{-1}P($  $)^2 + (q - q^{-1})P($  $)\right]$

which becomes

$$P(3_1) = \frac{a - a^{-1}}{q - q^{-1}}\left[a^{-2}q^2 + a^{-2}q^{-2} - a^{-4}\right].$$

The normalized HOMFLY-PT polynomial is simply the quantity in brackets. Specializing to $a = q^2$ gives the unnormalized Jones polynomial:

$$P_2(3_1) = \frac{q^2 - q^{-2}}{q - q^{-1}}\left[q^{-2} + q^{-6} - q^{-8}\right]. \tag{1.14}$$

Again, the normalized Jones polynomial is the factor in square brackets. Finally, we specialize to $a = q$, obtaining $P = 1$ in both the normalized and unnormalized cases! This is connected to the fact that $a = q$ corresponds to constructing the $\mathfrak{sl}(1)$ invariant, which must be vacuous since the Lie algebra is trivial. □

Remark 1.3. The study of this subject is made more difficult by the preponderance of various conventions in the literature. In particular, there is no agreement at all about standard usage with regard to the variables for polynomial invariants. Given ample forewarning, this should not cause too much confusion, but the reader must always be aware of the problem. In particular, it is extremely common for papers to differ from our conventions by the replacement

$$a \mapsto a^{1/2}, \quad q \mapsto q^{1/2}, \tag{1.15}$$

halving all powers that occur in knot polynomials. Some authors also make the change

$$a \mapsto a^{-1}, \quad q \mapsto q^{-1}, \tag{1.16}$$

and some make both.

We have by now seen a rich supply of knot polynomials, which can be straightforwardly computed by hand for simple enough diagrams, and are easy to write down and compare. One might then ask about the value of attempting to categorify at all. Given such simple and powerful invariants, why would one bother trying to replace them with much more complicated ones?

The simple answer is that the HOMFLY-PT polynomial and its relatives, while powerful, are not fully adequate for the job of classifying all knots up to ambient isotopy. Consider the two knot diagrams shown in Fig. 3, which represent the knots 5_1 and 10_{132} in the Rolfsen classification. While the knots are not equivalent, they have identical Alexander and Jones polynomials! In fact, we have

$$P_{a,q}(5_1) = P_{a,q}(10_{132}) \tag{1.17}$$

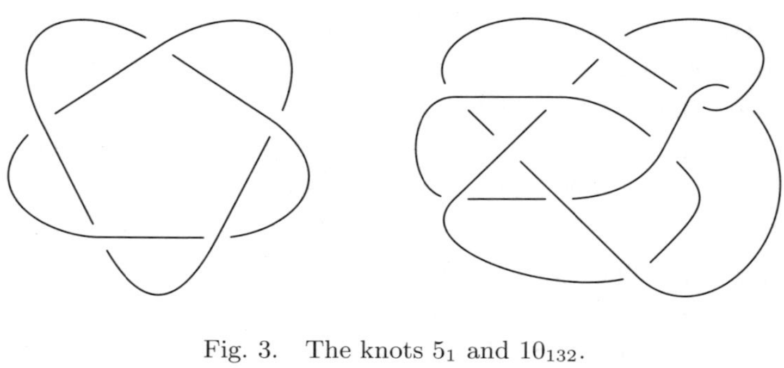

Fig. 3. The knots 5_1 and 10_{132}.

and, therefore, all specializations — including all $\mathfrak{sl}(N)$ invariants — will be identical for these two knots. Thus, even the HOMFLY-PT polynomial is not a perfect invariant and fails to distinguish between these two knots. This motivates us to search for a finer invariant. Categorification, as we shall see, provides one. Specifically, even though the Jones, Alexander, and HOMFLY-PT polynomials fail to distinguish the knots 5_1 and 10_{132} of our example, their respective categorifications do (cf. Fig. 9).

Before we step into the categorification era, let us make one more desperate attempt to gain power through polynomial knot invariants. To this end, let us introduce not one, but a whole sequence of knot polynomials $J_n(K;q) \in \mathbb{Z}[q,q^{-1}]$ called the *colored Jones polynomials*. For each non-negative integer n, the n-colored Jones polynomial of a knot K is the quantum group invariant associated to the decoration $\mathfrak{g} = \mathfrak{sl}(2)$ with its n-dimensional representation V_n. $J_2(K;q)$ is just the ordinary Jones polynomial. In *Chern–Simons theory* with gauge group $G = \mathrm{SU}(2)$, we can think of $J_n(K;q)$ as the expectation value of a Wilson loop operator on K, colored by the n-dimensional representation of SU(2) [2].

Moreover, the colored Jones polynomial obeys the following relations, known as *cabling formulas*, which follow directly from the rules of Chern–Simons TQFT:

$$\begin{aligned} J_{\bigoplus_i R_i}(K;q) &= \sum_i J_{R_i}(K;q), \\ J_R(K^n;q) &= J_{R^{\otimes n}}(K;q). \end{aligned} \tag{1.18}$$

Here K^n is the n-cabling of the knot K, obtained by taking the path of K and tracing it with a "cable" of n strands. These equations allow us to compute the n-colored Jones polynomial, given a way to compute the ordinary Jones polynomial and a little knowledge of representation theory. For instance, any knot K has $J_1(K;q) = 1$ and $J_2(K;q) = J(K;q)$, the ordinary Jones polynomial. Furthermore,

$$\begin{aligned} \mathbf{2}\otimes\mathbf{2} = \mathbf{1}\oplus\mathbf{3} &\implies J_3(K;q) = J(K^2;q) - 1, \\ \mathbf{2}\otimes\mathbf{2}\otimes\mathbf{2} = \mathbf{2}\oplus\mathbf{2}\oplus\mathbf{4} &\implies J_4(K;q) = J(K^3;q) - 2J(K;q), \end{aligned} \tag{1.19}$$

and so forth. We can switch to representations of lower dimension at the cost of considering more complicated links; however, the computability of the ordinary Jones polynomial means that this is still a good strategy for calculating colored Jones polynomials.

Example 1.4. Using the above formulas, it is easy to find n-colored Jones polynomial of the trefoil knot $K = 3_1$ for the first few values of n:

$$\begin{aligned} J_1 &= 1, \\ J_2 &= q + q^3 - q^4, \\ J_3 &= q^2 + q^5 - q^7 + q^8 - q^9 - q^{10} + q^{11}, \\ &\vdots \end{aligned} \tag{1.20}$$

where, for balance (and to keep the reader alert), we used the conventions which differ from (1.14) by the transformations (1.15) and (1.16).

Much like the ordinary Jones polynomial is a particular specialization (1.12) of the HOMFLY-PT polynomial, its colored version $J_n(K;q)$ can be obtained by the same specialization from the so-called *colored HOMFLY-PT polynomial* $P_n(K;a,q)$,

$$J_n(K;q) = P_n(K;a = q^2, q) \tag{1.21}$$

labeled by an integer n. More generally, the colored HOMFLY-PT polynomials $P^{\lambda}(K;a,q)$ are labeled by Young diagrams or 2d partitions λ. In these lectures, we shall consider only Young diagrams that consist of a single row (or a single column) and by Schur–Weyl duality correspond to totally symmetric (respectively totally anti-symmetric) representations. Thus, what we call $P_n(K;a,q)$ is the HOMFLY-PT polynomial of K colored by $\lambda = S^{n-1}$.

Even though $P_n(K;a,q)$ provide us with an infinite sequence of two-variable polynomial knot invariants, which can tell apart e.g. the two knots in (1.17), they are still not powerful enough to distinguish simple pairs of knots and links called *mutants*. The operation of *mutation* involves drawing a disc on a knot diagram

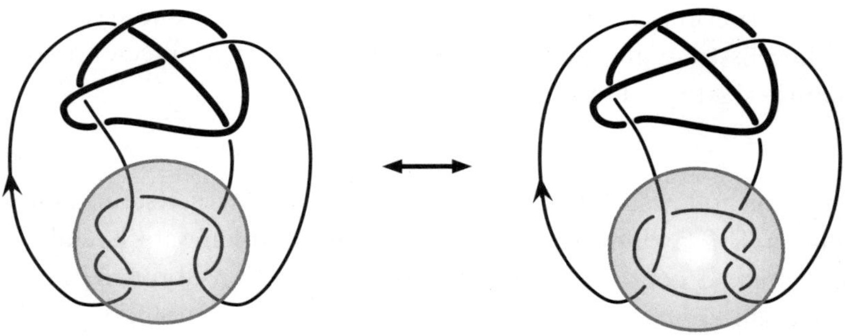

Fig. 4. Mutant knots.

such that two incoming and two outgoing strands pass its boundary, and then rotating the portion of the knot inside the disc by 180°. The Kinoshita–Terasaka and Conway knots shown in Fig. 4 are a famous pair of knots that are mutants of one another, but are nonetheless distinct; they can be distinguished by homological knot invariants, but not by any of the polynomial invariants we have discussed so far!

Theorem 1.1. *The colored Jones polynomial, the colored HOMFLY-PT polynomial, and the Alexander polynomial cannot distinguish mutant knots* [7], *while their respective categorifications can* [8–10].

2. The Classical *A*-Polynomial

In this section, we take a step back from quantum group invariants to discuss another classical invariant of knots: the so-called *A-polynomial.* Our introduction will be rather brief, intended to familiarize the reader with the general idea behind this invariant and catalogue some of its properties, rather than attempt a complete construction. For more information, we refer to the pioneering paper of Cooper *et al.* [12], in which the A-polynomial was first defined.

For a knot K, let $N(K) \subset S^3$ be an open tubular neighborhood of K. Then the *knot complement* is defined to be

$$M := S^3 \backslash N(K). \tag{2.1}$$

By construction, M is a 3-manifold with torus boundary, and our goal here is to explain that to every such manifold one can associate a planar algebraic curve

$$\mathscr{C} = \{(x, y) \in \mathbb{C}^2 : A(x, y) = 0\}, \tag{2.2}$$

defined as follows. The classical invariant of M is its fundamental group, $\pi_1(M)$, which in the case of knot complements is called the *knot group*. It contains a lot of useful information about M and can distinguish knots much better than any of the polynomial invariants we saw in Sec. 1.

Example 2.1. Consider the trefoil knot $K = 3_1$. Its knot group is the simplest example of a braid group:

$$\pi_1(M) = \langle a, b : aba = bab \rangle. \tag{2.3}$$

Although the knot group is a very good invariant, it is not easy to deal with due to its non-Abelian nature. To make life easier, while hopefully not giving up too much power, one can imagine considering representations of the knot group rather than the group itself. Thus, one can consider representations of $\pi_1(M)$ into a simple non-Abelian group, such as the group of 2×2 complex matrices,

$$\rho : \pi_1(M) \to \mathrm{SL}_2\mathbb{C}. \tag{2.4}$$

Associated to this construction is a polynomial invariant $A(x, y)$, whose zero locus (2.2) parametrizes in some sense the "space" of all such representations. Indeed, as we noted earlier, M is a 3-manifold with torus boundary,

$$\partial M = \partial N(K) \cong T^2. \tag{2.5}$$

Therefore, the fundamental group of ∂M is

$$\pi_1(\partial M) = \pi_1(T^2) = \mathbb{Z} \times \mathbb{Z}. \tag{2.6}$$

The generators of $\pi_1(\partial M)$ are the two basic cycles, which we will denote by m and ℓ (representing for *meridian* and *longitude*, respectively — see Fig. 5). m is the cycle that is contractible when considered as a loop in $N(K)$, and ℓ is the non-contractible cycle that follows the knot in $N(K)$. Of course, any representation

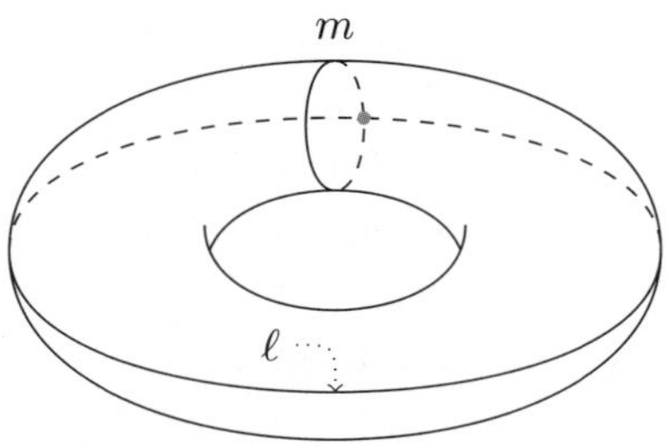

Fig. 5. The torus $T^2 = \partial N(K)$ for $K =$ unknot, with cycles m and ℓ.

$\pi_1(M) \to \mathrm{SL}_2\mathbb{C}$ restricts to a representation of $\pi_1(T^2 = \partial M)$; this gives a natural map of representations of $\pi_1(M)$ into the space of representations of $\pi_1(\partial M)$.

These cycles are represented in $\mathrm{SL}_2\mathbb{C}$ by 2×2 complex matrices $\rho(m)$ and $\rho(\ell)$ with determinant 1. Since the fundamental group of the torus is just $\mathbb{Z} \times \mathbb{Z}$, the matrices $\rho(m)$ and $\rho(\ell)$ commute, and can therefore be simultaneously brought to Jordan normal form by some change of basis, i.e. conjugacy by an element of $\mathrm{SL}_2\mathbb{C}$:

$$\rho(m) = \begin{pmatrix} x & \star \\ 0 & x^{-1} \end{pmatrix}, \quad \rho(\ell) = \begin{pmatrix} y & \star \\ 0 & y^{-1} \end{pmatrix}. \tag{2.7}$$

Therefore, we have a map that assigns two complex numbers to each representation of the knot group:

$$\begin{aligned} \mathrm{Hom}(\pi_1(M), \mathrm{SL}_2\mathbb{C})/\mathrm{conj.} &\to \mathbb{C}^\star \times \mathbb{C}^\star, \\ \rho &\mapsto (x, y), \end{aligned} \tag{2.8}$$

where x and y are the eigenvalues of $\rho(m)$ and $\rho(\ell)$, respectively. The image of this map is the *representation variety* $\mathscr{C} \subset \mathbb{C}^\star \times \mathbb{C}^\star$, whose defining polynomial is the A-polynomial of K. Note that this definition of the A-polynomial does not fix the overall numerical coefficient, which is usually chosen in such a way that $A(x, y)$ has integer coefficients (we return to this property below). For the same reason, the A-polynomial is only defined up to multiplication by arbitrary powers of x and y. Let us illustrate the idea of this construction with some specific examples.

Example 2.2. Let $K \subset S^3$ be the unknot. Then $N(K)$ and M are both homeomorphic to the solid torus $S^1 \times D^2$. Notice that m is contractible as a loop in $N(K)$ and ℓ is not, while the opposite is true in M: ℓ is contractible and m is not. Since ℓ is contractible in M, $\rho(\ell)$ must be the identity, and therefore we have $y = 1$ for all $(x, y) \in \mathscr{C}$. There is no restriction on x. Hence,

$$\mathscr{C}(\bigcirc) = \{(x, y) \in \mathbb{C}^\star \times \mathbb{C}^\star : y = 1\}, \tag{2.9}$$

and the A-polynomial of the unknot is therefore

$$A(x, y) = y - 1. \tag{2.10}$$

Example 2.3. Let $K \subset S^3$ be the trefoil knot 3_1. Then, as mentioned in (2.3), the knot group is given by

$$\pi_1(M) = \langle a, b : aba = bab \rangle, \tag{2.11}$$

where the meridian and longitude cycles can be identified as follows:

$$\begin{cases} m = a, \\ \ell = ba^2ba^{-4}. \end{cases} \tag{2.12}$$

Let us see what information we can get about the A-polynomial just by considering Abelian representations of $\pi_1(M)$, i.e. representations such that $\rho(a)$ and $\rho(b)$ commute. For such representations, the defining relations reduce to $a^2b = ab^2$ and therefore imply $a = b$. (Here, in a slight abuse of notation, we are simply writing a to refer to $\rho(a)$ and so forth.) Equation (2.12) then implies that $\ell = 1$ and $m = a$, so that $y = 1$ and x is unrestricted exactly as in Example 2.2. It follows that the A-polynomial contains $(y - 1)$ as a factor.

This example illustrates a more general phenomenon. Whenever M is a knot complement in S^3, it is true that the abelianization

$$\pi_1(M)^{\mathrm{ab}} = H_1(M) \cong \mathbb{Z}. \tag{2.13}$$

Therefore, the A-polynomial always contains $y - 1$ as a factor,

$$A(x, y) = (y - 1)(\cdots), \tag{2.14}$$

where the first piece carries information about Abelian representations, and any additional factors that occur arise from the non-Abelian representations. In the particular case $K = 3_1$, a similar analysis of non-Abelian representations of (2.3) into $\mathrm{SL}_2\mathbb{C}$ yields

$$A(x,y) = (y-1)(y+x^6). \tag{2.15}$$

To summarize, the algebraic curve $\mathscr{C}$ is (the closure of) the image of the representation variety of M in the representation variety $\mathbb{C}^\star \times \mathbb{C}^\star$ of its boundary torus ∂M. This image is always an affine algebraic variety of complex dimension 1, whose defining equation is precisely the A-polynomial [12].

This construction defines the A-polynomial as an invariant associated to any knot. However, extension to links requires extra care, since in that case $\partial N(L) \not\cong T^2$. Rather, the boundary of the link complement consists of several components, each of which is separately homeomorphic to a torus. Therefore, there will be more than two fundamental cycles to consider, and the analogous construction will generally produce a higher-dimensional character variety rather than a plane algebraic curve. One important consequence of this is that the A-polynomial cannot be computed by any known set of skein relations; as was made clear in Exercise 1.2, computations with skein relations require one to consider general links rather than just knots.

To conclude this brief introduction to the A-polynomial, we will list without proof several of its interesting properties:

- For any hyperbolic knot K,

$$A_K(x,y) \neq y-1. \tag{2.16}$$

  That is, the A-polynomial carries nontrivial information about non-Abelian representations of the knot group.
- Whenever K is a knot in a homology sphere, $A_K(x,y)$ contains only even powers of the variable x. Since in these lectures we shall only consider examples of this kind, we simplify expressions a bit by replacing x^2 with x. For instance, in these conventions the A-polynomial (2.15) of the trefoil knot looks like

$$A(x,y) = (y-1)(y+x^3). \tag{2.17}$$

- The A-polynomial is *reciprocal*: that is,

$$A(x,y) \sim A(x^{-1},y^{-1}), \tag{2.18}$$

where the equivalence is up to multiplication by powers of x and y. Such multiplications are irrelevant, because they do not change the zero locus of the A-polynomial in $\mathbb{C}^\star \times \mathbb{C}^\star$. This property can also be expressed by saying that the curve $\mathscr{C}$ lies in $(\mathbb{C}^\star \times \mathbb{C}^\star)/\mathbb{Z}_2$, where $\mathbb{Z}_2$ acts by $(x,y) \mapsto (x^{-1},y^{-1})$ and can be interpreted as the Weyl group of $\mathrm{SL}_2\mathbb{C}$.
- $A(x,y)$ is invariant under orientation reversal of the knot, but *not* under reversal of orientation in the ambient space. Therefore, it can distinguish mirror knots (knots related by the parity operation), such as the left- and right-handed versions of the trefoil. To be precise, if K' is the mirror of K, then

$$A_K(x,y) = 0 \iff A_{K'}(x^{-1},y) = 0. \tag{2.19}$$

- After multiplication by a constant, the A-polynomial can always be taken to have integer coefficients. It is then natural to ask: are these integers counting something, and if so, what? The integrality of the coefficients of $A(x,y)$ is a first hint of the deep connections with number theory. For instance, the following two properties, based on the *Newton polygon* of $A(x,y)$, illustrate this connection further.
- The A-polynomial is *tempered*: that is, the faces of the Newton polygon of $A(x,y)$ define cyclotomic polynomials in one variable. Examine, for example, the A-polynomial of the figure-eight knot:

$$A(x,y) = (y-1)(y^2 - (x^{-2} - x^{-1} - 2 - x + x^2)y + y^2). \tag{2.20}$$

- Furthermore, the slopes of the sides of the Newton polygon of $A(x,y)$ are boundary slopes of incompressible surfaces[b] in M.

---

[b] A proper embedding of a connected orientable surface $F \to M$ is called incompressible if the induced map $\pi_1(F) \to \pi_1(M)$ is injective. Its boundary slope is defined as follows. An incompressible surface $(F, \partial F)$ gives rise to a collection of parallel simple closed loops in ∂M. Choose one such loop and write its homology class as $\ell^p m^q$. Then, the boundary slope of $(F, \partial F)$ is defined as a rational number p/q.

While all of the above properties are interesting, and deserve to be explored much more fully, our next goal is to review the connection to physics [13], which explains known facts about the A-polynomial and leads to many new ones:

- The A-polynomial curve (2.2), though constructed as an algebraic curve, is most properly viewed as an object of symplectic geometry: specifically, a holomorphic Lagrangian submanifold.
- Its quantization with the symplectic form

$$\omega = \frac{dy}{y} \wedge \frac{dx}{x} \tag{2.21}$$

leads to interesting wave functions.
- The curve $\mathscr{C}$ has all the necessary attributes to be an analogue of the Seiberg–Witten curve for knots and 3-manifolds [14, 15].

As an appetizer and a simple example of what the physical interpretation of the A-polynomial has to offer, here we describe a curious property of the A-polynomial curve (2.2) that follows from this physical interpretation. For any closed cycle in the algebraic curve $\mathscr{C}$, the integral of the Liouville one-form (see (3.3) below) associated to the symplectic form (2.21) should be quantized [13]. Schematically,[c]

$$\oint_\Gamma \log y \frac{dx}{x} \in 2\pi^2 \cdot \mathbb{Q}. \tag{2.22}$$

This condition has an elegant interpretation in terms of algebraic K-theory and the Bloch group of $\bar{\mathbb{Q}}$. Moreover, it was conjectured in [16] that every curve of the form (2.2) — not necessarily describing the moduli of flat connections — is quantizable if and only if $\{x, y\} \in$

[c]To be more precise, all periods of the "real" and "imaginary" part of the Liouville one-form θ must obey

$$\oint_\Gamma \Big(\log|x| d(\arg y) - \log|y| d(\arg x) \Big) = 0,$$

$$\frac{1}{4\pi^2} \oint_\Gamma \Big(\log|x| d \log|y| + (\arg y) d(\arg x) \Big) \in \mathbb{Q}.$$

$K_2(\mathbb{C}(\mathscr{C}))$ is a torsion class. This generalization will be useful to us later, when we consider a refinement of the A-polynomial that has to do with categorification and homological knot invariants.

To see how stringent the condition (2.22) is, let us compare, for instance, the A-polynomial of the figure-eight knot (2.20):

$$A(x,y) = 1 - (x^{-4} - x^{-2} - 2 - x^2 + x^4)y + y^2 \tag{2.23}$$

and a similar polynomial

$$B(x,y) = 1 - (x^{-6} - x^{-2} - 2 - x^2 + x^6)y + y^2. \tag{2.24}$$

(Here the irreducible factor $(y-1)$, corresponding to Abelian representations, has been suppressed in both cases.) The second polynomial has all of the required symmetries of the A-polynomial, and is obtained from the A-polynomial of the figure-eight knot by a hardly noticeable modification. But $B(x,y)$ cannot occur as the A-polynomial of any knot since it violates the condition (2.22).

3. Quantization

Our next goal is to explain, following [13], how the physical interpretation of the A-polynomial in Chern–Simons theory can be used to provide a bridge between *quantum group invariants* of knots and *algebraic curves* that we discussed in Secs. 1 and 2, respectively. In particular, we shall see how quantization of Chern–Simons theory naturally leads to a *quantization* of the classical curve (2.2),

$$A(x,y) \rightsquigarrow \hat{A}(\hat{x},\hat{y};q), \tag{3.1}$$

i.e. a q-difference operator $\hat{A}(\hat{x},\hat{y};q)$ with many interesting properties. While this will require a crash course on basic tools of Quantum Mechanics, the payoff will be enormous and will lead to many generalizations and ramifications of the intriguing relations between quantum group invariants of knots, on the one hand, and algebraic curves, on the other. Thus, one such generalization will be the subject of Sec. 4, where we will discuss categorification and formulate a similar bridge between algebraic curves and *knot homologies*, finally explaining the title of these lecture notes.

We begin our discussion of the quantization problem with a lightning review of some mathematical aspects of classical mechanics. Part of our exposition here follows the earlier lecture notes [17] that we recommend as a complementary introduction to the subject. When it comes to Chern–Simons theory, besides the seminal paper [2], mathematically oriented readers may also want to consult the excellent books [18, 19].

As we discussed briefly in the Introduction, the description of a system in classical mechanics is most naturally formulated in the language of symplectic geometry. In the classical world, the state of a system at a particular instant in time is completely specified by giving $2N$ pieces of data: the values of the coordinates x_i and their conjugate momenta p_i, where $1 \leq i \leq N$. The $2N$-dimensional space parametrized by the x_i and p_i is the *phase space* $\mathscr{M}$ of the system. (For many typical systems, the space of possible configurations of the system is some manifold X, on which the x_i are coordinates, and the phase space is the cotangent bundle $\mathscr{M} = T^*X$.) Notice that, independent of the number N of generalized coordinates needed to specify the configuration of a system, the associated phase space is always of even dimension. In fact, phase space is always naturally equipped with the structure of a symplectic manifold, with a canonical symplectic form given by

$$\omega = dp \wedge dx. \tag{3.2}$$

(When the phase space is a cotangent bundle, (3.2) is just the canonical symplectic structure on any cotangent bundle, expressed in coordinates.) Recall that a symplectic form on a manifold is a closed, nondegenerate two-form, and that nondegeneracy immediately implies that any symplectic manifold must be of even dimension.

Since ω is closed, it locally admits a primitive form, the so-called *Liouville one-form*

$$\theta = p\, dx. \tag{3.3}$$

It should be apparent that $\omega = d\theta$, so that θ is indeed a primitive.

Let us now explore these ideas more concretely in the context of a simple example. As a model system, consider the 1-dimensional simple harmonic oscillator. The configuration space of this system is just $\mathbb{R}$ (with coordinate x), and the Hamiltonian is given by

$$H = \frac{1}{2}p^2 + \frac{1}{2}x^2. \tag{3.4}$$

Since $dH/dt = 0$, the energy is a conserved quantity, and since $N = 1$, this one conserved quantity serves to completely specify the classical trajectories of the system. They are curves in phase space of the form

$$\mathscr{C} : \frac{1}{2}(x^2 + p^2) - \mathscr{E} = 0, \tag{3.5}$$

for $\mathscr{E} \in \mathbb{R}_+$; these are concentric circles about the origin, with radius determined by the energy. Figure 6 shows the potential of this system, together with a typical trajectory in the phase space. The dashed line represents the lowest-energy wave function of the system, to which we will come in a moment.

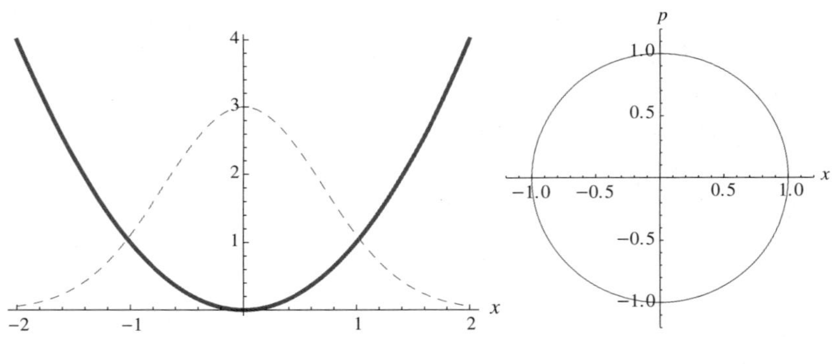

Fig. 6. On the left, the potential and lowest-energy wave function for the simple harmonic oscillator. On the right, the phase space of this system, with a typical classical trajectory.

Now, recall that a *Lagrangian submanifold* $\mathscr{C} \subset (\mathscr{M}, \omega)$ is a submanifold such that $\omega|_{\mathscr{C}} = 0$, having the maximal possible dimension, i.e. $\dim \mathscr{C} = \frac{1}{2} \dim \mathscr{M}$. (If $\mathscr{C}$ has dimension larger than half the dimension of $\mathscr{M}$, the symplectic form cannot be identically zero when restricted to $\mathscr{C}$, since it is nondegenerate on $\mathscr{M}$.) It should be clear that, in the above example, the classical trajectories (3.5) are Lagrangian submanifolds of the phase space.

Moreover, since in this example the degree of the symplectic form ω is equal to the dimension of the phase space, ω is a volume form — in fact, the standard volume form on $\mathbb{R}^2$. We can therefore compute the area encompassed by a trajectory of energy $\mathscr{E}$ by integrating ω over the region $x^2 + p^2 < 2\mathscr{E}$, obtaining

$$2\pi\mathscr{E} = \int_D dp \wedge dx, \tag{3.6}$$

where D is the disc enclosed by the trajectory $\mathscr{C}$. Therefore, classically, the energy of a trajectory is proportional to the area in phase space it encompasses.

How do these considerations relate to quantization of the system? It is well known that the energy levels of the simple harmonic oscillator are given by

$$\mathscr{E} = \frac{1}{2\pi} \int_D dp \wedge dx = \hbar \left(n + \frac{1}{2} \right) \tag{3.7}$$

when the system is quantized. Thus, we expect that, in quantizing a system, the number of quantum states contained in some region of phase space will be directly proportional to its area. Moreover, we interpret $\hbar$, which has the same units as area in phase space, as the amount of classical phase space per quantum state. Schematically,

$$\#\text{ states} \sim \text{area}/\hbar. \tag{3.8}$$

This relation has a long history in quantum physics; it is none other than the Bohr–Sommerfeld quantization condition.

Moreover, since ω admits a primitive, we can use the Stokes theorem to write

$$\mathscr{E} = \frac{1}{2\pi} \int_D \omega = \frac{1}{2\pi} \oint_{\mathscr{C}} \theta, \tag{3.9}$$

since $\mathscr{C} = \partial D$ and $d\theta = \omega$.

We have discussed counting quantum states; what about actually constructing them? In quantum mechanics, we expect the state to be a vector in a Hilbert space, which can be represented as a square-integrable wave function $Z(x)$. It turns out that, in the limit where

$\hbar$ is small, the wave function can be constructed to lowest order in a manner that bears a striking resemblance to (3.9):

$$\begin{aligned} Z(x) \xrightarrow[\hbar \to 0]{} & \exp\left[\frac{i}{\hbar}\int_0^x \theta + \cdots\right] \\ &= \exp\left[\frac{i}{\hbar}\int_0^x \sqrt{2\mathscr{E} - x^2}\, dx + \cdots\right]. \end{aligned} \tag{3.10}$$

Evaluating the wave function in this manner for the lowest-energy state of our system ($\mathscr{E} = \hbar/2$) yields

$$Z(x) \approx \exp\left[-\frac{1}{2\hbar}x^2 + \cdots\right]. \tag{3.11}$$

Indeed, $\exp(-x^2/2\hbar)$ is the exact expression for the $n = 0$ wave function.

We are slowly making progress towards understanding the quantization of our model system. The next step is to understand the transition between the classical and quantum notions of an *observable.* In the classical world, the observables x and p are coordinates in phase space — in other words, functions *on* the phase space:

$$x : \mathscr{M} \to \mathbb{R}, \quad (x, p) \mapsto x, \tag{3.12}$$

and so forth. General observables are functions of x and p, i.e. general elements of $C^\infty(\mathscr{M}, \mathbb{R})$.

In the quantum world, as is well known, x and p should be replaced by operators $\hat{x}$ and $\hat{p}$, obeying the canonical commutation relation

$$[\hat{p}, \hat{x}] = -i\hbar. \tag{3.13}$$

These operators now live in some noncommutative algebra, which is equipped with an action on the Hilbert space of states. In the position representation, for instance,

$$\hat{x}f(x) = xf(x), \quad \hat{p}f(x) = -i\hbar\frac{d}{dx}f(x), \tag{3.14}$$

where $f \in L^2(\mathbb{R})$. The constraint equation (3.5) that defines a classical trajectory is then replaced by the operator equation

$$\left[\frac{1}{2}(\hat{x}^2 + \hat{p}^2) - \mathscr{E}\right] Z(x) = 0, \tag{3.15}$$

which is just the familiar Schrödinger eigenvalue equation $\hat{H}Z = \mathscr{E}Z$. Now, unlike in the classical case, the solutions of (3.15) in the position representation will only be square-integrable (and therefore physically acceptable) for certain values of $\mathscr{E}$. These are precisely the familiar eigenvalues or allowed energy levels

$$\mathscr{E} = \hbar\left(n + \frac{1}{2}\right), \tag{3.16}$$

where $n = 0, 1, 2, \ldots$. Taking the lowest energy level ($n = 0$) as an example, the exact solution is $Z(x) = \exp(-x^2/2\hbar)$, just as we claimed above. The reader can easily verify this directly.

All of this discussion should be taken as illustrating our above claim that quantum mechanics should properly be understood as a "modern symplectic geometry," in which classical constraints are promoted to operator relations. We have constructed the following correspondence or dictionary between the elements of the classical and quantum descriptions of a system:

Table 1.

	Classical	Quantum
state space	symplectic manifold $(\mathscr{M}, \omega)$	Hilbert space $\mathscr{H}$
states	Lagrangian submanifolds $\mathscr{C} \subset \mathscr{M}$	vectors (wave functions) $Z \in \mathscr{H}$
observables	algebra of functions $f \in C^\infty(M)$	algebra of operators $\hat{f}$, acting on $\mathscr{H}$
constraints	$f_i = 0$	$\hat{f}_i Z = 0$

We now have a benchmark for what a successful quantization should accomplish: for a given classical system, it should construct the quantum counterpart for each element in the classical description, as summarized above. Of course, we would also like the correspondence principle to hold: that is, the quantum description should dovetail nicely with the classical one in some way when we take $\hbar \to 0$.

The correspondence between the classical and quantum descriptions is not quite as cut-and-dried as we have made it appear, and there are a few points that deserve further mention. Firstly, it should be apparent from our discussion of the harmonic oscillator that not every Lagrangian submanifold will have a quantum state associated to it; in particular, only a particular subset of these (obeying the Bohr–Sommerfeld quantization condition, or equivalently, corresponding to eigenvalues of the operator $\hat{H}$) will allow us to construct a square-integrable wave function $Z(x)$. There can be further constraints on *quantizable* Lagrangian submanifolds [20].

Secondly, let us briefly clarify why quantum state vectors correspond to Lagrangian submanifolds of the classical phase space and not to classical 1-dimensional trajectories, as one might naively think. (In our example of the harmonic oscillator we have $N = 1$ and, as a result, both Lagrangian submanifolds and classical trajectories are 1-dimensional.) The basic reason why Lagrangian submanifolds, rather than dimension-1 trajectories, are the correct objects to consider in attempting a quantization is the following. In quantum mechanics, we specify a state by giving the results of measurements of observables performed on that state. For this kind of information to be meaningful, the state must be a simultaneous eigenstate of all observables whose values we specify, which is only possible if all such observables mutually commute. As such, to describe the state space in quantum mechanics, we choose a "complete set of commuting observables" that gives a decomposition of $\mathscr{H}$ into 1-dimensional eigenspaces of these operators. For time-independent Hamiltonians, one of these operators will always be $\hat{H}$.

However, to the leading order in $\hbar$ the commutator of two quantum observables must be proportional to the Poisson bracket of the corresponding classical observables. Therefore, if $\hat{H}, \hat{f}_i$ form a complete set of commuting quantum-mechanical observables, we must have

$$\{H, f_i\}_{\text{P.B.}} = 0, \tag{3.17}$$

where $\{\cdot, \cdot\}_{\text{P.B.}}$ is the Poisson bracket. But we know that the classical time-evolution of the quantity f_i is determined by the equation

$$\frac{df_i}{dt} + \{H, f_i\}_{\text{P.B.}} = 0. \tag{3.18}$$

As such, the quantum-mechanical observables used in specifying the state must correspond to classically conserved quantities: *constants of the motion*. And it is well-known that the maximal possible number of classically conserved quantities is $N = \frac{1}{2}\dim \mathscr{M}$, corresponding to a completely integrable system; this follows from the nondegeneracy of the symplectic form on the classical phase space. For $N > 1$, then, specifying all of the constants of the motion does not completely pin down the classical trajectory; it specifies an N-dimensional submanifold $\mathscr{C} \subset \mathscr{M}$. However, it does give all the information it is possible for one to have about the quantum state. This is why Lagrangian submanifolds are the classical objects to which one attempts to associate quantum states.

We should also remark that it is still generically true that wave functions will be given to lowest order by

$$Z(x) = \exp\left[\frac{i}{\hbar}\int_{x_0}^{x}\theta + \cdots\right]. \tag{3.19}$$

This form fits all of the local requirements for $Z(x)$, although it may or may not produce a globally square-integrable wave function.

Finally, the quantum-mechanical algebra of operators is a non-commutative deformation or *q-deformation* of the algebra of functions $C^\infty(\mathscr{M})$, where the deformation is parametrized by

$$q := e^{\hbar}. \tag{3.20}$$

In the classical limit, $q \to 1$.

How are these general ideas about quantization implemented in the context of topological quantum field theories? To illustrate the connection, we will consider a specific example of a TQFT: the Chern–Simons gauge theory.

As in any gauge theory, the starting point of this theory is the choice of a gauge group G and the action functional, which in the present case is the Chern–Simons functional:

$$\frac{1}{\hbar}\int_M \mathrm{Tr}\left(A \wedge dA + \frac{2}{3}A \wedge A \wedge A\right). \tag{3.21}$$

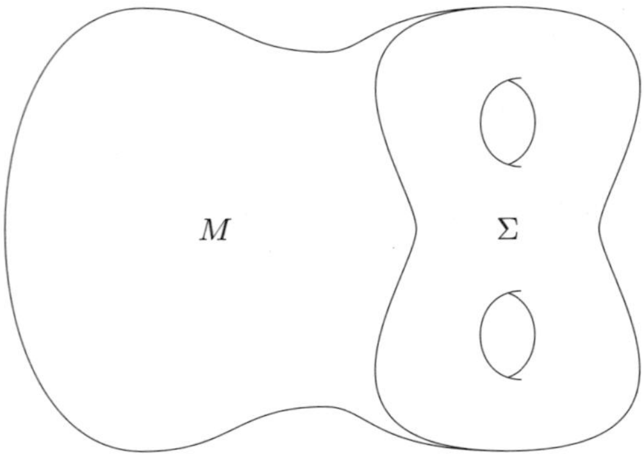

Fig. 7. The setup for Chern–Simons theory: an oriented 3-manifold M with boundary a 2-manifold Σ.

Here M is a 3-manifold, and the gauge field A is a connection on a principal G-bundle $E \to M$. The action functional (3.21) can be interpreted roughly as a Morse function on the space of gauge fields. We search for critical points of this functional by solving the equation of motion, which is the PDE

$$dA + A \wedge A = 0. \tag{3.22}$$

This equation says that A is a flat connection. How is this gauge theory formulation related to the picture of a TQFT as a functor, in the axiomatic language of Atiyah and Segal [18]?

The answer to this question is summarized in Table 2, and illustrates the way in which quantization plays a role. The action functional (3.21) defines a classical gauge theory; the classical phase space of this theory is the *moduli space of flat connections* $\mathscr{M} = \mathscr{M}_{\text{flat}}(G, \Sigma)$, where $\Sigma = \partial M$.

Now, let $\mathscr{M}_{\text{flat}}(G, M)$ be the moduli space of flat connections on M. There is a natural mapping

$$\mathscr{M}_{\text{flat}}(G, M) \hookrightarrow \mathscr{M}_{\text{flat}}(G, \Sigma) \tag{3.23}$$

induced by restriction to $\Sigma = \partial M$. The image of this map is the subspace of $\mathscr{M}$ consisting of flat connections on Σ that can be extended to M; this is a Lagrangian submanifold $\mathscr{C} \subset \mathscr{M}$.

We are now equipped with precisely the classical data referred to in our earlier discussion of the quantization problem. If we now quantize the classical Chern–Simons theory, the classical phase space $\mathscr{M}$ and the Lagrangian submanifold $\mathscr{C} \subset \mathscr{M}$ will be respectively replaced with a Hilbert space and a state vector in that Hilbert space. But these are precisely the objects that we expect a TQFT functor to associate to Σ and M!

To sum up, our situation is as follows:

Table 2.

Geometry	Classical CS	Quantum CS
2-manifold	symplectic manifold	vector space
Σ	$\mathscr{M} = \mathscr{M}_{\text{flat}}(G, \Sigma)$	$\mathscr{H}_\Sigma$
3-manifold	Lagrangian submanifold:	vector
$M\ (\partial M = \Sigma)$	connections extendible to M	$Z(M) \in \mathscr{H}_\Sigma$

To move from the first column to the second, we define the classical Chern–Simons theory. Moving from the second column to the third consists of a quantization of this theory. The usual picture of a TQFT as a functor is the composition of these two: it moves directly from the first to the third column, ignoring the second.

Let us discuss the phase space of classical Chern–Simons theory a little further. It is known that all flat connections on Riemann surfaces are described by their holonomies; that is, the moduli space consists of maps

$$\mathscr{M} = \text{Hom}(\pi_1(\Sigma) \to G)/\text{conjugation}. \tag{3.24}$$

As emphasized in the work of Atiyah and Bott [21], this space comes equipped with a natural symplectic form,

$$\omega = \frac{1}{4\pi^2} \int_\Sigma \text{Tr}\, \delta A \wedge \delta A, \tag{3.25}$$

where δ denotes the exterior derivative on $\mathscr{M}$, so that δA is a 1-form on Σ as well as on $\mathscr{M}$. The Lagrangian submanifold we are considering is then given by

$$\mathscr{C} = \text{Hom}(\pi_1(M) \to G)/\text{conjugation}, \tag{3.26}$$

and the inclusion map is induced by the natural map $\pi_1(\Sigma) \to \pi_1(M)$. This Lagrangian submanifold can be defined by classical constraint equations of the form

$$A_i = 0. \tag{3.27}$$

Quantization will then replace these with quantum constraints; that is, operator relations

$$\hat{A}_i Z = 0 \tag{3.28}$$

much like the classical constraint (3.5) was replaced by the operator equation (3.15) in our previous example.

Exercise 3.1. Verify that $\mathscr{C}$ is indeed Lagrangian with respect to the symplectic form (3.25). That is, show that

$$\omega|_{\mathscr{C} \subset \mathscr{M}} = 0. \tag{3.29}$$

Exercise 3.2. Let g be the genus of Σ. Show that, for $g > 1$,

$$\dim \mathscr{M} = (2g - 2) \dim G. \tag{3.30}$$

Solution. Consider the case where G is a simple group. The fundamental group $\pi_1(\Sigma)$ is generated by $2g$ elements A_i and B_i, $1 \leq i \leq g$, subject to the one relation

$$A_1 B_1 A_1^{-1} B_1^{-1} \cdots A_g B_g A_g^{-1} B_g^{-1} = 1. \tag{3.31}$$

After applying an element of $\mathrm{Hom}(\pi_1(\Sigma) \to G)$, the generators are mapped to group-valued matrices, and so we need $2g \cdot \dim G$ parameters to specify them all independently. However, there are constraints: the matrices must obey (3.31), one matrix equation which eliminates $\dim G$ degrees of freedom. Taking the quotient by conjugacy eliminates another $\dim G$ degrees of freedom, leaving

$$\dim \mathscr{M} = (2g - 2) \dim G, \tag{3.32}$$

as we expected. □

Let us now specialize this general discussion and consider the theory with gauge group $G = \mathrm{SL}_2\mathbb{C}$ on a 3-manifold that is a knot complement, $M = S^3 \backslash N(K)$. Then, of course, $\partial M = \Sigma \cong T^2$. It follows immediately that $\pi_1(\Sigma) = \mathbb{Z} \times \mathbb{Z}$, so that

$$\begin{aligned}\mathscr{M} &= \mathrm{Hom}(\mathbb{Z} \times \mathbb{Z} \to \mathrm{SL}_2\mathbb{C})/\text{conjugacy} \\ &= (\mathbb{C}^\star \times \mathbb{C}^\star)/\mathbb{Z}_2.\end{aligned} \tag{3.33}$$

This is exactly the space we considered in Sec. 2 in our discussion of the A-polynomial: it is the representation variety of the boundary torus of M! Moreover, the Lagrangian submanifold is in this case given by

$$\begin{aligned}\mathscr{C} &= \mathrm{Hom}(\pi_1(M) \to \mathrm{SL}_2\mathbb{C})/\text{conjugacy} \\ &= \{(x, y) \in (\mathbb{C}^\star \times \mathbb{C}^\star)/\mathbb{Z}_2 : A(x, y) = 0\},\end{aligned} \tag{3.34}$$

where $A(x, y)$ is a familiar polynomial in x and y, interpreted now as a classical observable giving the classical constraint relation that defines the submanifold $\mathscr{C} \subset \mathscr{M}$.

The appearance of the A-polynomial in this context clarifies two mysterious statements that were made in the previous section. Firstly, it makes apparent in what sense the zero locus of the A-polynomial is a natural object in symplectic geometry. Secondly, we can now make sense of the statement that one can "quantize" the A-polynomial. Having interpreted it as a classical constraint equation defining a Lagrangian submanifold of a classical phase space, it becomes obvious that quantization replaces the A-polynomial by an operator in a quantum constraint equation of the form (3.28).

What happens when we try to quantize the A-polynomial? The natural symplectic form (3.25) on the classical phase space takes the simple form [13]:

$$\omega = \frac{dy}{y} \wedge \frac{dx}{x} = d \ln y \wedge d \ln x. \tag{3.35}$$

The canonical commutation relation is therefore

$$\left[\widehat{\ln y}, \widehat{\ln x}\right] = \hbar, \tag{3.36}$$

which can be rewritten in the form

$$\hat{y}\hat{x} = q\hat{x}\hat{y} \tag{3.37}$$

with $q = e^{\hbar}$. Given this relation, what form do the operators $\hat{x}$ and $\hat{y}$ take in the position representation? Of course, we must have $\hat{x}f(x) = xf(x)$. Then the commutation relation becomes

$$q\hat{x}(\hat{y}f(x)) = \hat{y}(\hat{x}f(x)), \tag{3.38}$$

and implies that $\hat{y}$ should act as a *shift operator* $\hat{y}f(x) = f(qx)$. The reason for this name is the following. Notice that the symplectic form (3.35) has the canonical form in logarithms of x and y, rather than x and y themselves. Therefore, it is natural to introduce the logarithmic variable n by the relation $x = q^n$. Then, in terms of n the action of the operators $\hat{x}$ and $\hat{y}$ looks like

$$\hat{x}f(n) = q^n f(n), \quad \hat{y}f(n) = f(n+1). \tag{3.39}$$

The quantization of the polynomial $A(x,y) = \sum_k a_k(x)y^k$ will then be an operator of the form

$$\hat{A}(\hat{x},\hat{y};q) = \sum_k a_k(\hat{x};q)\hat{y}^k. \tag{3.40}$$

In general, quantization is a rather delicate and mysterious procedure [22] (see [20] for a recent discussion). However, for algebraic curves defined by classical constraint equations of the form $A(x,y) = 0$, recent progress in mathematical physics [23–27] has led to a systematic way of constructing the coefficients $a_k(\hat{x};q)$ of the quantum operator (3.40) entirely from the data of the classical A-polynomial [16] (see also [28]):

$$A(x,y) \rightsquigarrow \hat{A}(\hat{x},\hat{y};q). \tag{3.41}$$

In addition, in some cases the curve $A(x,y) = 0$ comes from extra data that may be very helpful in constructing its quantum analog. For instance, the construction [29] of the ordinary A-polynomial based on the triangulation data of a 3-manifold M admits a beautiful noncommutative lift [30]. However, since in what follows we need to apply the procedure (3.41) to arbitrary curves for which the extra data is not always available, we shall mainly focus on the so-called *topological recursion* approach that involves complex analysis and noncommutative algebra on $\mathscr{C}$.

In complex analysis, one of the basic ingredients associated to the curve $\mathscr{C} : A(x,y) = 0$ is the so-called *Bergman kernel*. It becomes the first brick in the foundation of the construction (3.41) based on the topological recursion, which after a few more systematic and completely rigorous steps builds the q-difference operator as a power series in $\hbar$:

$$A(x,y) \rightsquigarrow \hat{A}(\hat{x},\hat{y};q) = A(\hat{x},\hat{y}) + \hbar A_1(\hat{x},\hat{y}) + \cdots . \tag{3.42}$$

Even though we omit the intermediate steps due to constraints of space, the reader should simply be aware that a well-defined, systematic procedure exists. The existence and uniqueness of this procedure are well-motivated based on physical considerations; in fact, these form one of the basic premises of quantum mechanics.

By looking at (3.42) it would seem that we would therefore have to compute terms to arbitrarily high order in this series to write down the operator $\hat{A}$. However, in practice, this is not the case; we usually need to compute only one or sometimes two terms in the series to know $\hat{A}$ exactly! The trick is as follows: if we know, *a priori*, that the operator we construct can be written as a rational function of $q = e^{\hbar}$, then the higher order terms in the expansion in $\hbar$ must resum nicely into an expression of this form. We also have information about the classical limit $(q \to 1)$ of this expression. Armed with this information, it is usually pretty straightforward to construct the quantization of $A(x,y)$ in closed form.

For example, if we know both the classical term and the first quantum correction $\hbar A_1(\hat{x},\hat{y})$ in the expansion (3.42), there is a good chance we can reconstruct the quantum operator

$$\hat{A}(\hat{x},\hat{y};q) = \sum_{m,n} a_{m,n}\, q^{c_{m,n}}\, \hat{x}^m\, \hat{y}^n \tag{3.43}$$

simply from the data $\{a_{m,n}\}$ of the original polynomial $A(x,y) = \sum a_{m,n} x^m y^n$ and from the exponents $\{c_{m,n}\}$ determined by $\hbar A_1(\hat{x},\hat{y})$. This trick becomes especially useful for curves that come from knots and 3-manifolds. Indeed, in such examples the leading quantum correction is determined by the "classical" knot invariant $\Delta(q)$ called the twisted Alexander polynomial. Therefore, a simple mnemonic rule to remember what goes into the construction of the operator $\hat{A}(\hat{x}, \hat{y}; q)$ in such situations can be schematically expressed as [16]:

$$\boxed{\text{“}A(x,y) + \Delta(q) \;\Rightarrow\; \hat{A}(\hat{x},\hat{y};q)\text{”}}\,. \tag{3.44}$$

Concretely, the exponents $c_{m,n}$ in (3.43) can be determined by requiring that the relation

$$2\sum_{m,n} a_{m,n}\, c_{m,n}\, x^m y^n = \frac{\partial A}{\partial \ln x}\left(\frac{\partial A}{\partial \ln y}\right)^{-1}\frac{\partial^2 A}{(\partial \ln y)^2} + x\frac{\partial \Delta(x)}{\partial x}\frac{\partial A}{\partial \ln y} \tag{3.45}$$

holds for all values of x and y (along with $A(x, y) = 0$).

Example 3.3. Consider once more the trefoil knot $K = 3_1$, which has A-polynomial $A(x,y) = (y^{-1}-1)(y+x^3)$ and where, following our earlier agreement, we replaced x^2 by x to simplify the expressions, cf. (2.17). Notice that $A(x, y)$ in this example is a degree-2 polynomial in y. Quantization (3.42) then gives an operator which is also of degree 2 in $\hat{y}$

$$\hat{A}(\hat{x}, \hat{y}; q) = \alpha\hat{y}^{-1} + \beta + \gamma\hat{y}, \tag{3.46}$$

where

$$\begin{cases} \alpha = \dfrac{x^2(x-q)}{x^2-q}; \\[2ex] \beta = q\left(1 + x^{-1} - x + \dfrac{q-x}{x^2-q} - \dfrac{x-1}{x^2q-1}\right); \\[2ex] \gamma = \dfrac{q-x^{-1}}{1-qx^2}. \end{cases} \tag{3.47}$$

In the representation (3.39), our quantized constraint (3.28) then gives an operator relation that takes the form of a recurrence in the variable n:

$$\hat{A}Z = 0 \implies \alpha(q^n;q)Z_{n-1} + \beta(q^n;q)Z_n + \gamma(q^n;q)Z_{n+1} = 0, \tag{3.48}$$

where we recall that n was defined so that $x = q^n$.

Exercise 3.4. Solve this recurrence with the initial conditions

$$Z_n = 0 \quad \text{for } n \leq 0; \;\; Z_1 = 1. \tag{3.49}$$

That is, find the first few terms of the sequence $Z_n(q)$ for $n = 2, 3, \ldots$

Solution. Straightforward computation gives

$$\begin{aligned} Z_2(q) &= -\beta(q;q)/\gamma(q;q) \\ &= -\frac{1-q^3}{q-q^{-1}} \cdot q\left(1+q^{-1}-q-\frac{q-1}{q^3-1}\right) \\ &= -\frac{(1-q^3)(1+q-q^2)+q(q-1)}{q-q^{-1}} \\ &= \frac{-1+q^3+q^4-q^5}{q-q^{-1}} \\ &= q+q^3-q^4, \end{aligned} \tag{3.50}$$

as well as

$$\begin{aligned} Z_3(q) &= -(\alpha(q^2;q)+\beta(q^2;q)Z_2(q))/\gamma(q^2;q) \\ &= q^2+q^5-q^7+q^8-q^9-q^{10}+q^{11}, \end{aligned} \tag{3.51}$$

after a little manipulation. Notice that the Z_n all turn out to be polynomials! □

Now, we come to one of the punch lines of these lectures. The reader who has completed Exercise 1.2 and followed through the derivation of (1.20) may have noticed a startling coincidence: Z_n produced by our recurrence relation (3.48) is none other than the n-colored Jones polynomial; that is, the quantum group invariant of the knot decorated with extra data consisting of the Lie algebra $\mathfrak{g} = \mathfrak{sl}(2)$ and its n-dimensional representation $R = V_n$.

This is no coincidence, of course. As we reviewed in Sec. 1, the n-colored Jones polynomial is simply the partition function of Chern–Simons TQFT with gauge group $G = \mathrm{SU}(2)$. On the other hand, in this section we explained that the classical A-polynomial and its quantum, noncommutative version have a natural home in Chern–Simons TQFT with complex gauge group $G_{\mathbb{C}} = \mathrm{SL}_2\mathbb{C}$. In particular, we saw how the usual rules of quantum mechanics replace the classical constraint (3.34) with an operator relation (3.28),

$$\mathscr{C} : A(x, y) = 0 \rightsquigarrow \hat{A}(\hat{x}, \hat{y}; q) Z_{\mathrm{CS}}(M) = 0, \tag{3.52}$$

where $Z_{\mathrm{CS}}(M)$ is the state vector associated by quantization to the Lagrangian submanifold $\mathscr{C}$ (or, equivalently, associated by the Chern–Simons TQFT functor to the 3-manifold M). Since $G_{\mathbb{C}} = \mathrm{SL}_2\mathbb{C}$ is a complexification of $G = \mathrm{SU}(2)$, the partition functions in these two theories are closely related [31, 32]. In particular, it was argued in [13] that *both* SU(2) and $\mathrm{SL}_2\mathbb{C}$ partition functions must satisfy the quantum constraint equation (3.52). In the n-representation (3.39) it takes the form of a *recurrence relation*

$$A(x, y) = \sum_k a_k(x) y^k \rightsquigarrow \sum_k a_k(q^n; q) J_{n+k}(K; q) = 0, \tag{3.53}$$

which is precisely our q-difference equation (3.48) in the above example, where K was taken to be the trefoil knot. More generally, Eq. (3.53) is a q-difference equation, describing the behavior with respect to n, or "color dependence," of the n-colored Jones polynomial that is computed by Wilson loop operators in the SU(2) Chern–Simons theory.

The relation between the quantization of the A-polynomial and the quantum group invariants (3.53) that follows from Chern–Simons theory is the statement of the *quantum volume conjecture* [13] (see [17] for a review of earlier developments that led to it). This conjecture was independently proposed in [33] around the same time, and is also known as the *AJ-conjecture*. It provides a bridge between two seemingly distant areas of knot theory, the classical A-polynomial and the study of quantum group invariants. Before the discovery of this connection, the separate communities of knot theorists working on these two different types of invariants had very little contact with one another.

Do two knots having the same A-polynomial always have all the same n-colored Jones polynomials? Based on the above connection, we would expect an affirmative answer, given that the quantization procedure for the A-polynomial is essentially unique. This has been checked for knots up to large number of crossings, although there is as yet no formal proof. If it is true, then a single algebraic curve constructed without any reference to quantum groups encodes all the information about the whole tower of n-colored Jones polynomials:

$$A(x, y) \rightsquigarrow \hat{A}(\hat{x}, \hat{y}; q) \rightsquigarrow J_n(K; q) \,. \tag{3.54}$$

Nonetheless, even if all the n-colored Jones polynomials together carry no more information than the A-polynomial, their relation to quantum groups still makes them interesting objects of study in their own right. (It is also worth noting that the study of the colored Jones polynomial predates the discovery of the A-polynomial.)

Once we explained how to go, via quantization, from the classical A-polynomial to quantum group invariants (3.54) it is natural to ask whether there is a simple way to go back. The *generalized volume conjecture* [13] proposes an affirmative answer to this question and is also based on the fact that the analytic continuation of SU(2) is $\mathrm{SL}_2\mathbb{C}$. It states that, in the classical limit $q \to 1$ accompanied by the "large color" limit $n \to \infty$, the n-colored Jones polynomial, as a Wilson line in SU(2) Chern–Simons theory [2], exhibits the exponential behavior

$$J_n(K; q = e^{\hbar}) \underset{\substack{n \to \infty \\ \hbar \to 0}}{\sim} \exp\left(\frac{1}{\hbar} S_0(x) + \cdots\right), \tag{3.55}$$

where the limits are taken with $q^n = x$ held fixed. Here $S_0(x)$ is the classical action of $\mathrm{SL}_2\mathbb{C}$ Chern–Simons theory, which is

$$S_0(x) = \int \log y \frac{dx}{x} \tag{3.56}$$

evaluated on a path within the curve $\mathscr{C} : A(x, y) = 0$. Here, by an abuse of notation, the variable x stands in for a point on the Riemann surface; S_0 is actually a function on $\mathscr{C}$, and the integral in (3.56) is taken along a path in $\mathscr{C}$ from some fixed base point to the point at which S_0 is evaluated. Moreover, (3.56) is only well-defined if the integrality condition (2.22) holds! The change ΔS_0 that comes from composing the path used in our evaluation with an arbitrary closed cycle must be valued in $2\pi\mathbb{Z}$, so that the quantity e^{iS_0} is well-defined and independent of path; the integrality condition ensures that this is so.

To summarize, the generalized volume conjecture gives us two important ways of thinking about the A-polynomial: firstly, as a characteristic variety encoding information about $\mathrm{SL}_2\mathbb{C}$ flat connections, and secondly, as a *limit shape* in the limit of large color.

We have now begun to see how the seemingly disparate topics we have been discussing are connected to one another. Roughly speaking, there are four major themes in these lectures: quantum group invariants, the A-polynomial, quantization, and categorification. We have now seen how quantization relates the A-polynomial and quantum group invariants, providing a bridge between seemingly unrelated knot polynomials. In what remains, we will return to ideas of categorification, hoping to give at least a glimpse of how knot polynomials arise from deeper and more powerful homological invariants.

4. Categorification

Categorification is a powerful and flexible idea; it can mean different things in different contexts, and a given mathematical construction may admit more than one categorification depending on how one chooses to look at its structure. In the context of topological quantum field theories, however, categorification is manifested in a very natural way. The categorification of a 3-dimensional TQFT should be a 4-dimensional TQFT, from which the 3D theory is recovered by dimensional reduction, see e.g. [34, 35]. That is,

$$\text{3D TQFT} \underset{\text{dimensional reduction}}{\overset{\text{categorification}}{\rightleftarrows}} \text{4D TQFT}\,.$$

We can tabulate the information that each of these TQFTs should associate to geometrical objects in Table 3.

Thus, to a geometrical object of given dimension, a categorified TQFT associates objects of one higher categorical level than its decategorified counterpart. (The categorical level of the object associated by a TQFT to something in geometry corresponds to

Table 3.

Geometry	3D TQFT	4D TQFT
3-manifold M,	number $Z(M)$,	vector space $\mathscr{H}_K$
knot $K \subset M$	polynomial invariant $P(K)$	
2-manifold Σ	vector space $\mathscr{H}_\Sigma$	category $\mathbf{Cat}_\Sigma$

its *codimension*, so that a 4D TQFT assigns numerical invariants to 4-manifolds. Famous examples of these are given by Donaldson theory [36] and Seiberg–Witten theory [37].)

In 2000, Mikhail Khovanov [38] succeeded in constructing a categorification of the Jones polynomial. Like the Jones polynomial, it is associated to the extra data $\mathfrak{g} = \mathfrak{sl}(2)$ and its fundamental representation $R = V_2$. To give the barest outline, his construction associates a chain complex to a diagram of a link K. The homology of this chain complex can be shown to be invariant under the Reidemeister moves, and therefore to be an invariant of K. *Khovanov homology* $H_{i,j}(K)$ is doubly graded, and the Jones polynomial is its graded Euler characteristic, cf. (0.1),

$$J(q) = \sum_{i,j} (-1)^i q^j \dim H_{i,j}(K). \tag{4.1}$$

Sometimes it is convenient to encode information about the Khovanov homology in its Poincaré polynomial:

$$\mathrm{Kh}(q,t) = P_{\mathfrak{sl}(2),V_2}(q,t) = \sum_{i,j} t^i q^j \dim H_{i,j}(K). \tag{4.2}$$

The Jones polynomial is then recovered by making the evaluation at $t = -1$. As an example, the Poincaré polynomial of the trefoil knot is

$$\mathrm{Kh}(q,t;K=3_1) = q + q^3t^2 + q^4t^3. \tag{4.3}$$

It is easy to see that the evaluation at $t = -1$ indeed returns the normalized Jones polynomial of the trefoil knot (1.20) that we saw in Sec. 1. By definition, this version of the homology is called *reduced*. Its close cousin, the *unreduced knot homology* categorifies the *unnormalized* polynomial invariant. Thus, for the unnormalized

$i=3$	·	·	·	·	·	·	·	·	$\mathbb{Z}$
2	·	·	·	·	$\mathbb{Z}$	·	$\mathbb{Z}/2$	·	·
1	·	·	·	·	·	·	·	·	·
0	$\mathbb{Z}$	·	$\mathbb{Z}$	·	·	·	·	·	·
	1	2	3	4	5	6	7	8	$9=j$

Fig. 8. The Khovanov homology $H_{i,j}(K=3_1)$ of the trefoil knot.

Jones polynomial (1.14) of $K = 3_1$ the corresponding categorification is given by the unreduced Khovanov homology shown in Fig. 8.

Much like the Khovanov homology of a knot is a categorification of its Jones polynomial or quantum $\mathfrak{sl}(2)$ invariant, there exist generalizations [39–43] of the Khovanov homology categorifying the n-colored Jones polynomials for all n:

$$J_n(K;q) = P_n(K;q,t)|_{t=-1} = \sum_{i,j} q^i t^j \dim H_{i,j}^{\mathfrak{sl}(2),V_n}(K)\Big|_{t=-1}. \tag{4.4}$$

The n-colored $\mathfrak{sl}(2)$ knot homologies satisfy recursion relations, just like their decategorified versions, and exhibit beautiful asymptotic behavior in the limit of large n. Both of these behaviors are controlled by a *refined algebraic curve*, which is an analogue of the A-polynomial [44]:

$$\mathscr{C}^{\text{ref}}: \; A^{\text{ref}}(x,y;t) = 0. \tag{4.5}$$

This curve is a t-deformation of (the image of) the representation variety of a knot complement M in the classical phase space of the Chern–Simons theory, which is the moduli space $\mathscr{M}_{\text{flat}}(\text{SL}_2\mathbb{C},\Sigma)$ of flat connections. Here $\Sigma = \partial M$. Much like the representation variety (3.34) of M, its t-deformation (4.5) is a holomorphic Lagrangian submanifold with respect to the symplectic form (3.35).

Example 4.1. In Sec. 2 we derived the A-polynomial of the trefoil knot (2.17). Then, in Sec. 3 we discussed its quantization, or *noncommutative* q-deformation. In both cases, the result is a quadratic polynomial in y. Similarly, the *commutative* t-deformation of the A-polynomial for the trefoil knot is a quadratic polynomial in y,

$$A^{\mathrm{ref}}(x, y; t) = y^2 - \frac{1 - xt^2 + x^3t^5 + x^4t^6 + 2x^2t^2(t+1)}{1 + xt^3} y + \frac{(x-1)x^3t^4}{1 + xt^3} \tag{4.6}$$

which reduces to the ordinary A-polynomial (2.17) in the limit $t = -1$.

As in Sec. 3, quantization of $\mathscr{M}_{\mathrm{flat}}(\mathrm{SL}_2\mathbb{C}, \Sigma)$ with its natural symplectic form promotes x and y to operators obeying the commutation relation

$$\hat{y}\hat{x} = q\hat{x}\hat{y} \tag{4.7}$$

and turns the planar algebraic curve (4.5) into a q-difference recursion relation, cf. (3.53),

$$\boxed{\hat{A}^{\mathrm{ref}} P_\star(K; q, t) \simeq 0,} \tag{4.8}$$

where $\hat{x}P_n = q^n P_n$ and $\hat{y}P_n = P_{n+1}$. This recursion relation, called the *homological volume conjecture* in [44], provides a natural categorification of the generalized volume conjecture that was the subject of Sec. 3. Unlike the generalized volume conjecture, its homological version (4.8) is based on much more sophisticated physics that involves a physical interpretation of knot homologies in terms of refined BPS invariants [35, 45] and dynamics of supersymmetric gauge theories [11, 14, 15, 46]. The details of this physical framework go way beyond the scope of these lectures and we simply refer the interested reader to the original papers.

There also exists a homology theory categorifying the HOMFLY-PT polynomial [47, 48]. As should be obvious, this theory must be triply graded; the HOMFLY-PT polynomial is recovered by taking the graded Euler characteristic, cf. (4.1),

$$P_{a,q}(K) = \sum_{ijk} (-1)^i q^j a^k \dim \mathscr{H}_{ijk}(K). \tag{4.9}$$

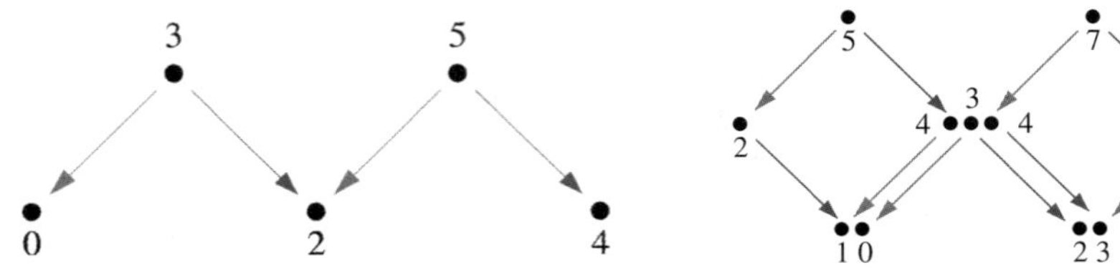

Fig. 9. The HOMFLY homology for knots 5_1 and 10_{132}. Each dot represents a generator, with its vertical and horizontal position indicating a- and q-degree respectively. The labels represent t-degree. The diagram can be thought of as similar to a Newton diagram for the Poincaré polynomial (4.10). (Images from [49].)

Just as we did for *Khovanov homology*, we can construct the Poincaré polynomial associated to the *HOMFLY homology*, which will encode information about the dimensions of its groups at each level:

$$\mathscr{P}(a,q,t) = \sum_{ijk} t^i q^j a^k \dim \mathscr{H}_{ijk}(K). \tag{4.10}$$

Then decategorification corresponds once more to evaluation at the value $t = -1$. It turns out that even the HOMFLY homology is not a complete invariant of knots; nonetheless, these homological invariant sare strictly finer and stronger than their decategorified counterparts. For instance, HOMFLY homology can distinguish between the knots 5_1 and 10_{132}, discussed earlier, that have identical Jones, Alexander, and HOMFLY-PT polynomials (1.17).

We should remark also that n-colored generalizations of HOMFLY homology can be constructed, and that the color dependence can be encoded in an algebraic curve, just as the zero locus of the A-polynomial encodes the information about color dependence of the n-colored Jones polynomial. We will return to this point and discuss the corresponding algebraic curve in much more detail in the final section of these lectures. Meanwhile, in the rest of this section we mostly focus on the ordinary, *uncolored* HOMFLY homology aiming to explain its structure and how to compute it in practice.

As we shall see, the structure of the homological knot invariants turns out to be so rich and so powerful that, once we learn enough about it, we will be able to compute, say, the Khovanov homology and the HOMFLY homology of the trefoil knot solely from the data of its Jones polynomial. In other words, in a moment we will learn powerful techniques that will allow us to reproduce (4.3) without even learning the definition of Khovanov homology. And, much of this structure is present — in fact, in a richer form! — in the *colored HOMFLY homology* as well [50].

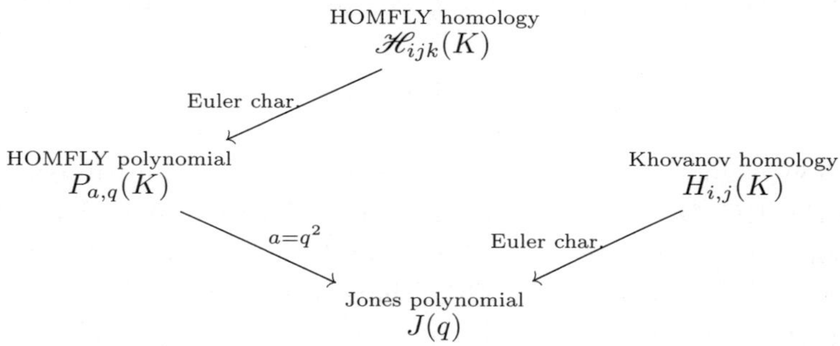

Fig. 10. A summary of relations between homological and polynomial invariants.

Let us start by summarizing the familiar relations (1.12), (4.1), (4.9) between homological and polynomial invariants diagramatically, as shown in Fig. 10. We would like to be able to fill in the missing fourth arrow, i.e. to have a way of recovering Khovanov homology directly from the HOMFLY homology. This, however, is rather delicate for a number of reasons. First, the specialization $a = q^2$ does not make sense in the context of the homology theories. At best one could try to complete the diagram by working with the Poincaré polynomials associated to these theories:

$$\begin{array}{ccccc} & & \mathscr{P}(a,q,t) & & \\ & \overset{t=-1}{\swarrow} & & \overset{a=q^2}{\searrow} & \\ P(a,q) & & & & \mathrm{Kh}(q,t) \\ & \underset{a=q^2}{\searrow} & & \underset{t=-1}{\swarrow} & \\ & & J(q) & & \end{array} \tag{4.11}$$

As we explain shortly, even this is too naive due to a simple, yet conceptual reason. Nevertheless, for a moment let us ignore this issue and proceed as if (4.11) were actually correct.

Example 4.2. Let us see if we can use the information in (4.11) to reconstruct $\mathscr{P}(a,q,t)$ for the trefoil knot. We know already that

$$\begin{cases} P(a,q) = aq^{-1} + aq - a^2, \\ \mathrm{Kh}(q,t) = q + q^3t^2 + q^4t^3. \end{cases} \tag{4.12}$$

We can attempt to guess $\mathscr{P}(a,q,t)$ just by comparing terms; this gives

$$\mathscr{P}(a,q,t) = aq^{-1} + aqt^2 + a^2t^3. \tag{4.13}$$

This naive guess turns out to be correct! Using only information from the HOMFLY-PT polynomial and Khovanov homology (both of which are easily computable), we have obtained information about the triply-graded HOMFLY homology theory, which encodes information about the $\mathfrak{sl}(N)$ homological invariants for all N.

In fact, one can even get to (4.13) without knowing the Khovanov homology! Our task is to assign a t-degree to each term in the HOMFLY-PT polynomial. We can do this using the following trick: From Exercise 1.2, the reader should know that evaluating $P(a,q)$ at $a = q$ yields a monomial (exactly which monomial depends on a simple knot invariant and a choice of normalization). This turns out to be true for any knot: the HOMFLY-PT polynomial will always become trivial, i.e. monomial, when evaluated at $a = q$. Therefore, to ensure the needed cancellation when the specialization $a = q$ is made, the normalized HOMFLY-PT polynomial for any knot must have the following schematic form:

$$P_{a,q} = 1 + (1 - a^{-1}q)Q(a,q), \tag{4.14}$$

where Q is some polynomial factor. The basic reason for this is that taking $a = q$ corresponds to asking about the $\mathfrak{sl}(1)$ polynomial invariant, which must always be trivial. A similar simplification happens in the case $a = q^{-1}$.

What about the $\mathfrak{sl}(1)$ *homological* invariant? Since $\mathscr{P}(a,q,t)$ has only positive coefficients, $\mathscr{P}(q,q,t)$ cannot be trivial — it must reduce to a monomial only because of cancellations that occur for $t = -1$. But we would not expect to be able to construct any nontrivial invariants with $\mathfrak{sl}(1)$, homological or otherwise. This is a clue that something more sophisticated must be happening in the way that one extracts Khovanov homology (generally, $\mathfrak{sl}(N)$ homology) from the HOMFLY homology.

The reason, to which we alluded earlier, is that when polynomial knot invariants are categorified one correspondingly needs to upgrade the specialization $a = q^N$ of Sec. 1 to homological level. In other words, trying to use the specialization $a = q^N$ as we did in diagram (4.11) is too naive and the suitable operation should also be from the world of homological algebra.

It turns out that the correct homological lift of the specialization $a = q^N$ involves a conceptually new ingredient, which has no analog at the (decategorified) polynomial level: a family of differentials $\{d_N\}$ on the HOMFLY homology, indexed by $N \in \mathbb{Z}$. These differentials endow HOMFLY homology with a structure that is much richer than what can be seen at the polynomial level and that is responsible for our claim that (4.13) can be derived even without the knowledge of the Khovanov homology. By viewing the triply-graded homology as a complex and taking its homology with respect to this differential, one recovers the doubly-graded Khovanov homology. Specifically, in the grading conventions of [50], the differentials have degree

$$\begin{aligned} d_{N>0}&: (-1, N, -1), \\ d_{N\leq 0}&: (-1, N, -3) \end{aligned} \tag{4.15}$$

with respect to (a, q, t) grading. The homology of $\mathscr{H}_\star$, viewed as a complex with differential d_N, returns the doubly-graded $\mathfrak{sl}(|N|)$ homology theory [47] or the *knot Floer homology* [51, 52] in the special case $N = 0$, see [49] for details. In particular, its homology with respect to the differentials d_1 and d_{-1} must be trivial.

For instance, in considering the reduction of HOMFLY homology to the $\mathfrak{sl}(1)$ homological invariant, almost all of the terms in the triply-graded HOMFLY homology will be killed by the differential d_1, leaving behind a "trivial" 1-dimensional space,

$$\dim(\mathscr{H}_\star, d_1) = 1. \tag{4.16}$$

As the differential d_1 has definite grading (4.15), the Poincaré polynomial of HOMFLY homology must therefore be of the following general form

$$\mathscr{P}(a,q,t) = 1 + (1 + a^{-1}qt^{-1})Q_+(a,q,t), \tag{4.17}$$

where the first term represents a contribution of the (trivial) $\mathfrak{sl}(1)$ knot homology, and $Q_+(a,q,t)$ is some polynomial with positive coefficients. Note that the Poincaré polynomial (4.17) necessarily has all of its coefficients non-negative. Similar structure follows from the existence of another *canceling differential* d_{-1} that also kills all but one generators of the HOMFLY homology. The physical interpretation of the differentials $\{d_N\}$ can be found in [50].

Now, just from the little we learned about the differentials d_1 and d_{-1}, we can reconstruct the HOMFLY homology of the trefoil knot. First, we can get information about the a- and q-degrees of nontrivial HOMFLY homology groups just from the HOMFLY-PT polynomial. For the trefoil knot, these are depicted below:

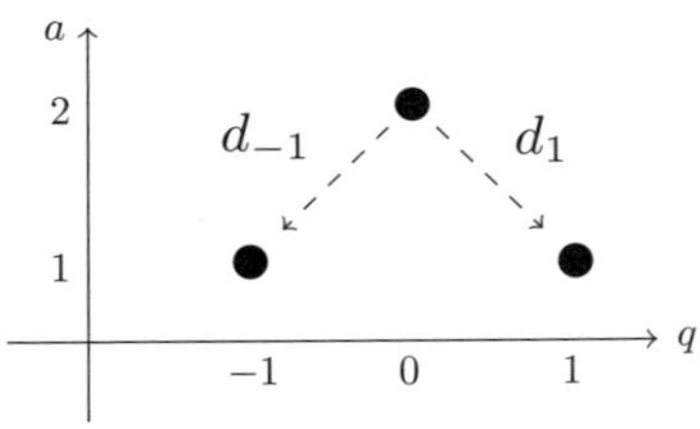

It is clear that each of the differentials $d_{\pm 1}$ can only act nontrivially in one place. From the condition that they give rise to trivial homology, each must be surjective; this determines the relative t-degree of each group. Taking the point with (a,q)-degree $(1,-1)$ to have $t = 0$, it immediately follows that the degrees of the other groups with respect to (a,q,t) degree are $(2,0,3)$ and $(1,1,2)$. We have now managed to extract this information without even computing Khovanov homology; the results of Exercise 1.2 and the above trick are all we need.

5. Epilogue: Super-A-polynomial

In this section, we give a somewhat deeper discussion of the connection between physics, homological knot invariants, and the quantization of the A-polynomial, constructing one final bridge between the ideas of quantization and categorification. This final section of the lectures can be seen as an addendum; based on recent progress [15, 53–55] it summarizes material that was covered in a talk given at the conference following the summer school, and so is somewhat more technical.

In these lectures, we saw several deformations of the classical A-polynomial $A(x,y)$ introduced in Sec. 2. In Sec. 3, we saw how quantization of $SL_2\mathbb{C}$ Chern–Simons theory leads to a *noncommutative* q-deformation (3.1). Then, in Sec. 4, we saw how more sophisticated physics based on refined BPS invariants leads to a categorification of the generalized volume conjecture and a *commutative* t-deformation (4.5).

These turn out to be special cases of a more general three-parameter "super-deformation" of the A-polynomial introduced in [54]. Two out of these three deformations are *commutative* and will be parametrized by a and t, while the third *noncommutative* deformation is produced essentially by the quantization procedure (3.42) of Sec. 3:

$$A^{\text{super}}(x,y;a,t) \rightsquigarrow \hat{A}^{\text{super}}(\hat{x},\hat{y};a,q,t). \tag{5.1}$$

What is the meaning of this *super-A-polynomial*?

The best way to answer this question is to consider an example. In fact, let us repeat the analogs of Example 3.3 and Exercise 3.4:

Example 5.1. For our favorite example, the trefoil knot $K = 3_1$, we know from our earlier discussion that the classical A-polynomial $A(x,y) = (y-1)(y+x^3)$ is quadratic in y, and so are its t-deformation (4.6) and q-deformation (3.46). The same is true of the super-A-polynomial of $K = 3_1$,

$$\begin{aligned} A^{\text{super}}(x,y;a,t) = y^2 &- \frac{a\left(1 - t^2x + 2t^2(1+at)x^2 + at^5x^3 + a^2t^6x^4\right)}{1+at^3x}y \\ &+ \frac{a^2t^4(x-1)x^3}{1+at^3x}, \end{aligned} \tag{5.2}$$

which clearly reduces to (4.6) upon setting $a = 1$ and to the ordinary A-polynomial (2.17) upon further specialization to $t = -1$. Moreover, the quantization procedure of Sec. 3 turns super-A-polynomial (5.2) into a q-difference operator, which can be interpreted as a recurrence relation, similar to (3.48),

$$\begin{aligned}\hat{A}^{\text{super}}(\hat{x}, \hat{y}; a, q, t) &= \alpha + \beta\hat{y} + \gamma\hat{y}^2 \\ &\Rightarrow \alpha\mathscr{P}_n + \beta\mathscr{P}_{n+1} + \gamma\mathscr{P}_{n+2} = 0. \end{aligned} \tag{5.3}$$

Here, the coefficients α, β, and γ are certain rational functions of the variables a, q, $x \equiv q^n$, and t, whose explicit form can be found in [54].

Exercise 5.2. As in Exercise 3.4, solve the recurrence (5.3) with the initial conditions

$$\mathscr{P}_n = 0 \quad \text{for } n \leq 0; \quad \mathscr{P}_1 = 1. \tag{5.4}$$

That is, find the first few terms of the sequence $\mathscr{P}_n(q)$ for $n = 2, 3, \ldots$.

Solution. Straightforward computation gives:

n	$\mathscr{P}_n(a, q, t)$
1	1
2	$aq^{-1} + aqt^2 + a^2t^3$
3	$a^2q^{-2} + a^2q(1+q)t^2 + a^3(1+q)t^3 + a^2q^4t^4 + a^3q^3(1+q)t^5 + a^4q^3t^6$
4	$a^3q^{-3} + a^3q(1+q+q^2)t^2 + a^4(1+q+q^2)t^3 + a^3q^5(1+q+q^2)t^4$ $+ a^4q^4(1+q)(1+q+q^2)t^5 + a^3q^4(a^2+a^2q+a^2q^2+q^5)t^6$ $+ a^4q^8(1+q+q^2)t^7 + a^5q^8(1+q+q^2)t^8 + a^6q^9t^9$

How should we interpret these polynomial invariants? The answer can be guessed from a couple of clues in the above table: firstly, all $\mathscr{P}_n(a, q, t)$ involve only positive integer coefficients. Secondly, we have seen $\mathscr{P}_2(a, q, t)$ before; it is the Poincaré polynomial (4.13) of the triply-graded HOMFLY homology of the trefoil knot! □

These considerations lead one to guess, correctly, that $\mathscr{P}_n(a, q, t)$ is the Poincaré polynomial of the n-colored generalization of the

HOMFLY homology:

$$\mathscr{P}_n(a,q,t) = \sum_{ijk} t^i q^j a^k \dim \mathscr{H}_{ijk}^{(n)}(K)\,. \tag{5.5}$$

Naively, one might expect that making the specialization $a = q^2$ in the polynomial $\mathscr{P}_n(a,q,t)$ should return the Poincaré polynomial for the n-colored $\mathfrak{sl}(2)$ homology in (4.4), and so forth. However, in the homological world, this specialization is a slightly bit more subtle. It turns out that, just as we saw earlier in Sec. 4, the colored homology $\mathscr{H}_{ijk}^{(n)}(K)$ comes naturally equipped with a family of differentials d_N; viewing $\mathscr{H}_{ijk}^{(n)}(K)$ as a complex and taking its homology with respect to the differential d_2 allows one to pass directly from the n-colored HOMFLY homology to the n-colored analog of the Khovanov homology.

To summarize, the super-A-polynomial encodes the "color dependence" of the *colored HOMFLY homology*, much like the ordinary A-polynomial and its t-*deformation* do for the colored Jones polynomial (3.53) and the colored $\mathfrak{sl}(2)$ homology (4.8), respectively:

$$\boxed{\hat{A}^{\text{super}}\mathscr{P}_\star(a,q,t) \simeq 0.} \tag{5.6}$$

Moreover, setting $q = 1$ gives the classical super-A-polynomial with two commutative parameters a and t. Its zero locus defines an algebraic curve

$$\mathscr{C}^{\text{super}}: \; A^{\text{super}}(x,y;a,t) = 0, \tag{5.7}$$

which in various limits reduces to the A-polynomial curve (2.2) and its "refined" version (4.5). This curve plays the same role for colored HOMFLY homology as the ordinary A-polynomial does for the colored Jones invariants. Specifically, there is an obvious analog of the generalized volume conjecture (3.55), which states that (5.7) is the *limit shape* for the S^n-colored HOMFLY homology in the large color limit $n \to \infty$ accompanied by $q \to 1$ [54].

A simple way to remember different specializations of the two-parameter "super-deformation" of the A-polynomial is via the following diagram:

$$
\begin{array}{ccccc}
 & & A^{\text{super}}(x,y;a,t) & & \\
 & \overset{a=1}{\swarrow} & & \overset{t=-1}{\searrow} & \\
A^{\text{ref}}(x,y;t) & & & & A^{\text{Q-def}}(x,y;a) \\
 & \underset{t=-1}{\searrow} & & \underset{a=1}{\swarrow} & \\
 & & A(x,y) & &
\end{array}
\tag{5.8}
$$

which should remind the reader of the diagram (4.11) expressing a similar relation between various polynomial and homological invariants discussed here. Indeed, each of the invariants in (4.11) has an n-colored analog, whose color dependence is controlled by the corresponding deformation of the A-polynomial in (5.8). In this diagram, we included yet another deformation of the A-polynomial, which can be obtained from the super-A-polynomial by setting $t = -1$. This so-called *Q-deformation* of the A-polynomial was recently studied in [53], where it was conjectured that $A^{\text{Q-def}}(x,y;a)$ agrees with the augmentation polynomial of *knot contact homology* [56–58].

As a closing remark, we should mention that the colored homological invariants have even more structure than we have so far discussed. One can also construct a family of *colored differentials*, which act by removing boxes from Young tableaux or reducing the dimension of the representation in the decoration of a link diagram [50]. For example,

$$(\mathscr{H}^{\square\square}, d_{\text{colored}}) \simeq \mathscr{H}^{\square}, \tag{5.8}$$

where $(\mathscr{H}^{\square\square}, d_{\text{colored}})$ denotes the homology of the complex with respect to the indicated differential. This can be expressed for the respective Poincaré polynomials by a relation of the form (4.17):

$$\mathscr{P}^{\square\square}(a,q,t) = a^s \mathscr{P}^{\square}(a,q^2,t) + (1+at)Q_+(a,q,t), \tag{5.9}$$

showing the color dependence of these invariants in the form that nicely integrates with the recursion (5.6).

In general, there are many more colored differentials, which altogether form a very rich and rigid structure [50]. To fully appreciate the beauty and the power of this structure one needs to consider *homologically thick knots*. Roughly speaking, these are the knots whose homological invariants contain a lot more new information compared to their polynomial predecessors. The knot $8_{19} = T^{(3,4)}$ — that can be equivalently viewed as a $(3,4)$ torus knot — is the first example of a homologically thick knot. Other examples of homologically thick knots and links include *mutants*.

In the case of n-colored HOMFLY homology that we discussed earlier, the colored differentials include the differentials d_N of Sec. 4 for special values of N in the range $-2n+3, \ldots, 1$. Note that in the uncolored theory ($n = 2$) this range contains only three differentials, $d_{\pm 1}$ and d_0, which play a very special role. Namely, the first two are canceling differentials, whereas d_0 is the differential that relates HOMFLY homology to *knot Floer homology* [49]. We emphasize that the last relation really requires the knowledge of how d_0 acts on HOMFLY homology, which is an extra data not contained in the Poincaré polynomial $\mathscr{P}(a,q,t)$. Curiously, this extra data is automatically contained in the colored version of the HOMFLY homology, so that *knot Floer homology* can be recovered directly from $\mathscr{P}_n(a,q,t)$, even for homologically thick knots!

The reason for this is that all three special differentials d_1, d_{-1} and d_0, have analogs in the n-colored theory. Moreover, they are part of the colored differentials d_N, with $N = -2n+3, \ldots, 1$. Specifically, in the n-colored HOMFLY homology the differentials d_1 and d_{1-n} are canceling, whereas d_{2-n} provides the relation to knot Floer homology [10, 50]. And the virtue of the colored theory is that the action of this latter differential can be deduced from the data of $\mathscr{P}_n(a,q,t)$ alone. In other words, what in the uncolored theory appears as a somewhat bizarre and irregular behavior at $N = -1, 0, +1$ becomes a natural and simple structure in the colored theory.

Acknowledgments

We would like to thank Tudor Dimofte, Hiroyuki Fuji, Lenhard Ng, Marko Stošić, Piotr Sułkowski, Cumrun Vafa, Edward Witten, Don Zagier for enlightening discussions and enjoyable collaborations on subjects considered in these notes. We are also grateful to Tudor Dimofte, Lenhard Ng, and Piotr Sułkowski for their comments on the draft. Moreover, special thanks are due to Stephan Stolz for hospitality during the 2012 Summer School on Topology and Field Theories at the Center for Mathematics, University of Notre Dame. This work is supported in part by DOE Grant DE-FG03-92-ER40701FG-02 and in part by NSF Grant PHY-0757647. Opinions and conclusions expressed here are those of the authors and do not necessarily reflect the views of funding agencies.

References

[1] N. Chriss and V. Ginzburg, *Representation Theory and Complex Geometry.* Birkhäuser, 1997.

[2] E. Witten, Quantum field theory and the Jones polynomial, *Comm. Math. Phys.* **121** (1989) 351–399.

[3] N. Reshetikhin and V. Turaev, Invariants of 3-manifolds via link polynomials and quantum groups, *Invent. Math.* **103** (1991) 547–597.

[4] P. Freyd, D. Yetter, J. Hoste, W. B. R. Lickorish, K. Millett and A. A. Ocneanu, A new polynomial invariant of knots and links, *Bull. Amer. Math. Soc.* **12** (1985) 239–246.

[5] D. Rolfsen, *Knots and Links.* AMS Chelsea Publishing, 1976.

[6] V. F. R. Jones, A polynomial invariant for knots via von Neumann algebras, *Bull. Amer. Math. Soc.* **12** (1985) 103–111.

[7] H. R. Morton and P. R. Cromwell, Distinguishing mutants by knot polynomials, *J. Knot Theory Ramification* **5** (1996) 225–238.

[8] P. Ozsvath and Z. Szabo, Knot Floer homology, genus bounds, and mutation, arXiv:math/0303225.

[9] S. Wehrli, Khovanov homology and Conway mutation, arXiv:math/0301312.

[10] T. power of refinement, "to appear,".

[11] T. Dimofte, D. Gaiotto and S. Gukov, 3-Manifolds and 3d Indices, arXiv:1112.5179.

[12] D. Cooper, M. Culler, H. Gillet, D. D. Long and P. B. Shalen, Plane curves associated to character varieties of 3-manifolds, *Invent. Math.* **118** (1994) 47–84.

[13] S. Gukov, Three-dimensional quantum gravity, Chern–Simons theory, and the A polynomial, *Comm. Math. Phys.* **255** (2005) 577–627.

[14] T. Dimofte, S. Gukov and L. Hollands, Vortex counting and Lagrangian 3-manifolds, *Lett. Math. Phys.* **98** (2011) 225–287.

[15] H. Fuji, S. Gukov, M. Stosic and P. Sulkowski, 3d analogs of Argyres–Douglas theories and knot homologies, *JHEP* **1301** (2013) 175.

[16] S. Gukov and P. Sułkowski, *JHEP* **1202** (2012) 070.

[17] T. Dimofte and S. Gukov, *Contemp. Math.* **541** (2011) 41–67.

[18] M. Atiyah, *The Geometry and Physics of Knots*, Cambridge University Press, 1990.

[19] T. Kohno, *Conformal Field Theory and Topology.* Iwanami Series in Modern Mathematics. Amer. Math. Soc., 1998.

[20] S. Gukov and E. Witten, Branes and quantization, arXiv:0809.0305.

[21] M. Atiyah and R. Bott, The Yang–Mills equations over Riemann surfaces, *Phil. Trans. R. Soc. London A* **308** (1982) 523–615.

[22] N. Woodhouse, *Geometric Quantization.* Oxford Univ. Press, 1997.

[23] A. S. Alexandrov, A. Mironov and A. Morozov, M-theory of matrix models, *Teor. Mat. Fiz.* **150** (2007) 179–192.

[24] M. Marino, Open string amplitudes and large order behavior in topological string theory, *JHEP* **0803** (2008) 060.

[25] B. Eynard and N. Orantin, Invariants of algebraic curves and topological expansion, arXiv:math-ph/0702045.

[26] V. Bouchard, A. Klemm, M. Marino, and S. Pasquetti, Remodeling the B-model, *Comm. Math. Phys.* **287** (2009) 117–178.

[27] R. Dijkgraaf, H. Fuji and M. Manabe, The volume conjecture, perturbative knot invariants, and recursion relations for topological strings, arXiv:1010.4542.

[28] G. Borot and B. Eynard, All-order asymptotics of hyperbolic knot invariants from non-perturbative topological recursion of A-polynomials, arXiv:1205.2261.

[29] W. D. Neumann and D. Zagier, Volumes of hyperbolic three-manifolds, *Topology* **24** (1985) 307–332.

[30] T. Dimofte, Quantum Riemann surfaces in Chern–Simons theory, arXiv:1102.4847.

[31] T. Dimofte, S. Gukov, J. Lenells and D. Zagier, Exact results for perturbative Chern–Simons theory with complex gauge group, *Comm. Num. Th. Phys.* **3** (2009) 363–443.

[32] E. Witten, Analytic continuation of Chern–Simons theory, arXiv:1001.2933.

[33] S. Garoufalidis, On the characteristic and deformation varieties of a knot, *Geom. Topol. Monographs* **7** (2004) 291–304.

[34] L. Crane and I. Frenkel, Four-dimensional topological quantum field theory, Hopf categories, and the canonical bases, *J. Math. Phys.* **35** (1994) 5136.

[35] S. Gukov, Gauge theory and knot homologies, *Fortsch. Phys.* **55** (2007) 473–490.

[36] E. Witten, Topological quantum field theory, *Comm. Math. Phys.* **117** (1988) 353.

[37] E. Witten, Monopoles and four manifolds, *Math. Res. Lett.* **1** (1994) 769–796.

[38] M. Khovanov, A categorification of the Jones polynomial, *Duke Math. J.* **101** (2000) 359–426.

[39] Y. Yonezawa, Quantum $(sl_n, \wedge V_n)$ link invariant and matrix factorizations, *Nagoya Math. J.* **204** (2011) 69–123.

[40] H. Wu, A colored $sl(N)$-homology for links in S^3, arXiv:0907.0695.
[41] B. Webster, Knot invariants and higher representation theory II: the categorification of quantum knot invariants, arXiv:1005.4559.
[42] B. Cooper and V. Krushkal, Categorification of the Jones–Wenzl projectors, arXiv:1005.5117.
[43] I. Frenkel, C. Stroppel and J. Sussan, Categorifying fractional Euler characteristics, Jones–Wenzl projector and $3j$-symbols, arXiv:1007.4680.
[44] H. Fuji, S. Gukov and P. Sułkowski, Volume conjecture: Refined and categorified (with an appendix by Hidetoshi Awata), arXiv:1203.2182.
[45] S. Gukov, A. Schwarz and C. Vafa, Khovanov–Rozansky homology and topological strings, *Lett. Math. Phys.* **74** (2005) 53–74.
[46] E. Witten, Fivebranes and knots, arXiv:1101.3216.
[47] M. Khovanov and L. Rozansky, Matrix factorizations and link homology, *Fund. Math.* **199** (2008) 1–91.
[48] M. Khovanov and L. Rozansky, Matrix factorizations and link homology II, *Geom. Topol.* **12** (2008) 1387–1425.
[49] N. M. Dunfield, S. Gukov and J. Rasmussen, The superpolynomial for knot homologies, arXiv:math/0505662.
[50] S. Gukov and M. Stosic, Homological algebra of knots and BPS states, arXiv:1112.0030.
[51] P. Ozsvath and Z. Szabo, Holomorphic disks and knot invariants, *Adv. Math.* **186** (2004) 58–116.
[52] J. Rasmussen, Floer homology and knot complements, arXiv:math/0306378.
[53] M. Aganagic and C. Vafa, Large N duality, mirror symmetry, and a Q-deformed A-polynomial for knots, arXiv:1204.4709.
[54] H. Fuji, S. Gukov and P. Sułkowski, Super-A-polynomial for knots and BPS states, arXiv:1205.1515.
[55] S. Nawata, P. Ramadevi, X. Sun and Zodinmawia, Super-A-polynomials for twist knots, *JHEP* **1211** (2012) 157.
[56] L. Ng, Framed knot contact homology, *Duke Math. J.* **141** (2008) 365–406.
[57] T. Ekholm, J. Etnyre, L. Ng and M. Sullivan, Filtrations on the knot contact homology of transverse knots, *Math. Ann.* **355** (2013) 1561–1591.
[58] L. Ng, Combinatorial knot contact homology and transverse knots, *Adv. Math.* **227** (2011) 2189–2219.

Dirac Operators in Gauge Theory

Andriy Haydys
Faculty of Mathematics, University of Bielefeld,
P. O. Box 100131, D-33501 Bielefeld, Germany
haydys@math.uni-bielefeld.de

This paper is a mixture of expository material and current research material. Among new results are examples of generalized harmonic spinors and their gauged version, the generalized Seiberg–Witten equations.

1. Introduction

A lot of advances in geometry and topology of low dimensional manifolds are intimately related to gauge theory. Recently a lot of interest attracted a variant of the anti-self-duality theory for higher dimensional manifolds equipped with metrics with special holonomies [7]. The anti-self-dual (asd) instantons on such manifolds can blow-up along certain subspaces of codimension four [36]. It is argued in [6] and [14] that in the limit one obtains certain "generalized harmonic spinors", which are harmonic sections of fiber bundles, the fibers of which are diffeomorphic to the moduli space of asd instantons on $\mathbb{R}^4$. One of the purposes of this paper is to present examples of generalized harmonic spinors and their gauged version, namely the generalized Seiberg–Witten equations.

The exposition is chosen so that the material should be accessible for a reader not familiar with the basics of spin geometry or the Seiberg–Witten theory. Throughout the focus is on examples.

Section 2 is a rapid introduction to Dirac operators and the Seiberg–Witten theory. We briefly introduce Dirac operators focusing on dimension four, which has the advantage that the corresponding spin group can be constructed "from scratch". The spin groups in low dimensions are intimately related to quaternions and the language of quaternions is emphasized throughout. This has a twofold purpose. First, for an unprepared reader this is a quick way to understand what Dirac operators are at least in low dimensions. Secondly, this prepares the ground for a certain generalization of Dirac operators considered later on. We finish the first part with the Seiberg–Witten equations emphasizing again the role of quaternions.

In Sec. 3, we introduce the generalized Dirac operator, whose zeros are the generalized harmonic spinors mentioned above. The generalized Dirac operator has its origins in physics [1] and was later considered also in mathematical literature [12, 30, 32]. The idea of the generalization is very simple. Recall that for the Euclidean space $\mathbb{R}^4$ the Dirac equation can be written in the form

$$\frac{\partial u}{\partial x_0} - i\frac{\partial u}{\partial x_1} - j\frac{\partial u}{\partial x_2} - k\frac{\partial u}{\partial x_3} = 0, \qquad u\colon \mathbb{R}^4 \to \mathbb{H}. \tag{1}$$

This clearly generalizes for maps $u\colon \mathbb{R}^4 \to M$ provided that M is a hypercomplex manifold. A generalization of (1) for non-flat source manifolds requires either some further restrictions on the target M or some additional structure on the source manifold. In this paper the first possibility is pursued, while a realization of the second one can be found for instance in [17].

In Sec. 4, examples of the generalized Seiberg–Witten equations are presented. With a suitable choice of the target space, these equations make sense for any four-manifold (or three-manifold). However, to omit technical details, only the case of $\mathbb{R}^4$ as the source manifold is considered.

2. Dirac Operators and the Seiberg–Witten Equations

2.1. *Clifford algebras and spin groups in low dimensions*

The purpose of this subsection is to recall briefly the notions of Clifford algebra and spin group focusing on low dimensions. More details can be found for instance in [21].

Since $\pi_1(\mathrm{SO}(n)) \cong \mathbb{Z}/2\mathbb{Z}$ for any $n \geq 3$, there is a simply connected Lie group denoted by $Spin(n)$ together with a homomorphism $Spin(n) \to \mathrm{SO}(n)$, which is a double covering. This characterizes $Spin(n)$ up to an isomorphism. The spin groups can be constructed explicitly with the help of Clifford algebras, however in low dimensions this can be done using the quaternions only. This is the approach taken here.

Let $\mathbb{H}$ denote the algebra of quaternions. Denote

$$Sp(1) = \{q \in \mathbb{H} \mid |q| = 1\} \cong S^3.$$

Clearly, this is a simply connected Lie group. Furthermore, identify $\operatorname{Im}\mathbb{H} = \{\bar{h} = -h\}$ with $\mathbb{R}^3$ and consider the homomorphism

$$\alpha\colon Sp(1) \to \mathrm{SO}(3), \quad q \mapsto A_q, \tag{2}$$

where $A_q h = qh\bar{q}$. It is easy to check that the corresponding Lie-algebra homomorphism is in fact an isomorphism. Since SO(3) is connected, α is surjective. Moreover, $\ker\alpha = \{\pm 1\}$. Hence, (2) is a nontrivial double covering, i.e. $Spin(3) \cong Sp(1)$.

To construct the group $Spin(4)$, first recall that the Hodge operator $*$ yields the splitting $\Lambda^2(\mathbb{R}^4)^* = \Lambda^2_+(\mathbb{R}^4)^* \oplus \Lambda^2_-(\mathbb{R}^4)^*$, where $\Lambda^2_\pm(\mathbb{R}^4)^* = \{\omega \mid *\omega = \pm\omega\}$. Since $\mathfrak{so}(4) \cong \Lambda^2(\mathbb{R}^4)^* = \Lambda^2_+(\mathbb{R}^4)^* \oplus \Lambda^2_-(\mathbb{R}^4)^* = \mathfrak{so}(3) \oplus \mathfrak{so}(3)$, the adjoint representation yields a homomorphism $\mathrm{SO}(4) \to \mathrm{SO}(3) \times \mathrm{SO}(3)$.

Identify $\mathbb{R}^4$ with $\mathbb{H}$ and consider the homomorphism[a]

$$\beta\colon Sp_+(1) \times Sp_-(1) \to \mathrm{SO}(4), \quad (q_+, q_-) \mapsto A_{q_+, q_-},$$

where $A_{q_+,q_-}h = q_+h\bar{q}_-$. An explicit computation shows that the composition $Sp_+(1) \times Sp_-(1) \to \mathrm{SO}(4) \to \mathrm{SO}(3) \times \mathrm{SO}(3)$ is given by $(q_+, q_-) \mapsto (A_{q_+}, A_{q_-})$. Hence, the Lie algebra homomorphism corresponding to β is an isomorphism and $\ker \beta$ is contained in $\{(\pm 1, \pm 1)\}$. As it is readily checked, $\ker \beta = \{\pm(1,1)\} \cong \mathbb{Z}/2\mathbb{Z}$. Hence, $Sp_+(1) \times Sp_-(1) \cong Spin(4)$.

Let U be a Euclidean vector space. Then the Clifford algebra $Cl(U)$ is the tensor algebra $TU = \mathbb{R} \oplus U \oplus U \otimes U \oplus \cdots$ modulo the ideal generated by elements $u \otimes u + |u|^2 \cdot 1$. In other words, $Cl(U)$ is generated by elements of U subject to the relations $u \cdot u = -|u|^2$. For instance, $Cl(\mathbb{R}^1) \cong \mathbb{R}[x]/(x^2+1) \cong \mathbb{C}$. The algebra $Cl(\mathbb{R}^2)$ is generated by $1, e_1, e_2$ subject to the relations $e_1^2 = -1 = e_2^2$ and $e_1 \cdot e_2 = -e_2 \cdot e_1$, which follows from $(e_1 + e_2)^2 = -2$. In other words, $Cl(\mathbb{R}^2) \cong \mathbb{H}$. In general, $Cl(\mathbb{R}^n)$ is generated by $1, e_1, \dots, e_n$ subject to the relations $e_i^2 = -1$ and $e_i \cdot e_j = -e_j \cdot e_i$ for $i \neq j$.

It is convenient to have some examples of modules over Clifford algebras. Such module is given by a vector space V together with a map

$$U \otimes V \to V, \quad u \otimes v \mapsto u \cdot v,$$

which satisfies $u \cdot (u \cdot v) = -|u|^2 v$ for all $u \in U$ and $v \in V$. An example of a $Cl(U)$-module is $V = \Lambda U^*$, where the $Cl(U)$-module structure is given by the map

$$u \otimes \varphi \mapsto \imath_u \varphi - \langle u, \cdot \rangle \wedge \varphi.$$

Let V be a quaternionic vector space. Then the quaternionic multiplication gives rise to the map $\operatorname{Im}\mathbb{H} \otimes V \to V$, $h \otimes v \mapsto h \cdot v$, which satisfies $h \cdot (h \cdot v) = -h\bar{h}v = -|h|^2 v$. Thus any quaternionic vector space is a $Cl(\mathbb{R}^3)$-module. In particular, the fundamental

[a]We adopt the common convention $Sp_\pm(1) = Sp(1)$. The significance of the subscripts "$\pm$" will be clear below.

representation $W \cong \mathbb{H}$ of $Sp(1) \cong Spin(3)$ with the action given by the left multiplication is a $Cl(\mathbb{R}^3)$-module.

Similarly, for any quaternionic vector space V the space $V \oplus V$ is a $Cl(\mathbb{R}^4)$-module. Indeed, the $Cl(\mathbb{R}^4)$-module structure is induced by the map

$$\begin{aligned} &\mathbb{H} \otimes_{\mathbb{R}} (V \oplus V) \to V \oplus V, \\ &h \otimes (v_1, v_2) \mapsto (hv_2, -\bar{h}v_1) = \begin{pmatrix} 0 & h \\ -\bar{h} & 0 \end{pmatrix} \begin{pmatrix} v_1 \\ v_2 \end{pmatrix}. \end{aligned} \tag{3}$$

In particular, the $Sp_+(1) \times Sp_-(1)$-representation $W^+ \oplus W^-$ is a $Cl(\mathbb{R}^4)$-module. Here, as the notation suggests, $W^\pm$ is the fundamental representation of $Sp_\pm(1)$.

2.2. *Dirac operators*

Let X be a Riemannian manifold. Denote by $Cl(X) \to X$ the bundle, whose fiber at a point $x \in X$ is $Cl(T_xX) \cong Cl(T_x^*X)$. Let $E \to X$ be a bundle of $Cl(X)$-modules, i.e. there is a morphism of vector bundles

$$Cl\colon TX \otimes E \to E, \quad (v, e) \mapsto v \cdot e,$$

such that $v \cdot (v \cdot e) = -|v|^2 e$. Then E is called a Dirac bundle if it is equipped with an Euclidean scalar product and a compatible connection ∇ such that the following conditions hold:

- $\langle v \cdot e_1, v \cdot e_2 \rangle = |v|^2 \langle e_1, e_2 \rangle$ for any $v \in T_xX$ and $e_1, e_2 \in E_x$;
- $\nabla(\varphi \cdot s) = (\nabla\varphi) \cdot s + \varphi \cdot \nabla s$ for any $\varphi \in \Gamma(Cl(X))$ and $s \in \Gamma(E)$.

Here $\nabla\varphi$ is obtained by extending the Levi-Cività connection from TX.

Definition 2.1. If E is a Dirac bundle, the operator

$$\mathcal{D}\colon \Gamma(E) \xrightarrow{\nabla} \Gamma(T^*X \otimes E) \xrightarrow{Cl} \Gamma(E)$$

is called the *Dirac operator* of E.

The Dirac operator is a (formally) self-adjoint first order elliptic partial differential operator. Elements of $\ker \mathcal{D}$ are called harmonic.

An example of a Dirac operator is given by choosing $E = \Lambda T^*X$, for which the corresponding Dirac operator is $\mathcal{D} = d + \delta$ [21, Theorem 5.12], where $\delta = \pm * d*$ and the sign depends on the dimension of the manifold and the degree of a form.

For the sake of simplicity, let us focus on a low dimensional case, say dimension four. Thus, from now on X denotes a Riemannian four-manifold. It is also convenient to assume that X is oriented.

As already mentioned above, the space $\mathbb{H} \oplus \mathbb{H}$ is a $Cl(\mathbb{R}^4)$-module. There are at least three ways to construct a Dirac bundle from this $Cl(\mathbb{R}^4)$-module.

One way is as follows. Denote by $P_{\mathrm{SO}} \to X$ the SO(4)-bundle of oriented orthonormal frames. Then X is called spin, if there is a $Spin(4)$-bundle $P_{Spin} \to X$, which is a fiberwise double covering of P_{SO}. Assume X is spin and choose a spin structure. Then, considering $\mathbb{H} \oplus \mathbb{H}$ as the representation $W^+ \oplus W^-$, one obtains the associated bundle still denoted by $W^+ \oplus W^-$. This bundle is called the spinor bundle of X and its sections are called spinors. The spinor bundle is a Dirac bundle and the corresponding Dirac operator is of the form

$$\mathcal{D} = \begin{pmatrix} 0 & \mathcal{D}^- \\ \mathcal{D}^+ & 0 \end{pmatrix}, \qquad \mathcal{D}^\pm \colon \Gamma(W^\pm) \to \Gamma(W^\mp).$$

The components $\mathcal{D}^\pm$ are also called Dirac operators. For instance, in the case $X = \mathbb{R}^4$ it readily follows from (3) that $\mathcal{D}^\pm \colon C^\infty(\mathbb{R}^4; \mathbb{H}) \to C^\infty(\mathbb{R}^4; \mathbb{H})$ can be written as

$$\mathcal{D}^+ = -\frac{\partial}{\partial x_0} + i\frac{\partial}{\partial x_1} + j\frac{\partial}{\partial x_2} + k\frac{\partial}{\partial x_3},$$
$$\mathcal{D}^- = \frac{\partial}{\partial x_0} + i\frac{\partial}{\partial x_1} + j\frac{\partial}{\partial x_2} + k\frac{\partial}{\partial x_3}.$$

The second way is a slight modification of the first one. Namely, denote by $Spin^c(4) = (Spin(4) \times S^1)/\pm 1$ and consider a $Spin^c(4)$-bundle $P_{Spin^c} \to X$ such that $P_{Spin^c}/S^1 \cong P_{\mathrm{SO}}$ (this isomorphism is fixed throughout). Unlike spin structures, *spin*^{c} structures always exist on oriented four-manifolds. Furthermore, denote by P_{det} the determinant bundle $P_{Spin^c}/\mathrm{SO}(4)$, which is a principal S^1-bundle. By observing that P_{Spin^c} is a double covering of $P_{\mathrm{SO}} \times P_{det}$, we obtain that a choice of connection a on P_{det} together with the Levi-Cività connection on P_{SO} induces a connection on P_{Spin^c}. Letting $Spin^c(4)$ act on $\mathbb{H} \oplus \mathbb{H}$ via

$$[q_+, q_-, z] \cdot (h_1, h_2) = (q_+ h_1 \bar{z}, q_- h_2 \bar{z})$$

we obtain the associated bundle still denoted by $W^+ \oplus W^-$, which is again a Dirac bundle. Hence, similarly to the case of spin four-manifolds, we obtain the *spin*^{c}-Dirac operators $\mathcal{D}_a^\pm \colon \Gamma(W^\pm) \to \Gamma(W^\mp)$.

The third way is to view $\mathbb{H} \oplus \mathbb{H}$ as the SO(4)-representation $(\Lambda_+^2(\mathbb{R}^4)^* \oplus \mathbb{R}) \oplus \mathbb{R}^4$. This leads to the following Dirac operator

$$\mathcal{D} = \begin{pmatrix} 0 & \mathcal{D}'' \\ \mathcal{D}' & 0 \end{pmatrix}, \quad \begin{aligned} \mathcal{D}' &= \delta^+ + d \colon \Omega_+^2(X) \oplus \Omega^0(X) \to \Omega^1(X); \\ \mathcal{D}'' &= d^+ + \delta \colon \Omega^1(X) \to \Omega_+^2(X) \oplus \Omega^0(X), \end{aligned}$$

where δ^+ is the restriction of $\delta = *d*$ to $\Omega_+^2(X)$.

We would like to mention briefly some applications of Dirac operators. An important property of Dirac operators is the Weitzenböck formula (also known as the Bochner identity) [21, Theorem 8.2], which can be written in the form

$$\mathcal{D}^2 = \nabla^* \nabla + \mathcal{R},$$

where $\nabla^*\nabla$ is the connection Laplacian and $\mathcal{R}$ is an expression, which depends algebraically on the curvature tensor. For instance, in the case $D = d + \delta$ the restriction of $\mathcal{R}$ to T^*X can be identified with the Ricci curvature. This implies in particular that for compact manifolds admitting a metric with positive Ricci curvature, the first Betti number vanishes.

For the spin-Dirac operator (not necessarily in dimension four) the curvature term $\mathcal{R}$ equals up to a constant to the scalar curvature. This implies that for a metric with positive scalar curvature there are no harmonic spinors. A consequence of this is that the signature of a spin four-manifold, which admits a metric with positive scalar curvature, vanishes (see, for instance, [24]).

A particular class of harmonic spinors is given by covariantly constant spinors (here, clearly, the dimension of the base manifold does not need to be four). For a complete simply connected irreducible manifold, the existence of a covariantly constant spinor is equivalent [38] to the holonomy group being one of $\mathrm{SU}(n)$, $Sp(n)$, G_2

or $Spin(7)$. A particularly interesting case for us is the last holonomy group, which can occur on eight-manifolds only. An eight-manifold M with holonomy in $Spin(7)$ is characterized by the existence of a particular closed four-form Ω called the Cayley form. A four-dimensional submanifold $X \subset M$ is called a Cayley-submanifold, if Ω restricted to X yields the volume form of the induced metric on X. Cayley-submanifolds are examples of calibrated submanifolds [15], hence they are volume minimizing in their homology classes. Cayley-submanifolds, possibly singular, also arise as blow-up loci [36] of $Spin(7)$-instantons, which are discussed in some details in Sec. 4.3 below. Finally, the space of infinitesimal deformations of Cayley-submanifolds is the space of harmonic spinors [23].

2.3. *The Seiberg–Witten equations*

Consider the map

$$\sigma\colon \mathbb{H} \to \operatorname{Im}\mathbb{H}, \quad x \mapsto xi\bar{x}.$$

Putting $x = z + jw$, this map can be written in a more common way, namely

$$\mathbb{C}^2 \to \mathfrak{su}(2), \quad (z,w) \mapsto \frac{1}{2}\begin{pmatrix} |z|^2 - |w|^2 & z\bar{w} \\ \bar{z}w & |w|^2 - |z|^2 \end{pmatrix}.$$

The map σ is $Spin^c(4)$-equivariant, if the source is regarded as the W^+-representation and the target as $\Lambda^2_+(\mathbb{R}^4)^* \cong \mathfrak{sp}_+(1)$. Choosing a *spin*^{c} structure P_{Spin^c} on an oriented Riemannian four-manifold X, we obtain an induced map, still denoted by σ, between the associated fiber bundles:

$$\sigma\colon W^+ \to \Lambda^2_+ T^*X.$$

The Seiberg–Witten equations [31] are

$$\begin{aligned} \mathcal{D}^+_a\varphi &= 0, \\ F^+_a &= \sigma(\varphi), \end{aligned} \qquad (a,\varphi) \in \mathcal{A}(P_{det}) \times \Gamma(W^+), \tag{4}$$

where $\mathcal{A}(P_{det})$ is the space of all connections on P_{det}. The space of solutions is invariant under the action of the gauge group $\mathcal{G} = \{g\colon X \to S^1\}$, which acts on a by the gauge transformations and on φ by the multiplication.

From now on assume that X is closed. Perturbing the second equation by a self-dual two-form one can achieve that for a generic choice of such perturbation the moduli space of solutions

$$\mathcal{M}_{SW} = \{(a,\varphi) \mid (a,\varphi) \text{ satisfies } (4)\}/\mathcal{G}$$

is a smooth oriented[b] compact manifold of dimension $d = (c_1(P_{det})^2 - 2\chi(X) - 3\,\mathrm{sign}(X))/4$, where χ and sign denote the Euler characteristic and the signature respectively.

Choose a basepoint $x_0 \in X$ and denote $\mathcal{G}_0 = \{g \in \mathcal{G} \mid g(x_0) = 1\}$. Then the space $\{(a,\varphi) \text{ satisfies } (4)\}/\mathcal{G}_0$ is a principal S^1-bundle over $\mathcal{M}_{SW}$. Let η denote the first Chern class of this bundle. Then, for a given *spin*^{c} structure the Seiberg–Witten invariant is the integer $\langle \eta^{\frac{d}{2}}, [\mathcal{M}_{SW}]\rangle$ provided d is even and 0 otherwise. This integer does not depend on the choice of the perturbation provided $b_2^+(X) > 1$. Thus, the Seiberg–Witten invariant is an integer-valued function on the space of all *spin*^{c} structures. Observe also that this space is an $H^2(X;\mathbb{Z})$-torsor.

A reader who wishes to learn more about basics of Seiberg–Witten theory is encouraged to consult [22, 24, 25, 29].

Let us give some sample applications of the Seiberg–Witten theory. The Seiberg–Witten invariant does not vanish on symplectic four-manifolds with $b_2^+ > 1$. Moreover, for such manifolds the Seiberg–Witten invariant coincides with a variant of the Gromov–Witten invariant [33]. Hence, there are strong restrictions on the smooth type of four-manifolds admitting symplectic structures.

The Seiberg–Witten invariant vanishes on connected sums of four-manifolds with $b_2^+ > 1$. There is however a refinement of the Seiberg–Witten invariant [3] not necessarily vanishing on connected sums. This refinement is based on the Seiberg–Witten *map* rather than on its zeros only. The Seiberg–Witten map was also used to prove [10] the "10/8-theorem", which is a strong restriction on the intersection form of a smooth four-manifold.

Seiberg–Witten theory also applies to the Riemannian geometry of four-manifolds. The Weitzenböck formula can be used to show

[b]This requires a choice of orientation on certain homology groups of X.

that on Riemannian four-manifolds with positive scalar curvature the Seiberg–Witten invariant vanishes [39]. Using Seiberg–Witten theory, LeBrun [20] showed that on four-manifolds there are obstructions to the existence of Einstein metrics besides the Hitchin–Thorpe inequality.

With the help of the three-dimensional variant of the Seiberg–Witten equations, Taubes proved [34] the Weinstein conjecture for three-manifolds. This states that for any closed three-manifold equipped with a contact form the associated Reeb vector field has a closed orbit.

3. Generalized Dirac Operators

3.1. *Generalized Dirac operators on four-manifolds*

Let (U, i_1, i_2, i_3) be a quaternionic vector space, where i_1, i_2, and i_3 are complex structures satisfying quaternionic relations. Let (V, I_1, I_2, I_3) be another quaternionic vector space. Considering (U, i_1) and (V, I_1) as complex vector spaces, we can decompose the space $Hom_{\mathbb{R}}(U, V)$ into two components consisting of complex linear and complex antilinear maps. Bringing i_2 and I_2 into consideration, we obtain that each component splits in turn into two subspaces (the remaining complex structures do not refine this splitting, since they are determined by the first two). This is shown schematically on the diagram:

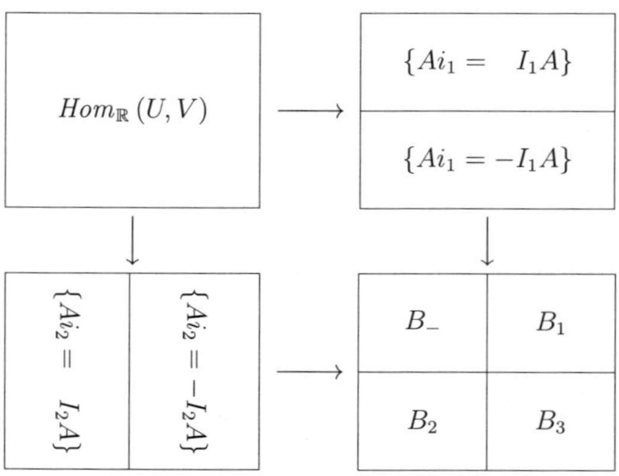

Here $B_- = Hom_{\mathbb{H}}(U,V)$ and $B_j = \{A \mid Ai_j = I_jA,\ Ai_k = -I_kA \text{ for } k \neq j\}$.

Notice that the group SO(3) acts on the space of quaternionic structures of any quaternionic vector space. Hence there is an induced action of SO(3) × SO(3) on $Hom_{\mathbb{R}}(U,V)$. Consider the action of the diagonally embedded SO(3). Then B_- is the trivial representation. Though each individual subspace $B_j, j = 1,2,3$, is not preserved by this action, their direct sum $B_1 \oplus B_2 \oplus B_3 = B_+$ is. To summarize, we have a splitting

$$Hom_{\mathbb{R}}(U,V) = B_- \oplus B_+ = Hom_{\mathbb{H}}(U,V) \oplus B_+, \tag{5}$$

which is invariant with respect to a simultaneous rotation of complex structures on both U and V.

Remark 3.1. It is easy to check that the map

$$A \mapsto \frac{1}{4}(A - I_1Ai_1 - I_2Ai_2 - I_2Ai_3) \tag{6}$$

is in fact the projection onto $Hom_{\mathbb{H}}(U,V)$. In particular, for $U = \mathbb{H}$ this projection can be written equivalently as

$$Hom_{\mathbb{R}}\left(\mathbb{R}^4, V\right) \to V \cong Hom_{\mathbb{H}}(\mathbb{H}, V),$$
$$A \mapsto \frac{1}{4}(Ae_0 - I_1Ae_1 - I_2Ae_2 - I_3Ae_3).$$

This in turn can be identified with the map $\mathbb{H} \otimes_{\mathbb{R}} V \to V,\ h \otimes v \mapsto \bar{h} \cdot v$ (cf. (3)).

Let $\tilde{G}$ be a Lie group together with a homomorphism $\mathbb{Z}/2\mathbb{Z} \to Z(\tilde{G})$. Denote

$$G = \tilde{G}/\pm 1 \quad \text{and} \quad \hat{G} = (Sp(1) \times \tilde{G})/\pm 1.$$

Clearly, there is a surjective Lie group homomorphism

$$(\rho_1, \rho_2)\colon \hat{G} \to \mathrm{SO}(3) \times G$$

with a finite kernel.

Let M be a hyper-Kähler[c] manifold. In particular, M comes equipped with a triple (I_1, I_2, I_3) of complex structures satisfying the quaternionic relations. Then for any purely imaginary quaternion $a = a_1 i + a_2 j + a_3 k$ of unit length denote $I_a = a_1 I_1 + a_2 I_2 + a_3 I_3$. We assume that $\hat{G}$ acts on M such that the following conditions hold:

(i) $(L_{\hat{g}})_* I_a (L_{\hat{g}^{-1}})_* = I_{\rho_1(\hat{g})a}$, where $L_{\hat{g}}\colon M \to M,\ m \mapsto \hat{g}\cdot m$;
(ii) $\tilde{G}$ preserves the hyper-Kähler structure of M, where $\tilde{G}$ is viewed as a subgroup of $\hat{G}$.

In the sequel, M is called the target manifold.

Let X^4 be an oriented Riemannian four-manifold. Denote by $P_\pm \xrightarrow{\pi_\pm} X$ the principal SO(3)-bundle of oriented orthonormal frames of $\Lambda^2_\pm T^*X$. For any $x \in X$ a quaternionic structure (i_1, i_2, i_3) on T_xX compatible with the scalar product and the orientation gives rise to a frame $(\omega_1^+, \omega_2^+, \omega_3^+)$ of $\Lambda^2_+ T^*_x X$, where $\omega_j^+ = g_x(i_j\cdot, \cdot)$. This correspondence allows us to interpret a point $p_+ \in P_+$ as a quaternionic structure on T_xX.

Let $\hat{P}$ be a principal $\hat{G}$-bundle. Then $\hat{P}/\tilde{G}$ is a principal SO(3)-bundle. We assume that $\hat{P}/\tilde{G}$ is isomorphic to P_+ and fix an isomorphism throughout. Similarly, $\hat{P}/Sp(1) = P_G$ is a principal G-bundle. Moreover, $\hat{P}$ is a finite covering of $P_+ \times P_G$. Hence, a connection a on P_G together with the Levi-Cività connection on P_+ determines a connection $\hat{a}$ on $\hat{P}$.

Let $\mathbb{M} = \hat{P} \times_{\hat{G}} M \xrightarrow{\pi} X$ be the associated bundle. Denote by $\mathcal{V} \to \mathbb{M}$ the vertical tangent bundle, i.e. $\mathcal{V} = \ker \pi_*$. For any section u of $\mathbb{M}$ the covariant derivative $\nabla^a u$ is a section of $Hom_{\mathbb{R}}(TX, u^*\mathcal{V})$. A consequence of Property (i) above is that for each $x \in X$ the vector space $\mathcal{V}_{u(x)}$ has a distinguished SO(3)-worth of quaternionic structures $\mathcal{I}_{u(x)}$, which can be canonically identified with $P_{+,x}$. Recalling that splitting (5) is SO(3)-invariant, we obtain

$$Hom_{\mathbb{R}}(TX, u^*\mathcal{V}) = Hom_{\mathbb{H}}(TX, u^*\mathcal{V}) \oplus Hom^+_{\mathbb{R}}(TX, u^*\mathcal{V}^+).$$

[c]The metric structure of M is non-essential for the purposes of this section but will play a role below.

Definition 3.2. We call the map $\mathcal{D}_a\colon u \mapsto -4(\nabla^a u)_{\mathbb{H}}$ the generalized Dirac operator of $\mathbb{M}$, where $(\nabla^a u)_{\mathbb{H}}$ is the $\mathbb{H}$-linear component of $\nabla^a u$.

Example 3.3. Let $P_{Spin^c} \to X$ be a *spin*^{c} structure. Recall that there is a short exact sequence

$$\{1\} \to Sp_-(1) \to Spin^c(4) \to (Sp_+(1) \times S^1)/\pm 1 \to \{1\}.$$

Hence, $\hat{P} = P_{Spin^c}/Sp_-(1)$ is a principal $\hat{G}$-bundle, where $\hat{G} = (Sp_+(1) \times S^1)/\pm 1 \cong \mathrm{U}(2)$. Furthermore, put $G = S^1$ and $M = \mathbb{H}$. Here $\mathbb{H}$ is viewed as being equipped with its left quaternionic structure and the $\hat{G}$-action is given by $[q_+, z] \cdot h = q_+ h\bar{z}$. It follows from Remark 3.1 that for these choices the "generalized" Dirac operator equals the *spin*^{c}-Dirac operator $\mathcal{D}_a^+$, where a is a connection on the S^1-bundle $\hat{P}/Sp_+(1) = P_{det}$.

Suitably modifying this example, one can also obtain the spin-Dirac operator $\mathcal{D}^+$ and $\delta^+ + d$. Details are left to the reader.

Remark 3.4. In the case $X = \mathbb{R}^4$, $G = \{1\}$, and $M = \mathbb{H}$ the equation $\mathcal{D}u = 0$ coincides with (1). This equation was studied by Fueter [9] in his attempts to construct a quaternionic version of the theory of holomorphic functions. Therefore, sometimes generalized harmonic spinors are also called Fueter-sections.

3.2. *Generalized Dirac operators on Kähler surfaces*

In this subsection a special case of the construction presented in the preceding subsection is studied. First, only the case $\hat{G} = \mathrm{SO}(3)$ is considered here. Secondly, X is assumed to be a Kähler surface.

Notice that the above assumptions imply in particular that $\hat{P} = P_+$. Moreover, for a Kähler surface the structure group of P_+ reduces to S^1. Concretely, one can think of $S^1 \subset \mathrm{SO}(3)$ as a stabilizer of a nonzero vector, say $(1, 0, 0) \in \mathbb{R}^3$. Let $P_{red} \subset P_+$ denote the corresponding S^1-subbundle. With our choices P_{red} is the principal bundle of the canonical bundle K_X. Then

$$\mathbb{M} = P_{red} \times_{S^1} M. \tag{7}$$

The induced S^1-action on M fixes I_1 and rotates the other two complex structures. This implies that the vertical bundle $\mathcal{V}$ carries a distinguished complex structure, which by a slight abuse of notation is also denoted by I_1. Then for $u \in \Gamma(\mathbb{M})$ we denote by $\partial u \in \Omega^{1,0}(X; u^*\mathcal{V})$ the $(1,0)$-component of the covariant derivative.

The proof of the next theorem is adapted from the proof of Proposition 4 in [12].

Theorem 3.5. *Let X be a compact Kähler surface. Then a spinor $u \in \Gamma(\mathbb{M})$ is harmonic, i.e. $\mathcal{D}u = 0$, if and only if $\partial u = 0$.*

Proof. First observe that the commutativity of S^1 together with (7) imply that $\mathbb{M}$ carries a fiberwise action of S^1, which in turn induces an S^1-action on $\Gamma(\mathbb{M})$.

For any $u \in \Gamma(\mathbb{M})$ we have the Weitzenböck-type formula [30]

$$\|\mathcal{D}u\|^2 = \|\nabla u\|^2 + \int_X \rho_0(u)\frac{s}{4}\, vol_X, \tag{8}$$

where s is the scalar curvature of X and ρ_0 is an S^1-invariant function. Here we used the fact that the self-dual part of the Weyl tensor vanishes on Kähler surfaces. Since the right-hand side of (8) is S^1-invariant, for any harmonic spinor u and any $z \in S^1$ the spinor $z \cdot u$ is also harmonic.

Pick a point $x \in X$ and choose a local trivialization of P_{red} on a neighborhood W of x. This trivialization gives rise to an almost quaternionic structure (i_1, i_2, i_3) on W, where i_1 is in fact the globally defined complex structure of X. Let $u_{loc}\colon W \to M$ be the local representation of u. Then by Remark 3.1 the harmonicity of u yields:

$$\nabla_v u_{loc} - I_1 \nabla_{i_1 v} u_{loc} - I_2 \nabla_{i_2 v} u_{loc} - I_3 \nabla_{i_3 v} u_{loc} = 0. \tag{9}$$

Here v is an arbitrary local vector field on W. Substituting $z \cdot u_{loc} = L_z u_{loc}$ instead of u_{loc} in (9), one obtains after a transformation:

$$\nabla_v u_{loc} - (L_{\bar z})_* I_1 (L_z)_* \nabla_{i_1 v} u_{loc} - (L_{\bar z})_* I_2 (L_z)_* \nabla_{i_2 v} u_{loc} - (L_{\bar z})_* I_3 (L_z)_* \nabla_{i_3 v} u_{loc} = 0. \quad (10)$$

Recall that the S^1-action preserves I_1 and rotates the other two complex structures. Hence, $(L_{\bar z})_* I_1 (L_z)_* = I_1$ for any $z \in S^1$ and there exists some $z \in S^1$ such that $(L_{\bar z})_* I_2 (L_z)_* = -I_2$ and $(L_{\bar z})_* I_3 (L_z)_* = -I_3$. Then (10) yields

$$\nabla_v u_{loc} - I_1 \nabla_{i_1 v} u_{loc} + I_2 \nabla_{i_2 v} u_{loc} + I_3 \nabla_{i_3 v} u_{loc} = 0.$$

Summing this with (9) leads to

$$\nabla_v u_{loc} - I_1 \nabla_{i_1 v} u_{loc} = 0 \Longleftrightarrow \nabla_{i_1 v} u_{loc} = -I_1 \nabla_v u_{loc}, \quad (11)$$

which means $\bar\partial u = 0$. On the other hand, it is easy to see that (11) implies (9). This finishes the proof. □

Holomorphic sections of bundles with fiber $T^*Gr_k(\mathbb{C}^n)$. Let (M_1, J_1, J_2, J_3) be a hyper-Kähler manifold equipped with an action of S^1, which fixes one complex structure, say J_1, and rotates the other two complex structures.[d] Assume also that there is an S^1-equivariant (J_1, I_1)-antiholomorphic map $\tau\colon M_1 \to M$. Then given a holomorphic section u_1 of $\mathbb{M}_1 = \hat P_{red} \times_{S^1} M_1$ we obtain a harmonic spinor $u \in \Gamma(\mathbb{M})$ by composing u_1 with the map $\mathbb{M}_1 \to \mathbb{M}$, $[\hat p, m_1] \mapsto [\hat p, \tau(m_1)]$. An example of this will be given below.

A large class of hyper-Kähler manifolds M_1 admitting S^1-action as described above was constructed in [8, 18], where M_1 is the cotangent bundle of a Kähler manifold Z (usually the hyper-Kähler metric is defined only in some neighborhood of the zero section). It is assumed in this case that $z \in S^1$ acts on T^*Z by the multiplication by z^p for some $p \in \mathbb{Z}$.

Identify $u_1 \in \Gamma(\mathbb{M}_1)$ with an equivariant map $\hat u_1\colon P_{red} \to M_1$. In the case $M_1 = T^*Z$ the map $\hat u_1$ can be composed with the projection $T^*Z \to Z$. The result is an S^1-invariant map $P_{red} \to Z$

[d]This action does not need to extend to an action of SO(3) as above.

or, equivalently, a map $v\colon X \to Z$. Writing $T^*Z \cong T^*Z \otimes_{\mathbb{C}} \mathbb{C}$ and letting S^1 act on $\mathbb{C}$ only, we see that the lift of v is given by a section ψ of $v^*T^*Z \otimes K_X^p$. Moreover, holomorphicity of u is equivalent to the holomorphicity of both v and ψ. Thus, in the case $M_1 = T^*Z$ we have

$$\{u_1 \in \Gamma(\mathbb{M}_1) \mid \bar{\partial} u_1 = 0\}$$
$$\cong \{(v, \psi) \mid v \in Map(X, Z), \psi \in \Gamma(v^*T^*Z \otimes K_X^p),\ \bar{\partial} v = 0,\ \bar{\partial}\psi = 0\}.$$

Let us consider the case $Z = Gr_n(\mathbb{C}^r)$ in some details. Recall that for a compact complex manifold X any holomorphic map $v\colon X \to Gr_n(\mathbb{C}^r)$ arises from an r-dimensional subspace $V \subset H^0(X; E)$ for some rank n holomorphic vector bundle $E \to X$ that is generated by holomorphic sections from V. Moreover, if $S \to Gr_n(\mathbb{C}^r)$ denotes the tautological vector bundle, then $E \cong v^*S$ and there is an embedding $E \hookrightarrow \underline{\mathbb{C}}^r$. Furthermore, $F = \underline{\mathbb{C}}^r/E$ is the pull-back of the canonical factor bundle Q on $Gr_n(\mathbb{C}^r)$. Since $T^*Gr_n(\mathbb{C}^r) \cong Q^\vee \otimes S$, it follows that $v^*T^*Gr_n(\mathbb{C}^r) \cong F^\vee \otimes E$. Hence we obtain the following result.

Proposition 3.6. *For a compact Kähler surface X, any holomorphic section of $P_{red} \times_{S^1} T^*Gr_n(\mathbb{C}^r) \to X$ can be constructed from the following data:*

- *A holomorphic rank n vector bundle E admitting r global holomorphic sections that generate E;*
- *A holomorphic section of $(\underline{\mathbb{C}}^r/E)^\vee \otimes E \otimes K_X^p$.*

Generalized harmonic spinors with values in the space of anti-self-dual instantons. Denote by $M_{n,r}(\mathbb{K})$ the space of matrices with n rows and r columns with entries from a ring $\mathbb{K} \in \{\mathbb{R}, \mathbb{C}, \mathbb{H}\}$. Consider the flat hyper-Kähler manifold

$$N = M_{n,n}(\mathbb{H}) \oplus M_{n,r}(\mathbb{H}) \cong M_{n,n}(\mathbb{C}) \oplus M_{n,n}(\mathbb{C}) \oplus M_{n,r}(\mathbb{C}) \oplus M_{r,n}(\mathbb{C}).$$

The group $U(n)$ acts on N as follows: $(B_1, B_2, C, D) \cdot g = (g^{-1}B_1,\ g^{-1}B_2,\ g^{-1}C,\ Dg)$. The corresponding moment map

$\mu\colon N \to \mathfrak{u}(n) \otimes \operatorname{Im}\mathbb{H}$, $\mu = \mu_{\mathbb{R}} i + \mu_{\mathbb{C}} j$, is given by

$$\mu_{\mathbb{R}} = \frac{i}{2}([B_1, \bar{B}_1^t] + [B_2, \bar{B}_2^t] + C\bar{C}^t - \bar{D}^t D), \quad \mu_{\mathbb{C}} = [B_1, B_2] + CD.$$

Denote

$$\mathcal{M}_0(r, n) = \{(B_1, B_2, C, D) \in N \mid \mu(B_1, B_2, C, D) = 0\}/U(n).$$

This space is called the hyper-Kähler reduction of N and carries itself a hyper-Kähler structure outside the singular locus [16]. Moreover, by the ADHM construction [2] there is a bijection between the non-singular part of $\mathcal{M}_0(r, n)$ and the moduli space of framed asd connections on a Hermitian bundle $E \to S^4$ of rank r and second Chern class n. Also, the following result holds.

Theorem 3.7. ([5, Corollary 3.4.10]) *There is a bijection between $\mathcal{M}_0(r, n)$ and the moduli space of framed* ideal *instantons on $\mathbb{R}^4$.*

Denote

$$\mathcal{M}_i(r, n) = \{(B_1, B_2, C, D) \in N \mid \mu(B_1, B_2, C, D) = i\}/U(n).$$

Clearly, $\mathcal{M}_i(r, n)$ is also a hyper-Kähler reduction of N but with respect to a different value of the moment map. This is a non-singular hyper-Kähler manifold, which was extensively studied by Nakajima [26–28]. In particular, $\mathcal{M}_i(r, n)$ is equipped with an S^1-action, which preserves one complex structure, say J_1, and rotates the other two complex structures. Moreover, there is a (J_1, I_1)-holomorphic map

$$\pi\colon \mathcal{M}_i(r, n) \to \mathcal{M}_0(r, n).$$

By putting $B_1 = 0 = B_2$ we see that $M_i(r, n)$ contains a hyper-Kähler submanifold $M_{n,r}(\mathbb{H})/\!/\!/_{\mu=i} U(n) = \{\mu(0, 0, C, D) = i\}/U(n)$, which is biholomorphic to $T^*Gr_n(\mathbb{C}^r)$ with respect to J_1 (details can be found for instance in [11, p. 303]). The complex conjugation on $M_{n,r}(\mathbb{C}) \oplus M_{r,n}(\mathbb{C}) \cong M_{n,r}(\mathbb{H})$ induces a J_1-antiholomorphic map

c on $T^*Gr_n(\mathbb{C}^r)$. Hence we obtain a (J_1, I_1)-antiholomorphic map

$$\tau\colon T^*Gr_n(\mathbb{C}^r) \xrightarrow{c} T^*Gr_n(\mathbb{C}^r) \hookrightarrow \mathcal{M}_i(r,n) \xrightarrow{\pi} \mathcal{M}_0(r,n).$$

Letting S^1 act on $\mathcal{M}_i(r,n)$ via

$$z \cdot (B_1, B_2, C, D) = (B_1, z^p B_2, C, z^p D)$$

the map τ becomes S^1-equivariant. Thus, recalling that composition of a holomorphic section with τ results in a harmonic spinor, we obtain the following result.

Theorem 3.8. *Let X be a compact Kähler surface. Then the same data as in Proposition* 3.6 *determine a harmonic section of* $\mathbb{M}_0(r,n) = P_{red} \times_{S^1} \mathcal{M}_0(r,n)$ *at least away from the singular locus.*

4. Generalized Seiberg–Witten Equations: Examples

In this section a generalization of the Seiberg–Witten equations first introduced in [30] and [32] is considered. This generalization makes sense for any four-manifold (or three-manifold) but for the sake of simplicity only the case $X = \mathbb{R}^4$ is considered here.

On $\mathbb{R}^4$ the generalized Seiberg–Witten equations can be defined as follows. Let M be a hyper-Kähler manifold equipped with a tri-Hamiltonian action of a Lie group G. Assume that $\mathfrak{g}$ is endowed with an *ad*-invariant scalar product and denote by $\mu\colon M \to \mathfrak{g} \otimes \operatorname{Im}\mathbb{H}$ the corresponding moment map. For a pair $(u, a) \in C^\infty(\mathbb{R}^4; M) \times \Omega^1(\mathbb{R}^4; \mathfrak{g})$, where a should be thought of as a connection on a trivial G-bundle, consider the following equations

$$\mathcal{D}_a u = 0, \quad F_a^+ = \mu \circ u, \tag{12}$$

which are called the generalized Seiberg–Witten equations. The first equation is already familiar from Sec. 3, while the second one needs a little explanation. Identifying $\Lambda^2_+(\mathbb{R}^4)^*$ with $\operatorname{Im}\mathbb{H}$, one can interpret $\mu \circ u$ as a self-dual two-form on $\mathbb{R}^4$ with values in $\mathfrak{g}$. This matches the term on the left-hand side.

A natural parameter of the construction is the target manifold M together with the G-action. Gauge theories, which can be obtained for different choices of M, are considered in some details below.

The simplest example is $M = \mathbb{H}$. For $G = S^1$, which acts by the multiplication on the right, one recovers the classical Seiberg–Witten equations. The details are left to the reader.

4.1. *Vafa–Witten equations*

Consider the flat hyper-Kähler manifold $M = \mathfrak{g} \otimes \mathbb{H}$ as the target manifold. Let G act on M by the "quaternization" of the adjoint action. The corresponding moment map is given by

$$\begin{aligned}\mu(\xi) &= ([\xi_2, \xi_3] + [\xi_0, \xi_1]) \otimes i + ([\xi_3, \xi_1] + [\xi_0, \xi_2]) \otimes j \\ &\quad + ([\xi_1, \xi_2] + [\xi_0, \xi_3]) \otimes k \\ &= \sigma(\operatorname{Im} \xi) + [\operatorname{Re} \xi, \operatorname{Im} \xi],\end{aligned}$$

where $\xi = \xi_0 + (\xi_1 i + \xi_2 j + \xi_3 k) = \operatorname{Re} \xi + \operatorname{Im} \xi$.

Furthermore, a map $u\colon \mathbb{R}^4 \to \mathfrak{g} \otimes \mathbb{H}$ can be identified with a pair $(c, b) \in \Omega^0(\mathbb{R}^4; \mathfrak{g}) \oplus \Omega^2_+(\mathbb{R}^4; \mathfrak{g})$, where c is the real part of u and b is obtained from the imaginary part of u via the identification $\operatorname{Im} \mathbb{H} \cong \Lambda^2_+(\mathbb{R}^4)^*$. The corresponding Dirac operator is $\mathcal{D}_a(c, b) = d_a c + \delta_a^+ b$. Hence, in the case $M = \mathfrak{g} \otimes \mathbb{H}$, the generalized Seiberg–Witten equations are

$$\begin{aligned} & d_a c + \delta_a^+ b = 0, \\ & F_a^+ - \sigma(b) + [b, c] = 0, \end{aligned} \qquad (a, b, c) \in \Omega^1(\mathbb{R}^4; \mathfrak{g}) \times \Omega^1(\mathbb{R}^4; \mathfrak{g}) \times \Omega^0(\mathbb{R}^4; \mathfrak{g}).$$

These equations first appeared in [37] and are known as the Vafa–Witten equations. Notice that the Vafa–Witten equations make sense on any (oriented Riemannian) four-manifold.

4.2. *Anti-self-duality equations on $G^{\mathbb{C}}$-bundles*

The previous example admits a different interpretation. Namely, the target manifold is chosen again to be $M = \mathfrak{g} \otimes \mathbb{H}$ but this time a map $u\colon \mathbb{R}^4 \to \mathfrak{g} \otimes \mathbb{H}$ is identified with some $b \in \Omega^1(\mathbb{R}^4; \mathfrak{g})$ according to the rule

$$u = \xi_0 + \xi_1 i + \xi_2 j + \xi_3 k \equiv \sum_{p=0}^{3} \xi_p dx_p = b.$$

The group G acts on M in the same manner as in the previous example and therefore the moment map is given by the same expression. However, this is also interpreted in a different way. Namely, for $b \in \Omega^1(\mathbb{R}^4; \mathfrak{g})$ a straightforward computation shows that the self-dual two-form $\mu \circ b$ is in fact $\frac{1}{2}[b \wedge b]^+$, where the symbol $[\cdot \wedge \cdot]$ stays for a combination of the wedge product and the Lie-brackets. The Dirac operator acting on one-forms was already mentioned above and equals $\delta_a + d_a^+$. Hence, in this case the generalized Seiberg–Witten equations take the following form:

$$\delta_a b = 0, \tag{13}$$

$$d_a^+ b = 0, \tag{14}$$

$$F_a^+ - \frac{1}{2}[b \wedge b]^+ = 0. \tag{15}$$

To see the geometric meaning of Eqs. (13)–(15), consider the one-form $A = a + ib \in \Omega^1(\mathbb{R}^4; \mathfrak{g}^{\mathbb{C}})$. Interpreting A as a connection on a trivial $G^{\mathbb{C}}$-bundle, one obtains

$$F_A^+ = \left(F_a + id_a b - \frac{1}{2}[b \wedge b] \right)^+ = F_a^+ - \frac{1}{2}[b \wedge b]^+ + id_a^+ b.$$

Hence, Eqs. (14) and (15) mean that $A = a + ib$ is anti-self-dual. Notice that these equations are invariant with respect to the *complex* gauge group $Map(\mathbb{R}^4; G^{\mathbb{C}})$.

It remains to clarify the meaning of (13). Notice that for a compact manifold X the space $\Omega^1(X; \mathfrak{g}^{\mathbb{C}})$ has a natural Kähler metric, which is preserved by the action of the *real* gauge group $Map(X; G)$. Then $(a, b) \mapsto \delta_a b$ is the moment map of this action. The same conclusion holds for $X = \mathbb{R}^4$ provided $\Omega^1(X; \mathfrak{g}^{\mathbb{C}})$ is replaced by a suitable Sobolev space. Thus, solutions of (13)–(15) are those anti-self-dual connections, which are in the zero level set of the moment map of the real gauge group. In other words, (13) is a "stability condition".

For the sake of brevity solutions of (13)–(15) are called (stable) complex anti-self-dual connections.

The moduli space of complex asd connections has some interesting properties, which are best seen from a more general perspective. For this reason it is convenient to deviate from the convention to work exclusively with $\mathbb{R}^4$ as the base four-manifold. Thus, let X be a closed Riemannian oriented four-manifold. Choose a principal

G-bundle $P \to X$ and denote $ad\,P = P \times_G \mathfrak{g}$. Equations (13)–(15) for a general four-manifold can be written in exactly the same form with the understanding that a stays for a connection on P and b is a one-form on X with values in $ad\,P$. The geometric meaning also remains valid in this case if $A = a + ib$ is interpreted as a connection on the corresponding principal $G^{\mathbb{C}}$-bundle $\mathcal{P} = P \times_G G^{\mathbb{C}}$.

Remark 4.1. Strictly speaking, on general four-manifold the complex anti-self-duality equations do not quite fit into the concept of [30]. The reason is, roughly speaking, that $Sp_-(1)$ must act nontrivially on the target manifold.

Let $\mathcal{M}_{casd}(\mathcal{P})$ denote the moduli space of complex asd connections. Clearly the moduli space of real asd connections $\mathcal{M}_{asd}(P)$ is contained in $\mathcal{M}_{casd}(\mathcal{P})$. By looking at the deformation complex of (13)–(15)

$$\begin{aligned} 0 &\to \Omega^0(ad\,P) \to \Omega^1(ad\,P) \oplus \Omega^1(ad\,P) \\ &\to \Omega^0(ad\,P) \oplus \Omega^2_+(ad\,P) \oplus \Omega^2_+(ad\,P) \to 0 \end{aligned}$$

it is easy to see that the expected dimension of $\mathcal{M}_{casd}(\mathcal{P})$ is twice the expected dimension of $\mathcal{M}_{asd}(P)$.

Theorem 4.2. *Let X be a closed oriented Riemannian four-manifold. Assume that both $\mathcal{M}_{asd}(P)$ and $\mathcal{M}_{casd}(\mathcal{P})$ are manifolds of expected dimensions. Then the following holds:*

(i) *$\mathcal{M}_{casd}(\mathcal{P})$ is Kähler;*
(ii) *$\mathcal{M}_{asd}(P)$ is a Lagrangian submanifold of $\mathcal{M}_{casd}(\mathcal{P})$;*
(iii) *If X is Kähler, then $\mathcal{M}_{casd}(\mathcal{P})$ is hyper-Kähler and $\mathcal{M}_{asd}(P)$ is a complex Lagrangian submanifold.*

The rest of this subsection is devoted to the sketch of the proof of Theorem 4.2. The configuration space $\mathcal{A}(P) \times \Omega^1(ad\,P) \cong T^*\mathcal{A}(P)$

is a flat infinite dimensional Kähler manifold.[e] Indeed, the Kähler structure is given explicitly by

$$\begin{aligned} I_1(\mathrm{v}, \mathrm{w}) &= (-\mathrm{w}, \mathrm{v}), \\ \omega_1((\mathrm{v}_1, \mathrm{w}_1), (\mathrm{v}_2, \mathrm{w}_2)) &= -\langle \mathrm{v}_2, \mathrm{w}_1 \rangle + \langle \mathrm{v}_1, \mathrm{w}_2 \rangle, \end{aligned} \tag{16}$$

where $\mathrm{v}, \mathrm{w} \in V = \Omega^1(ad\,P)$. As already mentioned above, the moment map of the real gauge group $\mathcal{G}(P)$ is given by $(a, b) \mapsto \delta_a b$. Moreover, $\mathcal{A}_{asd}(\mathcal{P}) = \{A \in \mathcal{A}(\mathcal{P}) \mid F_A^+ = 0\}$ is a complex subvariety of the configuration space. Then $\mathcal{M}_{casd}(\mathcal{P})$ is the Kähler reduction of $\mathcal{A}_{asd}(\mathcal{P})$ with respect to the action of the real gauge group, hence a Kähler manifold.

To see (ii), observe that the antisymplectic involution $(a, b) \mapsto (a, -b)$ on the configuration space induces an antisymplectic involution τ on $\mathcal{M}_{casd}(\mathcal{P})$. The fixed point set of τ is $\mathcal{M}_{asd}(P)$, whose dimension equals $\frac{1}{2} \dim \mathcal{M}_{casd}(\mathcal{P})$. Hence, $\mathcal{M}_{asd}(P)$ is a Lagrangian submanifold of $\mathcal{M}_{casd}(\mathcal{P})$.

It remains to show (iii). Recall that for a Kähler surface X there is the decomposition $\Omega^2_+(X; \mathbb{R}) \cong \Omega^0(X) \cdot \omega_X \oplus \Omega^{0,2}(X)$. Denote by $\Lambda\colon \Omega^2(X) \to \Omega^0(X)$ the adjoint operator of $L\colon \Omega^0(X) \to \Omega^2(X)$, $\alpha \mapsto \alpha\omega_X$. Then (13)–(15) can be written in the form

$$\delta_a b = 0, \quad \Lambda d_a^+ b = 0, \quad \Lambda\left(F_a^+ - \frac{1}{2}[b \wedge b]^+\right) = 0, \tag{17}$$

$$F_{a+bi}^{2,0} = 0, \quad F_{a+bi}^{0,2} = 0. \tag{18}$$

Furthermore, for a Hermitian vector space $(V, \langle\cdot,\cdot\rangle + i\omega(\cdot,\cdot))$ its complexification $V_{\mathbb{C}} \cong V \oplus V$ is a quaternion-Hermitian vector space. Explicitly, the quaternion-Hermitian structure is given by (16) together with

$$\begin{aligned} I_2(\mathrm{v}, \mathrm{w}) &= (I\mathrm{v}, -I\mathrm{w}), & \omega_2((\mathrm{v}_1, \mathrm{w}_1), (\mathrm{v}_2, \mathrm{w}_2)) &= \omega(\mathrm{v}_1, \mathrm{v}_2) - \omega(\mathrm{w}_1, \mathrm{w}_2), \\ I_3(\mathrm{v}, \mathrm{w}) &= (I\mathrm{w}, I\mathrm{v}), & \omega_3((\mathrm{v}_1, \mathrm{w}_1), (\mathrm{v}_2, \mathrm{w}_2)) &= \omega(\mathrm{w}_1, \mathrm{v}_2) + \omega(\mathrm{v}_1, \mathrm{w}_2). \end{aligned}$$

[e]Strictly speaking, one should pass to a suitable Sobolev space to get a Banach manifold structure; here and in the sequel, we work in a smooth category for the sake of simplicity of exposition.

This implies that $\mathcal{A}(\mathcal{P}) \cong \mathcal{A}(P) \times \Omega^1(ad\, P)$ is a flat hyper-Kähler manifold. A straightforward but somewhat lengthy computation shows that the action of the real gauge group $\mathcal{G}(P)$ preserves this hyper-Kähler structure and the zero level set of the corresponding hyper-Kähler moment map is given by solutions of (17). Furthermore, denote by $\mathcal{A}^{1,1}(\mathcal{P})$ the space of solutions of (18). Clearly, the tangent bundle of $\mathcal{A}^{1,1}(\mathcal{P})$ is preserved by I_1 and I_2. Therefore, $\mathcal{A}^{1,1}(\mathcal{P})$ is a hyper-Kähler submanifold of $\mathcal{A}(\mathcal{P})$. Thus, for a Kähler surface X the moduli space $\mathcal{M}_{casd}(\mathcal{P})$ is the hyper-Kähler reduction of $\mathcal{A}^{1,1}(\mathcal{P})$, hence a hyper-Kähler manifold. The remaining part of (iii) is shown in a similar manner to (ii).

4.3. *Spin*(7)*-instantons*

At first, it is convenient to recall the notion of a *Spin*(7)-instanton, which appeared in the mathematical literature in [7] for the first time. To do this, fix a splitting $\mathbb{R}^8 = U \oplus V$, where $U \cong \mathbb{H} \cong V$. Let θ (respectively η) denote the projection onto the first (respectively second) subspace. Think of θ and η as $\mathbb{H}$-valued one-forms on $\mathbb{R}^8$. The stabilizer of the Cayley form

$$\Omega = -\frac{1}{24}\mathrm{Re}(\theta \wedge \bar\theta \wedge \theta \wedge \bar\theta - 6\theta \wedge \bar\theta \wedge \eta \wedge \bar\eta + \eta \wedge \bar\eta \wedge \eta \wedge \bar\eta),$$

is [4] the subgroup $Spin(7) \subset \mathrm{SO}(8)$. The Cayley form gives rise to the linear map

$$\Lambda^2(\mathbb{R}^8)^* \to \Lambda^2(\mathbb{R}^8)^*, \quad \omega \mapsto -*(\omega \wedge \Omega),$$

which has two eigenvalues 3 and -1. The corresponding eigenspaces $\Lambda^2_+(\mathbb{R}^8)^*$ and $\Lambda^2_-(\mathbb{R}^8)^*$ are of dimensions 7 and 21 respectively. Hence, just like in four dimensions, there is the decomposition of the space of two-forms:

$$\Omega^2(\mathbb{R}^8) = \Omega^2_+(\mathbb{R}^8) \oplus \Omega^2_-(\mathbb{R}^8). \tag{19}$$

Then a connection A on a G-bundle over $\mathbb{R}^8$ is called a *Spin*(7)-instanton, if $F_A^+ = 0$.

Remark 4.3. In general, the base manifold $\mathbb{R}^8$ can be replaced by a Riemannian eight-manifold with holonomy *Spin*(7).

Notice also that the splitting $\mathbb{R}^8 = U \oplus V$ leads to the decomposition

$$\Lambda^k(\mathbb{R}^8)^* = \bigoplus_{p+q=k} \Lambda^{p,q}(\mathbb{R}^8)^*, \quad \text{where } \Lambda^{p,q}(\mathbb{R}^8)^* \cong \Lambda^p U^* \otimes \Lambda^q V^*.$$

Hence, there is a similar decomposition of differential forms on $\mathbb{R}^8$:

$$\Omega^k(\mathbb{R}^8) = \bigoplus_{p+q=k} \Omega^{p,q}(\mathbb{R}^8).$$

In particular, any connection on the trivial bundle $\underline{G} \to \mathbb{R}^8$ can be uniquely written as $A = a + b$, where $a \in \Omega^{1,0}(\mathbb{R}^8;\mathfrak{g})$ and $b \in \Omega^{0,1}(\mathbb{R}^8;\mathfrak{g})$. Think of b as a family of connections on $\mathbb{R}^4 \cong V \subset \mathbb{R}^8$ parametrized by U. Let F_b denote the corresponding family of curvatures. Then a computation [14] shows that A is a $Spin(7)$-instanton if and only if

$$(F_A^{1,1})^+ = 0 \quad \text{and} \quad F_a^+ = F_b^+.$$

Notice that the superscript "+" in the first equation is used in the sense of decomposition (19), while in the second one in the sense of the four-dimensional analogue of (19).

Putting aside $Spin(7)$-instantons for a while, consider the generalized Seiberg–Witten equations for the target hyper-Kähler manifold $\Omega^1(\mathbb{R}^4;\mathfrak{g})$, which is interpreted as the space of connections on the trivial bundle $\underline{G} \to \mathbb{R}^4$. The gauge group $Map(\mathbb{R}^4;G)$ acts on $\Omega^1(\mathbb{R}^4;\mathfrak{g})$ preserving its hyper-Kähler structure. The corresponding moment map is well-known to be $\mu(b) = F_b^+$. Hence, in this case the generalized Seiberg–Witten equations can be written as

$$\mathcal{D}_a b = 0, \quad F_a^+ = F_b^+.$$

A somewhat lengthy computation, which can be found in details in [14], shows that $\mathcal{D}_a b = (F_A^{1,1})^+$, where $A = a + b$. Thus, for $M = \Omega^1(\mathbb{R}^4;\mathfrak{g})$ the generalized Seiberg–Witten equations yield $Spin(7)$-instantons.

Remark 4.4. If $\mathbb{R}^4$ is replaced by a spin four-manifold X, the generalized Seiberg–Witten equations with the target manifold $M = \Omega^1(\mathbb{R}^4;\mathfrak{g})$ yield up to a zero-order term $Spin(7)$-instantons on the total space of the spinor bundle of X [14].

Remark 4.5. By using similar arguments, one can also show that solutions of the generalized Seiberg–Witten equations with the target manifold $M = \Omega^1(\mathbb{R}^3; \mathfrak{g}) \times \Omega^0(\mathbb{R}^3; \mathfrak{g})$ can be interpreted as G_2-monopoles on $\mathbb{R}^7$.

4.4. *Five-dimensional instantons*

Consider $M = \Omega^0(\mathbb{R}; \mathfrak{g} \otimes \mathbb{H})$ as the target manifold equipped with its flat hyper-Kähler structure, where $\mathfrak{g}$ is the Lie algebra of a Lie group G. The gauge group $\mathcal{G} = Map(\mathbb{R}; G)$ acts on $\Omega^0(\mathbb{R}; \mathfrak{g} \otimes \mathbb{H})$, namely $g \cdot T = gTg^{-1} - \dot{g}g^{-1}$, where $\dot{g}$ is the derivative of g with respect to the variable $t \in \mathbb{R}$. Then the hyper-Kähler moment map $\mu = \mu_1 i + \mu_2 j + \mu_3 k$ of this action is given by

$$\begin{aligned}\mu_1(T) &= \dot{T}_1 + [T_0, T_1] - [T_2, T_3],\\ \mu_2(T) &= \dot{T}_2 + [T_0, T_2] - [T_3, T_1],\\ \mu_3(T) &= \dot{T}_3 + [T_0, T_3] - [T_1, T_2],\end{aligned}$$

where $T = T_0 + T_1 i + T_2 j + T_3 k$. Recalling the identification $\mathbb{H} = \mathbb{R} \oplus \Lambda^2_+(\mathbb{R}^4)^*$, a map $u\colon \mathbb{R}^4 \to \Omega^0(\mathbb{R}; \mathfrak{g} \otimes \mathbb{H})$ can be interpreted as a map $(c, b)\colon \mathbb{R} \to \Omega^0(\mathbb{R}^4; \mathfrak{g}) \times \Omega^2_+(\mathbb{R}^4; \mathfrak{g})$. Similarly, a connection on $\underline{\mathcal{G}} \to \mathbb{R}^4$ can be interpreted as a map $a\colon \mathbb{R} \to \Omega^1(\mathbb{R}^4; \mathfrak{g})$. A little thought shows that the corresponding generalized Seiberg–Witten equations can be written in the form

$$\begin{aligned}\dot{a} &= \delta_a b + d_a c,\\ \dot{b} &= F_a^+ - \sigma(b) - [c, b].\end{aligned} \tag{20}$$

These are the five-dimensional instantons on $W^5 = \mathbb{R}^4 \times \mathbb{R}$ [13, Eq. (40)]. Equations (20) were discovered independently by Witten [40]. We actually obtained five-dimensional instantons for the first time along the lines outlined above. The reader can find some applications of five-dimensional instantons in the papers mentioned above.

Remark 4.6. One can replace $\Omega^0(\mathbb{R}; \mathfrak{g} \otimes \mathbb{H})$ in the above construction by the space $\tilde{\Omega}^0(I; \mathfrak{g} \otimes \mathbb{H})$ consisting of all maps $T\colon I = (-1, 1) \to \mathfrak{g} \otimes \mathbb{H}$ satisfying certain asymptotic conditions as $t \to \pm 1$. Then the hyper-Kähler reduction of $\tilde{\Omega}^0(I; \mathfrak{g} \otimes \mathbb{H})$ is the moduli space of magnetic monopoles on $\mathbb{R}^3$. This is also the hyper-Kähler reduction of the target manifold from Remark 4.5. Hence, it is naturally to expect that there is some relation between five-dimensional instantons and G_2-monopoles.

5. Remarks on Three-Manifolds

As already mentioned above, there is a generalization of the Dirac operator for three-manifolds due to Taubes [32]. Hence, one can also consider the generalized Seiberg–Witten equations in dimension three with the target manifolds as in the preceding section. It turns out that the analogues of the Vafa–Witten equations and the complex anti-self-duality equations coincide and yield stable flat $G^{\mathbb{C}}$-connections. The compactness property of the moduli space of flat $PSL(2;\mathbb{C})$-connections was recently studied by Taubes [35]. Finally, the construction of Sec. 4.3 leads to G_2-instantons, while that of Sec. 4.4 leads to the Kapustin–Witten equations [19].

Acknowledgment

I am thankful to an anonymous referee for helpful comments.

References

[1] D. Anselmi and P. Fré, Gauged hyperinstantons and monopole equations, *Phys. Lett. B* **347** (1995) 247–254.

[2] M. F. Atiyah, V. G. Drinfel′d, N. J. Hitchin and Yu. I. Manin, Construction of instantons, *Phys. Lett. A* **65** (1978) 185–187.

[3] S. Bauer and M. Furuta, A stable cohomotopy refinement of Seiberg–Witten invariants. I, *Invent. Math.* **155** (2004) 1–19.

[4] R. Bryant and S. Salamon, On the construction of some complete metrics with exceptional holonomy, *Duke Math. J.* **58** (1989) 829–850.

[5] S. K. Donaldson and P. B. Kronheimer, *The Geometry of Four-Manifolds*, Oxford Mathematical Monographs, The Clarendon Press, Oxford University Press, New York, 1990. Oxford Science Publications.

[6] S. Donaldson and E. Segal, Gauge theory in higher dimensions, II, in *Surveys in Differential Geometry, Vol. XVI. Geometry of Special Holonomy and*

Related Topics, volume 16 of Surv. Differ. Geom., Vol. 16, pp. 1–41. Int. Press, Somerville, MA, 2011.

[7] S. K. Donaldson and R. P. Thomas, Gauge theory in higher dimensions, in *The Geometric Universe* (Oxford, 1996), pp. 31–47. Oxford Univ. Press, Oxford, 1998.

[8] B. Feix, Hyper-Kähler metrics on cotangent bundles, *J. Reine Angew. Math.* **532** (2001) 33–46.

[9] R. Fueter, Die Funktionentheorie der Differentialgleichungen $\Delta u = 0$ und $\Delta\Delta u = 0$ mit vier reellen Variablen, *Comment. Math. Helv.* **7** (1934) 307–330.

[10] M. Furuta, Monopole equation and the $\frac{11}{8}$-conjecture, *Math. Res. Lett.* **8** (2001) 279–291.

[11] A. Haydys, Hyper-Kähler and quaternionic Kähler manifolds with S^1-symmetries, *J. Geom. Phys.* **58** (2008) 293–306.

[12] A. Haydys, Nonlinear Dirac operator and quaternionic analysis, *Comm. Math. Phys.* **281** (2008) 251–261.

[13] A. Haydys, Fukaya–Seidel category and gauge theory, *J. Symplectic Geom.*, to appear; arXiv:1010.2353.

[14] A. Haydys, Gauge theory, calibrated geometry and harmonic spinors, *J. Lond. Math. Soc.* **86**(2) (2012) 482–498.

[15] R. Harvey and B. Lawson, Calibrated geometries, *Acta Math.* **148** (1982) 47–157.

[16] N. J. Hitchin, A. Karlhede, U. Lindström and M. Roček, Hyper-Kähler metrics and supersymmetry, *Comm. Math. Phys.* **108** (1987) 535–589.

[17] S. Hohloch, G. Noetzel and D. A. Salamon, Hypercontact structures and Floer homology, *Geom. Topol.* **13** (2009) 2543–2617.

[18] D. Kaledin, A canonical hyper-Kähler metric on the total space of a cotangent bundle, in *Quaternionic Structures in Mathematics and Physics* (Rome, 1999), pp. 195–230. Univ. Studi Roma "La Sapienza", Rome, 1999.

[19] A. Kapustin and E. Witten, Electric-magnetic duality and the geometric Langlands program, *Commun. Number Theory Phys.* **1** (2007) 1–236.

[20] C. LeBrun, Four-manifolds without Einstein metrics, *Math. Res. Lett.* **3** (1996) 133–147.

[21] B. Lawson and M.-L. Michelsohn, *Spin Geometry*, Princeton Mathematical Series, Vol. 38, Princeton University Press, Princeton, NJ, 1989.

[22] M. Marcolli, *Seiberg-Witten Gauge Theory*, Texts and Readings in Mathematics, Hindustan Book Agency, New Delhi, 1999. With an appendix by the author and Erion J. Clark.

[23] R. McLean, Deformations of calibrated submanifolds, *Comm. Anal. Geom.* **6** (1998) 705–747.

[24] J. Moore, *Lectures on Seiberg–Witten Invariants*, of Lecture Notes in Mathematics, Vol. 1629, Springer-Verlag, Berlin, 1996.

[25] J. Morgan, *The Seiberg–Witten Equations and Applications to the Topology of Smooth Four-Manifolds*, Mathematical Notes, Vol. 44, Princeton University Press, Princeton, NJ, 1996.

[26] H. Nakajima, Instantons on ALE spaces, quiver varieties, and Kac–Moody algebras, *Duke Math. J.* **76** (1994) 365–416.

[27] H. Nakajima, Resolutions of moduli spaces of ideal instantons on $\mathbf{R}^4$, in *Topology, Geometry and Field Theory*, pp. 129–136. World Scientific, River Edge, NJ, 1994.

[28] H. Nakajima, *Lectures on Hilbert Schemes of Points on Surfaces*, of *University Lecture Series*, Vol. 18, American Mathematical Society, Providence, RI, 1999.

[29] L. Nicolaescu, *Notes on Seiberg–Witten Theory*, Graduate Studies in Mathematics, Vol. 28, American Mathematical Society, Providence, RI, 2000.

[30] V. Ya. Pidstrygach, Hyper-Kähler manifolds and the Seiberg–Witten equations, *Tr. Mat. Inst. Steklova* **246** (Algebr. Geom. Metody, Svyazi i Prilozh.) (2004) 263–276.

[31] N. Seiberg and E. Witten, Electric-magnetic duality, monopole condensation, and confinement in $N = 2$ supersymmetric Yang–Mills theory, *Nucl. Phys. B* **426** (1994) 19–52.

[32] C. Taubes, Nonlinear generalizations of a 3-manifold's Dirac operator, in *Trends in Mathematical Physics* (Knoxville, TN, 1998), AMS/IP Stud. Adv. Math. Vol. 13, pp. 475–486. *Amer. Math. Soc.*, Providence, RI, 1999.

[33] C. Taubes, *Seiberg–Witten and Gromov Invariants for Symplectic 4-Manifolds*, First International Press Lecture Series, Vol. 2, Int. Press, Somerville, MA, 2000. Edited by Richard Wentworth.

[34] C. Taubes, The Seiberg–Witten equations and the Weinstein conjecture, *Geom. Topol.* **11** (2007) 2117–2202.

[35] C. H. Taubes, PSL(2;C) connections on 3-manifolds with L2 bounds on curvature, arXiv:1205.0514, 2012.

[36] G. Tian, Gauge theory and calibrated geometry. I. *Ann. of Math. (2)* **151** (2000) 193–268.

[37] C. Vafa and E. Witten, A strong coupling test of S-duality, *Nucl. Phys. B* **431** (1994) 3–77.

[38] M. Wang, Parallel spinors and parallel forms, *Ann. Global Anal. Geom.* **7** (1989) 59–68.

[39] E. Witten, Monopoles and four-manifolds, *Math. Res. Lett.* **1** (1994) 769–796.

[40] E. Witten, Fivebranes and knots, *Quantum Topol.* **3** (2012) 1–137.

Graph-Links: The State of the Art

D. P. Ilyutko[*,†,¶], V. O. Manturov[†,‡] and I. M. Nikonov[*,†,§]

[*]*Department of Mechanics and Mathematics, Moscow State University, Moscow, Russia*

[†]*Delone Laboratory of Discrete and Computational Geometry, Yaroslavl State University, Yaroslavl, Russia*

[‡]*Department of Fundamental Sciences, Bauman Moscow State Technical University Moscow, Russia*

[‡]*Laboratory of Quantum Topology, Chelyabinsk State University, Chelyabinsk, Russia*

[§]*Faculty of Management, National Research University Higher School of Economics, Moscow, Russia*

[¶]*ilyutko@yandex.ru*

[‡]*vomanturov@yandex.ru*

[§]*nikonov@mech.math.msu.su*

In this paper we discuss the latest achievements in graph-link theory created by the first two authors in 2008–2009. This theory is a far-reaching generalization of classical and virtual knot theory and originates from the notion of the intersection graph (of a chord diagram). For a given graph a chord diagram having this graph as its intersection graph may not exist, however many invariants of classical and virtual knots can be extended to the theory of graph-links, in particular, Khovanov homology. Here we construct invariants of graph-links, many of them are based on the notion of parity due to the second author. We derive new properties of graph-links, which do not exist in the realizable case, and present non-realizable graph-links (i.e. such graph-links which have no representative realizable by a chord diagram), define an orientation for graph-links and construct the theory of Khovanov homology for graph-links.

1. Introduction

Classical knots can be encoded by four-valent graphs (with some additional structure) and moves on them, but not all four-valent graphs with such structure are planar. This simple paradigm led Louis Kauffman in the late 1990s to the notion of *virtual knot* and *virtual link* [15]. If a four-valent graph with such a structure (see below) cannot be embedded in $\mathbb{R}^2$, one can *immerse* it (new intersections and self-intersections are marked as virtual crossings). Then we define *virtual knots* as equivalence classes of such immersions by *generalized Reidemeister moves.* The latter consist of the classical Reidemeister moves Ω_1, Ω_2, Ω_3 and the *detour move* representing the following: A branch of the diagram containing several consecutive virtual crossings but not containing classical crossings can be transformed into any other branch with the same endpoints, see Fig. 1.

Another way of thinking of virtual knots (not links) is as follows. Take a (classical) knot diagram and consider it as an immersion of the standard circle S^1 in $\mathbb{R}^2$. On this circle, connect by arrows those points which have the same image (arrows directed from the preimage of the undercrossing to the preimage of the overcrossing). Moreover, let us endow each arrow with a sign coinciding with the sign of the crossing (the *writhe number* of a crossing), i.e. it equals +1 if we have

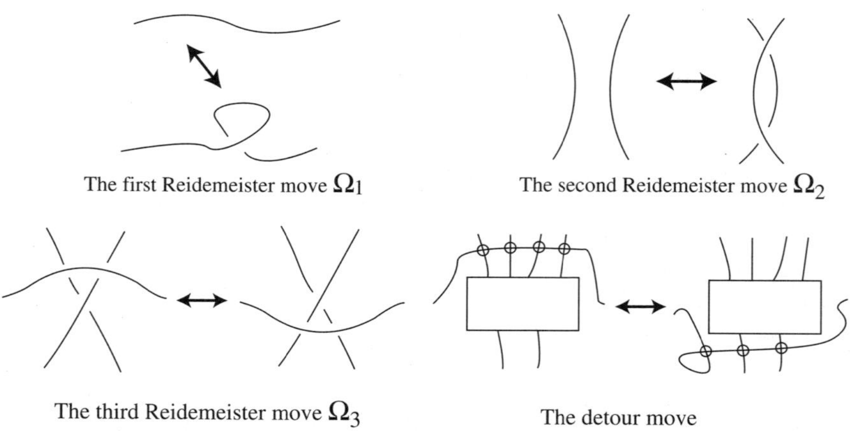

Fig. 1. The generalized Reidemeister moves.

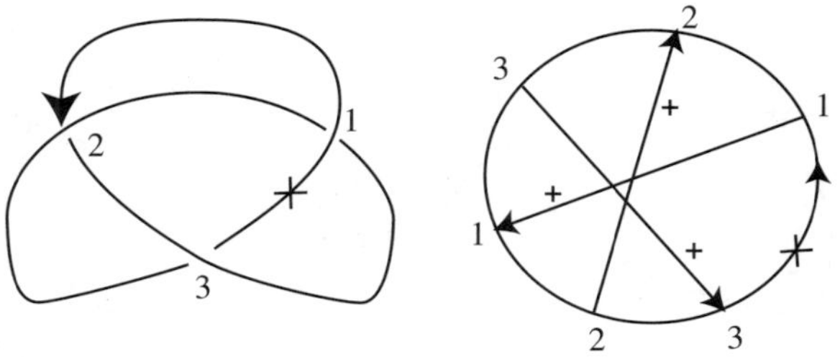

Fig. 2. The Gauss diagram of the right-handed trefoil.

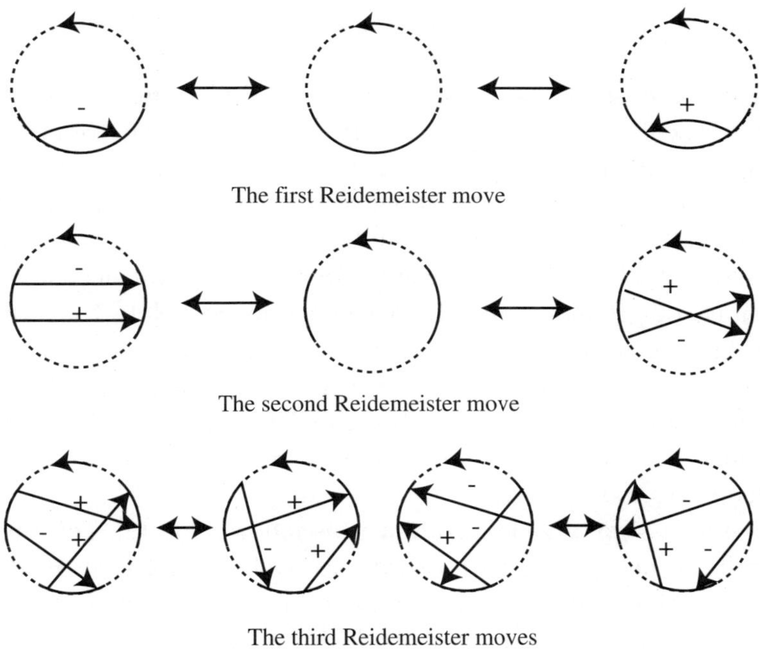

Fig. 3. Moves on Gauss diagrams.

⨯ and -1 for ⨯. Thus we obtain the *Gauss diagram* of the knot, see Fig. 2. For Gauss diagrams, one can easily rewrite Reidemeister moves, see, e.g. [7, 31] and Fig. 3 (the other versions of the third Reidemeister move are combinations of the moves depicted in the figure, not only third ones).

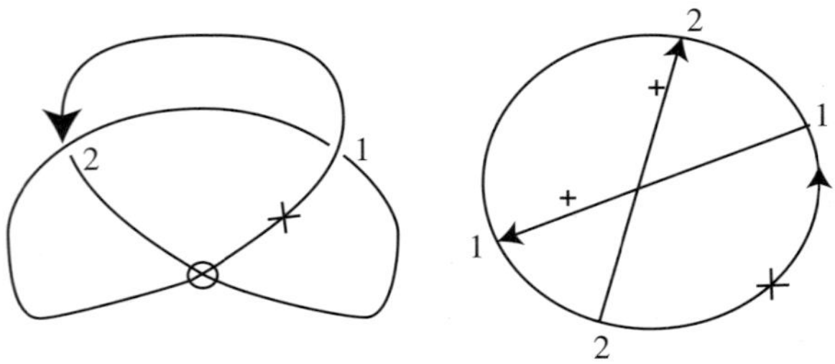

Fig. 4. The Gauss diagram of a virtual knot.

However, if one tries to define classical knots by Gauss diagrams and Reidemeister moves on them, one faces the problem of non-realizability of some Gauss diagrams by knot diagrams. Indeed, if we try to draw a curve on the plane with prescribed order of crossings and ways of connecting them, in some cases we fail, see Fig. 4.

From this point of view, classical knots are the equivalence classes of realizable Gauss diagrams modulo Reidemeister moves which do not violate realizability, and *virtual knots* are equivalence classes of all Gauss diagrams modulo Reidemeister moves. Note that virtual crossings are not present in chord diagrams, and a detour move applied to a planar diagram of a virtual knot does not affect the corresponding chord diagram. So, the list of Reidemeister moves for Gauss diagrams consists of moves corresponding to the classical Reidemeister moves only.

The next step of generalization is as follows [11, 12]. For a chord diagram D, i.e. a circle with chords, see Definition 2.1, let us define its intersection graph $G(D)$, as follows. We say that two chords a, b of D are *linked* if two endpoints of a lie in two different components of the complement to the endpoints of b. Otherwise, we say that chords are *unlinked*. Every chord is thought to be unlinked with itself.

Now, the vertices of $G(D)$ correspond to the chords of D, and two vertices are connected by an edge whenever the two corresponding chords are linked, see Fig. 5. It turns out that the Reidemeister moves for Gauss diagrams can be rewritten in the language of intersection graphs (with the loss of some information), see [37] and Fig. 6 ("dashed" means that we consider two cases: the graph with the

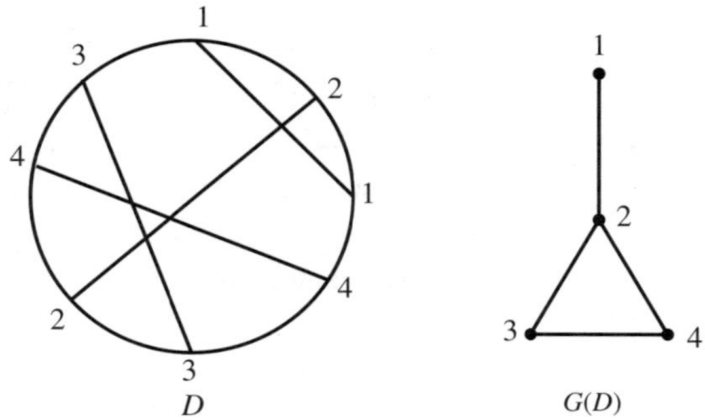

Fig. 5. A chord diagram D and its intersection graph $G(D)$.

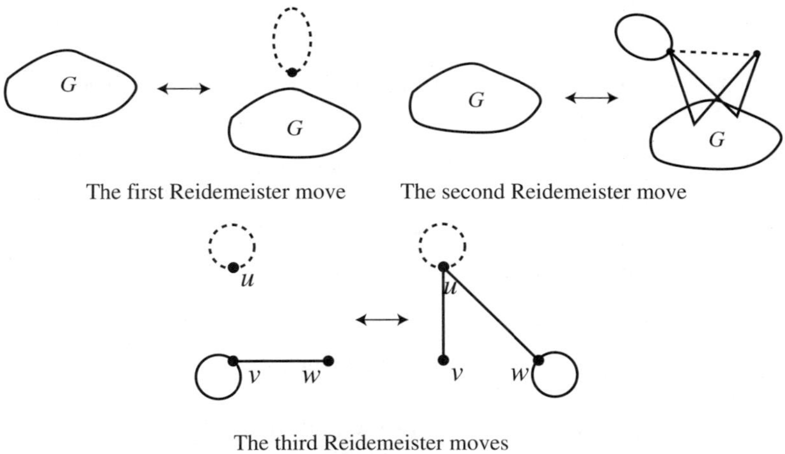

Fig. 6. Moves on intersection graphs for Gauss diagrams.

dashed curve and the graph without one). We add a loop to any vertex of the intersection graph, which has a negative writhe number, and forget the orientation of each arrow (so we lose some information about the knot). For more details and precise description of these moves see ahead.

It is easy to see that not all graphs are realizable as intersection graphs of chord diagrams, see [5] and Fig. 7. Moreover, if a graph is realizable, then its realization might not be unique, see Fig. 8. This non-uniqueness usually corresponds to so-called mutations of virtual

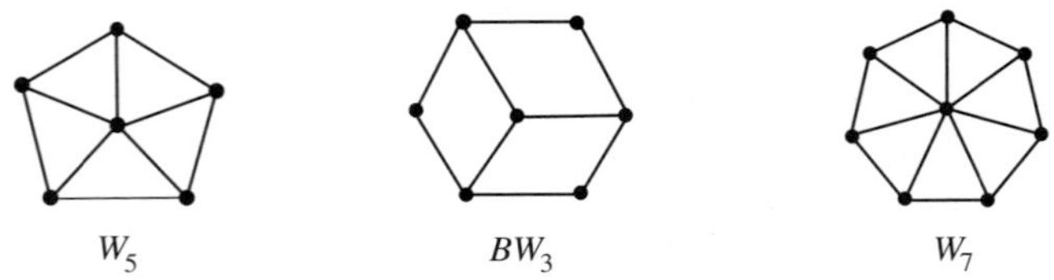

Fig. 7. Bouchet graphs.

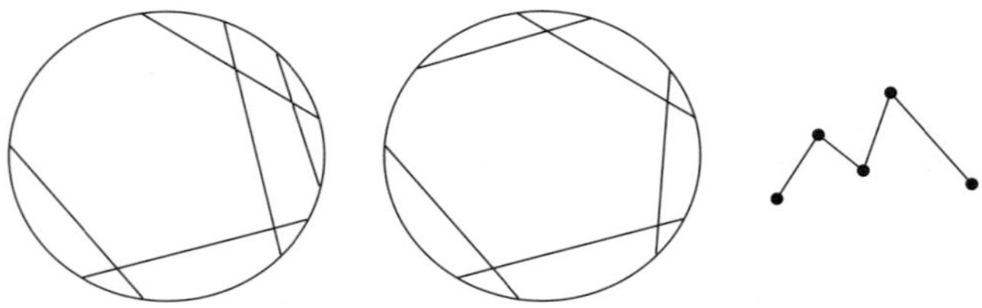

Fig. 8. A graph not uniquely represented by chord diagrams.

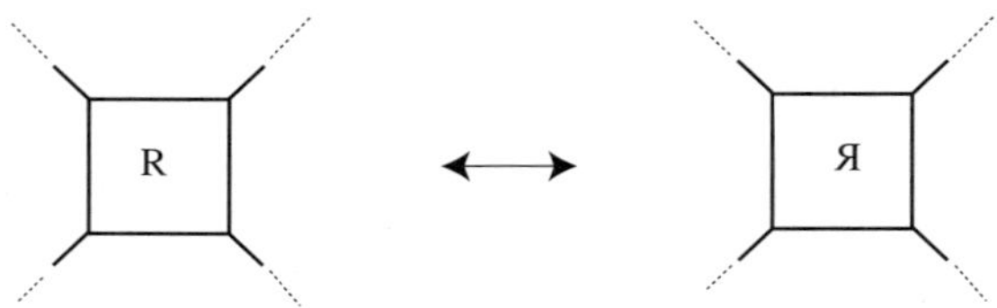

Fig. 9. Mutation.

knots. The *mutation operation* (shown in Fig. 9) cuts a piece of a knot diagram inside a box, turns it by a half-twist and pastes it to the other part. Considering all graphs without multiple edges (loops are allowed) as intersection graphs and moves on them which are obtained from the Reidemeister moves on Gauss diagrams, we get a new theory: the theory of homotopy classes of looped interlacement graphs. The graph-link theory is constructed analogously, but for links, we consider chord diagrams in some other sense, not Gauss diagrams.

Graph-link theory can be treated as a theory of links whose diagrams have meaningful crossings but are *not drawable*: we have crossings which correspond to vertices of the intersection graph (or chords of the chord diagram, if such a diagram exists) but if we can say how these crossings are connected to each other, then we have a chord diagram and get a bare virtual knot.

Thus, this theory goes far beyond virtual knot theory. However, if we have two realizable Gauss diagrams connected by a Reidemeister move, then their intersection graphs are connected by the corresponding moves.

It turns out that many classical invariants of virtual and classical knots which do not change under mutation can be read from the corresponding graph-links and these invariants can be generalized for graph-links. The reason comes from the significant power of the combinatorial data of the intersection graph.

Probably, the simplest evidence that one can get some information out of the intersection graph is the number of circles one has in a certain state after a smoothing, see Fig. 10. The simplest example shows that the number of circles obtained after smoothings in two crossings gives three circles if the corresponding chords are unlinked or one circle if they are not. The general theorem (the Circuit-Nullity Formula, see [8, 26, 33, 35]) allows one to count the number of circles in Kauffman's states out of the intersection graph, and this number is equal to the corank (or nullity) of the adjacency matrix of the intersection graph plus one (matrices are considered over $\mathbb{Z}_2$). In particular, this means that graphs not necessarily corresponding to any knot admit a way of generalizing the Kauffman bracket, which coincides with the usual Kauffman bracket when the graph is realizable by a knot.

Thus, having an abstract graph G on n vertices (possibly, unrelated to any concrete chord diagram and having nothing to do

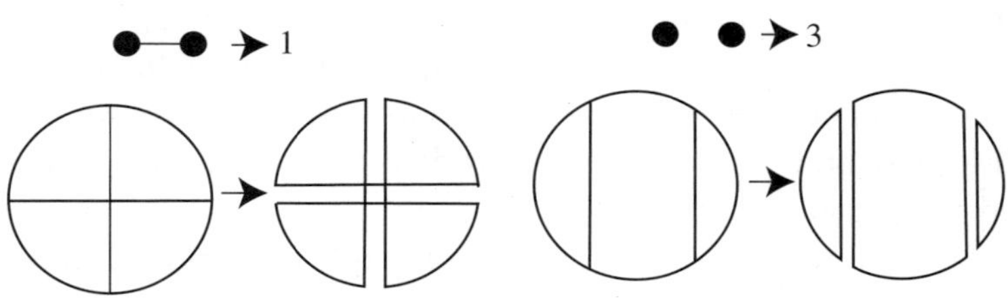

Fig. 10. Surgery along two chords results in one or three circles.

with any classical or virtual knot diagram), we may count the number of non-existing circles in its 2^n *Kauffman states*. Here by a *Kauffman state* we mean any subset of the set of vertices. With a bit more effort, we may count the number of circles in 3^n states, where each crossings is smoothed in either way ( $\to$  or  $\to$ ) or is unchanged ( $\to$ ). It turns out that this data (3^n numbers) is extremely powerful for constructing many invariants.

The paper is organized as follows.

In the next section, we give main definitions and reformulate our setting to be able to work with multi-component *graph-links* rather than just *graph-knots*. This is done by changing the way of coding (classical and virtual) diagrams by chord diagrams so as to adopt it to the case of links with many components. This approach deals with *rotating circuits* where we pass at each classical vertex from an edge to a non-opposite edge. This allows one to encode all four-valent graphs (diagrams of "links"), however, it causes some ambiguity in the choice of a circuit. This ambiguity yields one more move on graphs, see ahead.

In both cases, when we deal with Gauss diagram approach and when we deal with rotating circuit approach, we obtain some "non-realizable" knot theories: the theory of *homotopy classes of looped interlacement graphs* and the graph-link theory. The first approach was originally developed by L. Traldi and L. Zulli [37] and the second one was invented by D. P. Ilyutko and V. O. Manturov [11, 12]. As it was proved by D. P. Ilyutko [10], these two theories lead to the same objects in the case of one-component links.

In Sec. 3, we realize the program of "counting non-existing circles" for the construction of the Kauffman bracket and Jones polynomial for graph-knots. According to Sec. 2, the theory graph-knots and the theory of homotopy classes of looped interlacement graphs are equivalent. Therefore, in some definition and theorems we consider graph-knots, and in other cases we deal with homotopy classes of looped interlacement graphs.

In Sec. 4, we construct a first example of a graph-link with many components which is a non-realizable graph-link (note that the first example of a non-realizable graph-link was constructed by

V. O. Manturov and it was a graph-knot). This example is based on the following contradiction. For a four-component link, there are six types of mixed crossings: each mixed crossing type corresponds to two components it belongs to. However, there is a graph-link whose "component count" gives four, and for every diagram of it, there are seven mixed crossings all of different types. In addition, in this section we introduce the notion of oriented graph-link and construct the Jones polynomial for graph-links with many components.

A breakthrough in virtual knot theory was undertaken by the second-named author who introduced the crucial notion of parity, see [22]. The next section is devoted to the parity theory and its generalizations applied to graph-links (cf. [20–22, 24]).

The simplest Gauss diagram which does not correspond to any classical knot is the diagram consisting of two linked chords. Both these chords are *odd* in the sense that the corresponding vertices of the intersection graph have odd valency. It turns out that the existence of *odd* chords in virtual knot theory leads to various important consequences, one of which can be formulated as follows: *if a diagram K of a knot (link) is odd, i.e. all chords are odd, and irreducible (no decreasing second Reidemeister move can be applied to it), then every diagram of the same link contains K as a pattern.* This allows one to see many nontrivial properties of virtual knots by looking at their diagrams. Later on the above principle was generalized for diagrams with even crossings so that *if a diagram K is complicated enough then it reproduces itself (appears as a subdiagram of every K' equivalent to K).*

We shall formalize this statement in Sec. 5; with some effort, this statement can be easily reformulated in terms of graph-links [13]. In particular, it leads to many graphs (diagrams of graph-links) which are *non-realizable, irreducible and odd*, see, for example, W_5 in Fig. 7. This means that whatever another diagram of the graph-link we take, it will contain the initial diagram as a *persistent pattern.* In particular, this means that if the initial diagram is not realizable (not drawable), then none of the diagrams of the graph-link, it represents, is drawable. This led to the historically first example of a non-realizable graph-link.

Section 6 is devoted to the construction of Khovanov homology theory [1, 16, 17] for graph-links. Being a very powerful invariant (detects the unknot among the classical knots) and defined in a completely combinatorial way, Khovanov homology requires more knowledge of the knot structure than just the Kauffman bracket: besides the state circle count, it also requires the information of how the crossings interfere when passing from a state to a neighboring state. In the setup of graph-links, it is possible to count *how many non-existing circles* one has in each state, however, it is impossible to explain *how these non-existing circles interfere*. This problem was solved by changing a point of view by I. M. Nikonov in the case of $\mathbb{Z}_2$ coefficients [28, 29].

2. Two Approaches to Graph-Link Theory

2.1. *Graph-links and homotopy classes of looped interlacement graphs*

A virtual diagram is called *connected* if the underlying graph (classical crossings play a role of vertices and virtual crossings are just intersection points of edges, see Fig. 11) is connected as a graph. Since any two equivalent (in the class of all virtual diagrams) connected virtual diagrams are equivalent in the class of connected virtual diagrams [11], without loss of generality, all virtual diagrams are assumed to be connected and contain at least one classical crossing.

Note that the underlying graph of a virtual diagram is a four-valent graph with a *cross structure*. The latter means that at each vertex we know what two edges being adjacent to the vertex are

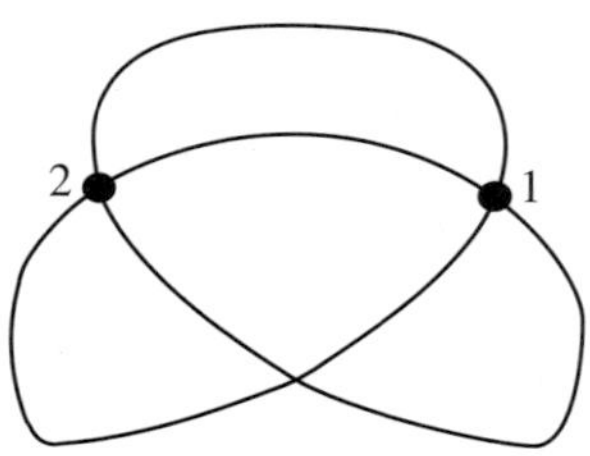

Fig. 11. The underlying graph of the virtual trefoil, see Fig. 4.

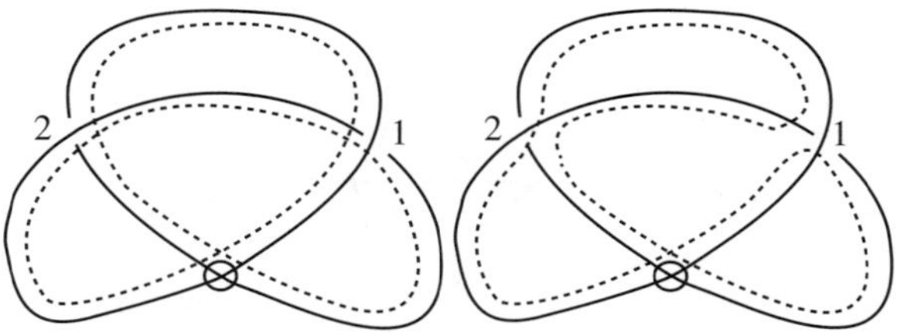

Fig. 12. A Gauss and a rotating circuits on the virtual trefoil, see Fig. 4.

"opposite to each other". A four-valent graph with a cross structure is also called a *framed graph*, see [9–13]. Since the underlying graph can contain loops, we consider each edge consisting of two half-edges constituting this edge. An Euler tour on a four-valent graph with a cross structure is called a *Gauss circuit* if traveling along the tour at each vertex we pass from an edge to the edge being opposite to it, see Fig. 12. If at each vertex we pass from an edge to a non-opposite edge, then the Euler tour is called a *rotating circuit*, see Fig. 12.

Definition 2.1. By a *chord diagram* we mean a cubic graph consisting of one selected non-oriented Hamiltonian cycle[a] (a *core circle* or *circle*) and a set of non-oriented edges (*chords*) connecting points on the cycle.

A chord diagram is *labeled* if every chord is endowed with a label (a, α), where $a \in \{0, 1\}$ is the *framing* of the chord, and $\alpha \in \{\pm\}$ is the *sign* of the chord. We shall also consider chord diagrams whose chords have only one bit of information, a label 0 or 1. We call such diagrams *framed.*

Remark 2.2. As a rule, a chord diagram is depicted on the plane as the standard Euclidean circle with a collection of chords (intersection points of chords which appear as artifacts of drawing chords are not considered as vertices). Chords with framing 0 (respectively, 1) are depicted with thick lines (respectively, dashed lines).

We also deal with *oriented chord diagrams*, i.e. chord diagrams with its core circle to be oriented. In that case we consider chord

[a] A cycle passing through all vertices of the graph exactly once.

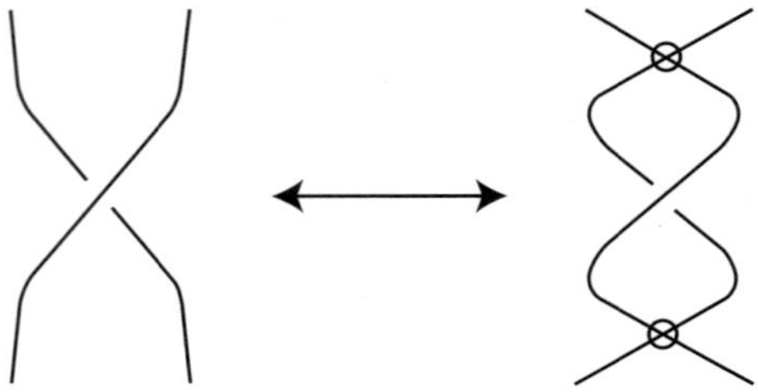

Fig. 13. Virtualization.

diagrams up to isomorphisms of graphs preserving the orientation of the core circle.

Let D be a labeled chord diagram. One can construct a virtual link diagram $K(D)$ (up to *virtualization*, also called Z*-move*, see Fig. 13) in such a way that the chord diagram D coincides with the chord diagram of a rotating circuit on $K(D)$. Let us immerse the diagram D in $\mathbb{R}^2$ by taking an embedding of the core circle and placing some chords inside the circle and the others outside the circle. After that we remove neighborhoods of each of the chord ends and replace the chords with a pair of lines connecting four points on the circle which are obtained after removing neighborhoods. The new chords lead to a classical crossing only (with each other) if the chord is framed by 0, and form classical and virtual crossings if the chord is framed by 1, see Fig. 14 (intersections of chords from different pair form virtual crossings). We also require that the initial piece of the circle corresponds to the A-smoothing  $\to$  if the chord is positive and to the B-smoothing  $\to$ 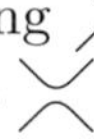 if it is negative.

Conversely, having a connected virtual diagram K, one can get a labeled chord diagram $D_C(K)$. Indeed, one takes a rotating circuit C on K (more precisely, on the underlying graph of K) and constructs the labeled chord diagram, see Fig. 15. The sign of the chord is $+$ (respectively, $-$) if the circuit locally agrees with the A-smoothing (respectively, the B-smoothing), and the framing of a chord is 0 (respectively, 1) if two opposite half-edges have the opposite (respectively, the same) orientations. It can be easily checked that this operation is indeed the inverse to the operation of

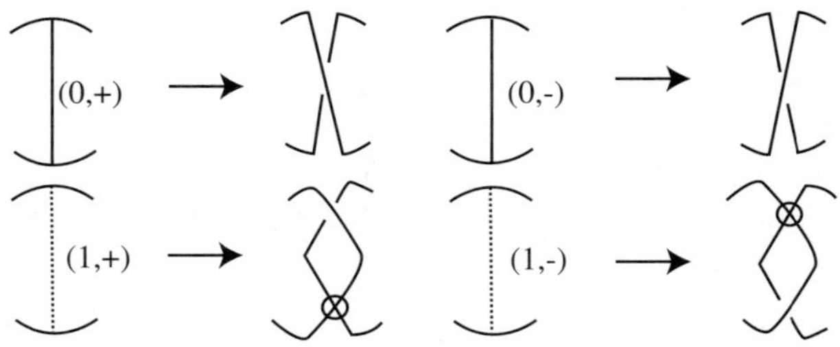

Fig. 14. Replacing a chord with a pair of lines.

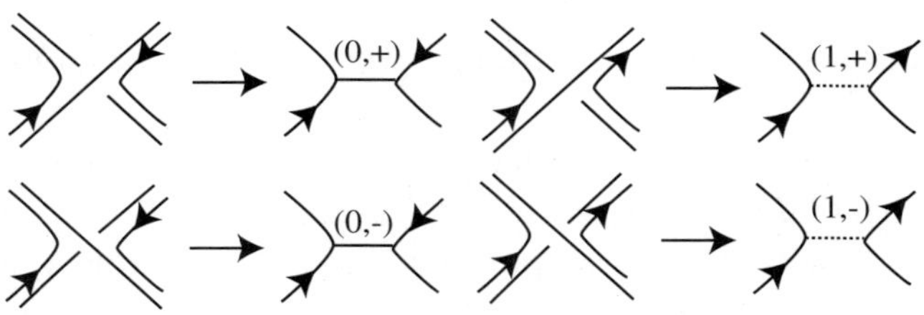

Fig. 15. Replacing a classical crossing with the labeled chord.

constructing a virtual link diagram out of a chord diagram: If we take a chord diagram D, and construct a virtual diagram $K(D)$ out of it, then for some circuit C the chord diagram $D_C(K(D))$ will coincide with D. This proves the following.

Theorem 2.3. (see [19]) *For any connected virtual diagram K' there is a certain labeled chord diagram D such that $K' = K(D)$.*

Now we are describing moves on graphs, obtained from the Reidemeister moves on virtual diagrams by using rotating circuits [11, 12] and the Gauss circuit [37]. These moves in both cases will correspond to "real" Reidemeister moves when applied to realizable graphs. Then we shall extend these moves to all graphs (not only realizable ones). As a result, we get two new objects: a *graph-link* and a *homotopy class of looped interlacement graphs*. Thus, virtual diagrams are represented (with loss of some information) by graphs. Note that we have obtained two new objects in a way similar to the generalization of classical knots to virtual knots: The passage from realizable Gauss diagrams (classical knots) to arbitrary chord diagrams leads to the concept of a virtual knot, and the passage from realizable (by means of chord diagrams) graphs to arbitrary graphs leads to the concept of a graph-link and a homotopy class of looped interlacement graphs (here "looped" corresponds to the writhe number, if the writhe number is -1, then the corresponding vertex has a loop). To construct the first object we shall use simple labeled graphs, and for the second one we shall use (unlabeled) graphs without multiple edges, but loops are allowed.

Definition 2.4. A graph is *labeled* if every vertex v of it is endowed with a pair (a, α), where $a \in \{0, 1\}$ is the *framing* of v, and $\alpha \in \{\pm\}$ is the *sign* of v.

Let D be a labeled chord diagram. The *labeled intersection graph* (cf. [6, 32]) $G(D)$ of D is the intersection graph whose vertices are endowed with the corresponding labels. A simple graph H is called *realizable* if there is a chord diagram D such that $H = G(D)$. Otherwise, a graph is called *non-realizable.*

Remark 2.5. We shall also consider simple graphs whose vertices have only one label, 0 or 1. We call these graphs *framed.*

In the realizable case, framed graphs are intersection graphs of framed chord diagrams.

The following lemma is evident.

Lemma 2.6. *A simple graph is realizable if and only if each of its connected components is realizable.*

Definition 2.7. Let G be a graph with the set of vertices $\mathcal{V}(G)$ and let $v \in \mathcal{V}(G)$. The set of all vertices adjacent to v is called the *neighborhood* of v and denoted by $N(v)$ or $N_G(v)$.

Let us define two operations on simple unlabeled graphs. In the realizable case these operations correspond to transformations of a rotating circuit into another rotating circuit.

Definition 2.8. Let G be a graph. The *local complementation* of G at $v \in \mathcal{V}(G)$ is the operation which toggles adjacencies between

a, $b \in N(v)$, $a \neq b$, and does not change the rest of G. Denote the graph obtained from G by the local complementation at a vertex v by $\mathrm{LC}(G; v)$.

Definition 2.9. Let G be a graph with distinct vertices u and v. The *pivoting operation* of G at u and v is the operation which toggles adjacencies between x, y such that x, $y \notin \{u, v\}$, $x \in N(u)$, $y \in N(v)$ and either $x \notin N(v)$ or $y \notin N(u)$, and does not change the rest of G. Denote the graph obtained from G by the pivoting operation at the vertices u and v by $\mathrm{piv}(G; u, v)$.

It is not difficult to prove the following lemma.

Lemma 2.10. *If u and v are adjacent, then there is an isomorphism*

$$\mathrm{piv}(G; u, v) \cong \mathrm{LC}(\mathrm{LC}(\mathrm{LC}(G; u); v); u) \cong \mathrm{LC}(\mathrm{LC}(\mathrm{LC}(G; v); u); v).$$

Let us define *graph-moves*, i.e. moves on labeled graphs. We consider labeled chord diagrams constructed by using rotating circuits and moves on them which originate from "real" Reidemeister moves on virtual diagrams. Then we extend these moves to arbitrary labeled graphs by using intersection graphs of chord diagrams. These moves were defined in [11, 12].

Definition 2.11. $\Omega_g 1$. *The first graph-move* is an addition/removal of an isolated vertex labeled $(0, \alpha)$, $\alpha \in \{\pm\}$.

$\Omega_g 2$. *The second graph-move* is an addition/removal of two non-adjacent (respectively, adjacent) vertices labeled $(0, \pm\alpha)$ (respectively, $(1, \pm\alpha)$) and having the same adjacencies with the other vertices.

$\Omega_g 3$. *The third graph-move* is defined as follows. Let u, v, w be three vertices of G all having the label $(0, -)$ so that u is adjacent only to v and w in G, and v and w are not adjacent to each other. Then we only change the adjacencies of u with the vertices v, w and $t \in \big(N(v) \setminus N(w)\big) \cup \big(N(w) \setminus N(v)\big)$ (for the other pairs of vertices we do not change their adjacencies). In addition, we switch the sign of v and w to $+$. The inverse operation is also called the third Reidemeister graph-move.

$\Omega_g 4$. *The fourth graph-move* for G is defined as follows. We take two adjacent vertices u and v labeled $(0, \alpha)$ and $(0, \beta)$, respectively.

Replace G with $\mathrm{piv}(G; u, v)$ and change the signs of u and v so that the sign of u becomes $-\beta$ and the sign of v becomes $-\alpha$.

$\Omega_g 4'$. In this fourth graph-move we take a vertex v with the label $(1, \alpha)$. Replace G with $\mathrm{LC}(G; v)$ and change the sign of v and the framing for each $u \in N(v)$.

Remark 2.12. The third Reidemeister graph-move does not exhaust all the possibilities for representing the third Reidemeister move on chord diagrams constructed by rotating circuits. The other versions of the third Reidemeister move are combinations of the second, third and fourth graph-moves.

Remark 2.13. The fourth graph-moves $\Omega_g 4$ and $\Omega_g 4'$ in the realizable case correspond to a rotating circuit change on a virtual diagram. Sometimes, applying these graph-moves we just say that we change the circuit.

The comparison of the graph-moves with the Reidemeister moves yields the following theorem.

Theorem 2.14. *Let G_1 and G_2 be two labeled intersection graphs corresponding to virtual diagrams K_1 and K_2, respectively. If K_1 and K_2 are equivalent, then G_1 and G_2 are obtained from one another by a sequence of $\Omega_g 1 - \Omega_g 4'$.*

Definition 2.15. A *graph-link* is an equivalence class of simple labeled graphs modulo $\Omega_g 1 - \Omega_g 4'$ graph-moves.

Remark 2.16. If we consider framed graphs, then graph-moves for them are obtained from the graph-moves $\Omega_g 1 - \Omega_g 4'$ by forgetting sign of vertices, i.e. the second components of labels. In this case we use the same notation.

Definition 2.17. A *free framed graph* is an equivalence class of framed graphs modulo $\Omega_g 4$ and $\Omega_g 4'$ graph-moves up to signs of labels. A *free graph-link* is an equivalence class of free framed graphs modulo $\Omega_g 1 - \Omega_g 3$.

Let us consider another approach based on Gauss circuits. Let $D_G(K)$ be the Gauss diagram of a virtual diagram K. Let us construct the graph obtained from the intersection graph of $D_G(K)$ by adding loops to vertices corresponding to chords with the negative writhe number [37]. We refer to this graph as a *looped interlacement graph* or *looped graph*. Let us describe the set of moves on looped graphs, see Fig. 6. These moves are similar to the moves for graph-links and also correspond to "real" Reidemeister moves on virtual diagrams.

Definition 2.18. $\Omega 1$. *The first move* for looped interlacement graphs is an addition/removal of an isolated looped or unlooped vertex.

$\Omega 2$. *The second move* for looped interlacement graphs is an addition/removal of two vertices having the same adjacencies with other vertices and, moreover, one of which is looped and the other one is unlooped.

$\Omega 3$. *The third move* for looped interlacement graphs is defined as follows. Let u, v, w be three vertices such that v is looped, w is unlooped, v and w are adjacent, u is adjacent to neither v nor w, and every vertex $x \notin \{u, v, w\}$ is adjacent to either 0 or precisely two of u, v, w. Then we only toggle adjacencies between three vertices u, v and w. The inverse operation is also called the third Reidemeister move.

Remark 2.19. The two third Reidemeister moves (the first one is the move with the third vertex being looped, and the second one is the move with the third vertex being unlooped) do not exhaust all the possibilities for representing the third Reidemeister move on Gauss diagrams [37]. It can be shown that all the other versions of the third Reidemeister move are combinations of the second and third Reidemeister moves described in Definition 2.18, see [31] for details.

Definition 2.20. We call an equivalence class of graphs (without multiple edges, but loops are allowed) modulo the three moves listed in Definition 2.18 a *homotopy class* of looped graphs. A *free homotopy* *class* of looped graphs is an equivalence class of simple graphs modulo the moves up to loops, i.e. we forget about loops.

Remark 2.21. The equivalence relation from Definition 2.20 is called the *Reidemeister equivalence* in [37], and it differs from the classical homotopy of links.

One may generalize the two approaches to graph-link theory by considering any Euler tour of a virtual diagram. This was initiated by Traldi [36], where he introduced the notion of a marked graph. But we think that there is an equivalence between the theory of graph-links and the theory of marked graphs.

2.2. *An equivalence between two approaches*

Let G be a labeled graph on $\mathcal{V}(G) = \{v_1, \ldots, v_n\}$.

Definition 2.22. The *adjacency matrix* $A(G)$ of a labeled graph G is the matrix over $\mathbb{Z}_2$ defined as follows: a_{ii} is equal to the framing of v_i, $a_{ij} = 1$, $i \neq j$, if and only if v_i is adjacent to v_j, and $a_{ij} = 0$ otherwise.

Theorem 2.23. (see [11, 12]) *If G and G' represent the same graph-link, then* $\mathrm{corank}_{\mathbb{Z}_2}(A(G) + E) = \mathrm{corank}_{\mathbb{Z}_2}(A(G') + E)$, *where E is the identity matrix and the corank of a matrix equals the difference between the size of the matrix and its rank. Thus, the number* $\mathrm{corank}_{\mathbb{Z}_2}(A(G) + E)$, *where G is a representative of a graph-link $\mathfrak{F}$, is an invariant of the graph-link $\mathfrak{F}$.*

Remark 2.24. All matrices are considered over $\mathbb{Z}_2$ and the determinant, rank and corank are also calculated over $\mathbb{Z}_2$.

Definition 2.25. The *number of components* in a graph-link $\mathfrak{F}$ is $\mathrm{corank}_{\mathbb{Z}_2}(A(G) + E) + 1$, where G is a representative of $\mathfrak{F}$. A graph-link $\mathfrak{F}$ with $\mathrm{corank}_{\mathbb{Z}_2}(A(G) + E) = 0$ for any representative G of $\mathfrak{F}$ is called a *graph-knot*.

Assume $\mathrm{corank}_{\mathbb{Z}_2}(A(G)+E) = 0$, and set $B_i(G) = A(G)+E+E_{ii}$ for each vertex $v_i \in \mathcal{V}(G)$; here E_{ii} is the matrix with the only one nonzero element equal to 1 in the ith column and ith row. The *writhe number* w_i of G at v_i is $w_i = (-1)^{\mathrm{corank}_{\mathbb{Z}_2} B_i(G)}\,\mathrm{sign}(v_i)$, and the *writhe number* of G is

$$w(G) = \sum_{i=1}^{n} w_i.$$

Remark 2.26. Note if G is a realizable graph, then $\mathrm{corank}_{\mathbb{Z}_2}(A(G)+E)+1$ and w_i are the "real" number of components and writhe number of the crossing corresponding to v_i, respectively.

Let L be a looped graph with enumerated set of vertices.

Definition 2.27. Define the *adjacency matrix* $A(L) = (a_{ij})$ over $\mathbb{Z}_2$ as: $a_{ii} = 1$ if and only if the vertex numbered i is looped, and $a_{ii} = 0$ otherwise; $a_{ij} = 1$, $i \neq j$, if and only if the vertex with the number i is adjacent to the vertex with the number j, and $a_{ij} = 0$ otherwise.

There is an equivalence between the set of homotopy classes of looped graphs and the set of graph-knots. Namely, the following theorem holds.

Theorem 2.28. (see [25]) *There is a one-to-one correspondence between the set of all looped graphs and the set of all equivalence classes of labeled graphs G with* $\mathrm{corank}_{\mathbb{Z}_2}(A(G)+E) = 0$ *under two fourth graph-moves. This correspondence generates an equivalence between the set of all homotopy classes of looped graphs and the set of all graph-knots. Moreover, if K is a virtual knot diagram, and $\mathfrak{F}$ is the graph-knot constructed from K and $\mathfrak{L}$ is the homotopy class of looped graphs constructed from K, then $\mathfrak{F}$ and $\mathfrak{L}$ are related by this equivalence.*

Sketch of the Proof. Let us construct a map χ from the set of all graph-knots to the set of all homotopy classes of looped graphs. This map will have an inverse map.

Let G be a representative of a graph-knot $\mathfrak{F}$. Let us consider the simple graph H having the adjacency matrix coinciding with $(A(G)+E)^{-1}$ up to diagonal elements. Here we say that two matrices $A = (a_{ij})$ and $B = (b_{kl})$ coincide (or are equal to each other) *up to diagonal elements* if $a_{ij} = b_{ij}$ for any $i \neq j$. Let us construct the graph $L(G)$ from H by just adding loops to any vertex of H corresponding to the vertex of G with the negative writhe number (in the sense of Definition 2.25). By definition, put $\chi(\mathfrak{F}) = \mathfrak{L}$, where $\mathfrak{L}$ is the homotopy class of $L(G)$.

Let us define the map ψ from the set of all homotopy classes of looped graphs to the set of all graph-knots. Let $\mathfrak{L}$ be the homotopy class of L. It is not difficult to prove that we can construct a symmetric matrix $A = (a_{ij})$ over $\mathbb{Z}_2$ coinciding with the adjacency matrix of L up to diagonal elements and having $\det A = 1$. Let $G(L)$ be the labeled simple graph having the matrix $(A^{-1} + E)$ as its adjacency matrix. Therefore, the first component of the label of a vertex is equal to the corresponding diagonal element of $(A^{-1} + E)$, the second component of the label of the vertex with the number i is $w_i(1 - 2a_{ii})$, where $w_i = 1$ if the vertex of L with the number i does not have a loop, and $w_i = -1$ otherwise. Set $\psi(\mathfrak{L}) = \mathfrak{F}$, where $G(L)$ is a representative of $\mathfrak{F}$.

It turns out that the maps χ and ψ are well-defined and inverse to each other. This fact proves the theorem. □

3. The Kauffman Bracket and Jones Polynomial

In the previous section we constructed an invariant of graph-links, this invariant was the number of components of a graph-link. Now we are going to construct a polynomial invariant and formulate some theorems of minimality by using this invariant. We construct this polynomial invariant for graph-knots. In Sec. 4 we generalize this polynomial for the case of any graph-link. In Sec. 5 we get minimality theorems from parity.

Let G be a labeled graph. Let $s \subset \mathcal{V}(G)$ be a subset of the set $\mathcal{V}(G)$ of vertices of G. Set $G(s)$ to be the induced subgraph of the graph G with the set of vertices $\mathcal{V}(G(s)) = s$ and the set of edges $E(G(s))$ such that $\{u, v\} \in E(G(s))$, where $u, v \in s$, if and only if $\{u, v\} \in E(G)$.

Definition 3.1. We call a subset of $\mathcal{V}(G)$ a *state* of G. The *A-state* is the state consisting of all the vertices of G labeled $(a, -)$, $a \in \{0, 1\}$, and no vertex labeled $(b, +)$, $b \in \{0, 1\}$. Analogously, the *B-state* is the state consisting of all vertices of G labeled $(b, +)$ and no vertex labeled $(a, -)$.

Two states are called *neighboring* if they differ only in one vertex, which belongs to one state and not to the other state. We define the *number of circles* in a state s as $\mathrm{corank}_{\mathbb{Z}_2} A(G(s)) + 1$.

The *Kauffman bracket polynomial* of a labeled graph G is

$$\langle G\rangle(t) = \sum_s t^{\alpha(s)-\beta(s)}(-t^2 - t^{-2})^{\mathrm{corank}_{\mathbb{Z}_2} A(G(s))},$$

where the sum is taken over all states s of G, $\alpha(s)$ is equal to the sum of the vertices labeled $(a, -)$ from s and the vertices labeled $(b, +)$ from $\mathcal{V}(G)\setminus s$, $\beta(s) = |\mathcal{V}(G)| - \alpha(s)$.

Theorem 3.2. (see [11, 12]) *The Kauffman bracket polynomial of a labeled graph is invariant under $\Omega_g 2 - \Omega_g 4'$ and gets multiplied by $(-t^{\pm 3})$ under $\Omega_g 1$.*

Definition 3.3. Define the *Jones polynomial* for a labeled graph G with $\mathrm{corank}_{\mathbb{Z}_2}(A(G) + E) = 0$ as $X(G)(t) = (-t)^{-3w(G)}\langle G\rangle(t)$.

Remark 3.4. The Jones polynomial can be defined for any graph-link, but first we have to define the notion of an "oriented" graph-link. We do this in Sec. 4.

Theorem 3.5. (see [11, 12]) *The Jones polynomial is an invariant of graph-knots.*

The main results concerning minimality theorems in the classical case come from the well-known Kauffman–Murasugi–Thistlethwaite theorem [18, 19, 27, 34], see also [18, 19] for generalizing this result for the case of virtual diagrams.

Let us generalize the notions used for Kauffman–Murasugi–Thistlethwaite theorem for graph-links.

Definition 3.6. A labeled graph G on n vertices is *alternating* if $k + l = n + 2$, where k is the number of circles in the A-state s_1, i.e. $k = \mathrm{corank}_{\mathbb{Z}_2} A(G(s_1)) + 1$, and l is the number of circles in the B-state s_2 of G, i.e. $l = \mathrm{corank}_{\mathbb{Z}_2} A(G(s_2)) + 1$.

The *atom genus* of G is $1 - (k + l - n)/2$.

A labeled graph G is *non-split* if it has no isolated vertices.

In the definition of the atom genus we have just used $\chi = 2 - 2g$, where χ is the Euler characteristic, and counted χ by using the number of crossings n, number of edges $2n$ and the number of 2-cells (A-state circles and B-state circles).

Proposition 3.7. *A labeled graph G is alternating if and only if its atom genus is equal to* 0.

Proposition 3.8. *For any labeled graph G on n vertices we have*

$$\mathrm{span}\langle G\rangle \leq 4n - 4g(G),$$

where $g(G)$ is the genus of the corresponding atom.

Theorem 3.9. (see [11, 12]) *An alternating non-split labeled graph is minimal.*

Example 3.10. Consider the second graph Bouchet BW_3, see Fig. 7: i is connected to j if and only if $i - j \equiv \pm 1 \pmod 6$, $i, j = 1, \ldots, 6$, and 7 is connected to 2, 4, 6. Label all vertices with even numbers by $(0, +)$, and label all the other vertices by $(0, -)$. BW_3 is alternating. By Theorem 3.9, BW_3 is minimal. In the next section we shall show that this graph-link is non-realizable.

4. A Non-Realizable Graph-Link and a Way to Define an Orientation on Graph-Links

In this section we deal with graph-links with many components. The first question considered in this section is whether there exists a *non-realizable* graph-link, see Definition 4.1? To show that the answer to this question is affirmative we construct an invariant and a graph-link such that the value of the invariant for this graph-link is equal to the number which cannot be realized in the case of classical and virtual links.

The second part of the section is devoted to a way to define an orientation on graph-links in such a way that we could generalize the Jones polynomial to the case of graph-links. It is known that the writhe number does not depend on an orientation of a knot, and depends on the "mutual" orientations of two components. Therefore, to define an orientation on a graph-link we construct a graph-knot from this graph-link and this graph-knot gives us an orientation on the graph-link. As usual, we have in mind a realizable case and mimic all the necessary definitions for general case.

4.1. *A non-realizable graph-link*

Let us give the main definitions and present an example of a non-realizable graph-link.

Definition 4.1. We call a (free) graph-link (respectively, free framed graph, (free) homotopy class of looped graphs) *realizable* if it has a representative which can be realized by a chord diagram. Otherwise, a (free) graph-link (respectively, (free) homotopy class of looped graphs) is called *non-realizable.*

Remark 4.2. It is not difficult to show that if a free framed graph is realizable, then each representative of it is realized by a chord diagram (we just redraw the picture).

Theorem 4.3. *The graph-link generated by the labeled graph from Example* 3.10 *is non-realizable.*

Here we give a sketch of the proof of this fact.

Before proving this result, we give some definitions and formulate some assertions.

Let $\mathfrak{F}$ be a graph-link with more than one component. Let G be any representative of $\mathfrak{F}$ with $\mathcal{V}(G) = \{v_1, \dots, v_n\}$.

Definition 4.4. We say that a vertex $v_i \in \mathcal{V}(G)$ *lies on one component* of G if $\operatorname{corank}_{\mathbb{Z}_2} B_i(G) \neq \operatorname{corank}_{\mathbb{Z}_2} \widehat{B}_i(G)$, where $B_i(G) = A(G) + E + E_{ii}$ and $\widehat{B}_i(G)$ is the matrix obtained from the matrix $B(G) = A(G)+E$ by deleting the ith column and ith row. Otherwise, we say that v_i *belongs to two different components.*

Remark 4.5. Let $G' = G \setminus \{v_i\}$, i.e. G' is obtained from G by deleting the vertex v_i and all the edges incident to this vertex. In the realizable case, this operation corresponds to a smoothing of the crossing corresponding to v_i. Since $\widehat{B}_i(G) = B(G')$, we get that $\operatorname{corank}_{\mathbb{Z}_2} \widehat{B}_i(G)$ is equal to the number of components of the link after this smoothing. It is not difficult to see that $\operatorname{corank}_{\mathbb{Z}_2} B_i(G)$ equals the number of components of the link obtained by smoothing the crossing corresponding to v_i the other way.

Therefore, in the realizable case, Definition 4.4 means that v_i lies on one component if after two smoothings we have links with different numbers of components. This definition coincides with the "real" definition of a vertex which lies on one component.

Let us define the following relation on the set of all vertices belonging to two different components.

Definition 4.6. Let v_i and v_j be two vertices from $\mathcal{V}(G)$ belonging to two different components. We say that *two components meet* at these vertices if either $v_i = v_j$, or v_i lies on one component of the labeled graph $G \setminus \{v_j\}$.

Remark 4.7. In the realizable case, Definition 4.6 means that after a smoothing of one vertex, the other one lies on the component obtained by "joining" two components sharing the first vertex.

It is not difficult to prove the following lemma.

Lemma 4.8. *The relation from Definition* 4.6 *is an equivalence relation on the set of vertices belonging to two different components.*

Let us consider equivalence classes in the set of vertices belonging to two different components modulo the relation from Definition 4.6 and define the number $\vartheta(G)$ to be the number of equivalence classes having an odd number of vertices.

Theorem 4.9. *The number $\vartheta(G)$ is invariant under graph-moves.*

Theorem 4.3 can now be proved by following the main principle. In the realizable case, the claim of the theorem follows straightforwardly from definitions, and in the general case, we have the following definitions reformulated in the language of adjacency matrices.

It is easy to show that the graph-link generated by the labeled graph BW_3 from Example 3.10 has four components and $\vartheta(BW_3) = 7$, and for any realizable graph-link with four components we have ϑ to be strictly less than 7.

4.2. *An oriented graph-link and the Jones polynomial*

Let G be a labeled graph with k components and $\mathcal{V}(G) = \{v_1, \ldots, v_n\}$. It is not difficult to show that there exists a sequence of vertices $(v_{i_1}, \ldots, v_{i_{k-1}})$ consisting of $k-1$ vertices $v_{i_1}, \ldots, v_{i_{k-1}}$ such that the graph $(\cdots((G\backslash\{v_{i_1}\})\backslash\{v_{i_2}\})\cdots)\backslash\{v_{i_{k-1}}\}$ obtained from G by deleting all the vertices of the sequence in consecutive order has only one component.

Lemma 4.10. *For any sequence $(\alpha_{i_1}, \ldots, \alpha_{i_{k-1}})$ of signs $\alpha_{i_j} \in \{\pm\}$ there exists a labeled graph $G(v_{i_1}, \ldots, v_{i_{k-1}})$ obtained from G by fourth graph-moves such that the labeled graph $G' = G(v_{i_1}, \ldots, v_{i_{k-1}})\backslash \{v_{i_1}, \ldots, v_{i_{k-1}}\}$ has one component and the signs of the vertices of $G(v_{i_1}, \ldots, v_{i_{k-1}})$, which correspond to the vertices v_{i_j}, $j = 1, \ldots, k-1$, of G, coincide with α_{i_j}.*

Proof. First note that fourth graph-moves change signs of only those vertices which these moves are applied to. Further, since deleting a vertex from a graph has no influence on signs and adjacencies of the remaining vertices, then the lemma can be proved by the induction. □

Since G' has one component, we can define the *writhe number* w_i of each v_i' of its vertices by putting $w_i = (-1)^{\operatorname{corank} B_i(G')} \operatorname{sign}(v_i')$ and the *writhe number* of G' itself by the formula $w(G') = \sum_{i=1}^{n-k+1} w_i$.

Definition 4.11. The *writhe number* $w_i(G)$ of the graph G at a vertex $v_i \in \mathcal{V}(G)$ with respect to a sequence of vertices $(v_{i_1}, \ldots, v_{i_{k-1}})$ with signs $(\alpha_{i_1}, \ldots, \alpha_{i_{k-1}})$ is the writhe number of the vertex of G', which corresponds to the vertex v_i if $i \neq i_j$, and $w_{i_j}(G) = \alpha_{i_j}$ otherwise. The *writhe number* of G with respect to the sequence of vertices $(v_{i_1}, \ldots, v_{i_{k-1}})$ with signs $(\alpha_{i_1}, \ldots, \alpha_{i_{k-1}})$ is

$$w(G) = \sum_{i=1}^{n} w_i(G).$$

It is easy to show that a writhe number is well defined, i.e. it does not depend on a choice of the graph $G(v_{i_1}, \ldots, v_{i_{k-1}})$, but only does depend on the signs of the vertices being deleted.

Definition 4.12. Let two sequences of vertices $(v_{i_1}, \ldots, v_{i_{k-1}})$ and $(v_{j_1}, \ldots, v_{j_{k-1}})$ with signs $(\alpha_{i_1}, \ldots, \alpha_{i_{k-1}})$ and $(\alpha_{j_1}, \ldots, \alpha_{j_{k-1}})$ respectively, be given. Assume that after the deletion of these vertices from the corresponding graphs $G(v_{i_1}, \ldots, v_{i_{k-1}})$ and $G(v_{j_1}, \ldots, v_{j_{k-1}})$ we have the two graphs with one component each. We say that these sequences are *equivalent* if the writhe number at vertices v_{i_p} (respectively, v_{j_p}), $p = 1, \ldots, k-1$, with respect to these sequences coincide.

Definition 4.13. We say that a labeled graph G is *oriented* if an equivalence class of sequences of vertices with signs after the deletion of which we get a labeled graph with one component is fixed.

If we have a graph G with an orientation, i.e. a sequence of vertices $(v_{i_1}, \ldots, v_{i_{k-1}})$ with signs $(\alpha_{i_1}, \ldots, \alpha_{i_{k-1}})$ is fixed, we can define an orientation on any graph $\widetilde{G}$ obtained from G by applying a single graph-move. Note that after applying a first, third, fourth graph-move or a second graph-move increasing the number of vertices of G, we can define the orientation on $\widetilde{G}$ by the sequence of vertices corresponding to the sequence $(v_{i_1}, \ldots, v_{i_{k-1}})$ with the same signs. But if we apply a second graph-move decreasing the number of vertices of G, then we first choose a sequence of vertices $(v_{j_1}, \ldots, v_{j_{k-1}})$ with signs $(\alpha_{j_1}, \ldots, \alpha_{j_{k-1}})$ equivalent to the sequence $(v_{i_1}, \ldots, v_{i_{k-1}})$ with signs $(\alpha_{i_1}, \ldots, \alpha_{i_{k-1}})$ such that the second graph-move has no effect on $v_{j_1}, \ldots, v_{j_{k-1}}$, and then the orientation on $\widetilde{G}$ is generated by the sequence of vertices corresponding to $(v_{j_1}, \ldots, v_{j_{k-1}})$ with signs $(\alpha_{j_1}, \ldots, \alpha_{j_{k-1}})$. We say that the graphs G and $\widetilde{G}$ *have the same orientation.*

Lemma 4.14. *The writhe number of an oriented labeled graph is changed by* ± 1 *under the first graph-move* $\Omega_g 1$*. More precisely, it is changed by* -1 *if we add the vertex with the positive sign, and by* $+1$ *if we add the vertex with the negative sign.*

The writhe number of an oriented labeled graph is invariant under graph-moves $\Omega_g 2 - \Omega_g 4'$.

Definition 4.15. We say that a graph-link $\mathfrak{F}$ is *oriented* if all its representatives are oriented and for any two representatives G' and G'' of $\mathfrak{F}$ there exists a sequence $G_1 = G', G_2, \ldots, G_s = G''$ such that the graphs G_p and G_{p+1}, $p = 1, \ldots, s-1$, are obtained from each other by one graph-move and have the same orientation.

Definition 4.16. Let G be an oriented graph. Define the *Jones polynomial* as $X(G)(t) = (-t)^{-3w(G)}\langle G\rangle(t)$, where $\langle G\rangle(t)$ is the Kauffman bracket polynomial.

Theorem 4.17. *The Jones polynomial* $X(G)(t)$ *is an invariant of oriented graph-links.*

5. Parity and Other Invariants. Persisting Configurations

In this section we consider the parity for free graph-knots and free graph-links in the spirit of [20–22, 24]. Having a parity we can, for example, strengthen some invariants, construct non-realizable graph-knots and functorial maps.

Since we have constructed the equivalence between the set of graph-knots and the set of homotopy classes of looped interlacement graphs, it is sufficient to construct a parity for free homotopy classes of looped interlacement graphs and for graph-links with more than one component.

5.1. *Smoothing operations and Turaev's* Δ

Let us define a smoothing operation for free graph-links. As usual we first mimic the definition of a smoothing operation on four-valent graphs with a cross structure [20, 21, 23] for the case of realizable free graph-links, and then we use the same definition for all graphs.

Let $\mathcal{G}$ be a free framed graph, i.e. an equivalence class of framed graphs modulo the fourth graph-moves. All representatives of a free framed graph have the same number of vertices. Therefore, the notion of a vertex for a free framed graph is well defined.

Let v be a vertex of $\mathcal{G}$. Let us consider two cases. In the first case, there exists a representative H of $\mathcal{G}$ for which v has either a framing 1 or a degree greater than 0. It is not difficult to see that v has the same property for each representative of $\mathcal{G}$, and there are two representatives H_1 and H_2 of $\mathcal{G}$ which differ from each other by $\Omega_g 4$ or $\Omega_g 4'$ at v. By a *smoothing* of the free framed graph $\mathcal{G}$ at the vertex v we mean any of the two free framed graphs having the representatives $H_1 \setminus \{v\}$ and $H_2 \setminus \{v\}$, respectively. In the realizable case this means that the four-valent graph with a cross structure has a rotating circuit having any of the two possible connection types at the vertex. Then the smoothing at the vertex corresponds to the removal of the chord of the framed chord diagram (the vertex of the intersection graph) corresponding to the vertex. If a smoothing of a four-valent graph with a cross structure leads to a disconnected graph, this may be repaired by taking another representative of the same graph-link. We get the second case: the vertex v has a framing 0 and is isolated for a representative, and, therefore, for any representative of $\mathcal{G}$. Let H be a representative of $\mathcal{G}$. Let us construct the new graph H' obtained from H by adding a new vertex u with framing 0 to H which is adjacent only to v, see Fig. 16 for the case of realizable graphs (the dashed line is a rotating circuit). By a *smoothing* of the free framed graph $\mathcal{G}$ at the vertex v we mean any of the two free framed graphs having the representatives $H \setminus \{v\}$ and H', respectively.

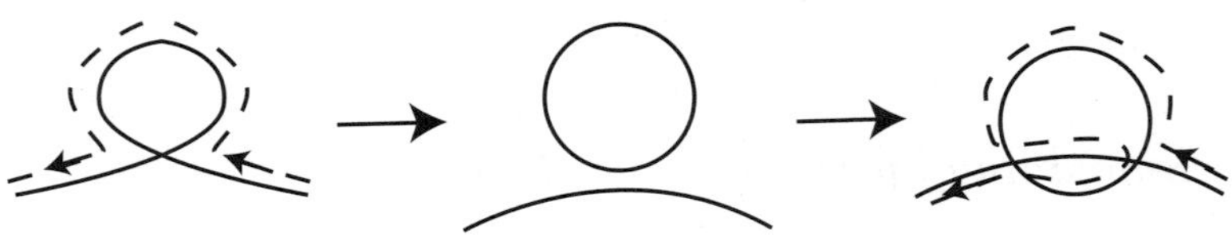

Fig. 16. One of the two smoothings at an isolated vertex.

Generally, a *smoothing* of a free framed graph in a collection of vertices is a free framed graph obtained by a sequence of smoothings.

Remark 5.1. Sometimes after applying a smoothing, additional vertices can appear, which were absent in the original graph. We do not apply any smoothing to these new vertices.

Proposition 5.2. (see [25]) *Assume $\mathcal{G}$, $\mathcal{G}'$ are free framed graphs, and $\mathcal{G}$ can be obtained from $\mathcal{G}'$ as a result of smoothings at some vertices. If $\mathcal{G}'$ is realizable by a chord diagram, then so is $\mathcal{G}$.*

Let $i > 1$ be a natural number. Define the set $\mathbb{Z}_2\mathcal{G}_i$ to be the $\mathbb{Z}_2$-linear space generated by the set $\mathcal{G}_i$ of free framed graphs having i components (in the sense of Definition 2.25) modulo the following relations:

(1) the second Reidemeister graph-move,
(2) $\mathcal{G} = 0$, if $\mathcal{G}$ has two vertices with framing 0 which are adjacent only to each other.

The meaning of (2) is that a free framed graph equals zero if it has a "component not linked with others" (see Definition 2.25). More precisely, if a free framed graph has two vertices with framing 0 which are adjacent only to each other, then it has a unicursal component not linked with others.

For $i = 1$, we define $\mathbb{Z}_2\mathcal{G}_1$ analogously with respect to equivalence (1) and not (2).

Let us now define a map $\Delta\colon \mathbb{Z}_2\mathcal{G}_1 \to \mathbb{Z}_2\mathcal{G}_2$. We take a free framed graph $\mathcal{G}$ with $\mathrm{corank}_{\mathbb{Z}_2}(A(\mathcal{G})+E) = 0$ and construct an element $\Delta(\mathcal{G})$ from $\mathbb{Z}_2\mathcal{G}_2$ as follows. For each vertex v of $\mathcal{G}$, there are two ways of smoothing it. One way gives us a graph from $\mathbb{Z}_2\mathcal{G}_1$, and the other smoothing gives us a free framed graph $\mathcal{G}_v$ from $\mathbb{Z}_2\mathcal{G}_2$. We take $\mathcal{G}_v$ and set

$$\Delta(\mathcal{G}) = \sum_v \mathcal{G}_v \in \mathbb{Z}_2\mathcal{G}_2.$$

Theorem 5.3. (see [14]) *Δ is a well-defined mapping from $\mathbb{Z}_2\mathcal{G}_1$ to $\mathbb{Z}_2\mathcal{G}_2$.*

Definition 5.4. We call a vertex v_i of a free framed graph $\mathcal{G}$ *oriented* if it lies on one component.

Using the notion of oriented vertex we can define the map Δ^i (the iteration of the map Δ) by considering smoothings at oriented vertices and taking that smoothing which has more components than the other in the previous step.

Corollary 5.5. *Δ^i is a well-defined mapping from $\mathbb{Z}_2\mathcal{G}_1$ to $\mathbb{Z}_2\mathcal{G}_{i+1}$.*

5.2. *Definition of parity*

Let $\mathfrak{L}$ (respectively, $\mathfrak{F}$) be a free homotopy class of graphs or homotopy class of looped interlacement graphs (respectively, a free graph-link or graph-link).

Let us define the category $\mathfrak{C}(\mathfrak{L})$ (respectively, $\mathfrak{C}(\mathfrak{F})$) of graphs of the (free) homotopy class $\mathfrak{L}$ (respectively, the (free) graph-link $\mathfrak{F}$). The objects of $\mathfrak{C}(\mathfrak{L})$ (respectively, $\mathfrak{C}(\mathfrak{F})$) are graphs of $\mathfrak{L}$ (respectively, labeled graphs of $\mathfrak{F}$) and morphisms of the category are (formal) compositions of *elementary morphisms*. By an elementary morphism we mean

- an isomorphism of graphs;
- a Reidemeister move.

Definition 5.6. A *partial bijection* of sets X and Y is a triple $(\widetilde{X}, \widetilde{Y}, \phi)$, where $\widetilde{X} \subset X$, $\widetilde{Y} \subset Y$ and $\phi\colon \widetilde{X} \to \widetilde{Y}$ is a bijection.

Let us denote by $\mathcal{V}$ the vertex functor on $\mathfrak{C}(\mathfrak{L})$ (respectively, $\mathfrak{C}(\mathfrak{F})$), i.e. the functor from $\mathfrak{C}(\mathfrak{L})$ (respectively, $\mathfrak{C}(\mathfrak{F})$) to the category; objects of which are finite sets and morphisms are partial bijections. For each graph G we define $\mathcal{V}(G)$ to be the set of the vertices of G. Any elementary morphism $f\colon G \to G'$ naturally induces a partial bijection $f_*\colon \mathcal{V}(G) \to \mathcal{V}(G')$.

Let A be an abelian group.

Definition 5.7. A *parity* p on graphs of the (free) homotopy class $\mathfrak{L}$ (respectively, the (free) graph-link $\mathfrak{F}$) with coefficients in A is a family of maps $p_G\colon \mathcal{V}(G) \to A$, where $G \in \mathrm{ob}(\mathfrak{C}(\mathfrak{L}))$ (respectively, $G \in$

$\mathrm{ob}(\mathfrak{C}(\mathfrak{F})))$ is an object of the category, such that for any elementary morphism $f\colon G \to G'$ the following conditions hold:

(1) $p_{G'}(f_*(v)) = p_G(v)$ provided that $v \in \mathcal{V}(G)$ and there exists $f_*(v) \in \mathcal{V}(G')$, i.e. the parity of the corresponding vertices is the same;
(2) $p_G(v_1) + p_G(v_2) = 0$ if f is a decreasing second Reidemeister move and v_1, v_2 are the disappearing vertices;
(3) $p_G(v_1) + p_G(v_2) + p_G(v_3) = 0$ if f is a third Reidemeister move and v_1, v_2, v_3 are the vertices participating in this move.

One can prove the following lemma.

Lemma 5.8. *Let p be any parity and G be a (labeled) graph. Then $p_G(v) = 0$ if f is a decreasing first Reidemeister move applied to G and v is the disappearing vertex of G.*

Let us consider two examples of parity with coefficients from $\mathbb{Z}_2$.

Example 5.9. Let $\mathfrak{L}$ be a (free) homotopy class of looped graphs, and L be its representative.

Definition 5.10. Define the map $\mathrm{gp}_L\colon \mathcal{V}(L) \to \mathbb{Z}_2$ by putting $\mathrm{gp}_L(v) = 0$ if the degree of v is even (*an even vertex*), and $\mathrm{gp}_G(v) = 1$ otherwise (*an odd vertex*).

Lemma 5.11. (see [13]) *The map* gp *is a parity for $\mathfrak{L}$.*

The parity gp is called the *Gaussian parity.*

Example 5.12. Let $\mathfrak{F}$ be a (free) two-component graph-link, and G be its representative. Using Definition 5.10, we can define the notion of oriented vertex for G.

Definition 5.13. Define the map $p_G\colon \mathcal{V}(G) \to \mathbb{Z}_2$ by putting $p_G(v) = 0$ if v is an oriented vertex (*an even vertex*), and $p_G(v) = 1$ otherwise (*an odd vertex*).

Lemma 5.14. (see [13]) *The map p is a parity for $\mathfrak{F}$.*

Definition 5.15. A parity p_u with coefficients in an abelian group A_u is called *universal* if for any parity p with coefficients in an abelian group A there exists a unique homomorphism of groups $\rho\colon A_u \to A$ such that $p_G = \rho \circ (p_u)_G$ for any graph G.

Let us describe the construction of the universal parity in the general case.

Let G be a graph, and let v be its vertex. Denote by $1_{G,v}$ the generator of the direct summand in the group $\bigoplus_G \bigoplus_{v\in\mathcal{V}(G)} \mathbb{Z}$ corresponding to the vertex v of G.

Let A_u be the group

$$A_u = \left(\bigoplus_G \bigoplus_{v\in\mathcal{V}(G)} \mathbb{Z} \right) \Big/ \mathcal{R},$$

where $\mathcal{R}$ is the set of relations of four types:

(1) $1_{G',f_*(v)} = 1_{G,v}$ if $v \in \mathcal{V}(G)$ and there exists $f_*(v) \in \mathcal{V}(G')$;
(2) $1_{G,v_1} + 1_{G,v_2} = 0$ if f is a decreasing second Reidemeister move and v_1, v_2 are the disappearing crossings;
(3) $1_{G,v_1} + 1_{G,v_2} + 1_{G,v_3} = 0$ if f is a third Reidemeister move and v_1, v_2, v_3 are the crossings participating in this move.

The map $(p_u)_G$ for each graph G is defined by the formula $(p_u)_G(v) = 1_{G,v}$, $v \in \mathcal{V}(G)$.

If p is a parity with coefficients in a group A, one defines the map $\rho\colon A_u \to A$ in the following way:

$$\rho\left(\sum_{G,v\in\mathcal{V}(G)} \lambda_{G,v} 1_{G,v} \right) = \sum_{G,v\in\mathcal{V}(G)} \lambda_{G,v} p_G(v), \qquad \lambda_{G,v} \in \mathbb{Z}.$$

The main theorem of the present subsection is the following.

Theorem 5.16. (see [25]) *Let a free homotopy class be given. Then the Gaussian parity on its graphs is the universal parity.*

5.3. *Minimality*

Using the Gaussian parity and the corresponding constructions from the theory of free knots and links, see [20–24], we can define the map Δ^i_{odd}, where the sum is taken over all odd oriented vertices, or Δ^i_{even}, where the sum is taken over all even oriented vertices. We have to define the notion of *even* and *odd* vertex for free framed graphs with many components. As usual, we mimic the realizable case.

Definition 5.17. We call a vertex v of a free framed graph $\mathcal{G}$ with one component *even* (respectively, *odd*) if the vertex corresponding to v of the looped interlacement graph $\chi(\mathcal{G})$ is even (respectively, odd).

Let $\mathcal{G}_{v_1,\ldots,v_{k-1}}$ be the free framed graph with k components which is obtained from $\mathcal{G}$ by smoothing $\mathcal{G}$ consequently at $v_1,\ldots,v_{k-1}$, where v_1 is oriented in $\mathcal{G}$ and v_i, $i = 2,\ldots,k-1$, is an oriented vertex in $\mathcal{G}_{v_1,\ldots,v_{i-1}}$. An oriented vertex u of $\mathcal{G}_{v_1,\ldots,v_{k-1}}$ is *even with respect to the smoothing at* $v_1,\ldots,v_{k-1}$ (respectively, *odd with respect to the smoothing at* $v_1,\ldots,v_{k-1}$) if the number of oriented vertices in $\mathcal{G}_{v_1,\ldots,v_{k-1}}$ which are incident to u in $\chi(\mathcal{G})$ is even (respectively, odd).

Remark 5.18. We have defined even vertices only for those free framed graphs with many components which originate from free framed graphs with one component. First, it suffices to define the iteration $\Delta^i_{\text{even(odd)}}$. Second, it is possible to define even vertices in the general case, but we do not want to complicate the construction.

Proposition 5.19. Δ^i_{odd} *is a well-defined mapping from* $\mathbb{Z}_2\mathcal{G}_1$ *to* $\mathbb{Z}_2\mathcal{G}_{i+1}$.

Now, we can define brackets $[\cdot]$ and $\{\cdot\}$ for graph-knots and graph-links, respectively. For a graph G representing a free graph-knot and for a graph H representing a free two-component graph-link consider the following sums:

$$[G] = \sum_{s \text{ even,1 comp.}} G_s \in \mathbb{Z}_2\mathcal{G}_1 \quad \text{and} \quad \{H\} = \sum_{s \text{ even, nontrivial}} H_s \in \mathbb{Z}_2\mathcal{G}_2.$$

In the first formula, the sums are taken over all smoothings at all even vertices, and only those summands are taken into account where $\mathrm{corank}_{\mathbb{Z}_2}(A(G_s)+E) = 0$. In the second formula, the sums are taken over all smoothings at all even vertices, and only those summands are taken into account where H_s is not equivalent to any simple graph having two vertices with framing 0 and adjacent only to each other (here, the equivalence relation is generated by second Reidemeister moves). Thus, if G has k even vertices, then $[G]$ will contain at most 2^k summands, and if all vertices of G are odd, then we shall have exactly one summand, the graph G itself. The same is true for H and $\{H\}$.

Theorem 5.20. (see [25]) *If G and G' represent the same free graph-knot, then the following equality holds*: $[G] = [G']$. *If H and H' represent the same free graph-link with two components, then* $\{H\} = \{H'\}$.

Definition 5.21. We call a labeled graph G (respectively, a looped graph L) *minimal* if there is no representative of the graph-link corresponding to G (respectively, the homotopy class of L) having strictly smaller number of vertices than G (respectively, L) has.

Theorem 5.22. (see [25]) *Let G (respectively, H) be a simple labeled graph representing a free graph-knot (respectively, a two-component graph-link) with all odd vertices in the sense of Definition* 5.10 (*respectively, Definitions* 5.13), *such that no decreasing second Reidemeister move is applicable to G (respectively, H). Then G (respectively, H) is minimal.*

As a consequence of this theorem we may deduce the following corollaries.

Corollary 5.23. (cf. [20]) *The graph W_5 shown in Fig.* 7 *with each vertex having the framing* 1 *is minimal for the free graph-knot. In particular, this free graph-knot is nontrivial and non-realizable.*

Corollary 5.24. *The free two-component graph-link generated by W_5 with each vertex having the framing* 0 *is non-realizable.*

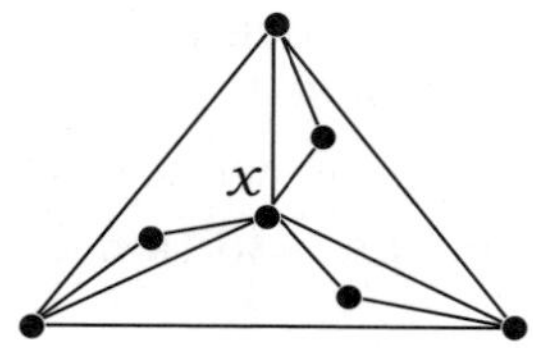

Fig. 17. An even non-realizable graph representing a non-realizable homotopy class.

Let us consider one more example. Note that the graph from this example has all even vertices, but using the operation Δ, we obtain a graph with all odd vertices which is non-realizable.

Proposition 5.25. *The looped graph* K *shown in Fig.* 17 *is minimal, and the corresponding homotopy class of looped graphs is non-realizable.*

The proof consists of the following steps.

First, note that $\Delta(K)$ consists of seven summands $L + \sum_i L_i$, where only one summand (corresponding to the vertex x) is a two-component free graph-link with all *odd* vertices; for each of the remaining summands L_i, there is at least one even vertex.

Now, the two-component free graph-link generated by L has the representative W_5; all framings of the vertices are 0. To see it, one should apply the map from Theorem 2.28.

Now, consider the bracket $\{\Delta(K)\} = L + \sum_i \{L_i\}$. Note that all the graph-links generated by summands $\{L_i\}$ have representatives with strictly less than six vertices since each of L_i has at least one even vertex; on the other hand, the graph-link generated by L has no representative with less than six crossings; so, this element L is not canceled in the sum. Since it is not realizable, the free framed knot K is not realizable either.

6. Graph-Link Homology

In this section we describe a homological invariant of graph-links: (odd) Khovanov homology. Our construction generalizes the definition of odd Khovanov homology of knots given by J. Bloom [2].

The construction allows one to define homology for all graph-links in the case when the coefficient field has the characteristic 2. As for integral Khovanov homology, one needs to slightly modify the theory of graph-links and impose an additional restriction — principal unimodularity.

6.1. *Khovanov homology of graph-links with coefficients in $\mathbb{Z}_2$*

Let G be a simple labeled graph with n vertices, $\mathcal{V} = \mathcal{V}(G)$ be the set of its vertices and $A = A(G)$ be the adjacency matrix of G.

Let $s \subset \mathcal{V} = \mathcal{V}(G)$ be a state of G. Denote $A(s) = A(G(s))$. Consider the vector space

$$V(s) = \mathbb{Z}_2\langle x_1, \dots, x_n \,|\, r_1^s, \dots, r_n^s\rangle$$

with generators $x_1, \dots, x_n$ and relations $r_1^s, \dots, r_n^s$, where the relations are given by the formula

$$r_i^s = \begin{cases} x_i + \sum_{\{j\,|\,v_j \in s\}} a_{ij}x_j & \text{if } v_i \notin s, \\ \sum_{\{j\,|\,v_j \in s\}} a_{ij}x_j & \text{if } v_i \in s. \end{cases}$$

The dimension of $V(s)$ is equal to $\operatorname{corank}_{\mathbb{Z}_2} A(s)$.

There is a natural bijection between states $s \subset \mathcal{V}$ and vertices of the hypercube $\{0,1\}^n$: the state s corresponds to the vector $(\alpha_1, \dots, \alpha_n)$, where $\alpha_i = 0$ if $v_i \in s$ and $\operatorname{sign}(v_i) = 1$ or if $v_i \notin s$ and $\operatorname{sign}(v_i) = -1$. Otherwise we set $\alpha_i = 1$.

Every edge of the hypercube connects states s and $s \oplus i$, where $s \oplus i$ denotes $s \cup \{v_i\}$ if $v_i \notin s$ and $s \setminus \{v_i\}$ if $v_i \in s$. We shall denote such edge as $s \to s \oplus i$. We orient the arrow so that $v_i \notin s$ if $\operatorname{sign}(v_i) = -1$ and $v_i \in s$ if $\operatorname{sign}(v_i) = 1$.

We assign to each edge $s \to s \oplus i$ the map

$$\partial^s_{s\oplus i} \colon \bigwedge^* V(s) \to \bigwedge^* V(s \oplus i)$$

of exterior algebras defined by the formula

$$\partial^s_{s\oplus i}(u) = \begin{cases} x_i \wedge u & \text{if } x_i = 0 \in V(s), \\ u & \text{if } x_i \neq 0 \in V(s). \end{cases}$$

Consider the graded vector space

$$C(G) = \bigoplus_{s \subset \mathcal{V}} \bigwedge^{*} V(s)$$

and the map ∂ on it

$$\partial(u) = \sum_{\{s,s' \subset \mathcal{V} \,|\, s \to s'\}} \partial^{s}_{s'}(u).$$

It turns out that the map ∂ is well defined and turns the space $C(G)$ into a chain complex.

Definition 6.1. We call homology $\mathrm{Kh}(G)$ of the complex $(C(G), \partial)$ the *reduced (odd) Khovanov homology* of the labeled simple graph G.

Theorem 6.2. (see [28]) *Khovanov homology* $\mathrm{Kh}(G)$ *is an invariant of graph-links.*

Example 6.3. Let us consider the Bouchet graph W_5 (Fig. 7 on the left) whose vertices are labeled with $(0, +)$. Khovanov homology of this labeled graph W_5 was calculated by J. Bloom: $\mathrm{Kh}_1(W_5) = \mathrm{Kh}_2(W_5) = \mathrm{Kh}_4(W_5) = \mathrm{Kh}_5(W_5) = \mathbb{Z}_2$, $\mathrm{Kh}_3(W_5) = \mathbb{Z}_2 \oplus \mathbb{Z}_2$, the rest homology groups vanish. This result implies that the corresponding graph-link is not trivial. Earlier (see Corollary 5.23) we used parity to prove a stronger statement that the graph-link is non-realizable.

6.2. *PU-graph-links and integral odd Khovanov homology*

The definition of integer-valued odd Khovanov homology for graph-links faces the difficulty that the signs of the integer intersection matrix of an oriented chord diagram does not survive in general after mutations. But we can keep track of the signs if one is able to distinguish outer and inner chords. This reason confines us to bipartite graph-links. There is another restriction, which is necessary to retain the orientation after Reidemeister moves, — *principal unimodularity* [4]. So the definition of integer odd Khovanov homology is given for bipartite principally unimodular graph-links.

6.2.1. *Principally unimodular bipartite graph-links*

Let G be an oriented bipartite graph without loops and multiple edges and $\mathcal{V} = \mathcal{V}(G)$ be the set of its vertices. We assume G is labeled with signs, i.e. there is a map sign: $\mathcal{V} \to \{-1, 1\}$.

Fix an enumeration of vertices for G. We define the *oriented adjacency matrix* $A(G) = (a_{ij})_{i,j=1,\dots,n}$ over $\mathbb{Z}$ as follows: $a_{ij} = 1$ and $a_{ji} = -1$ if and only if v_i is the beginning and v_j is the end of an edge in the graph G (we shall denote this situation as $v_i \to v_j$), and $a_{ij} = 0$ if and only if v_i and v_j are not adjacent. Besides we set $a_{ii} = 0$.

For any state $s \subset \mathcal{V}$ we denote by $G(s)$ the induced subgraph in G with the set of vertices s and put $A(s) = A(G(s))$. Since G is bipartite, the set of vertices splits into a disjoint sum $\mathcal{V} = \mathcal{V}_0 \sqcup \mathcal{V}_1$.

Below we define an oriented version of the Reidemeister moves on oriented bipartite graphs.

R. For a given vertex $v \in \mathcal{V}(G)$ we change the direction of all the edges incident to v.

Ω_1. The first Reidemeister move is an addition/removal of an isolated vertex labeled $+$ or $-$.

Ω_2. The second Reidemeister move is an addition/removal of two non-adjacent vertices u and v having the different signs and the same neighborhoods so that the new graph remains bipartite. We require the orientations of the new edges to be compatible: for any vertex $w \in \mathcal{V}(G)$ we have $u \to w$ (respectively, $u \leftarrow w$) whenever $v \to w$ (respectively, $v \leftarrow w$).

Ω_3. The third Reidemeister move is defined as follows. Let u, v, w be three vertices of G with the sign $-$ and u be adjacent only to v and w so that $u \to v$ and $u \to w$. Then we disconnect u from v and w. We set $u \to t$ (respectively, $u \leftarrow t$) for all t such that $v \to t$ (respectively, $v \leftarrow t$) and set $u \to t$ (respectively, $u \leftarrow t$) if $w \leftarrow t$ (respectively, $w \to t$). In addition, we change the signs of v and w to $+$. The inverse operation is also called the third Reidemeister move.

Ω_4. The fourth Reidemeister move is defined as follows. We take two adjacent vertices u labeled α and v labeled β. Then we change the label of u to $-\beta$ and the label of v to $-\alpha$ and change also the orientation of the edge uv. After that we change the adjacency for each pair (t, w) of vertices where $t \in N(u)$ and $w \in N(v)$. We set the orientation of a new edge tw so that the square $utwv$ is *even*, i.e. the number of codirectional edges in the round $utwv$ is even (see examples below).

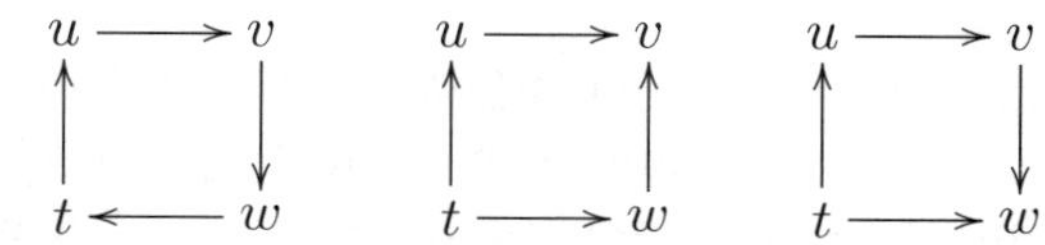

even square *even square* *odd square*

The move Ω_4 defined above allows one to change orientations of edges of the graph in an arbitrary way by adding and removing additional vertices with the move Ω_2.

Proposition 6.4. (see [29]) *Let G be an oriented bipartite labeled graph and $\widetilde{G}$ differ from G by orientation of edges. Then we can obtain $\widetilde{G}$ by applying the moves Ω_2 and Ω_4 to the graph G.*

This proposition shows that the theory of oriented bipartite graphs with moves $R, \Omega_1, \ldots, \Omega_4$ is in fact the theory of *undirected* labeled bipartite graphs with the usual Reidemeister moves of graph-links which preserve the bipartite structure of the graphs. So we have to impose some additional constraints to make orientations of graphs significant.

Definition 6.5. Let G be an oriented bipartite labeled graph. We call the orientation of G *principally unimodular* if for each state $s \subset \mathcal{V}$ we have $\det A(s)$ is equal to either 0 or 1. We call the graph G *PU-oriented.*

Any bipartite graph, which is realizable as the intersection graph of a chord diagram, is PU-oriented [3]. The inverse statement is not true, the graph Θ below has a PU-orientation but is not realizable:

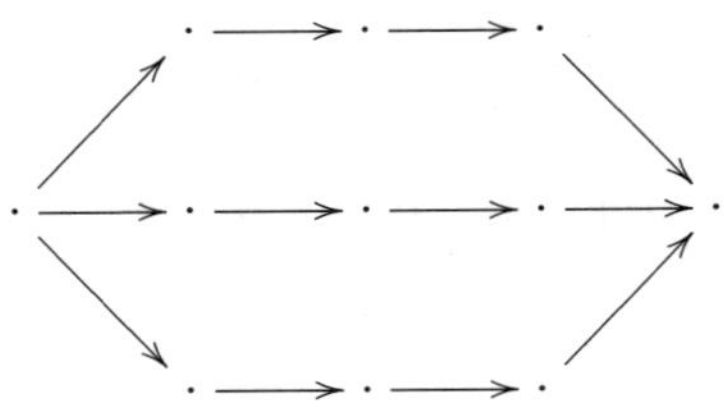

Indeed, any proper principal minor of the adjacency matrix is equal to either 0 or 1 since any proper subgraph in Θ is realizable. The determinant of the adjacency matrix is zero because Θ has no perfect matching. On the other hand, consider any of the three cycles in Θ. This subgraph is the intersection graph of some chord diagram which is uniquely defined. It is easy to see that one cannot expand this chord diagram to a diagram with the intersection graph Θ.

There are bipartite graphs which are not PU-orientable. For example, the Bouchet graph BW_3 (Fig. 7) has no principally unimodular orientation. Tutte showed in [38] that the graph BW_3 was the unique obstruction to PU-orientability; in other words, any bipartite graph is not PU-orientable if and only if it can be transformed by pivoting operations to a graph that contains BW_3.

The question whether there exists a bipartite PU-oriented graph that cannot be transformed by Reidemeister moves into a realizable graph is still open.

The following statement shows that principal unimodularity is compatible with almost all Reidemeister moves.

Proposition 6.6. (see [29]) *Let G be PU-oriented. Then*

(1) *any complete subgraph of G is PU-oriented*;
(2) *if G' is obtained from G by applying the moves $R, \Omega_1, \Omega_3, \Omega_4$, then G' is PU-oriented.*

The set of PU-oriented graphs is not stable under the second Reidemeister move. So we define the *principally unimodular second Reidemeister move* Ω_2^{PU} by requiring the result to be a PU-oriented bipartite graph.

Definition 6.7. A *PU-oriented graph-link* is an equivalence class of PU-oriented bipartite labeled graphs modulo moves $R, \Omega_1, \Omega_2^{PU}, \Omega_3, \Omega_4$.

Let us consider some properties of PU-oriented graphs.

The following proposition justifies the introduction of principal unimodularity in the definition of oriented graph-links since it secures against appearance of odd squares after Reidemeister moves. As we have seen above (Proposition 6.4) odd squares eliminate significance of graph orientation.

Proposition 6.8. (see [29]) *Let G be an oriented bipartite labeled graph. Then G is PU-oriented if and only if any graph G' obtained from G by a sequence of moves Ω_4 does not contain odd squares (4-cycles).*

Proposition 6.9. (see [29]) *Any two PU-orientations of a bipartite graph G coincide up to reversions R.*

Thus the theory of PU-oriented graph-links is in fact a theory of *PU-orientable* graph-links. We shall call graphs admitting PU-orientation *principally unimodular graphs* (or *PU-graphs*) and call the corresponding graph-links *PU-graphs-links.*

6.2.2. *Odd Khovanov homology of PU-graph-links*

Let G be a PU-oriented bipartite labeled graph with n vertices and $A = A(G)$ be its adjacency matrix.

Suppose $s \subset \mathcal{V} = \mathcal{V}(G)$. Consider the vector space

$$V(s) = \mathbb{Z}\langle x_1, \dots, x_n \,|\, r_1^s, \dots, r_n^s \rangle,$$

where the relations $r_1^s, \dots, r_n^s$ are given by the formula

$$r_i^s = \begin{cases} x_i - \sum_{\{j \,|\, v_j \in s\}} \operatorname{sign}(v_j) a_{ij} x_j & \text{if } v_i \notin s, \\ -\sum_{\{j \,|\, v_j \in s\}} \operatorname{sign}(v_j) a_{ij} x_j & \text{if } v_i \in s. \end{cases}$$

The rank of $V(s)$ is equal to corank $A(s)$.

There is a natural bijection between states $s \subset \mathcal{V}$ and vertices of the hypercube $\{0,1\}^n$. Every edge of the hypercube is of the type $s \to s \oplus i$, where $s \oplus i$ denotes $s \cup \{v_i\}$ if $v_i \notin s$ and $s \backslash \{v_i\}$ if $v_i \in s$. We orient the arrow so that $v_i \notin s$ if $\mathrm{sign}(v_i) = -1$ and $v_i \in s$ if $\mathrm{sign}(v_i) = 1$.

We assign to every edge $s \to s \oplus i$ the map

$$\partial^s_{s\oplus i} \colon \bigwedge^* V(s) \to \bigwedge^* V(s \oplus i)$$

of exterior algebras defined by the formula

$$\partial^s_{s\oplus i}(u) = \begin{cases} x_i \wedge u & \text{if } x_i = 0 \in V(s), \\ u & \text{if } x_i \neq 0 \in V(s). \end{cases}$$

The element u is considered as a polynomial of generators x_i and can be regarded as an element of the external algebra $\bigwedge^* V(s)$ as well as the algebra $\bigwedge^* V(s \oplus i)$.

The construction of the differential like a (signed) sum of partial differentials needs the partial differentials to commute or anticommute. In other words, two-dimensional faces of the state cube must be (anti)commutative.

Lemma 6.10. (see [29]) *Any 2-face of the hypercube of states is commutative or anticommutative.*

Thus, we can classify the 2-faces of the state cube as anticommutative faces (type A), commutative faces (type C) and zero faces (types Z). Further we split zero faces into two types: X and Y.

Definition 6.11. We call a vertex $v \in V$ *inner* if $v \in \mathcal{V}_0$ and $\mathrm{sign}(v) = -1$ or $v \in \mathcal{V}_1$ and $\mathrm{sign}(v) = 1$. Otherwise v is *outer*.

Remark 6.12. The terms "inner" and "outer" come from realizable situation. Consider any connected diagram of a classical link. Then the chord diagram that corresponds to some rotating circuit (see Fig. 15) is embedded in the plane. Hence, some chords lie inside the region bounded by the core circle of the chord diagram and some chord lie in the outer region. A partition of the chords into "inner" and "outer" ones defines a bipartite structure of the intersection graph of the chord diagram.

It turns out that among the two vertices, which determine a 2-face of type Z, there is one inner and one outer [29].

Let us consider a 2-face of type Z determined by vertices v_i and v_j:

$$\begin{array}{ccc} \bigwedge^* V(s\oplus j) & \xrightarrow{\partial^{s\oplus j}_{s\oplus i\oplus j}} & \bigwedge^* V(s\oplus i\oplus j) \\ {\scriptstyle \partial^s_{s\oplus j}}\uparrow & & \uparrow{\scriptstyle \partial^{s\oplus i}_{s\oplus i\oplus j}} \\ \bigwedge^* V(s) & \xrightarrow[\partial^s_{s\oplus i}]{} & \bigwedge^* V(s\oplus i). \end{array}$$

Let v_i be an inner vertex of the face. We assign the face to the *type* X if $x_i = \operatorname{sign}(v_j)x_j \in V(s\oplus i)$ and assign it to the *type* Y if $x_i = -\operatorname{sign}(v_j)x_j \in V(s\oplus i)$.

Note that definition of types X and Y relies essentially on bipartite structure of the graph. This provides an obstruction to extension of the construction of odd Khovanov homology to virtual links (even with orientable atoms).

Let us denote the set of the edges in the hypercube as $\mathcal{E}$. We call any map $\varepsilon\colon \mathcal{E} \to \{\pm 1\}$ an *edge assignment* (see [30]). A 2-face is called *even* (respectively, *odd*) if it contains even (respectively, odd) number of edges e with $\varepsilon(e) = -1$. A *type X edge assignment* is an edge assignment such that all faces of type A and X are even and all faces of type C and Y are odd. Similarly, *type Y edge assignment* is an edge assignment for which faces of type A and Y are even and faces of type C and X are odd.

Lemma 6.13. (see [29]) *Any PU-oriented bipartite labeled graph has an edge assignment of type X and one of type Y.*

Given a type X or type Y edge assignment ε we define the chain complex

$$C(G) = \bigoplus_{s\subset\mathcal{V}} \bigwedge^* V(s)$$

with a differential

$$\partial_\varepsilon(u) = \sum_{\{s,s'\subset\mathcal{V}\,|\,s\to s'=\varepsilon\in\mathcal{E}\}} \varepsilon(e)\partial^s_{s'}(u).$$

Definition 6.14. Homology $\mathrm{Kh}'(G)$ of the complex $(C(G),\partial)$ is called *reduced odd Khovanov homology* of the labeled simple graph G.

Theorem 6.15. (see [29]) *Khovanov homology* $\mathrm{Kh}'(G)$ *is an invariant of bipartite PU-graph-links, i.e. it is invariant under* $R, \Omega_1, \Omega_2^{PU}$, Ω_3, Ω_4 *moves and does not depend on the choice of edge assignment.*

6.3. *Jones polynomial of graph-links and Khovanov homology of graph-links*

Let G be a simple labeled graph with n vertices. Khovanov complex $C(G)$ has two gradings: the homological grading M_0 and the algebraic grading deg of the graded algebra $\bigwedge^* V(s)$. The differential is not homogeneous with respect to deg but it is compatible with the grading Q_0, where for any element $u \in \bigwedge^r(s)$ we define

$$Q_0(u) = \dim_{\mathbb{Z}_2} V(s) - 2r + M_0(s).$$

The differential increases the grading M_0 by 1 and leaves the grading Q_0 unchanged. Then

$$\mathrm{Kh}(G) = \bigoplus_{m,q\in\mathbb{Z}} \mathrm{Kh}(G)_{(m,q)}.$$

The following theorem shows that the Euler characteristic of Khovanov homology coincides (up to a multiplier) with Kauffman bracket of the graph G.

Theorem 6.16. (see [28]) *The following equality holds*:

$$\sum_{m,q\in\mathbb{Z}} (-1)^m \dim_{\mathbb{Z}_2} \mathrm{Kh}(G)_{(m,q)} \cdot t^q = (-it^{1/2})^n \langle G\rangle(it^{-1/2}).$$

Proof. The left term of the equality coincides with the Euler characteristic of the chain complex $C(G)$. Each state $s \subset \mathcal{V}(G)$ has the homological grading $\beta(s)$ and the corresponding chain space $\bigwedge^* V(s)$ contributes $(-1)^{\beta(s)}(t+t^{-1})^{\dim V(s)}t^{\beta(s)}$ to the Euler characteristic. Since $\dim V(s) = \operatorname{corank} A(s)$ and $\beta(s) = -\frac{1}{2}(\alpha(s) - \beta(s) - n)$ the sum over all states yields the right term of the equality. □

The gradings M_0, Q_0 of the Khovanov complex are not preserved by the Reidemeister moves as shown in the following table

Table 1.

	M_0	Q_0
Ω_1^-	0	−1
Ω_1^+	1	2
Ω_2	1	1
Ω_3	0	0
Ω_4	0	0
Ω_4'	0	0

Here $\Omega_1^{\pm}$ denotes an addition of an isolated vertex, and Ω_2 denotes an addition of two vertices. The entries of the cells are the shifts of the gradings after the corresponding move.

Thus the groups $\mathrm{Kh}(G)_{(m,q)}$ are not invariants of graph-links. Nevertheless, we can normalize the gradings for the graph-knots or oriented graph-links.

The writhe number of a graph-knot (oriented graph-link) is invariant under the moves $\Omega_2 - \Omega_4$. The move $\Omega_1^{\pm}$ changes the writhe number by ∓ 1. Using this fact and Table 1, we construct two gradings which are invariant under the Reidemeister moves

$$M = M_0 + \frac{w(G) + n(G)}{2}, \quad Q = Q_0 + \frac{3w(G) - n(G)}{2},$$

where $n(G)$ is the number of vertices in G.

Let us denote the homogeneous part of $\mathrm{Kh}(G)$ with the gradings $M = m$ and $Q = q$ by $\mathrm{Kh}_{m,q}(G)$. Invariance of the gradings under Reidemeister moves implies the following statement.

Theorem 6.17. (see [28]) *The groups* $\mathrm{Kh}_{m,q}(G)$ *are invariants of graph-knots (oriented graph-links) which categorifies the Jones* *polynomial*:

$$\sum_{m,q\in\mathbb{Z}} (-1)^m \dim_{\mathbb{Z}_2} \mathrm{Kh}_{m,q}(G) \cdot t^q = X(G)(it^{-1/2}).$$

Analogous results hold for odd Khovanov homology of PU-oriented graph-links.

7. Unsolved Problems

We conclude this paper with the significant one question which seems to be very interesting for us: is there a graph-link having two representatives realizable by chord diagrams but they are equivalent in the graph-link only by means of non-realizable graphs?

Acknowledgments

D.P.I. and I.M.N. were partially supported by grants of the Russian Government 11.G34.31.0053, RF President NSh — 1410.2012.1, RFBR 12-01-31432, 13-01-00664-a, 13-01-00830-a, 14-01-91161 and 14-01-31288. V.O.M. was partially supported by grants of the Russian Government 11.G34.31.0053 and 14.Z50.31.0020, RF President NSh — 1410.2012.1, RFBR 12-01-31432, 13-01-00830-a 14-01-91161 and 14-01-31288.

References

[1] D. Bar-Natan, On Khovanov's categorification of the Jones polynomial, *Algebr. Geom. Topol.* **2** (2002) 337–370.
[2] J. Bloom, Odd khovanov homology is mutation invariant, *Math. Res. Lett.* **17** (2010) 1–10.
[3] A. Bouchet, Unimodularity and circle graphs, *Discrete Math.* **66** (1987) 203–208.
[4] A. Bouchet, W. H. Cunningham and J. F. Geelen, Principally unimodular skew-symmetric matrices, *Combinatorica* **18** (1998) 461–486.
[5] A. Bouchet, Circle graph obstructions, *J. Combin. Theory Ser. B* **60** (1994) 107–144.
[6] S. V. Chmutov, S. V. Duzhin and S. K. Lando, Vassiliev knot invariants. I, II, III, *Adv. Sov. Math.* **21** (1994) 117–147.

[7] S. Chmutov, S. Duzhin and J. Mostovoy, *Introduction to Vassiliev Knot Invariants* (Cambridge Univ. Press, 2012).

[8] M. Cohn and A. Lempel, Cycle decomposition by disjoint transpositions, *J. Combin. Theory Ser. A* **13** (1972) 83–89.

[9] D. P. Ilyutko, Framed 4-valent graphs: Euler tours, Gauss circuits and rotating circuits, *Sb. Math.* **202** (2011) 1303–1326 [*Mat. Sb.* **202** (2011) 53–76 (in Russian)].

[10] D. P. Ilyutko, An equivalence between the set of graph-knots and the set of homotopy classes of looped graphs, *J. Knot Theory Ramifications* **21** (2012). DOI:10.1142/S0218216512500010.

[11] D. P. Ilyutko and V. O. Manturov, Introduction to graph-link theory, *J. Knot Theory Ramifications* **18** (2009) 791–823.

[12] D. P. Ilyutko and V. O. Manturov, Graph-links, *Dokl. Math.* **80** (2009) 739–742 [*Dokl. Akad. Nauk* **428** (2009) 591–594 (in Russian)].

[13] D. P. Ilyutko and V. O. Manturov, Graph-links, in *Introductory Lectures on Knot Theory, Selected Lectures Presented at the Advanced School and Conference on Knot Theory and Its Applications to Physics and Biology*, pp. 135–161. Series of Knots and Everything, Vol. 46 (World Scientific, 2012).

[14] D. P. Ilyutko, V. O. Manturov and I. M. Nikonov, Parity in knot theory and graph-links, *J. Math. Sci.* **193** (2013) 809–965.

[15] L. H. Kauffman, Virtual knots, talks at *MSRI Meeting*, January 1997 and *AMS Meeting at University of Maryland*, College Park, March 1997.

[16] M. Khovanov, A categorification of the Jones polynomial, *Duke Math. J.* **101** (1997) 359–426.

[17] M. Khovanov, Link homology and Frobenius extensions, arXiv:math.GT/0411447.

[18] V. O. Manturov, Atoms and minimal diagrams of virtual links, *Dokl. Math.* **68** (2003) 37–39 [*Dokl. Akad. Nauk* **391** (2003) 166–168 (in Russian)].

[19] V. O. Manturov, *Teoriya Uzlov* (*Knot Theory*), Moscow–Izhevsk: RCD, 512 pp (in Russian).

[20] V. O. Manturov, On free knots, arXiv:math.GT/0901.2214.

[21] V. O. Manturov, On free knots and links, arXiv:math.GT/0902.0127.

[22] V. O. Manturov, Parity in knot theory, *Sb. Math.* **201** (2010) 693–733 [*Mat. Sb.* **201** (2010) 65–110 (in Russian)].

[23] V. O. Manturov, Free knots and parity, in *Introductory Lectures on Knot Theory, Selected Lectures Presented at the Advanced School and Conference on Knot Theory and Its Applications to Physics and Biology*, pp. 321–345. Series of Knots and Everything, Vol. 46 (World Scientific, 2012).

[24] V. O. Manturov, Parity in virtual and classical knot theory, *ibid.*

[25] V. O. Manturov and D. P. Ilyutko, *Virtual Knots: The State of the Art*, World Scientific (2013).

[26] G. Moran, Chords in a circle and linear algebra over GF(2), *J. Combin. Theory Ser. A* **37** (1984) 239–247.

[27] K. Murasugi, The Jones polynomial and classical conjectures in knot theory, *Topology* **26** (1987) 187–194.

[28] I. Nikonov, Khovanov homology of graph-links, arXiv:math.GT/1005.2812.
[29] I. Nikonov, Odd Khovanov homology of principally unimodular bipartite graph-links, arXiv:math.GT/1006.0161.
[30] P. Ozsvath, J. Rasmussen and Z. Szabo, Odd Khovanov homology, arXiv:math.QA/0710.4300.
[31] O.-P. Östlund, Invariants of knot diagrams and relations among Reidemeister moves, arXiv:math.GT/0005108.
[32] R. C. Read and P. Rosenstiehl, *On the Gauss Crossing Problem*, Colloq. Math. Soc. Janos Bolyai, (North-Holland, 1976), pp. 843–876.
[33] E. Soboleva, Vassiliev knot invariants coming from Lie algebras and 4-invariants, *J. Knot Theory Ramifications* **10** (2001) 161–169.
[34] M. Thistlethwaite, A spanning tree expansion for the Jones polynonial, *Topology* **26** (1987) 297–309.
[35] L. Traldi, Binary nullity, Euler circuits and interlace polynomials, arXiv:math.CO/0903.4405.
[36] L. Traldi, A bracket polynomial for graphs. II. Links, Euler circuits and marked graphs, *J. Knot Theory Ramifications* **19** (2010) 547–586.
[37] L. Traldi and L. Zulli, A bracket polynomial for graphs, *J. Knot Theory Ramifications* **18** (2009) 1681–1709.
[38] W. T. Tutte, A homotopy theorem for matroids I, II, *Trans. Amer. Math. Soc.* **88** (1958) 144–174.

A Survey of Heegaard Floer Homology

András Juhász
Mathematical Institute,
University of Oxford, Andrew Wiles Building,
Woodstock Road, Oxford, OX2 6GG, UK
juhasza@maths.ox.ac.uk

Since its inception in 2001, Heegaard Floer homology has developed into such a large area of low-dimensional topology that it has become impossible to overview all of its applications and ramifications in a single paper. For the state of affairs in 2004, see the excellent survey article of Ozsváth and Szabó [70]. A decade later, this work has two goals. The first is to provide a conceptual introduction to the theory for graduate students and interested researchers, the second is to survey the current state of the field, without aiming for completeness.

After reviewing the structure of Heegaard Floer homology, treating it as a black box, we list some of its most important applications. Many of these are purely topological results, not referring to Heegaard Floer homology itself. Then, we briefly outline the construction of Lagrangian intersection Floer homology, as defined by Fukaya, Oh, Ono, and Ohta [16]. Given a strongly $\mathfrak{s}$-admissible based Heegaard diagram $(\Sigma, \boldsymbol{\alpha}, \boldsymbol{\beta}, z)$ of the Spin^c 3-manifold $(Y, \mathfrak{s})$, we construct the Heegaard Floer chain complex $CF^\infty(\Sigma, \boldsymbol{\alpha}, \boldsymbol{\beta}, z, \mathfrak{s})$ as a special case of the above, and try to motivate the role of the various seemingly ad hoc features such as admissibility, the choice of basepoint, and Spin^c-structures. We also discuss the proof of invariance of the homology $HF^\infty(\Sigma, \boldsymbol{\alpha}, \boldsymbol{\beta}, \mathfrak{s})$ up to isomorphism under all the choices made, and how to define $HF^\infty(Y, \mathfrak{s})$ using this in a functorial way (naturality). Next, we explain why Heegaard Floer homology is computable, and how it lends itself to the various combinatorial descriptions that most students encounter first during their studies. The last chapter gives an overview of the definition and applications of sutured Floer homology, which includes sketches of some of the key proofs.

Throughout, we have tried to collect some of the important open conjectures in the area. For example, a positive answer to two of these would give a new proof of the Poincaré conjecture.

1. Background

Four-manifold topology was revolutionized in 1982 by the work of Donaldson, who pioneered techniques coming from theoretical physics, namely gauge theory, to study smooth 4-manifolds. In order to obtain an interesting invariant of the smooth structure, the idea is to fix some additional geometric structure on the manifold such as a Riemannian metric, write as a nonlinear PDE, and then study the topology of the moduli space of solutions. If one is lucky enough, this does not depend on the additional choices made, only on the smooth structure. The Donaldson polynomial invariants [11] arise from the cohomology of a certain compactification of the moduli space of SU(2) Yang–Mills instantons over a Riemannian 4-manifold. They are independent of the choice of metric, but do depend on the smooth structure. Unfortunately, the lack of compactness often makes it difficult to work with.

Based on arguments coming from string theory, Seiberg and Witten wrote down a different set of equations whose solution spaces are usually compact, and hence easier to work with. They also wrote down a conjectural relationship between their theory and the Donaldson polynomial invariants.

Heegaard Floer homology was defined by Ozsváth and Szabó [71, 74]. It grew out of an attempt to better understand the Seiberg–Witten invariant of closed 4-manifolds, so that it becomes more computable. It consists of a package of invariants of closed oriented 3-manifolds, maps induced on these by cobordisms, and a 4-manifold invariant obtained via mixing the various flavors. Simultaneously, an analogous theory was developed by Kronheimer and Mrowka, called monopole Floer homology, based directly on the Seiberg–Witten monopole equations. The two theories have been shown to be equivalent. The proof passes through a third type of invariant of 3-manifolds called embedded contact homology (ECH), due to Hutchings. This is defined in terms of a contact structure on the 3-manifold, but turns out to be independent of this choice. The Heegaard Floer, monopole, and ECH invariants of 3-manifolds are equivalent, but they are adapted to different aspects of 3-manifold and contact topology.

The motivation for the definition of *HF* was provided by Atiyah's [4] topological quantum field theory (TQFT) picture and the Atiyah–Floer conjecture [3]. A $(3+1)$-dimensional TQFT over $\mathbb{Z}$ assigns to a closed oriented 3-manifold Y a finitely generated Abelian group $Z(Y)$, and to an oriented smooth cobordism W from Y to Y' a homomorphism $F_W \colon Z(Y) \to Z(Y')$. This assignment satisfies certain axioms. For example, it is functorial from the cobordism category of oriented 3-manifolds to the category of finitely generated Abelian groups, and $Z(\varnothing) = \mathbb{Z}$. Given a smooth oriented 4-manifold X, one can view it as a cobordism from $\varnothing$ to $\varnothing$. Then $F_X \colon \mathbb{Z} \to \mathbb{Z}$ is multiplication by some integer $n(X)$, which is an invariant of the smooth 4-manifold X.

The Seiberg–Witten invariant of a smooth Spin^c 4-manifold $(X, \mathfrak{s})$ with $b_2^+(X) \geq 2$ is an integer $SW(X, \mathfrak{s})$ obtained by choosing a Riemannian metric g on X, and considering the moduli space of solutions to the so-called monopole equations, up to gauge equivalence. This moduli space is a compact oriented manifold of dimension

$$d(\mathfrak{s}) = \frac{c_1(\mathfrak{s})^2 - 2\chi(X) - 3\sigma(X)}{4}$$

for a generic g. The moduli spaces corresponding to different generic metrics are cobordant. Hence, when $d(\mathfrak{s}) = 0$, the signed count of elements in this 0-dimensional moduli space is independent of the choice of g, giving rise to the invariant $SW(X, \mathfrak{s})$.

Unfortunately, the Seiberg–Witten invariant does not quite fit in the above TQFT picture, but similar constructions are available due to Kronheimer and Mrowka (monopole Floer homology, *HM*) and to Ozsváth and Szabó (*HF*). For several years, it was conjectured that these two theories are isomorphic, and this has recently been settled by the work of several people. Surprisingly, the equivalence between HM and HF passes through embedded contact homology (ECH), defined by Hutchings [31] and Hutchings–Taubes [32, 33]. To define ECH, one starts out with a contact 3-manifold (Y, α) (i.e. α is a one-form such that $\alpha \wedge d\alpha > 0$), and the chain complex is generated by certain periodic Reeb orbits lying in a given homology class $\Gamma \in H_1(Y)$. Taubes [92] proved that

$$\widecheck{HM}(Y, \mathfrak{s}_\alpha + PD[\Gamma]) \cong ECH(-Y, \alpha, \Gamma),$$

where $\mathfrak{s}_\alpha$ is the Spin^c-structure given by $\ker(\alpha)$. This establishes that $ECH(Y, \alpha)$ only depends on the 3-manifold Y. As of now, there is no intrinsic proof of this fact, and even showing that $ECH(Y, \alpha)$ only depends on the contact 2-plane field $\ker(\alpha)$ is a formidable task. Recently, Kutluhan–Lee–Taubes [45–48] proved that

$$\widecheck{HM}(Y, \mathfrak{s}) \cong HF^+(Y, \mathfrak{s}),$$

passing through a version of embedded contact homology for stable Hamiltonian structures generalizing contact structures. Meanwhile, Colin–Ghiggini–Honda [7–9] showed that

$$ECH(-Y, \alpha, \Gamma) \cong HF^+(Y, \mathfrak{s}_\alpha + PD(\Gamma)),$$

which, together with the isomorphism constructed by Taubes, also gives

$$\widecheck{HM}(Y, \mathfrak{s}) \cong HF^+(Y, \mathfrak{s}).$$

2. Overview of the Structure of *HF*

Given a closed, connected, oriented 3-manifold Y and a Spin^c-structure $\mathfrak{s} \in \text{Spin}^c(Y)$, Heegaard Floer homology assigns to the pair $(Y, \mathfrak{s})$ a finitely generated Abelian group $\widehat{HF}(Y, \mathfrak{s})$, and $\mathbb{Z}[U]$-modules $HF^\infty(Y, \mathfrak{s})$, $HF^+(Y, \mathfrak{s})$, and $HF^-(Y, \mathfrak{s})$, where $\mathbb{Z}[U]$ is the polynomial ring in the formal variable U. In the future, when we write HF°, we mean one of these four flavors of Heegaard Floer homology. Furthermore, let

$$HF^\circ(Y) = \bigoplus_{\mathfrak{s}\in\mathrm{Spin}^c(Y)} HF^\circ(Y,\mathfrak{s}).$$

Each of these groups carries a relative $\mathbb{Z}_{d(c_1(\mathfrak{s}))}$-grading, where $d(c_1(\mathfrak{s}))$ is the divisibility of the cohomology class $c_1(\mathfrak{s}) \in H^2(Y)$. (If A is a finitely generated Abelian group, the divisibility $d(a)$ of $a \in A$ is 0 if a is torsion, and otherwise the image of a in A/Tors is $d(a)$ times a primitive element.) When $c_1(\mathfrak{s})$ is torsion, then $d(c_1(\mathfrak{s})) = 0$, and the relative $\mathbb{Z}$-grading can be lifted to an absolute $\mathbb{Q}$-grading. Furthermore, each group $HF^\circ(Y)$ carries an absolute $\mathbb{Z}_2$-grading. If $b_1(Y) > 0$ and $\mathfrak{s}$ is a non-torsion Spin^c-structure, and one takes the Euler characteristic of $HF^+(Y,\mathfrak{s})$ with respect to this $\mathbb{Z}_2$-grading, then one recovers the Turaev torsion of Y in the Spin^c-structure $\mathfrak{s}$, cf. Turaev [94] (when $b_1(Y) = 1$, the torsion is calculated in the "chamber" containing $c_1(\mathfrak{s})$). On the other hand,

$$\chi(\widehat{HF}(Y,\mathfrak{s})) = \begin{cases} 1 & \text{if } b_1(Y) = 0 \\ 0 & \text{if } b_1(Y) > 0 \end{cases}$$

for every $\mathfrak{s} \in \mathrm{Spin}^c(Y)$.

The three flavors of Heegaard Floer homology are related by the exact sequence

$$\cdots \longrightarrow HF^-(Y,\mathfrak{s}) \xrightarrow{\iota} HF^\infty(Y,\mathfrak{s}) \xrightarrow{\pi} HF^+(Y,\mathfrak{s}) \xrightarrow{\delta} \cdots. \tag{1}$$

This gives rise to the invariant

$$HF^+_{\mathrm{red}}(Y,\mathfrak{s}) = \mathrm{coker}(\pi) \cong \ker(\iota) = HF^-_{\mathrm{red}}(Y,\mathfrak{s}),$$

where the isomorphism is given by the coboundary map δ.

Furthermore, we have an exact sequence

$$\cdots \longrightarrow \widehat{HF}(Y,\mathfrak{s}) \xrightarrow{i} HF^+(Y,\mathfrak{s}) \xrightarrow{U^+} HF^+(Y,\mathfrak{s}) \xrightarrow{p} \cdots, \tag{2}$$

and similarly,

$$\cdots \longrightarrow HF^-(Y,\mathfrak{s}) \xrightarrow{U^-} HF^-(Y,\mathfrak{s}) \longrightarrow \widehat{HF}(Y,\mathfrak{s}) \longrightarrow \cdots.$$

There is one more piece of algebraic structure on $HF^\circ(Y,\mathfrak{s})$, which is an action of the group $\Lambda^*(H_1(Y)/\mathrm{Tors})$.

When Y is a rational homology sphere, $HF^+(Y,\mathfrak{s})$ is absolutely $\mathbb{Q}$-graded for every $\mathfrak{s} \in \text{Spin}^c(Y)$. The *correction term* $d(Y,\mathfrak{s})$, introduced by Ozsváth and Szabó [67], is the minimal grading of any non-torsion element in the image of $HF^\infty(Y,\mathfrak{s})$ in $HF^+(Y,\mathfrak{s})$. This probably coincides with the gauge-theoretic invariant of Frøyshov [15].

Ozsváth and Szabó [74] showed that HF° is well-defined up to isomorphism, and checked some naturality properties in [79]. The assignment $Y \mapsto HF^\circ(Y)$ was made completely functorial by Thurston and the author [39], where we also showed that the mapping class group of Y acts on $HF^\circ(Y)$. Naturality is necessary to be able to talk about maps between HF groups, and to be able to talk about concrete elements. It turns out that the $+$, $-$ and ∞ versions are indeed natural (in analogy with the corresponding flavors of monopole Floer homology, where there is no basepoint dependence). However, $\widehat{HF}$ is only functorial on the category of based 3-manifolds and basepoint preserving diffeomorphisms (this is work in progress, joint with Ozsváth and Thurston). Indeed, let γ be a loop in Y passing through the basepoint p, and consider the automorphism d of (Y,p) which is a finger move along γ. For $x \in \widehat{HF}(Y,p)$, we have

$$d_*(x) = x + p \circ i([\gamma] \cdot x),$$

where p and i are the maps in the exact sequence (2), while $[\gamma]$ is the class of the curve γ in $H_1(Y)/\text{Tors}$. This map is nontrivial for example when

$$Y = \Sigma(2,3,7)\#(S^1 \times S^2)$$

and $\gamma = S^1 \times \{\text{pt}\}$, the basepoint being an arbitrary element of γ.

Heegaard Floer homology enjoys various symmetry properties. There is an involution on the set $\text{Spin}^c(Y)$, denoted by $\mathfrak{s} \mapsto \bar{\mathfrak{s}}$. If $\mathfrak{s}$ is represented by a vector field v, then $\bar{\mathfrak{s}}$ is represented by $-v$. Ozsváth and Szabó [71] showed that

$$HF^\circ(Y,\mathfrak{s}) \cong HF^\circ(Y,\bar{\mathfrak{s}})$$

as $\mathbb{Z}[U] \otimes_{\mathbb{Z}} \Lambda^*(H_1(Y)/\text{Tors})$-modules. The Heegaard Floer chain complexes gives rise to both homology and cohomology theories. We denote by $\widehat{HF}_*$, HF^+_*, and HF^-_* the homologies, and $\widehat{HF}^*$, HF^*_+, and HF^*_- the corresponding cohomologies, respectively. If $-Y$ denotes Y with its orientation reversed, then

$$\widehat{HF}^*(Y,\mathfrak{s}) \cong \widehat{HF}_*(-Y,\mathfrak{s}) \quad \text{and} \quad HF^*_\pm(Y,\mathfrak{s}) \cong HF^\mp_*(-Y,\mathfrak{s}).$$

As in a TQFT, cobordisms of 3-manifolds induce homomorphisms. More precisely, if $(W,\mathfrak{s})$ is a Spin^c-cobordism from $(Y_0,\mathfrak{s}_0)$ to $(Y_1,\mathfrak{s}_1)$, then Ozsváth and Szabó [79] associate to it a map

$$F^\circ_{W,\mathfrak{s}}\colon HF^\circ(Y_0,s_0) \to HF^\circ(Y_1,\mathfrak{s}_1).$$

When $\mathfrak{s}_1$ and $\mathfrak{s}_2$ are both torsion, this homomorphism shifts the absolute $\mathbb{Q}$-grading by the number

$$d(\mathfrak{s}) = \frac{c_1(\mathfrak{s})^2 - 2\chi(W) - 3\sigma(W)}{4}$$

(note the "coincidence" with the dimension of the Seiberg–Witten moduli space). More generally, there is also a map

$$F^\circ_{W,\mathfrak{s}}\colon HF^\circ(Y_0,\mathfrak{s}_0) \otimes \Lambda^*(H_1(W)/\mathrm{Tors}) \to HF^\circ(Y_1,\mathfrak{s}_1).$$

When $b_2^+(W) > 0$, then $F^\infty_{W,\mathfrak{s}} = 0$.

Let $W\colon Y_0 \to Y_1$ be a cobordism with $b_2^+(W) \geq 2$. An *admissible cut* of the cobordisms W is a 3-manifold $N \subset W$ such that

- N divides W into two cobordisms $W_1\colon Y_0 \to N$ and $W_2\colon N \to Y_1$,
- $b_2^+(W_1) \geq 1$ and $b_2^+(W_2) \geq 1$,
- $\delta H^1(N) = 0$ in $H^2(W,\partial W)$.

An admissible cut always exists. Since $F^\infty_{W_1,\mathfrak{s}} = 0$, the long exact sequence (1) implies that the image of the map

$$F^-_{W_1,\mathfrak{s}|_{W_1}}\colon HF^-(Y_0,\mathfrak{s}|_{Y_0}) \to HF^-(N,\mathfrak{s}|_N)$$

lies in $HF^-_{\mathrm{red}}(N,\mathfrak{s}|_N)$. Similarly, the map

$$F^+_{W,\mathfrak{s}|_{W_2}}\colon HF^+(N,\mathfrak{s}|_N) \to HF^+(Y_1,\mathfrak{s}|_{Y_1})$$

factors through the projection of $HF^+(N, \mathfrak{s}|_N)$ to $HF^+_{\text{red}}(N, \mathfrak{s}|_N)$. We define the *mixed invariant*

$$F^{\text{mix}}_{W,\mathfrak{s}}\colon HF^-(Y_0, \mathfrak{s}|_{Y_0}) \otimes_{\mathbb{Z}} \Lambda^*(H_1(W)/\text{Tors}) \to HF^+(Y_1, \mathfrak{s}|_{Y_1})$$

using the formula

$$F^{\text{mix}}_{W,\mathfrak{s}} = F^+_{W_1,\mathfrak{s}|_{W_1}} \circ \tau^{-1} \circ F^-_{W_0,\mathfrak{s}|_{W_0}},$$

where $\tau\colon HF^-_{\text{red}}(N, \mathfrak{s}|_N) \to HF^+_{\text{red}}(N, \mathfrak{s}|_N)$ is the isomorphism induced by the coboundary map δ in the long exact sequence (1). As the notation suggests, the map $F^{\text{mix}}_{W,\mathfrak{s}}$ is independent of the choice of admissible cut N.

Given a closed oriented smooth 4-manifold X with $b_2^+(W) \geq 2$ and a Spin^c-structure $\mathfrak{s} \in \text{Spin}^c(X)$, we define the *absolute invariant* $\Phi_{X,\mathfrak{s}}$ as follows. First, let $W = X \backslash 2B^4$, this can be viewed as a cobordism from S^3 to S^3. We will also write $\mathfrak{s}$ for the restriction of $\mathfrak{s}$ to W. Note that there is a unique Spin^c-structure $\mathfrak{s}_0$ on S^3, and that $HF^-(S^3, \mathfrak{s}_0) \cong \mathbb{Z}[U]$, while

$$HF^+(S^3, \mathfrak{s}_0) \cong \mathbb{Z}[U, U^{-1}]/U\mathbb{Z}[U].$$

We write Θ_- for a generator of the 0-degree part of $HF^-(S^3, \mathfrak{s}_0)$ and Θ_+ for a generator of the 0-degree part of $HF^+(S^3, \mathfrak{s}_0)$, these are both well-defined up to sign. Then the map

$$\Phi_{X,\mathfrak{s}}\colon \mathbb{Z}[U] \otimes \Lambda^*(H_1(X)/\text{Tors}) \to \mathbb{Z}/\pm 1$$

is defined by taking $\Phi_{X,\mathfrak{s}}(U^n \otimes \zeta)$ to be the coefficient of Θ_+ in $F^{\text{mix}}_{W,\mathfrak{s}}(U^n \cdot \Theta_- \otimes \zeta)$. Note that $\Phi_{X,\mathfrak{s}}$ vanishes on those homogeneous elements whose degree is different from $d(\mathfrak{s})$. The sign ambiguity comes from the choice of $\Theta_\pm$. Ozsváth and Szabó conjectured that one can recover the Seiberg–Witten invariants from this as follows.

Conjecture 2.1. *Take a basis $b_1, \ldots, b_k$ of $H_1(X)/Tors$, and let n be such that the degree of $U^n \otimes (b_1 \wedge \cdots \wedge b_k)$ is $d(\mathfrak{s})$. Then*

$$SW(X, \mathfrak{s}) = \Phi_{X,\mathfrak{s}}(U^n \otimes (b_1 \wedge \cdots \wedge b_k)).$$

Note that there is a more general version of SW that is obtained by integrating different elements of the cohomology of the configuration space over the Seiberg–Witten moduli space, that is conjectured to agree with $\Phi_{X,\mathfrak{s}}$.

The Heegaard Floer package also contains a knot invariant, called *knot Floer homology*. Given a null-homologous knot or link K in a closed, connected, oriented 3-manifold Y, Ozsváth and Szabó [73], and independently Rasmussen [83], assigned to it a finitely generated Abelian group $\widehat{HFK}(Y,K)$. This refines $\widehat{HF}(Y)$ in the sense that there is a filtration on the chain complex defining $\widehat{HF}(Y)$ such that the homology of the associated graded object is $\widehat{HFK}(Y,K)$. Consequently, there is a spectral sequence from $\widehat{HFK}(Y,K)$ converging to $\widehat{HF}(Y)$. When $Y = S^3$, the smallest filtration level for which the inclusion map on homology is nonzero into $\widehat{HF}(S^3) \cong \mathbb{Z}$ is denoted by $\tau(K)$, see [68].

In the case of $Y = S^3$, the group $\widehat{HFK}(K) = \widehat{HFK}(S^3,K)$ is bi-graded; i.e.

$$\widehat{HFK}(K) = \bigoplus_{i,j\in\mathbb{Z}} \widehat{HFK}_j(K,i).$$

Here i is called the Alexander grading and j is the homological grading. This is justified by the fact that

$$\sum_{i,j\in\mathbb{Z}} (-1)^j \cdot \mathrm{rk}(\widehat{HFK}_j(K,i))t^i$$

is the symmetrized Alexander polynomial $\Delta_K(t)$ of K. The proof relies on the fact that knot Floer homology satisfies an unoriented skein exact sequence. Another way of saying this is that knot Floer homology categorifies the Alexander polynomial, just like Khovanov homology is a categorification of the Jones polynomial. According to a conjecture of Rasmussen [12, 84], the two theories are related.

Conjecture 2.2. *There is a spectral sequence starting from the reduced Khovanov homology of K and converging to $\widehat{HFK}(K)$. In particular,*

$$\mathrm{rk}(\widetilde{Kh}(K)) \geq \mathrm{rk}(\widehat{HFK}(K)).$$

The author showed in [37] that knot cobordisms induce maps on knot Floer homology, making the categorification complete. However, to make knot Floer homology functorial, one needs to work with based knots, according to the work of Sarkar [88].

It is a classical result that the degree of $\Delta_K(t)$ provides a lower bound on the Seifert genus $g(K)$ (which is the minimal genus of an oriented surface in S^3 bounded by K). Ozsváth and Szabó showed in [72] that in fact knot Floer homology detects the Seifert genus in the sense that

$$g(K) = \max\{i \in \mathbb{Z} : \widehat{HFK}_*(K,i) \neq 0\}.$$

Furthermore, by work of Ghiggini–Ni [22, 60, 61] and the author [35, 38], the knot K is fibered if and only if

$$\widehat{HFK}_*(K, g(K)) \cong \mathbb{Z}.$$

A generalization of knot Floer homology, also due to Ozsváth and Szabó [80], is called *link Floer homology*. Given a link L in S^3, this invariant is denoted by $\widehat{HFL}(L)$, and is graded by $H_1(S^3 \backslash L)$. Its graded Euler characteristic gives rise to the multivariable Alexander polynomial of L, and it detects the Thurston norm of the link complement.

Since knot Floer homology detects the genus, it is sensitive to Conway mutation. E.g., it distinguishes the Conway and the Kinoshita–Teresaka knots, as the first one has genus 3, while the second one has genus 2. The δ-grading on $\widehat{HFK}(K)$ is defined as the difference of the Alexander and the homological gradings. Then we have the following conjecture, communicated to me by Zoltán Szabó, and supported by computational evidence.

Conjecture 2.3. *The rank of knot Floer homology is unchanged by Conway mutation in each δ-grading.*

So far, $\widehat{HFL}(L)$ has proved to be torsion-free in each example computed.

Conjecture 2.4. *The group $\widehat{HFL}(L)$ is torsion free for every link L.*

Knot and link Floer homology are invariant of the link complement. It is natural to ask whether they are particular cases of some more general invariant for 3-manifolds with boundary. There are two existing theories, namely, sutured Floer homology (SFH) due to the author [34], and bordered Floer homology due to Lipshitz, Ozsváth, and Thurston [51]. However, for both, one needs more structure on the boundary. SFH is defined for sutured manifolds, which were introduced by Gabai [17, 18]. A *sutured manifold* is a pair (M, γ), where M is a compact oriented 3-manifold with boundary, and $\gamma \subset \partial M$ can be thought of as a thickened oriented one-manifold that divides ∂M into subsurfaces $R_-(\gamma)$ and $R_+(\gamma)$. The components of γ are called the *sutures*. Then the sutured Floer homology $SFH(M, \gamma)$ is a finitely generated Abelian group.

If $p \in Y$ is a point, then the sutured manifold $Y(p)$ is obtained by removing a ball around p and putting a single suture on the boundary. For this, we have

$$SFH(Y(p)) \cong \widehat{HF}(Y).$$

Furthermore, if L is a link in Y, then $Y(L)$ denotes the sutured manifold where $M = Y \setminus N(L)$, and on each boundary torus, we have two oppositely oriented meridional sutures. Then

$$SFH(S^3(L)) \cong \widehat{HFL}(L).$$

So SFH is a common generalization of both $\widehat{HF}$ and $\widehat{HFL}$. Note that one can define $Y(p)$ and $Y(L)$ canonically using the real blow-up construction, cf. [39, Definitions 2.4 and 2.5].

The bordered Floer complex $\widehat{CFD}(M)$ is defined for a compact 3-manifold M with connected parametrized boundary. The parametrization amounts to fixing a handle decomposition of ∂M with a single 0-handle. Here, $\widehat{CFD}(M)$ is a differential graded $\mathcal{A}(\partial M)$-module, and $\mathcal{A}(\partial M)$ is a differential graded algebra that depends on the handle decomposition. If S is a surface in the 3-manifold Y that cuts it into pieces $-M_1$ and M_2, then we have the gluing formula

$$\widehat{HF}(Y) \cong \mathrm{Mor}_{\mathcal{A}(S)}(\widehat{CFD}(M_1), \widehat{CFD}(M_2)).$$

Bordered Floer homology is currently being developed at a rapid pace, and due to space constraints, we refer the interested reader to the survey article of Lipshitz, Ozsváth, and Thurston [53].

It is worth pointing out the relationship between SFH and the bordered theory. Given a 3-manifold M with parametrized boundary, for a set of sutures γ, there is an associated $\mathcal{A}(\partial M)$ module such that tensoring with it one obtains $SFH(M, \gamma)$, see [52]. On the other hand, bordered Floer homology can also be recovered from $SFH(M, \gamma)$ for all γ and certain cobordism maps between these.

Heegaard Floer homology and monopole Floer homology are equivalent. One definite advantage of the former is that it can be computed algorithmically. The breakthrough results in this direction are due to Sarkar and Wang [89], who gave an algorithm for computing $\widehat{HF}(Y)$ for an arbitrary closed, connected, oriented 3-manifold Y, and to Manolescu, Ozsváth, and Sarkar [57], who gave a combinatorial characterization of knot Floer homology $\widehat{HFK}(S^3, K)$, where the input data is a grid diagram for K. The latter led to an invariant of Legendrian and transverse knots in contact 3-manifolds, cf. Ozsváth, Szabó, and Thurston [81].

The hat version of HF is considerably simpler to compute than the other flavors. It took Manolescu, Ozsváth, and Thurston [56, 58] several years to bring the grid diagram approach to fruition and show that all flavors of HF, including the 4-manifold invariants, are algorithmically computable. The input data in these is a surgery presentation of the 3-manifold, where the link on which we do integral surgery is given by a grid diagram. The combinatorial theory of Heegaard Floer homology has grown into a large area that we do not intend to cover here, instead, we refer the reader to the survey article of Manolescu [55]. It is important to note that the above-mentioned algorithms are all far from being polynomial time and are unsuitable for even computing the knot Floer homology of slightly larger knots. Also, so far, these theories have shed very little light on the geometry of 3- and 4-manifolds. One notable result is due to Sarkar [87], which is the second completely combinatorial proof of the Milnor conjecture on the slice genera of torus knots (following Rasmussen's proof [85] via Khovanov homology).

Heegaard Floer homology has been really fruitful in the study of contact 3-manifolds, and Legendrian and transverse knots. Given a contact 3-manifold (Y, ξ), Ozsváth and Szabó [75] associate to it an element

$$c(\xi) \in \widehat{HF}(-Y)/\pm 1.$$

This captures a lot of geometric information about the contact structure, as we shall see in the following section.

3. Applications

The goal of this section is to showcase some of the many applications of Heegaard Floer homology, with an emphasis on results whose statements are purely topological and do not refer to *HF* itself. We also list results that show that *HF* contains very deep geometric information.

Let $g^*(K)$ denote the 4-ball genus of a knot $K \subset S^3$; i.e. the minimal genus of a smooth oriented surface bounded by K in D^4. Currently, there is no algorithm known for computing $g^*(K)$, or even to determine whether $g^*(K) = 0$ (such knots are called *slice*). For example, it is not known whether the 11-crossing Conway knot is slice (whereas its mutant, the Kinoshita–Teresaka knot, is slice). Using knot Floer homology, Ozsváth and Szabó [68] constructed an invariant $\tau(K) \in \mathbb{Z}$ such that

$$|\tau(K)| \leq g^*(K). \tag{3}$$

Moreover, τ descends to a group homomorphism from the concordance group of knots in S^3 to $\mathbb{Z}$. This allowed them to give another proof of the Milnor conjecture, originally proved by Kronheimer and Mrowka [40] using Donaldson invariants. This states that for the torus knot $T_{p,q}$, one has

$$g^*(T_{p,q}) = \frac{(p-1)(q-1)}{2}.$$

As alluded to in the previous section, Sarkar reproved inequality (3) using grid diagrams and combinatorics, giving a purely elementary proof of the Milnor conjecture. Note that Rasmussen introduced the s-invariant using Khovanov homology, which also gives a lower bound on the 4-ball genus, as $|s(K)| \leq 2g^*(K)$. Rasmussen conjectured that $s(K) = 2\tau(K)$, which was then disproved by Hedden and Ording [29]. Rasmussen [85] gave the first purely combinatorial proof of the Milnor conjecture using his s-invariant.

There are many examples of knots K with trivial Alexander polynomial — and due to Freedman, these are topologically slice — but for which $\tau(K) \neq 0$, and are hence not smoothly slice. Such knots can be used to construct exotic smooth structures on $\mathbb{R}^4$.

We already mentioned that knot Floer homology detects the Seifert genus and fiberedness of a knot. Ozsváth and Szabó [72] proved that $\widehat{HF}(Y)$ detects the Thurston norm of Y, while Ni [62] showed that it also detects fiberedness of Y. It is an interesting question how much geometric information is contained by the Heegaard Floer groups. For example, currently no relationship is known between $\pi_1(Y)$ and $\widehat{HF}(Y)$. We state an important conjecture that would make progress in this direction. But first, we need two definitions.

Note that for a rational homology 3-sphere Y, we always have

$$\operatorname{rk} \widehat{HF}(Y) \geq |H_1(Y)|.$$

Indeed, for every Spin^c-structure $\mathfrak{s} \in \mathrm{Spin}^c(Y)$, the Euler characteristic of $\widehat{HF}(Y, \mathfrak{s})$ with respect to the absolute $\mathbb{Z}_2$-grading is 1 for a rational homology 3-sphere. So $\operatorname{rk} \widehat{HF}(Y, \mathfrak{s}) \geq 1$ for every $\mathfrak{s} \in \mathrm{Spin}^c(Y)$. A rational homology 3-sphere Y is called an *L-space* if

$$\operatorname{rk} \widehat{HF}(Y) = |H_1(Y)|;$$

i.e. if its Heegaard Floer homology is as simple as possible. This is equivalent to saying that for every Spin^c-structure $\mathfrak{s} \in \mathrm{Spin}^c(Y)$, we have $\operatorname{rk} \widehat{HF}(Y, \mathfrak{s}) = 1$. The terminology originates from the fact that every lens-space is an L-space. However, there are many more: Ozsváth and Szabó [78] showed that the double cover of S^3 branched over any non-split alternating link is an L-space. Furthermore, they proved [77] that every 3-manifold with elliptic geometry is an L-space. To extend the notion of L-spaces from rational homology 3-spheres to arbitrary 3-manifolds, we need to look beyond the hat version of Heegaard Floer homology. We say that a 3-manifold Y is an L-space if $HF^+_{\mathrm{red}}(Y) = 0$.

A group G is called *left-orderable* if it is nontrivial, and it can be endowed with a linear order such that if $g < f$ for g, $f \in G$, then $hg < hf$ for every $h \in G$.

Conjecture 2.5. *Let Y be an irreducible rational homology 3-sphere. Then the following three statements are equivalent:*

(a) *Y is an L-space,*
(b) *$\pi_1(Y)$ is not left-orderable,*
(c) *Y carries no taut foliation.*

The conjecture that (a) and (b) are equivalent is due to Boyer, Gordon, and Watson [6]. Ozsváth and Szabó [72] proved that (a) implies (c). Now we state another conjecture, originally due to Ozsváth and Szabó, cf. Hedden and Ording [28].

Conjecture 2.6. *If Y is an irreducible homology sphere that is an L-space, then Y is homeomorphic to either S^3 or the Poincaré homology sphere.*

Observe that the implication (b) $\Rightarrow$ (a) in Conjecture 2.5, together with Conjecture 2.6, would imply the *Poincaré conjecture*. Indeed, if Y is a simply-connected 3-manifold, then it is an irreducible homology sphere. Since $\pi_1(Y) = 1$ is not left-orderable, Y is an L-space by Conjecture 2.5. Using Conjecture 2.6, we get that Y is homeomorphic to S^3, as the Poincaré homology sphere is not simply-connected.

Heegaard Floer homology has been particularly successful in tackling problems on Dehn surgery. The main tool is the following surgery exact triangle.

Theorem 2.7. *Let K be a knot in the closed oriented 3-manifold Y, together with framings f, $g \in H_1(\partial N(K))$ such that $m \cdot f = f \cdot g = g \cdot m = 1$, where m denotes the class of the meridian. Then there is an exact sequence*

$$\cdots \longrightarrow \widehat{HF}(Y) \longrightarrow \widehat{HF}(Y_f(K)) \longrightarrow \widehat{HF}(Y_g(K)) \longrightarrow \cdots .$$

In particular, when Y is a homology 3-sphere, then we can take $Y_f(K) = Y_0(K)$ and $Y_g(K) = Y_1(K)$.

The following theorem was originally proved by Kronheimer, Mrowka, Ozsváth, and Szabó [44] using monopole Floer homology, but now we know that it is isomorphic to HF.

Theorem 2.8. *If the knot $K \subset S^3$ is not the unknot, then*

$$S^3_{p/q}(K) \not\approx \mathbb{R}P^3.$$

More generally, they obtain the following.

Theorem 2.9. *Let K be a knot in S^3. If there is an orientation preserving homeomorphism between $S^3_{p/q}(K)$ and the lens space $S^3_{p/q}(U)$, then $K = U$.*

The lens space realization problem, asking which lens spaces can be obtained by Dehn-surgery along a nontrivial knot in S^3, has recently been settled via Heegaard Floer homology by Greene [25]. Note that the cyclic surgery theorem of Culler, Gordon, Luecke, and Shalen [10] ensures that if a surgery on a nontrivial knot yields a lens space, then the surgery coefficient has to be an integer. It is still an open question of Berge exactly which knots yield lens space surgeries. These are conjectured to be the doubly primitive knots that were classified by Berge. What Greene showed is that the lens spaces that can be realized by surgery on a nontrivial knot are exactly the ones obtained by surgery on a Berge knot. He also exhibits that such a knot has to have the same knot Floer homology as a Berge knot.

Another area where Heegaard Floer homology has been very successful is deciding whether a knot has unknotting number one. Given a knot K in S^3, we denote by $u(K)$ its *unknotting number*, which is the minimal number of times K intersects itself during a regular homotopy to the unknot. Currently, no algorithm is known for computing $u(K)$, or even to decide whether $u(K) = 1$. Of course, it is easy to give an upper bound on $u(K)$ by exhibiting a concrete unknotting sequence. A classical lower bound is provided for example by the knot signature. It is easy to prove that $g^*(K) \leq u(K)$, and hence $|\tau(K)|$ also provides a lower bound. However, $g^*(K)$ can be zero, while $u(K)$ is large. Note that recent work of Lackenby (in progress) gives an algorithm for telling whether a non-split link with at least two components and satisfying some mild restrictions has *unlinking number* one.

The main tool for detecting unknotting one knots is the *Montesinos trick*. Given a knot K in S^3, let $\Sigma(K)$ denote the double cover of S^3, branched along K. If $u(K) = 1$, then there exists a knot C in S^3 and an integer n such that

$$\Sigma(K) = S^3_{n/2}(C).$$

Now the surgery exact triangle can be used to give an obstruction to $\Sigma(K)$ being a half-integral Dehn surgery on a knot. Using this method, Ozsváth and Szabó [76] showed that, for example, $u(8_{10}) = 2$.

The links L and L' in S^3 are related by *Conway mutation* if there exists an embedded 2-sphere $S \subset S^2$ that intersects L in four points $L \cap S = L' \cap S$, and we can obtain L' from L by removing the ball bounded by S and regluing it using an involution that fixes two points on S. Viro showed that if L and L' are mutants, then $\Sigma(L) \approx \Sigma(L')$. Greene [24] showed that the converse also holds for alternating links. More precisely, he proved that the following are equivalent:

(a) L and L' are mutants,
(b) $\Sigma(L) \approx \Sigma(L')$,
(c) $\widehat{HF}(\Sigma(L)) \cong \widehat{HF}(\Sigma(L'))$ as absolutely graded, relatively Spin^c-graded groups.

The proof that (a) and (b) are equivalent passes through *HF* in an essential way.

The correction terms $d(Y, \mathfrak{s})$, introduced by Ozsváth and Szabó [67], have numerous important applications. For example, they provide alternate proofs of Donaldson's diagonalizability theorem and the Thom conjecture for $\mathbb{C}P^2$. Due to lack of space, we do not pursue this 4-manifold topological direction any further.

Finally, we mention some of the results in contact topology obtained using *HF*. First a few definitions. Given a contact 3-manifold (Y, ξ), we say that $\Delta \subset Y$ is an *overtwisted disk* if ξ is tangent to Δ along $\partial\Delta$. The contact structure ξ is called *overtwisted* if it contains an overtwisted disk, and is *tight* otherwise. By a result of Eliashberg, overtwisted contact structures satisfy an h-principle, and there is a unique one up to isotopy in each homotopy class of oriented 2-plane fields. In general, it is very difficult to classify tight contact structures on 3-manifolds.

We say that (Y, ξ) is *Stein fillable* if Y bounds a complex surface S such that the 2-plane field ξ consists of complex lines. Furthermore, S admits a Morse function $f\colon S \to I$ with $Y = \partial S = f^{-1}(1)$, and for every regular value t of f, the complex tangencies to $f^{-1}(t)$ form a contact structure. If (Y, ξ) is Stein fillable, then it is necessarily tight.

Given a contact 3-manifold (Y, ξ), Ozsváth and Szabó [75] assign to it an invariant $c(\xi) \in \widehat{HF}(-Y)$ using open book decompositions. They proved that if (Y, ξ) is overtwisted, then $c(\xi) = 0$. In other words, $c(\xi)$ can be used to detect tightness. However, the converse is not true, as shown by Ghiggini [20, 21]: there are tight contact structures with $c(\xi) = 0$. But at least we have the following nonvanishing result of Ozsváth and Szabó: if ξ is Stein fillable, then $c(\xi) \neq 0$.

The above result has been strengthened by Ghiggini [21]. We say that (Y, ξ) is *strongly symplectically fillable* if there is a contact form α such that $\xi = \ker(\alpha)$ and a symplectic 4-manifold (X, ω) such that $\partial X = Y$ and $\omega|_Y = d\alpha$. For example, every Stein fillable contact 3-manifold is strongly symplectically fillable. Furthermore, the contact 3-manifold (Y, ξ) is called *weakly symplectically fillable* if there is a symplectic manifold (X, ω) such that $\partial X = Y$ and $\omega|_\xi > 0$. Every strongly fillable contact manifold is weakly fillable, and every weakly fillable contact manifold is tight. Ghiggini showed that if ξ is strongly symplectically fillable, then $c(\xi) \neq 0$. He also proved that for a particular family $\{(Y, \xi_n) \colon n \in \mathbb{Z}_+\}$ of weakly symplectically fillable contact 3-manifolds constructed by Giroux, one has $c(\xi_n) = 0$ for every $n \in \mathbb{Z}_+$. In particular, each (Y, ξ_n) is weakly fillable (and hence tight), but not strongly fillable (and hence not Stein fillable). It was also Ghiggini [19] who constructed the first family of strongly fillable but not Stein fillable contact structures, and this work relies on Heegaard Floer homology as well.

Using the contact invariant, Lisca and Stipsicz [54] proved that every closed, oriented Seifert fibred 3-manifold carries a tight contact structure, except if it arises as $(2n-1)$-surgery on the torus knot $T_{2,2n+1}$. Furthermore, Ghiggini, Lisca, and Stipsicz [23] used the contact invariant to classify all tight contact structures on certain small Seifert fibered spaces.

4. Lagrangian Floer Homology

The machinery underlying the construction of Heegaard Floer homology is Lagrangian intersection Floer theory, first introduced by Floer himself [14]. We will follow the threatment of Fukaya *et al.* [16, Chap. 2].

First, recall that a *symplectic $\mathbb{R}$-vector space* is a pair (V, ω), where V is an $\mathbb{R}$-vector space and ω, called the *symplectic form*, is a non-degenerate antisymmetric bilinear form on V. The existence of such an ω implies that V is necessarily of some even dimension $2n$. A subspace W of V is called *Lagrangian* if it is of dimension n, and $\omega(v, w) = 0$ for every v, $w \in W$. If $V = \mathbb{R}^{2n}$ with standard basis $(v_1, \ldots, v_{2n})$, then there is a canonical symplectic form ω_0 such that $\omega_0(v_k, v_{n+k}) = 1$ and $\omega_0(v_{n+k}, v_k) = -1$ for every $k \in \{1, \ldots, n\}$, and such that ω_0 vanishes for every other pair of basis vectors. We denote the Grassmannian of Lagrangian subspaces in $(\mathbb{R}^{2n}, \omega_0)$ by $\Lambda(n)$. Note that $\Lambda(n) = U(n)/O(n)$, and $\pi_1(\Lambda(n)) = \mathbb{Z}$.

A *symplectic manifold* is a pair (M, ω) such that M is a $2n$-manifold for some positive integer n, and ω is a closed non-degenerate differential 2-form on M; i.e. $d\omega = 0$ and ω^n is nowhere zero. In other words, (T_pM, ω_p) is a symplectic vector space for every point $p \in M$ and $d\omega = 0$. We say that L is a *Lagrangian submanifold* of (M, ω) if L is a smooth submanifold of M such that T_pL is a Lagrangian subspace of (T_pM, ω_p) for every $p \in L$.

Note that every symplectic vector bundle over S^1 is symplectically trivial; i.e. isomorphic to the product $S^1 \times \mathbb{R}^{2n}$ with symplectic form ω_0 on each fiber. Hence, given a curve $\gamma\colon S^1 \to L$, the bundle γ^*TM is symplectically trivial; let

$$\Phi\colon \gamma^*TM \to S^1 \times \mathbb{R}^{2n}$$

be one such trivialization. We denote by p the projection from $S^1 \times \mathbb{R}^{2n} \to \mathbb{R}^{2n}$. For every $\tau \in S^1$, consider the Lagrangian subspace $p \circ \Phi(T_{\gamma(\tau)}L)$ of $(\mathbb{R}^{2n}, \omega_0)$, this is an element of $\Lambda(n)$. So we have obtained a loop $S^1 \to \Lambda(n)$ whose homotopy class, considered as an element of $\mathbb{Z}$, is called the Maslov index of γ in the trivialization Φ, and is denoted by $\mu_\Phi(\gamma)$.

This gives rise to the Maslov index homomorphism

$$\mu_L \colon \pi_2(M, L) \to \mathbb{Z}.$$

Given a continuous map $u\colon (D^2, \partial D^2) \to (M, L)$, we let $\mu_L([u]) = \mu_\Phi(\gamma)$, where $\gamma = u|_{\partial D^2}$, and Φ is the trivialization of γ^*TM that extend to u^*TM.

Suppose we have a symplectic $2n$-manifold (M, ω), together with two compact and connected Lagrangian submanifolds L_0 and L_1 that intersect transversely. Then fix a path ℓ_0 from L_0 to L_1. Let $\Omega(L_0, L_1)$ be the space of paths $\ell\colon I \to M$ such that $\ell(0) \in L_0$ and $\ell(1) \in L_1$, and we denote by $\Omega_{\ell_0}(L_0, L_1)$ the component of ℓ_0. The universal covering space of $\Omega_{\ell_0}(L_0, L_1)$ is homeomorphic to the space of pairs $(\ell, [w])$, where $\ell \in \Omega_{\ell_0}(L_0, L_1)$ and $[w]$ is a homotopy class of maps $w\colon I \times I \to M$ such that $w(0,t) = \ell_0(t)$ and $w(1,t) = \ell(t)$ for every $t \in I$, while $w(\tau, 0) \in L_0$ and $w(\tau, 1) \in L_1$ for every $\tau \in I$.

Next, we define another covering space of $\Omega_{\ell_0}(L_0, L_1)$. Let $c\colon S^1 \to \Omega(L_0, L_1)$ be a closed loop. This can be thought of as a map $c\colon S^1 \times [0,1] \to M$ such that $c_0 = c(\cdot, 0)$ is a curve on L_0 and $c_1 = c(\cdot, 1)$ is a curve on L_1. Since every symplectic vector bundle over S^1 is trivial, the bundle c^*TM is symplectically trivial. Fix such a trivialization $\Psi\colon c^*TM \to S^1 \times I \times \mathbb{R}^{2n}$, and let Ψ_i be the induced trivialization of c_i^*TM for $i \in \{0,1\}$. Then one can show that $\mu_{\Psi_0}(c_0) - \mu_{\Psi_1}(c_1)$ is independent of the choice of trivialization Ψ and only depends on the homotopy class of the loop c; we denote it by $\mu(c)$. In fact, it defines a homomorphism

$$\mu\colon \pi_1(\Omega_{\ell_0}(L_0, L_1), \ell_0) \to \mathbb{Z}.$$

Given two paths w and w' in $\Omega_{\ell_0}(L_0, L_1)$ from ℓ_0 to ℓ, we denote by $w\#\overline{w}'$ the concatenation of w with the reverse of w'. This is a closed loop in $\Omega_{\ell_0}(L_0, L_1)$ based at ℓ_0. We say that (ℓ, w) and (ℓ, w') are equivalent, and write $(\ell, w) \sim (\ell, w')$, if $\int_w \omega = \int_{w'} \omega$; i.e. if $\int_{w\#\overline{w}'} \omega = 0$, and if $\mu(w\#\overline{w}') = 0$. We denote by $[\ell, w]$ the equivalence class of (ℓ, w). Then the space $\widetilde{\Omega}_{\ell_0}(L_0, L_1)$ is defined to be the set of equivalence classes of pairs (ℓ, w); this is a covering space of $\Omega_{\ell_0}(L_0, L_1)$.

Loosely speaking, Lagrangian intersection Floer homology of the pair (L_0, L_1) is the homology of the path space $\widetilde{\Omega}_{\ell_0}(L_0, L_1)$. One can compute the homology of a finite-dimensional manifold M using Morse theory: First, one has to choose a Morse function f and a Riemannian metric on M whose gradient satisfies the Morse–Smale condition. The chain complex is freely generated by the critical points of f, and the differential counts gradient flow-lines between critical points of index difference one. However, on an infinite-dimensional manifold, it does not make sense to talk about the Morse index, and there are other technical difficulties. It was Floer's observation that in our case it suffices to define the index difference for a pair of critical points, and one can still obtain homology groups under certain hypothesis on the pair (L_0, L_1). These hypothese were relaxed by Fukaya *et al.* [16], but in some cases the differential does not square to zero, and the theory is obstructed in an essential way.

For our "Morse function," we take the functional $\mathcal{A} = \mathcal{A}_{\ell_0} : \widetilde{\Omega}_{\ell_0}(L_0, L_1) \to \mathbb{R}$, given by the formula

$$\mathcal{A}([\ell, w]) = \int w^* \omega.$$

The critical points of $\mathcal{A}$ correspond to the intersection points of L_0 and L_1; i.e. they are of the form $[\ell_p, w]$, where $\ell_p \colon I \to M$ is the constant path at some $p \in L_0 \cap L_1$. We denote by $C(\mathcal{A})$ the set of critical points of $\mathcal{A}$.

To define a metric on $\widetilde{\Omega}_{\ell_0}(L_0, L_1)$, we first choose a t-dependent family $J = \{J_t \colon t \in I\}$ of almost complex structures on M compatible with ω, and consider the induced family of Riemannian metrics $g_J = \omega(\cdot, J\cdot)$ on M. This will, in turn, induce an L^2-metric on $\widetilde{\Omega}_{\ell_0}(L_0, L_1)$ by

$$\langle \xi, \eta \rangle = \int_0^1 g_{J_t}(\xi(t), \eta(t))dt,$$

where ξ and η are vector fields along some path $\ell \in \Omega_{\ell_0}(L_0, L_1)$.

It turns out that the moduli space of gradient flow-lines of $\mathcal{A}$ between critical points $[\ell_p, w]$ and $[\ell_q, w']$ is the set $\mathcal{M}_J([\ell_p, w], [\ell_q, w'])$ of maps $u\colon \mathbb{R} \times I \to M$ such that

(a) $u(\mathbb{R} \times \{j\}) \subset L_j$ for $j \in \{0, 1\}$,
(b) $\partial u/\partial \tau + J(t, u)\partial u/\partial t = 0$ (i.e. u is *J-holomorphic*),
(c) $\lim_{\tau \to -\infty} u(\tau, t) = p$ and $\lim_{\tau \to +\infty} u(\tau, t) = q$,
(d) $w \# u \sim w'$, where the concatenation is taken along ℓ_p.

Note that the t-dependence of J is to ensure that the above moduli spaces are cut out transversely, and are hence smooth manifolds. From now on, we will suppress J in the notation whenever possible. Observe that $\mathbb{R} \times I$ is conformally equivalent with $D^2 \backslash \{\pm i\}$. So, instead, one can consider J-holomorphic Whitney disks $u\colon D^2 \to M$ connecting p and q (i.e. $u(\partial D^2 \cap \{\Re \leq 0\}) \subset L_0$, $u(\partial D^2 \cap \{\Re \geq 0\}) \subset L_1$, $u(-i) = p$, and $u(i) = q$) in the relative homotopy class given by $w \# u \sim w'$. This is often the viewpoint taken in Heegaard Floer homology.

We next define a consistent grading on $C(\mathcal{A})$. For this, we have to fix a section λ^0 of $\ell_0^* \Lambda M$, where ΛM is the bundle of Lagrangian Grassmanians of TM, such that $\lambda^0(0) = T_{\ell_0(0)} L_0$ and $\lambda^0(1) = T_{\ell_0(1)} L_1$. Let $[\ell_p, w] \in C(\mathcal{A})$ be an element corresponding to the intersection point $p \in L_0 \cap L_1$. Choose a trivialization

$$\Phi\colon w^*TM \to I^2 \times \mathbb{R}^{2n} \to \mathbb{R}^{2n}$$

such that

- $\Phi(0, t) \circ w^*(\lambda^0(t)) \equiv \mathbb{R}^n$,
- $\Phi(1, 0) \circ w^*(T_p L_0) = \mathbb{R}^n$, and
- $\Phi(1, 1) \circ w^*(T_p L_1) = i \cdot \mathbb{R}^n$.

This Φ will induce a loop $\lambda_\Phi\colon \partial I^2 \to \Lambda(n)$, such that

- $\lambda_\Phi(s, 0) = \Phi \circ w^*(T_{w(s,0)} L_0)$,
- $\lambda_\Phi(1, t) = e^{\frac{\pi i t}{2}} \mathbb{R}^n$,
- $\lambda_\Phi(s, 1) = \Phi \circ w^*(T_{w(s,1)} L_1)$, and
- $\lambda_\Phi(0, t) = \mathbb{R}^n$

for every $s, t \in I$. We define the degree of $[\ell_p, w]$ by

$$\mu_{\lambda_0}([\ell_p, w]) = \mu(\lambda_\Phi).$$

The above number is independent of the choice of trivialization Φ. So we obtain a grading on $C(\mathcal{A})$, provided the based path ℓ_0 and the section λ_0 of ΛM over ℓ_0 are chosen.

We say that the pair of Lagrangians (L_0, L_1) is *relatively spin* if there is a class $st \in H^2(M; \mathbb{Z}_2)$ such that $st|_{L_i} = w_2(L_i)$ for $i = 0, 1$. A *relative spin structure* on (L_0, L_1) consists of a class $st \in H^2(M; \mathbb{Z}_2)$, an oriented vector bundle ξ over the 3-skeleton of M such that $w_2(\xi) = st$, and Spin-structures on $TL_i \oplus \xi$ over the 2-skeleton of L_i.

The following transversality result was obtained by Floer [14], with some missing details filled in by Oh [63, 66].

Theorem 4.1. *Let $[\ell_p, w]$, $[\ell_q, w']$ be critical points of the functional $\mathcal{A}$. Then the space $\widetilde{\mathcal{M}}([\ell_p, w], [\ell_q, w'])$ is a smooth manifold of dimension $\mu([\ell_p, w']) - \mu([\ell_q, w])$. If we also assume that the pair (L_0, L_1) is relatively spin, then the space will carry an orientation.*

Since $J = \{J_t\}$ does not depend on τ, there is a natural $\mathbb{R}$-action on the moduli space $\mathcal{M}([\ell_p, w], [\ell_q, w'])$ defined by translating along the τ-direction. We put

$$\widetilde{\mathcal{M}}([\ell_p, w], [\ell_q, w']) = \mathcal{M}([\ell_p, w], [\ell_q, w'])/\mathbb{R}.$$

A Lagrangian L is called *monotonic* if the symplectic area of every pseudo-holomorphic disk with boundary on L is positively proportional to the Maslov index μ_L. From now on, we assume that L_0 and L_1 are monotonic. When the moduli space $\widetilde{\mathcal{M}}([\ell_p, w], [\ell_q, w'])$ is 0-dimensional; i.e. when $\mu([\ell_q, w']) - \mu([\ell_p, w]) = 0$, then it is a compact oriented 0-dimensional manifold, hence we can consider the algebraic count $\#\widetilde{\mathcal{M}}([\ell_p, w], [\ell_q, w']) \in \mathbb{Z}$ of its points.

Definition 4.2. The Floer chain complex $(CF(L_0, L_1; \ell_0), \partial_0)$ is defined as

$$CF^k(L_0, L_1) = CF^k(L_0, L_1; \ell_0) = \widehat{\bigoplus_{\substack{[\ell_p, w] \in C(\mathcal{A}) \\ \mu([\ell_p, w]) = k}}} \mathbb{Q}[\ell_p, w],$$

where $\widehat{\bigoplus}$ means a certain algebraic completion, and the differential of $[\ell_p, w] \in CF^k(L_0, L_1)$ is given by

$$\partial_0[\ell_p, w] = \sum_{[\ell_q, w'] \in CF^{k-1}(L_0, L_1)} \#\widetilde{\mathcal{M}}([\ell_p, w], [\ell_q, w']) \cdot [\ell_q, w'].$$

We briefly explain the completion used in the above definition. The elements of $CF^k(L_0, L_1)$ are the (infinite) sums

$$\sum_{[\ell_p, w]} a_{[\ell_p, w]}[\ell_p, w]$$

such that $a_{[\ell_p, w]} \in \mathbb{Q}$, and for each $C \in \mathbb{R}$, the set

$$\{[\ell_p, w] \colon \mathcal{A}(\ell_p, w) \leq C,\ a_{[\ell_p, w]} \neq 0\}$$

is finite.

Let $G(L_0, L_1)$ be the group of deck transformations of the covering

$$\widetilde{\Omega}_{\ell_0}(L_0, L_1) \to \Omega_{\ell_0}(L_0, L_1).$$

Then $CF(L_0, L_1)$ is a module over the Novikov ring $\Lambda(L_0, L_1)$, where $\Lambda^k(L_0, L_1)$ is the set of all (infinite) sums $\sum_{g \in G(L_0, L_1),\, \mu(g) = k} a_g[g]$ such that $a_g \in \mathbb{Q}$, and for each $C \in \mathbb{R}$, the set

$$\{g \in G(L_0, L_1) \colon E(g) \leq C,\ a_g \neq 0\}$$

is finite.

The minimal Maslov number of a Lagrangian L is the nonnegative integer Σ_L defined by $\mathrm{Im}(\mu_L) = \Sigma_L \mathbb{Z}$, where $\mu_L \colon \pi_2(M, L) \to \mathbb{Z}$ is the Maslov index homomorphism. Building on work of Floer [14], Oh [63, 64] proved that $\partial_0 \circ \partial_0 = 0$ for monotone Lagrangian submanifolds, where the minimal Maslov number is bigger than two, with some topological restrictions on the pair (L_0, L_1). As in the case of Morse homology, the proof relies on showing that the ends of the compactification of 1-dimensional moduli spaces $\mathcal{M}([\ell_p, w], [\ell_q, w'])$ are only broken flow-lines.

The reader might be wondering why one has to consider the covering space $\widetilde{\Omega}_{\ell_0}(L_0, L_1)$ instead of trying to do Morse theory on $\Omega_{\ell_0}(L_0, L_1)$. If one did that, one would run into the problem of having infinitely many pseudo-holomorphic disks (gradient flow-lines on the path space) between two intersection points (critical points of the functional), and the boundary map would not be defined. If one instead passed to the whole universal cover, this problem would be resolved, and we would get the twisted (co)homology of $\Omega_{\ell_0}(L_0, L_1)$ with coefficients lying in some completion of $\mathbb{Z}[\pi_1(\Omega_{\ell_0}(L_0, L_1))]$. This is unfortunate since the coefficient ring depends on the pair of Lagrangians. Since there are only finitely many disks with bounded symplectic area, it suffices to pass to the cover $\widetilde{\Omega}_{\ell_0}(L_0, L_1)$, and we can work over a Novikov ring independent of the Lagrangians. Twisting with the Maslov index allows one to define a homological $\mathbb{Z}$-grading, solving the issue of having disks between two intersection points of different Maslov indices (there would still be a $\mathbb{Z}_2$-grading in this case).

5. Definition of Heegaard Floer Homology

Definition 5.1. Let Y be a closed, connected, oriented 3-manifold. A *Heegaard diagram* for Y is a triple $(\Sigma, \boldsymbol{\alpha}, \boldsymbol{\beta})$ such that

- Σ is a closed oriented genus g surface embedded in Y,
- $\boldsymbol{\alpha} \subset \Sigma$ is a properly embedded 1-manifold with components $\alpha_1, \dots, \alpha_g$ that are linearly independent in $H_1(\Sigma)$, and each of which bounds a disk in Y to the negative side of Σ,
- $\boldsymbol{\beta} \subset \Sigma$ is a properly embedded 1-manifold with components $\beta_1, \dots, \beta_g$ that are linearly independent in $H_1(\Sigma)$, and each of which bounds a disk to the positive side of Σ.

A Heegaard diagram of Y arises from a self-indexing Morse function $f\colon Y \to \mathbb{R}$ with a single minimum and maximum, together with a Riemannian metric on Y. If p is a critical point of f, then we denote by $W^s(p)$ and $W^u(p)$ the stable and unstable manifolds of p under the gradient flow of f. We take Σ to be $f^{-1}(3/2)$. The α-curves are of the form $\Sigma \cap W^u(p)$, where p is an index-1 critical point of f, and the β-curves are of the form $W^s(q) \cap \Sigma$, where q is an index-2 critical point of f.

Let $\mathcal{H} = (\Sigma, \boldsymbol{\alpha}, \boldsymbol{\beta})$ and $\mathcal{H}' = (\Sigma', \boldsymbol{\alpha}', \boldsymbol{\beta}')$ be Heegaard diagrams of Y. Then we say that $\mathcal{H}'$ is obtained from $\mathcal{H}$ by a *stabilization* if $\Sigma' \backslash \Sigma$ is a punctured torus, and $\boldsymbol{\alpha}' = \boldsymbol{\alpha} \cup \alpha$ and $\boldsymbol{\beta}' = \boldsymbol{\beta} \cup \beta$, where α and β are two curves in $\Sigma' \backslash \Sigma$ that intersect each other transversely in a single point. In this case, we also say that $\mathcal{H}$ is obtained from $\mathcal{H}'$ via a *destabilization.*

The diagram $\mathcal{H}'$ is obtained from $\mathcal{H}$ by *handlesliding* α_i over α_j along the arc $a \subset \Sigma$ connecting α_i and α_j if $\Sigma = \Sigma'$ and $\boldsymbol{\beta} = \boldsymbol{\beta}'$, while $\boldsymbol{\alpha}' = (\boldsymbol{\alpha} \backslash \alpha_i) \cup \alpha_i'$, where α_i' is the boundary component of a thin regular neighborhood $N(\alpha_i \cup a \cup \alpha_j)$ distinct from α_i and α_j.

According to a theorem of Reidemeister and Singer [86, 91], any two Heegaard diagrams of the 3-manifold Y become diffeomorphic after a sequence of (de)stabilizations and handleslides. Slightly more is true, any two diagrams become isotopic in Y after a suitable sequence of (de)stabilizations and handleslides. However, this isotopy is far from unique, even homotopically. This will cause some serious difficulties when trying to prove the naturality of Heegaard Floer homology, and the failure of naturality for the hat version.

For the construction of Heegaard Floer homology, it is essential to fix a basepoint $z \in \Sigma \backslash (\boldsymbol{\alpha} \cup \boldsymbol{\beta})$. Furthermore, we choose a complex structure $\mathfrak{j}$ on Σ. Consider the symmetric product

$$\mathrm{Sym}^g(\Sigma) = \overbrace{\Sigma \times \cdots \times \Sigma}^{g} / S_g,$$

this is the space of unordered g-tuples of points in Σ. To see that this is a manifold, observe that $\mathrm{Sym}^g(\mathbb{C}) \cong \mathbb{C}^g$ by the fundamental theorem of algebra: Given an ordered g-tuple $(a_1, \ldots, a_g) \in \mathbb{C}^g$, the roots of the polynomial $z^g + a_1 z^{g-1} + \cdots + a_g$ give an unordered g-tuple. Note that the smooth structure on $\mathrm{Sym}^g(\Sigma)$ depends on $\mathfrak{j}$, but the diffeomorphism type is independent. The complex structure $\mathfrak{j}$ induces a complex structure $\mathrm{Sym}^g(\mathfrak{j})$ on the symmetric product $\mathrm{Sym}^g(\Sigma)$. Let

$$\mathbb{T}_\alpha = \alpha_1 \times \cdots \times \alpha_g / S_g,$$

and similarly,

$$\mathbb{T}_\beta = \beta_1 \times \cdots \times \beta_g / S_g,$$

these tori are totally real submanifolds of $\mathrm{Sym}^g(\Sigma)$; i.e. they contain no complex lines. As shown by Ozsváth and Szabó, one can study intersection Floer homology for the pair $(\mathbb{T}_\alpha, \mathbb{T}_\beta)$ in a manner analogous to the Lagrangian case. If one chooses a symplectic structure on Σ, the induced symplectic structure on $\mathrm{Sym}^g(\Sigma)$ is singular. Perutz [82] constructed a symplectic structure on $\mathrm{Sym}^g(\Sigma)$ compatible with the complex structure that makes $\mathbb{T}_\alpha$ and $\mathbb{T}_\beta$ Lagrangian and monotonic with minimal Maslov number two. Furthermore, he proved that a handleslide can be realized by a Hamiltonian isotopy, however, the pair might cease to be monotonic in the meantime (this is related to admissibility that we discuss next).

Monotonicity of $\mathbb{T}_\alpha$ and $\mathbb{T}_\beta$ suffices to define the boundary map ∂ for the Lagrangian intersection Floer homology of the pair $(\mathbb{T}_\alpha, \mathbb{T}_\beta)$ in $\mathrm{Sym}^g(\Sigma)$, but since the minimal Maslov number is two, one cannot simply refer to the work of Floer [14] and Oh [63] to claim that $\partial^2 = 0$ in this setting. However, Ozsváth and Szabó [74, Theorem 3.15] prove that in the boundary of the compactification of the 1-dimensional moduli spaces, the algebraic count of configurations that are not broken flow-lines is zero (these are so-called boundary bubbles, and they might actually appear), and hence indeed we have $\partial^2 = 0$. In the language of Fukaya *et al.* [16], the Lagrangian intersection Floer homology of the pair $(\mathbb{T}_\alpha, \mathbb{T}_\beta)$ is unobstructed.

However, the intersection Floer homology of $(\mathbb{T}_\alpha, \mathbb{T}_\beta)$ is *not* an invariant of the underlying 3-manifold. Indeed, consider the two diagrams of $S^1 \times S^2$ in Fig. 1. Both of them are genus one, so there is a single α-curve α and a single β-curve β, and both of them are

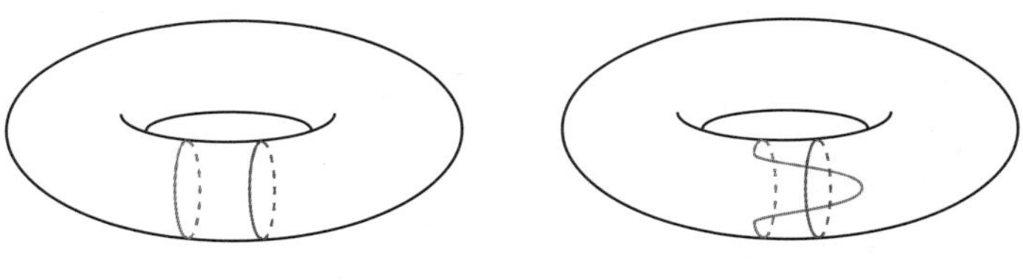

Fig. 1. A non-admissible diagram of $S^1 \times S^2$ on the left, and an admissible one on the right.

meridians. In the first diagram, α and β are disjoint, hence the Floer homology group is generated by the empty set, in the second, they intersect in two points transversely and the boundary map is zero. The two homologies have different ranks.

To make the construction work, Ozsváth and Szabó introduced two additional ingredients. One is the choice of a basepoint z in $\Sigma\backslash(\boldsymbol{\alpha} \cup \boldsymbol{\beta})$, and we record the intersection number of every holomorphic disk with the subvariety

$$V_z = \{z\} \times \mathrm{Sym}^{g-1}(\Sigma).$$

More precisely, we introduce an additional formal variable U, and deform the Floer complex according to the intersection number with V_z. There are several different ways of doing this, giving rise to the different flavors of Heegaard Floer homology. Cf. Seidel [90], who first introduced such an idea in a different context. Without fixing the basepoint, the invariant obtained would only capture the algebraic topology of Y. Furthermore, it would not be natural, as illustrated by [39, Example 3.3].

The second ingredient is that, when $b_1(Y) > 0$, we only consider so-called *admissible* diagrams. In the case of $\widehat{HF}$ and HF^+, we use weakly admissibile diagrams, but to define HF^∞ and HF^-, the diagram has to be strongly admissible in reference to a fixed Spin^c-structure on Y. We will discuss admissibility later.

In fact, Heegaard Floer homology is an invariant of a Spin^c 3-manifold. As explained in the previous section, Lagrangian intersection Floer homology is Morse theory on a *component* of the space $\Omega(\mathbb{T}_\alpha, \mathbb{T}_\beta)$ of paths connecting $\mathbb{T}_\alpha$ and $\mathbb{T}_\beta$. The choice of basepoint z sets up a bijection between the components of this path space with Spin^c-structures on Y. We explain this construction next.

The group $\mathrm{Spin}^c(n)$ is defined to be

$$(\mathrm{Spin}(n) \times S^1)/\langle(-1,-1)\rangle;$$

i.e. it is the complexification of the group $\mathrm{Spin}(n)$. It fits into the exact sequence

$$1 \to S^1 \to \mathrm{Spin}^c(n) \to \mathrm{SO}(n) \to 1.$$

The first map is given by $z \mapsto [1, z]$, while the second is $[g, z] \mapsto p(g)$, where $p\colon \mathrm{Spin}(n) \to \mathrm{SO}(n)$ is the covering map.

Given an oriented Riemannian n-manifold M, a Spin^c-structure on M is the reduction/lift of the structure group of TM, the principal SO(n)-bundle P_M of oriented orthonormal frames on M, to Spin$^c(n)$, considered up to equivalence. For us, an equivalent description of Spin^c-structures on 3-manifolds due to Turaev [93] will be more relevant, which we now review. Let Y be a closed, connected, oriented 3-manifold. We say that the nowhere vanishing vector fields v and w on Y are *homologous*, and we write $v \sim w$, if they are homotopic in the complement of a ball. A Spin^c-structure on the 3-manifold Y is the homology class of nowhere vanishing vector fields, and we denote the set of these by Spin$^c(Y)$. Note that we can think of a Spin^c-structure as a homotopy class of nowhere vanishing vector fields over the 2-skeleton of Y, hence the obstruction to homotoping v to w over the 2-skeleton (or equivalently, in the complement of a ball), is an element of $H^2(Y) \cong H_1(Y)$. This way, we can view Spin$^c(Y)$ as an affine space over $H_1(Y)$. The concrete action of a class $a \in H_1(M)$ on $\mathfrak{s} \in$ Spin$^c(Y)$ is given by Reeb turbularization: we represent a as an embedded oriented 1-manifold, and then "turbularize" the vector field v along it. A homomorphism

$$c_1 \colon \mathrm{Spin}^c(Y) \to H^2(Y)$$

is given by taking the first Chern class (or equivalently, Euler class) of the oriented 2-plane field $v^{\perp}$, where v is a nowhere vanishing vector field representing $\mathfrak{s}$. Then we have the formula

$$\mathfrak{s} - \mathfrak{s}' = 2(c_1(\mathfrak{s}) - c_1(\mathfrak{s}')).$$

In particular, the homomorphism c_1 is injective if and only if there is no 2-torsion in $H^2(Y)$.

Let $(\Sigma, \boldsymbol{\alpha}, \boldsymbol{\beta})$ be a Heegaard diagram, and consider the tori $\mathbb{T}_\alpha$ and $\mathbb{T}_\beta$ in Sym$^g(\Sigma)$. We view D^2 as the unit disk in $\mathbb{C}$. Given intersection points $\mathbf{x}, \mathbf{y} \in \mathbb{T}_\alpha \cap \mathbb{T}_\beta$, a topological Whitney disk from $\mathbf{x}$ to $\mathbf{y}$ is a continuous map $u \colon D^2 \to \mathrm{Sym}^g(\Sigma)$ such that

- $u(-i) = \mathbf{x}$ and $u(i) = \mathbf{y}$,
- $u(\partial D^2 \cap \{\Re \leq 0\}) \subset \mathbb{T}_\alpha$,
- $u(\partial D^2 \cap \{\Re \geq 0\}) \subset \mathbb{T}_\beta$.

A topological Whitney disk can be viewed as a path in $\Omega(\mathbb{T}_\alpha, \mathbb{T}_\beta)$ connecting the constant paths $\ell_{\mathbf{x}}$ and $\ell_{\mathbf{y}}$. We say that u and u' are *homotopic* if they are homotopic through topological Whitney disks. We denote by $\pi_2(\mathbf{x}, \mathbf{y})$ the set of homotopy classes of Whitney disks from $\mathbf{x}$ to $\mathbf{y}$. This is non-empty if and only if $\ell_{\mathbf{x}}$ and $\ell_{\mathbf{y}}$ lie in the same component of $\Omega(\mathbb{T}_\alpha, \mathbb{T}_\beta)$.

Given a based Heegaard diagram $\mathcal{H} = (\Sigma, \boldsymbol{\alpha}, \boldsymbol{\beta}, z)$ of Y, we can associate a Spin^c-structure $\mathfrak{s}(\mathbf{x})$ to every intersection point $\mathbf{x} \in \mathbb{T}_\alpha \cap \mathbb{T}_\beta$ as follows. Pick a self-indexing Morse function f on Y with a unique minimum and maximum, and a Riemannian metric g, inducing $\mathcal{H}$. Let v be the gradient of f with respect to g, and consider the flow of v. Then let N be a regular neighborhood of the union of the flow-lines of v passing through the points of $\mathbf{x}$ and through z. Then $v|_{Y \setminus N}$ extends to a nowhere vanishing vector field v' on Y since each component of N contains two singularities of v of opposite indices. Then $\mathfrak{s}(\mathbf{x})$ is defined to be the homology class of v'. Given intersection points $\mathbf{x}$, $\mathbf{y} \in \mathbb{T}_\alpha \cap \mathbb{T}_\beta$, there is a topological Whitney disk connecting $\mathbf{x}$ and $\mathbf{y}$ if and only if $\mathfrak{s}(\mathbf{x}) = \mathfrak{s}(\mathbf{y})$. Note that $\mathfrak{s}(\mathbf{x})$ depends on the choice of basepoint z.

Now we explain the notion of admissibility. One advantage of looking at admissible diagrams is that the sums appearing in the definition of the boundary map are finite; we do not need to use the completion of Fukaya *et al.* and Novikov coefficients. Furthermore, it rules out issues such as the one in the example in Fig. 1, as the diagram on the left-hand side is not admissible.

Again, let Y be a closed, connected, oriented 3-manifold, and $\mathcal{H} = (\Sigma, \boldsymbol{\alpha}, \boldsymbol{\beta}, z)$ a based Heegaard diagram of Y. A *domain* is a $\mathbb{Z}$-linear combination of components of $\Sigma \setminus (\boldsymbol{\alpha} \cup \boldsymbol{\beta})$ (we call these *regions*), this can be viewed as a 2-chain on Σ. We denote by $D(\Sigma, \boldsymbol{\alpha}, \boldsymbol{\beta})$ the set of domains, this is a free $\mathbb{Z}$-module. We define a map

$$\mathcal{D} \colon \pi_2(\mathbf{x}, \mathbf{y}) \to D(\Sigma, \boldsymbol{\alpha}, \boldsymbol{\beta})$$

as follows: Given a homotopy class of topological Whitney disks $\phi \in \pi_2(\mathbb{T}_\alpha, \mathbb{T}_\beta)$, the coefficient of $\mathcal{D}(\phi)$ at $p \in \Sigma \setminus (\boldsymbol{\alpha} \cup \boldsymbol{\beta})$ is

$$n_p(\phi) = \#(\phi \cap V_p),$$

where $V_p = \{p\} \times \mathrm{Sym}^{g-1}(\Sigma)$. We say that a domain $\mathcal{D} \in D(\Sigma, \boldsymbol{\alpha}, \boldsymbol{\beta})$ connects $\mathbf{x}$ and $\mathbf{y}$ if $\partial\mathcal{D} \cap \boldsymbol{\alpha}$ is a 1-chain with boundary $\mathbf{x} - \mathbf{y}$, and $\partial\mathcal{D} \cap \boldsymbol{\beta}$ is a 1-chain with boundary $\mathbf{y} - \mathbf{x}$. Let $D(\mathbf{x}, \mathbf{y})$ be the set of such domains. Essentially, the map $\mathcal{D}$ gives a bijection between $\pi_2(\mathbf{x}, \mathbf{y})$ and $D(\mathbf{x}, \mathbf{y})$. It is important to note that $\pi_2(\mathrm{Sym}^g(\Sigma)) = \mathbb{Z}$, and if we view the generator as an element ϕ of $\pi_2(\mathbf{x}, \mathbf{x})$, then $\mathcal{D}(\phi) = [\Sigma]$; i.e. it has multiplicity one everywhere.

A *periodic domain* $\mathcal{P}$ in $\mathcal{H}$ is an element of $D(\Sigma, \boldsymbol{\alpha}, \boldsymbol{\beta})$ such that $n_z(\mathcal{P}) = 0$, and $\partial\mathcal{P}$ is a linear combination of α-circles and β-circles (as opposed to having α-arcs or β-arcs in $\partial\mathcal{P}$). For every $\mathbf{x} \in \mathbb{T}_\alpha \cap \mathbb{T}_\beta$ and $\mathcal{D} \in D(\mathbf{x}, \mathbf{x})$, the domain $\mathcal{D} - n_z(\mathcal{D}) \cdot [\Sigma]$ is a periodic domain. In the symmetric product $\mathrm{Sym}^g(\Sigma)$, periodic domains correspond to homology classes in $H_2(\mathrm{Sym}^g(\Sigma), \mathbb{T}_\alpha \cup \mathbb{T}_\beta)$ disjoint from V_z, or equivalently, to elements of $\pi_1(\Omega(\mathbb{T}_\alpha, \mathbb{T}_\beta))$ disjoint from V_z. Given a periodic domain $\mathcal{P}$, we can cap off the boundary components of $\mathcal{P}$ in Y by the disks bounded by the $\boldsymbol{\alpha}$- and $\boldsymbol{\beta}$-curves in the two handlebodies, and obtain a 2-cycle in Y whose homology class we denote by $H(\mathcal{P}) \in H_2(Y)$. We say that a diagram is *weakly admissible* if every nonzero periodic domain has both positive and negative coefficients. This suffices for the definition of $\widehat{HF}(Y)$ and $HF^+(Y)$. We can always isotope the α- and β-curves in an arbitrary Heegaard diagram to make it admissible. Ozsváth and Szabó [74] showed that a diagram is weakly admissible if and only if Σ can be endowed with a volume form for which each periodic domain has total signed area equal to zero.

To define the other flavors of Heegaard Floer homology, we need *strong admissibility*, which is in reference to a Spin^c-structure $\mathfrak{s} \in \mathrm{Spin}^c(Y)$. A pointed Heegaard diagram is called strongly admissible for the Spin^c structure $\mathfrak{s}$ if for every nontrivial periodic domain $\mathcal{P}$ with

$$\langle c_1(\mathfrak{s}), H(\mathcal{P})\rangle = 2n \geq 0,$$

the domain $\mathcal{P}$ has some coefficient bigger than n.

If a diagram is strongly admissible for a single Spin^c-structure, then it is weakly admissible. Furthermore, when $H^2(Y)$ is torsion; i.e. when $b_1(Y) = 0$, then the notions of weak and strong admissibility coincide. According to Lekili [49, Proposition 30] (the same proof also covers the case $k = 0$), the diagram $(\Sigma, \boldsymbol{\alpha}, \boldsymbol{\beta}, z)$ is strongly admissible for the Spin^c-structure $\mathfrak{s}$ if and only if the component of the path space $\Omega(\mathbb{T}_\alpha, \mathbb{T}_\beta)$ corresponding to $\mathfrak{s}$ is monotonic. More precisely, if $\mathbf{x} \in \mathbb{T}_\alpha \cap \mathbb{T}_\beta$ satisfies $\mathfrak{s}(\mathbf{x}) = \mathfrak{s}$, then there exists an area form ξ on Σ such that for an induced symplectic form ω_ξ on $\mathrm{Sym}^g(\Sigma)$ that makes $\mathbb{T}_\alpha$ and $\mathbb{T}_\beta$ Lagrangian, and the ω_ξ-area and the index maps

$$\pi_1(\Omega(\mathbb{T}_\alpha, \mathbb{T}_\beta), \mathbf{x}) \to \mathbb{R}$$

are proportional. Recall that monotonicity ensures that ∂ is well-defined, and that the Lagrangian intersection Floer homology is unobstructed; i.e. that $\partial^2 = 0$ (assuming the minimal Maslov number is greater than two, or if the contributions of boundary bubbles to ∂^2 is zero, which is the case for HF). We can always isotope the α- and β-curves to make the diagram strongly admissible for a given Spin^c-structure, but as shown by the example of $S^1 \times S^2$, we cannot necessarily make it strongly admissible for all of them simultaneously.

Let $(\Sigma, \boldsymbol{\alpha}, \boldsymbol{\beta}, z)$ be strongly $\mathfrak{s}$-admissible, and suppose that $\mathfrak{s}(\mathbf{x}_0) = \mathfrak{s}$. Let $\Omega_{\mathbf{x}_0}(\mathbb{T}_\alpha, \mathbb{T}_\beta)$ the component of $\ell_{\mathbf{x}_0}$. Recall that to define Lagrangian intersection Floer homology, we need to consider the cover $\widetilde{\Omega}_{\mathbf{x}_0}(\mathbb{T}_\alpha, \mathbb{T}_\beta)$, and generators are certain equivalence classes $[\ell_{\mathbf{x}}, w]$, where w is a path of paths from $\ell_{\mathbf{x}_0}$ to $\ell_{\mathbf{x}}$. Then we can view w as a topological Whitney disk from $\mathbf{x}_0$ to $\mathbf{x}$. Given another such disk w', recall that $w \sim w'$ if they have the same Maslov index and symplectic area. Equivalently, if $\phi = w \# \overline{w}'$ has Maslov index and symplectic area zero. The domain $\mathcal{D}(\phi)$ can be written as $\mathcal{P} + n_z(\phi) \cdot [\Sigma]$. By the strong admissibility condition, we can choose the symplectic form on Σ such that the signed area of $\mathcal{P}$ is zero, hence the symplectic volume of the corresponding element of $\pi_2(\mathbf{x}_0, \mathbf{x}_0)$ is also zero. By the result of Lekili, this also has Maslov index zero (this also follows from a Maslov index formula of Ozsváth and Szabó). The component of ϕ corresponding to $n_z(\phi) \cdot [\Sigma]$ is $n_z(\phi)$ times the generator of $\pi_2(\mathrm{Sym}^g(\Sigma))$. This has Maslov index two. We conclude that $w \sim w'$ if and only if $n_z(w) = n_z(w')$, and the map

$$\pi_{\mathbf{x}_0} \colon \widetilde{\Omega}_{\mathbf{x}_0}(\mathbb{T}_\alpha, \mathbb{T}_\beta) \to \Omega_{\mathbf{x}_0}(\mathbb{T}_\alpha, \mathbb{T}_\beta)$$

is a trivial $\mathbb{Z}$-covering. So strong $\mathfrak{s}$-admissibility ensures that the covering $\pi_{\mathbf{x}_0}$ of the component of the path-space corresponding to $\mathfrak{s}$ is trivial, and the trivialization is given by the choice of basepoint z. In particular, we can identify a generator $[\ell_{\mathbf{x}}, w]$ of the Floer chain complex with the pair $[\mathbf{x}, n_z(w)] \in (\mathbb{T}_\alpha \cap \mathbb{T}_\beta) \times \mathbb{Z}$. These are exactly the generators of the chain complex $CF^\infty(\boldsymbol{\alpha}, \boldsymbol{\beta}, z)$, as defined by Ozsváth and Szabó. The group of deck transformations of the covering $\pi_{\mathbf{x}_0}$ is $\mathbb{Z}$, Ozsváth and Szabó denote the generator by U. To obtain $U[\ell_{\mathbf{x}}, w]$, we multiply w with the generator of $\pi_2(\mathrm{Sym}^g(\Sigma))$. This corresponds to $U[\mathbf{x}, i] = [\mathbf{x}, i-1]$.

Given intersection points $\mathbf{x}$, $\mathbf{y} \in \mathbb{T}_\alpha \cap \mathbb{T}_\beta$ with $\mathfrak{s}(\mathbf{x}) = \mathfrak{s}(\mathbf{y})$, there is a homotopy class ϕ of topological Whitney disks from $\mathbf{x}$ to $\mathbf{y}$, and as in the case of Lagrangian intersection Floer homology, we can consider the moduli space $\mathcal{M}(\phi)$ of pseudo-holomorphic representatives of ϕ. This has an $\mathbb{R}$-action, and we write

$$\widetilde{\mathcal{M}}(\phi) = \mathcal{M}(\phi)/\mathbb{R}.$$

The expected dimension of $\mathcal{M}(\phi)$ is given by the Maslov index $\mu(\phi)$. For the definition of the Heegaard Floer differential, we are going to count rigid pseudo-holomorphic discs; i.e. ones that have Maslov index one. If $\mu(\phi) = 1$, then $\widetilde{\mathcal{M}}(\phi)$ is a finite collection of points. We can either count the number of these points modulo 2, or after an appropriate choice of orientations, the points of $\widetilde{\mathcal{M}}(\phi)$ come with signs, and we can take the algebraic number of them, $\#\widetilde{\mathcal{M}}(\phi)$. As was mentioned before, all the moduli spaces are oriented if the pair of Lagrangians $(\mathbb{T}_\alpha, \mathbb{T}_\beta)$ is relatively spin. In particular, it suffices to endow $\mathbb{T}_\alpha$ and $\mathbb{T}_\beta$ with a Spin-structure.

Now we are ready to define the Heegaard Floer chain complex $CF^\infty(\mathcal{H}, \mathfrak{s})$ for a strongly $\mathfrak{s}$-admissible pointed Heegaard diagram $\mathcal{H} = (\Sigma, \boldsymbol{\alpha}, \boldsymbol{\beta}, z)$. It is the free $\mathbb{Z}$-module generated by pairs $[\mathbf{x}, i]$, where $\mathbf{x} \in \mathbb{T}_\alpha \cap \mathbb{T}_\beta$ is an intersection point, and $i \in \mathbb{Z}$ is an integer. The grading is defined by

$$\mathrm{gr}([\mathbf{x}, i], [\mathbf{y}, j]) = \mu(\phi) - 2n_z(\phi) + 2i - 2j,$$

where ϕ is a topological Whitney disk from $\mathbf{x}$ to $\mathbf{y}$, and $n_z(\phi)$ is the algebraic intersection number of ϕ and V_z. The boundary map ∂^∞ is given by the formula

$$\partial^\infty[\mathbf{x}, i] = \sum_{\mathbf{y}\in\mathbb{T}_\alpha\cap\mathbb{T}_\beta} \sum_{\substack{\phi\in\pi_2(\mathbf{x},\mathbf{y})\\ \mu(\Phi)=1}} \#\widetilde{\mathcal{M}}(\phi)\cdot[\mathbf{y}, i - n_z(\phi)].$$

Note that if $\mathfrak{s}(\mathbf{y}) \neq \mathfrak{s}$, then the second sum is automatically zero since there is not even a topological Whitney disk from $\mathbf{x}$ to $\mathbf{y}$. Hence, on the right, we only obtain generators $[\mathbf{y}, j]$ such that $\mathfrak{s}(\mathbf{y}) = \mathfrak{s}$. The fact that this sum is finite and we do not have to resort to Novikov coefficients is again a consequence of admissibility. Indeed, Ozsváth and Szabó showed [74] that if the diagram is strongly $\mathfrak{s}$-admissible, then, given intersection points $\mathbf{x}$, $\mathbf{y} \in \mathbb{T}_\alpha \cap \mathbb{T}_\beta$, there are only finitely many homotopy classes $\phi \in \pi_2(\mathbf{x}, \mathbf{y})$ for which $\mu(\phi) = d$ and $\mathcal{D}(\phi) \geq 0$ (i.e. all of its coefficients are nonnegative). Note that if ϕ has a pseudo-holomorphic representative, then $\mathcal{D}(\phi) \geq 0$ by positivity of intersection with the hypersurfaces V_p for at least one point p in each component of $\Sigma\backslash(\boldsymbol{\alpha} \cup \boldsymbol{\beta})$ (we restrict ourselves to almost complex structures on $\mathrm{Sym}^g(\Sigma)$ that are holomorphic around these V_p).

If we choose a coherent system of orientations for the moduli spaces, we get that $\partial^\infty \circ \partial^\infty = 0$. The appropriate choice of Spin-structures on $\mathbb{T}_\alpha$ and $\mathbb{T}_\beta$ are obtained by picking the non-fillable Spin-structure on each α_i and β_j, and then taking the product of these. This is how we canonically orient all the moduli spaces.

There is a chain map

$$U \colon CF^\infty(\mathcal{H}, \mathfrak{s}) \to CF^\infty(\mathcal{H}, \mathfrak{s})$$

defined by $U[\mathbf{x}, i] = [\mathbf{x}, i - 1]$, which lowers degree by two. As we mentioned above, this corresponds to a deck transformation in the Floer theory.

We obtain the chain complexes CF^-, CF^+, and $\widehat{CF}$ from CF^∞ as follows. Let CF^- be the subcomplex of CF^∞ generated by pairs $[\mathbf{x}, i]$ such that $i < 0$. Then CF^+ is the quotient complex CF^∞/CF^-. So there is a short exact sequence

$$0 \longrightarrow CF^-(\mathcal{H}, \mathfrak{s}) \longrightarrow CF^\infty(\mathcal{H}, \mathfrak{s}) \longrightarrow CF^+(\mathcal{H}, \mathfrak{s}) \longrightarrow 0.$$

The endomorphism U on CF^∞ restricts to an endomorphism U^- on CF^-, and hence we also obtain an endomorphism U^+ on the quotient CF^+. We denote the kernel of U^+ by $\widehat{CF}$. We can also define $\widehat{CF}(\Sigma, \boldsymbol{\alpha}, \boldsymbol{\beta}, z, \mathfrak{s})$ by taking the free $\mathbb{Z}$-module generated by those intersection points $\mathbf{x} \in \mathbb{T}_\alpha \cap \mathbb{T}_\beta$ for which $\mathfrak{s}(\mathbf{x}) = \mathfrak{s}$, and the boundary map $\widehat{\partial}$ is given by

$$\widehat{\partial}\mathbf{x} = \sum_{y \in \mathbb{T}_\alpha \cap \mathbb{T}_\beta} \sum_{\substack{\phi \in \pi_2(\mathbf{x},\mathbf{y}) \\ \mu(\phi)=1,\, n_z(\phi)=0}} \#\widetilde{\mathcal{M}}(\phi) \cdot \mathbf{y}.$$

This give rise to the short exact sequence

$$0 \longrightarrow \widehat{CF}(\mathcal{H}, \mathfrak{s}) \longrightarrow CF^+(\mathcal{H}, \mathfrak{s}) \xrightarrow{U^+} CF^+(\mathcal{H}, \mathfrak{s}) \longrightarrow 0.$$

By taking homology, we obtain the $\mathbb{Z}[U]$-modules

$$HF^\infty(\mathcal{H}, \mathfrak{s}),\ HF^+(\mathcal{H}, \mathfrak{s}),\ HF^-(\mathcal{H}, \mathfrak{s}) \quad \text{and} \quad \widehat{HF}(\mathcal{H}, \mathfrak{s}),$$

where the U-action is trivial on $\widehat{HF}(\mathcal{H}, \mathfrak{s})$. These are related by long exact sequences induced by the above short exact sequences of chain complexes.

5.1. *Invariance and naturality*

Let Y be a closed, connected, oriented 3-manifold, and $\mathfrak{s} \in \mathrm{Spin}^c(Y)$ a Spin^c-structure on Y. Ozsváth and Szabó [74] show that, given any two strongly $\mathfrak{s}$-admissible pointed Heegaard diagrams $\mathcal{H} = (\Sigma, \boldsymbol{\alpha}, \boldsymbol{\beta}, z)$ and $\mathcal{H}' = (\Sigma', \boldsymbol{\alpha}', \boldsymbol{\beta}', z')$ of Y, one has

$$HF^\circ(\mathcal{H}, \mathfrak{s}) \cong HF^\circ(\mathcal{H}', \mathfrak{s}),$$

where $\circ$ is one of the four flavors hat, $+$, $-$, and ∞. As mentioned before, by the Reidemeister–Singer theorem [86, 91], the diagrams $\mathcal{H}$ and $\mathcal{H}'$ become *diffeomorphic* after a sequence of Heegaard moves; i.e. isotopies of the α- and β-curves, α- and β-handleslides, stabilizations, and destabilizations. In fact, Ozsváth and Szabó show that there is also a sequence of Heegaard moves that passes through strongly $\mathfrak{s}$-admissible diagrams. Then what remains to show is that HF is invariant under changing the complex structure $\mathfrak{j}$ on Σ and the 1-parameter family J_t of perturbations of $\mathrm{Sym}^g(\mathfrak{j})$ on $\mathrm{Sym}^g(\Sigma)$, and under any Heegaard move that preserves admissibility.

To assign a concrete group $HF^\circ(Y, p, \mathfrak{s})$ to the based Spin^c 3-manifold $(Y, p, \mathfrak{s})$, one needs more. The first steps toward naturality were made by Ozsváth and Szabó [79], and completely worked out by Thurston and the author [39] over $\mathbb{Z}_2$. What one needs to construct is an isomorphism

$$F^{\mathfrak{s}}_{\mathcal{H},\mathcal{H}'}\colon HF^\circ(\mathcal{H},\mathfrak{s}) \to HF^\circ(\mathcal{H}',\mathfrak{s})$$

for any pair of admissible based Heegaard diagrams $\mathcal{H}$, $\mathcal{H}'$ of the based 3-manifold (Y,p) that satisfy the following two properties:

(a) $F^{\mathfrak{s}}_{\mathcal{H},\mathcal{H}} = \mathrm{Id}_{HF^\circ(\mathcal{H},\mathfrak{s})}$ for any admissible diagram $\mathcal{H}$,
(b) $F^{\mathfrak{s}}_{\mathcal{H}',\mathcal{H}''} \circ F^{\mathfrak{s}}_{\mathcal{H},\mathcal{H}'} = F^{\mathfrak{s}}_{\mathcal{H},\mathcal{H}''}$ for any admissible diagrams $\mathcal{H}$, $\mathcal{H}'$, and $\mathcal{H}''$ of (Y,p).

This is an instance of a *transitive system of groups*, as defined by Eilenberg and Steenrod [13, Definition 6.1]. We will call the maps $F_{\mathcal{H},\mathcal{H}'}$ *canonical isomorphisms.* We would like to warn the reader of the widespread practice of using the word "canonical" for any well-defined map, without checking property (b). Given such a transitive system, we obtain $HF^\circ(Y,p)$ by taking the product of all the groups $HF^\circ(\mathcal{H})$, where $\mathcal{H}$ is an admissible diagram of (Y,p) (note that these form a set as Σ is a subset of Y), and take elements x in this product such that $F^{\mathfrak{s}}_{\mathcal{H},\mathcal{H}'}(x(\mathcal{H})) = x(\mathcal{H}')$ for any pair $\mathcal{H}$, $\mathcal{H}'$. For every admissible diagram $\mathcal{H}$, we have an isomorphism

$$P^{\mathfrak{s}}_{\mathcal{H}}\colon HF^\circ(Y,p,\mathfrak{s}) \to HF^\circ(\mathcal{H},\mathfrak{s}).$$

One first constructs canonical isomorphisms for changing (j,J_t) using continuation maps. Next, suppose that $(\Sigma,\boldsymbol{\alpha},\boldsymbol{\beta},z)$ and $(\Sigma,\boldsymbol{\alpha},\boldsymbol{\beta}',z)$ are diagrams of (Y,p) (in particular, $z=p$), and that the triple diagram $(\Sigma,\boldsymbol{\alpha},\boldsymbol{\beta},\boldsymbol{\beta}',z)$ is admissible. Note that $(\Sigma,\boldsymbol{\beta},\boldsymbol{\beta}')$ is a diagram of $\#^g(S^1\times S^2)$, and that the group $\widehat{HF}(\Sigma,\boldsymbol{\beta},\boldsymbol{\beta}',z,\mathfrak{s}_0)$ is isomorphic with $H_*(T^g;\mathbb{Z}_2)$. The "fundamental class" is denoted by $\Theta_{\beta,\beta'}$. Then we obtain a map

$$\Psi^{\boldsymbol{\alpha}}_{\boldsymbol{\beta}\to\boldsymbol{\beta}'}\colon HF^\circ(\Sigma,\boldsymbol{\alpha},\boldsymbol{\beta},z) \to HF^\circ(\Sigma,\boldsymbol{\alpha},\boldsymbol{\beta}',z)$$

by counting rigid pseudo-holomorphic triangles in $\mathrm{Sym}^g(\Sigma)$ with edges lying on $\mathbb{T}_\alpha$, $\mathbb{T}_\beta$, and $\mathbb{T}_{\beta'}$, and one corner mapping to $\Theta_{\beta,\beta'}$. Given admissible diagrams $\mathcal{H}=(\Sigma,\boldsymbol{\alpha},\boldsymbol{\beta},z)$ and $\mathcal{H}'=(\Sigma,\boldsymbol{\alpha}',\boldsymbol{\beta}',z)$

of (Y,p), we construct $F_{\mathcal{H},\mathcal{H}'}$ as follows. First, assume that the quadruple diagram $(\Sigma,\boldsymbol{\alpha},\boldsymbol{\alpha}',\boldsymbol{\beta},\boldsymbol{\beta}')$ is admissible. Then let

$$\Psi^{\boldsymbol{\alpha}\to\boldsymbol{\alpha}'}_{\boldsymbol{\beta}\to\boldsymbol{\beta}'} = \Psi^{\boldsymbol{\alpha}\to\boldsymbol{\alpha}'}_{\boldsymbol{\beta}'} \circ \Psi^{\boldsymbol{\alpha}}_{\boldsymbol{\beta}\to\boldsymbol{\beta}'}.$$

In the general case, we pick two sets of attaching curves $\overline{\boldsymbol{\alpha}}$ and $\overline{\boldsymbol{\beta}}$ such that the quadruple diagrams $(\Sigma,\boldsymbol{\alpha},\overline{\boldsymbol{\alpha}},\boldsymbol{\beta},\overline{\boldsymbol{\beta}})$ and $(\Sigma,\boldsymbol{\alpha}',\overline{\boldsymbol{\alpha}},\boldsymbol{\beta}',\overline{\boldsymbol{\beta}})$ are both admissible (this is always possible). Then let

$$F_{\mathcal{H},\mathcal{H}'} = \Psi^{\overline{\boldsymbol{\alpha}}\to\boldsymbol{\alpha}'}_{\overline{\boldsymbol{\beta}}\to\boldsymbol{\beta}'} \circ \Psi^{\boldsymbol{\alpha}\to\overline{\boldsymbol{\alpha}}}_{\boldsymbol{\beta}\to\overline{\boldsymbol{\beta}}}.$$

Now suppose that $\mathcal{H}'$ is obtained from $\mathcal{H}$ by a stabilization. In this case, $\Sigma'\backslash\Sigma$ is a punctured torus T, with a single α-curve α_0 and a single β-curve β_0 that intersect in a unique point θ. We obtain a map

$$CF^\circ(\mathcal{H}) \to CF^\circ(\mathcal{H}')$$

by mapping the generator $\mathbf{x}$ to $\mathbf{x}\times\theta$. This is a chain map when the complex structure is chosen such that the connected sum neck along ∂T is very long. The map induced on the homology is $F_{\mathcal{H},\mathcal{H}'}$. Similarly, when $\mathcal{H}'$ is obtained from $\mathcal{H}$ by a destabilization, then we take $F_{\mathcal{H},\mathcal{H}'} = F^{-1}_{\mathcal{H}',\mathcal{H}}$.

Finally, we define $F_{\mathcal{H},\mathcal{H}'}$ when $\mathcal{H}$ and $\mathcal{H}'$ are related by a diffeomorphism. In this case, d induces a symplectomorphism between $\mathrm{Sym}^g(\Sigma)$ and $\mathrm{Sym}^g(\Sigma')$ that maps $\mathbb{T}_\alpha$ to $\mathbb{T}_{\alpha'}$ and $\mathbb{T}_\beta$ to $\mathbb{T}_{\beta'}$. If we use a complex structure $\mathfrak{j}$ on Σ and $d_*(\mathfrak{j})$ on Σ', then d tautologically induces a map between $HF^\circ(\mathcal{H})$ and $HF^\circ(\mathcal{H}')$.

To obtain $F_{\mathcal{H},\mathcal{H}'}$ for an arbitrary pair of diagrams of (Y,p), we take a sequence of diagrams $\mathcal{H}_0,\dots,\mathcal{H}_n$ such that $\mathcal{H}_0=\mathcal{H}$, $\mathcal{H}_n=\mathcal{H}'$, and $\mathcal{H}_i$ and $\mathcal{H}_{i+1}$ are related by changing the α- and β-curves, a stabilization or destabilization, or a diffeomorphism that is isotopic to the identity in Y fixing p. Then we let

$$F_{\mathcal{H},\mathcal{H}'} = F_{\mathcal{H}_{n-1},\mathcal{H}_n} \circ \cdots \circ F_{\mathcal{H}_0,\mathcal{H}_1}.$$

The main result of our paper with Thurston [39] is that this is independent of the choice of sequence $\mathcal{H}_0,\dots,\mathcal{H}_n$. The idea of the proof is the following. Consider the space of gradient vector fields v on Y, these are vector fields that arise as the gradient of a smooth function on Y with respect to some Riemannian metric. Given a generic gradient vector field v, there is a contractible space of overcomplete Heegaard diagrams of Y compatible with v in the sense that Σ is transverse to v and contains the index zero and one critical points on its negative side, while the index two and three critical points on its positive side. The α-curves are obtained by taking the intersection of Σ with the unstable manifolds of the index one critical points, while the β-curves are obtained by intersecting the stable manifolds of the index two critical points with Σ. We call the diagram overcomplete because the number of α- and β-curves might exceed $g(\Sigma)$.

Given a generic 1-parameter family v_t of gradient vector fields, we can deform the associated diagram smoothly as long as v_t is generic. We then study what happens to the associated diagram as one passes a bifurcation value of t. These correspond to handleslides, a generalized form of (de)stabilization, and creation or cancellation of 0-homologous α- or β-curves. By writing the generalized (de)stabilizations as a sequence of regular Heegaard moves, and choosing suitable subdiagrams, this shows that any pair of diagrams can be connected by a sequence of Heegaard moves and isotoping the diagram in Y.

To show that any two sequences of moves give the same map $F_{\mathcal{H},\mathcal{H}'}$, it suffices to prove that if we compose the elementary maps along an arbitrary loop of diagrams, we get the identity. Corresponding to this loop of moves, we can construct a loop of gradient vector fields along ∂D^2. Then we extend this to a generic 2-parameter vector field over D^2. We subdivide the disk into small polygons, and show that the composition around each small polygon is zero. For this end, one has to understand the different types of codimension-two bifurcations that appear in such a 2-parameter family, and then translate these to loops of diagrams. There are many cases which make the discussion rather complicated. Along the codimension-one strata, one has to resolve the generalized (de)stabilizations, pick suitable subdiagrams of the overcomplete diagrams, and then write each small loop as a product of elementary loops. Finally, we check that HF has no monodromy for each such elementary loop. Most of the elementary loops were already considered by Ozsváth and Szabó in [79], except for a new type of loop called a simple handleswap.

As mentioned earlier, $HF^\circ(Y,p)$ is independent of the choice of basepoint p when $\circ$ is one of $+$, $-$, or ∞. However, $\widehat{HF}(Y,p)$ does depend on p, but the action of $\pi_1(Y,p)$ factors through $H_1(Y)/\text{Tors}$.

5.2. *Cobordism maps*

Given a Spin^c-cobordism $(W,\mathfrak{s})$ from $(Y_0,\mathfrak{s}_0)$ to $(Y_1,\mathfrak{s}_1)$, Ozsváth and Szabó associate to it a homomorphism

$$F^\circ_{W,\mathfrak{s}}\colon HF^\circ(Y_0,\mathfrak{s}_0) \to HF^\circ(Y_1,\mathfrak{s}_1).$$

This is defined in terms of a relative handle decomposition of W built on $Y_0 \times I$ with no 0- and 4-handles, and associating maps to each i-handle attachment for $i \in \{1,2,3\}$. The 1-handle maps are similar to the stabilization isomorphisms used in the proof of invariance: One takes the connected sum of the diagram with a torus with a new α- and a new β-curve that intersect in a single point θ, and map each intersection point $\mathbf{x} \in \mathbb{T}_\alpha \cap \mathbb{T}_\beta$ to $\mathbf{x} \times \{\theta\}$. This is a chain map if the complex structure on Σ is chosen such that connected sum neck is very long. The 3-handle map is the inverse of the one-handle map.

The key ingredient is the map associated to a 2-handle attachment. Suppose that W is obtained by attaching a 2-handle to Y_0 along a framed knot K. Take a diagram

$$(\Sigma, \boldsymbol{\alpha} = \{\alpha_1,\dots,\alpha_g\}, \boldsymbol{\beta}_K = \{\beta_1,\dots,\beta_{g-1}\})$$

of $Y_0 \backslash N(K)$. Then β_g is chosen such that it represents the meridian of K, while β'_g is given by the surgery slope along K, and we require that $|\beta_g \cap \beta'_g| = 1$. We write $\boldsymbol{\beta} = \boldsymbol{\beta}_K \cup \{\beta_g\}$ and $\boldsymbol{\beta}' = \boldsymbol{\beta}'_K \cup \{\beta'_g\}$, where $\boldsymbol{\beta}'_K = \{\beta'_1,\dots,\beta'_{g-1}\}$ is a small isotopic copy of $\boldsymbol{\beta}_K$ such that $|\beta_i \cap \beta'_i| = 2$ for every $i \in \{1,\dots,g-1\}$. Then $(\Sigma,\boldsymbol{\alpha},\boldsymbol{\beta})$ is a diagram of Y_0, while $(\Sigma,\boldsymbol{\alpha},\boldsymbol{\beta}')$ is a diagram of Y_1. Furthermore, $(\Sigma,\boldsymbol{\beta},\boldsymbol{\beta}')$ is a diagram of $\#^{g-1}(S^1 \times S^2)$, and there is a distinguished "top" generator $\theta \in \mathbb{T}_\beta \cap \mathbb{T}_{\beta'}$. The cobordism map is defined by counting pseudo-holomorphic triangles in $\mathrm{Sym}^g(\Sigma)$ with edges mapping to $\mathbb{T}_\alpha$, $\mathbb{T}_\beta$, and $\mathbb{T}_{\beta'}$, respectively, and such that one corner goes to θ, and whose homotopy class corresponds to $\mathfrak{s}$ in a suitable sense.

The difficult part is showing that the composition of all the maps associated to the handle attachments is independent of the choices made, including the choice of handle decomposition. This composition is then denoted $F^\circ_{W,\mathfrak{s}}$. For further details, we refer the reader to [37, 79].

5.3. *Computing HF*

What makes Heegaard Floer homology computable is the following observation. Let $(\Sigma, \boldsymbol{\alpha}, \boldsymbol{\beta})$ be a Heegaard diagram, $\mathfrak{j}$ a complex structure on Σ, and let

$$u\colon D^2 \to \mathrm{Sym}^g(\Sigma)$$

be a pseudo-holomorphic disk with respect to $\mathrm{Sym}^g(\mathfrak{j})$ connecting $\mathbf{x}, \mathbf{y} \in \mathbb{T}_\alpha \cap \mathbb{T}_\beta$. Then we can pull back the g-fold branched covering

$$\Sigma \times \mathrm{Sym}^{g-1}(\Sigma) \to \mathrm{Sym}^g(\Sigma)$$

to obtain a g-fold branched covering $p\colon S \to D^2$; i.e. the following diagram is commutative:

$$\begin{array}{ccccc} S & \xrightarrow{\overline{u}} & \Sigma \times \mathrm{Sym}^{g-1}(\Sigma) & \xrightarrow{\pi_1} & \Sigma \\ \downarrow{\scriptstyle p} & & \downarrow & & \\ D^2 & \xrightarrow{u} & \mathrm{Sym}^g(\Sigma). & & \end{array}$$

Let $\pi_1 : \Sigma \times \mathrm{Sym}^{g-1}(S) \to \Sigma$ be the projection onto the first factor. Then

$$f = \pi_1 \circ \overline{u}\colon S \to \Sigma$$

is a holomorphic map. The pair (p, f) completely determines u. Indeed, for $x \in D^2$, if $p^{-1}(x) = \{s_1, \ldots, s_g\} \subset S$ (this is a multiset, with pre-images of branched points counted with multiplicity), then

$$u(x) = \{f(s_1), \ldots, f(s_g)\} \in \mathrm{Sym}^g(\Sigma).$$

In the opposite direction, given a g-fold branched cover $p\colon S \to D^2$ and a holomorphic map $f\colon S \to \Sigma$ such that $f(p^{-1}(-i)) = \mathbf{x}$, $f(p^{-1}(i)) = \mathbf{y}$, and the arcs in between $p^{-1}(-i)$ and $p^{-1}(i)$ alternatingly map to $\boldsymbol{\alpha}$ and $\boldsymbol{\beta}$, the above formula defines a holomorphic representative of $\pi_2(\mathbf{x}, \mathbf{y})$.

So finding the holomorphic representatives of a homotopy class $\phi \in \pi_2(\mathbf{x}, \mathbf{y})$ is equivalent to finding g-fold branched coverings $p\colon S \to D^2$, together with a holomorphic map $f\colon S \to \Sigma$ such that $f_*[S]$ is the 2-chain $\mathcal{D}(\phi)$, and which satisfies the appropriate boundary conditions. Such maps f can often be found using the Riemann mapping theorem, but in general there is no algorithm known for completely determining their moduli space. Since the homology of the Heegaard Floer chain complex (and more generally, its chain homotopy type) is independent of the choice of j up to isomorphism, in concrete computations it is often helpful to choose a degenerate complex structure. If one knows some of the moduli spaces, together with the fact that $\partial^2 = 0$, it is sometimes possible to work out ∂ completely.

The above viewpoint was developed by Lipshitz [50] into a self-contained definition of Heegaard Floer homology that does not refer to Lagrangian intersection Floer homology. He called this the "cylindrical reformulation" of Heegaard Floer homology. Here, one studies pseudo-holomorphic curves in the 4-manifold $\Sigma \times \mathbb{R} \times I$ with suitable boundary conditions. Indeed, a holomorphic map $v\colon S \to \Sigma \times \mathbb{R} \times I$ can be projected to the factors Σ and $\mathbb{R} \times I$ (which is conformally equivalent to $D^2 \backslash \{\pm i\}$) to obtain a pair (p, f), and vice versa.

An important result in his paper is a formula for computing the Maslov index of an arbitrary domain $\mathcal{D}$ connecting intersection points $\mathbf{x}$ and $\mathbf{y}$. It consists of three terms. One is the Euler measure $e(\mathcal{D})$ of $\mathcal{D}$. To define this, assume that $\boldsymbol{\alpha}$ and $\boldsymbol{\beta}$ meet at right angles. If S is a surface with k acute right-angled corners and l obtuse right-angled corners, then

$$e(S) = \chi(S) - k/4 + l/4.$$

Every domain is a linear combination of such surfaces, and we extend e to domains linearly. The other two terms are the averages $n_{\mathbf{x}}(\mathcal{D})$ and $n_{\mathbf{y}}(\mathcal{D})$ of the coefficients of $\mathcal{D}$ at the points of $\mathbf{x}$ and $\mathbf{y}$, respectively (called point measures). So the Maslov index of $\mathcal{D}$ is given by

$$\mu(\mathcal{D}) = n_{\mathbf{x}}(\mathcal{D}) + n_{\mathbf{y}}(\mathcal{D}) + e(\mathcal{D}).$$

A homotopy class of Whitney disks ϕ is called α-injective if all of the multiplicities of its domain $\mathcal{D}(\phi)$ are 0 or 1, if its interior (i.e. the interior of the region with multiplicity 1) is disjoint from $\boldsymbol{\alpha}$, and its boundary contains intervals in each α_i. Using work of Oh [65], Ozsváth and Szabó proved [74, Proposition 3.9] that if a homotopy class of Whitney disks ϕ is α-injective, then we can perturb the α-curves such that the moduli space $\mathcal{M}(\phi)$ of $\mathrm{Sym}^g(\mathrm{j})$-holomorphic disks is smoothly cut out by its defining equation. So we can use the unperturbed complex structure $\mathrm{Sym}^g(\mathrm{j})$ on $\mathrm{Sym}^g(\Sigma)$ in that case. This holds for example when the domain of $\phi \in \pi_2(\mathbf{x}, \mathbf{y})$ has coefficient one in an embedded rectangle or bigon, and is zero elsewhere. In both cases, $\#\widetilde{\mathcal{M}}(\phi) = \pm 1$. This is the basis of the algorithm of Sarkar and Wang [89] for computing $\widehat{HF}$: One first performs isotopies on the α-curves in the Heegaard diagram $(\Sigma, \boldsymbol{\alpha}, \boldsymbol{\beta}, z)$ until all the components of $\Sigma \backslash (\boldsymbol{\alpha} \cup \boldsymbol{\beta})$ disjoint from z become bigons and rectangles. Then they show that the domain of every rigid holomorphic disk with multiplicity zero at z is an embedded bigon or rectangle, and we saw above that each of these contributes ± 1 to the boundary map. Hence, the differential for the hat version becomes easy to compute, without having to resort to complex analysis.

Manolescu, Ozsváth, and Sarkar [57] realized that one can use grid diagrams to algorithmically compute knot Floer homology. A grid diagram is a multi-pointed Heegaard diagram on the torus, where every α-curve is a longitude and every β-curve is a meridian, and these form a rectangular grid. In each row and in each column there is exactly one z and one w basepoint. This can be thought of as a sutured diagram for the knot complement, with several sutures on each boundary torus. The Floer homology of this computes a stabilized version of $\widehat{HFK}$, there is a factor of $\mathbb{Z}^2$ for each additional pair of basepoints on each link component. Since every domain having multiplicity zero at the basepoints is a rectangle, the boundary map is completely combinatorial.

A more efficient algorithm for computing knot Floer homology is currently being developed by Szabó, which is based on ideas coming from bordered Floer homology. This can effectively compute the knot Floer homology of knots of 13 crossings, possibly even more.

Ozsváth and Szabó [69] introduced an efficient algorithm for computing Heegaard Floer homology for 3-manifolds obtained by plumbing spheres along certain graphs. These include, for example, all Seifert fibered rational homology spheres. Motivated by this algorithm, Némethi [59] defined an invariant, called *lattice homology*, for any negative definite plumbed 3-manifold. He conjectured that this agrees with HF^- for rational homology 3-spheres:

Conjecture 5.2. *For any negative definite plumbing tree G, the lattice homology $\mathbb{HF}^-(G)$ agrees with the Heegaard Floer homology $HF^-(Y_G)$ of the corresponding 3-manifold Y_G.*

6. Sutured Floer Homology

Sutured Floer homology, defined by the author [34], is an invariant of certain 3-manifolds with boundary. It generalizes the hat version of Heegaard Floer homology, basically by allowing multiple basepoints, and letting the number of α- and β-curves differ from the genus of the Heegaard surface. We now review the necessary definitions.

Sutured manifolds were introduced by Gabai [17], mainly to prove the Property R conjecture: If zero framed Dehn surgery on a knot K in S^3 gives $S^1 \times S^2$, then K is the unknot. A sutured manifold is a pair (M, γ), where M is a compact oriented 3-manifold with boundary, and $\gamma \subset \partial M$, the sutures, is an oriented 1-manifold that divides the boundary into two subsurfaces $R_+(\gamma)$ and $R_-(\gamma)$. The orientation of $R_+(\gamma)$ agrees with that of ∂M, while $R_-(\gamma)$ and ∂M are oriented oppositely. Often, γ is considered to be a closed regular neighborhood of the sutures, while the sutures themselves are denoted by $s(\gamma)$. In the latter case, one should think of M as having corners along $\partial\gamma$. We write $R(\gamma) = R_+(\gamma) \cup R_-(\gamma)$. A sutured manifold (M, γ) is *taut* if M is irreducible, and $R(\gamma)$ is incompressible and Thurston norm minimizing in its homology class in $H_2(M, \gamma)$.

We say that (M, γ) is a *product sutured manifold* if there is a compact oriented surface R such that $M = R \times I$ and $\gamma = \partial R \times I$. Recall that after Conjecture 2.4, we defined the sutured manifolds $Y(p)$ and $Y(L)$ for a based 3-manifold (Y, p) and a link $L \subset Y$, respectively. The sutured manifold $Y(p)$ is obtained from Y by removing a ball about p, and putting a single suture on the resulting S^2 boundary component. The sutured manifold $Y(L)$ is obtained from Y by removing a tubular neighborhood of L, and the sutures consist of two oppositely oriented meridians on each torus boundary component.

Gabai used sutured manifolds to study taut foliations on 3-manifolds. A foliation of a sutured manifold (M, γ) is a codimension-one, transversely oriented foliation that is transverse to γ, and such that each component of $R(\gamma)$ is a leaf. Such a foliation $\mathcal{F}$ is called *taut* if there is a properly embedded curve or arc in M that is transverse to the foliation, and which intersects each leaf of $\mathcal{F}$ at least once. A deep theorem of Gabai [17] states that a sutured manifold (M, γ) carries a taut foliation if and only if it is taut. The difficult direction is showing that if (M, γ) is taut, then it carries a taut foliation. The idea of the proof is the following. One can define a complexity for sutured manifolds, which is minimal for products. Gabai showed that if (M, γ) is taut and not a product, then there is always a properly embedded, oriented surface S such that if we cut M along S and add the negative side of S to R_- and the positive side to R_+, then the resulting sutured manifold (M', γ') is of strictly smaller complexity. This operation is called a sutured manifold decomposition, and is denoted by

$$(M, \gamma) \rightsquigarrow^{S} (M', \gamma').$$

Hence one obtains a sequence of decompositions

$$(M, \gamma) \rightsquigarrow^{S_1} (M_1, \gamma_1) \rightsquigarrow^{S_2} \cdots \rightsquigarrow^{S_n} (M_n, \gamma_n)$$

resulting in a product sutured manifold (M_n, γ_n); he calls such a sequence a *sutured manifold hierarchy.* A product sutured manifold carries an obvious taut foliation, namely the product foliation. Given a sutured manifold decomposition $(M, \gamma) \rightsquigarrow^S (M', \gamma')$ and a taut foliation $\mathcal{F}'$ on (M', γ'), Gabai constructs a taut foliation $\mathcal{F}$ on (M, γ), assuming that the surface S is *well-groomed.* This is a technical condition on S, and the sutured manifold hierarchy starting with (M, γ) and ending in a product can be chosen such that each decomposing surface S_i is well-groomed. Starting with the product foliation on (M_n, γ_n), we end up with a taut foliation on (M, γ).

A sutured manifold (M, γ) is called *balanced* if $\chi(R_+(\gamma)) = \chi(R_-(\gamma))$, the manifold M has no closed components, and each component of ∂M has at least one suture. Note that the first condition automatically holds when (M, γ) is taut. The author [34] defined sutured Floer homology, which assigns a finitely generated Abelian group $SFH(M, \gamma)$ to every balanced sutured manifold (M, γ). It splits along *relative* Spin^c*-structures.* We define these next. Let v_0 be a vector field that points into M along $R_-(\gamma)$, points out of M along $R_+(\gamma)$, and which is tangent to ∂M and points from $R_-(\gamma)$ to $R_+(\gamma)$ along the sutures. A relative Spin^c-structure on (M, γ) is the homology class of a nowhere zero vector field v on M such that $v|_{\partial M} = v_0$, where two such vector fields v and v' are said to be homologous if they are homotopic through nowhere zero vector fields relative to ∂M in the complement of a ball in the interior of M. We denote the set of Spin^c-structures by $\text{Spin}^c(M, \gamma)$, this is an affine space over $H_1(M)$. As claimed above, we have a splitting

$$SFH(M, \gamma) = \bigoplus_{\mathfrak{s} \in \text{Spin}^c(M,\gamma)} SFH(M, \gamma, \mathfrak{s}).$$

The group $SFH(M, \gamma)$ is constructed in the spirit of the hat version of Heegaard Floer homology for closed 3-manifolds. We can represent every balanced sutured manifold (M, γ) by a multi-pointed Heegaard diagram $(\overline{\Sigma}, \boldsymbol{\alpha}, \boldsymbol{\beta}, \underline{z})$, where $\overline{\Sigma}$ is a closed oriented surface, $\boldsymbol{\alpha} = \{\alpha_1, \ldots, \alpha_d\}$ and $\boldsymbol{\beta} = \{\beta_1, \ldots, \beta_d\}$ are sets of pairwise disjoint oriented simple closed curves on $\overline{\Sigma}$, and $\underline{z} = \{z_1, \ldots, z_k\}$ is a set of points on $\overline{\Sigma} \backslash (\boldsymbol{\alpha} \cup \boldsymbol{\beta})$. Given such a multi-pointed diagram, we can associate to it a sutured manifold as follows. Take $\Sigma = \overline{\Sigma} \backslash N(\underline{z})$, and let M be the 3-manifold obtained by attaching 3-dimensional

2-handles to $\Sigma \times I$ along $\alpha_i \times \{0\}$ and $\beta_j \times \{1\}$ for every $i, j \in \{1, \ldots, d\}$. Finally, we take $\gamma = \partial\Sigma \times I$ and $s(\gamma) = \partial\Sigma \times \{1/2\}$. The subsurface $R_-(\gamma)$ of ∂M is obtained from $\Sigma \times \{0\}$ by compressing it along the curves $\alpha_i \times \{0\}$. Similarly, $R_+(\gamma)$ is obtained from $\Sigma \times \{1\}$ by compressing it along the curves $\beta_j \times \{1\}$. The condition that $\chi(R_-(\gamma)) = \chi(R_+(\gamma))$ is hence equivalent to having the same number of α- and β-curves. There will be at least one suture on each component of ∂M if and only if there is at least one basepoint in each component of $\overline{\Sigma}\backslash\boldsymbol{\alpha}$ and $\overline{\Sigma}\backslash\boldsymbol{\beta}$. For naturality purposes, we consider Σ to be a subsurface of M that divides M into two "sutured compression bodies." Each α-curve and β-curve bounds a disk in one of these two sutured compression bodies.

To define $SFH(M, \gamma)$, we consider the tori $\mathbb{T}_\alpha = \alpha_1 \times \cdots \times \alpha_d$ and $\mathbb{T}_\beta = \beta_1 \times \cdots \times \beta_d$ inside the symmetric product $\mathrm{Sym}^d(\overline{\Sigma})$. So, in the definition of $\widehat{HF}(Y)$ for a closed 3-manifold Y, it was just a coincidence that the exponent of the symmetric product was given by the genus of the Heegaard surface, in general, it is the number of the α- and β-curves. The balanced condition on (M, γ) (in particular, that $\chi(R_+(\gamma)) = \chi(R_-(\gamma)))$ is essential in ensuring that $\mathbb{T}_\alpha$ and $\mathbb{T}_\beta$ are both half-dimensional submanifolds of $\mathrm{Sym}^d(\overline{\Sigma})$. For a suitable symplectic form on $\mathrm{Sym}^d(\overline{\Sigma})$, the tori $\mathbb{T}_\alpha$ and $\mathbb{T}_\beta$ will be Lagrangian. Then $SFH(M, \gamma)$ is defined to be the Lagrangian intersection Floer homology of $\mathbb{T}_\alpha$ and $\mathbb{T}_\beta$ in $\mathrm{Sym}^d(\overline{\Sigma})$, where the boundary map counts disks disjoint from the hypersurfaces

$$V_{z_i} = \{z_i\} \times \mathrm{Sym}^{d-1}(\overline{\Sigma})$$

for $i \in \{1, \ldots, k\}$. More concretely, given intersection points $\mathbf{x}$, $\mathbf{y} \in \mathbb{T}_\alpha \cap \mathbb{T}_\beta$, the coefficient of $\mathbf{y}$ in $\partial\mathbf{x}$ is given by counting the algebraic number of points in the moduli space $\widetilde{\mathcal{M}}(\phi)$, where ϕ is a homotopy class of pseudo-holomorphic Whitney disks connecting $\mathbf{x}$ and $\mathbf{y}$ of Maslov-index one, and intersecting each V_{z_i} algebraically zero times. This way we obtain the chain complex $CF(\overline{\Sigma}, \boldsymbol{\alpha}, \boldsymbol{\beta}, \underline{z})$, whose homology is $SFH(\overline{\Sigma}, \boldsymbol{\alpha}, \boldsymbol{\beta}, \underline{z})$. For different diagrams, in [39], we construct canonical isomorphisms, and the limit of the arising transitive system of groups is $SFH(M, \gamma)$.

Just like in the case of $\widehat{HF}$, we need to assume that the diagram satisfies a weak admissibility condition, namely, that every nonzero periodic domain has a positive and a negative coefficient. This translates to a monotonicity condition in the language of Floer homology. Since every component of $\overline{\Sigma}\backslash\boldsymbol{\alpha}$ and $\overline{\Sigma}\backslash\boldsymbol{\beta}$ contains a basepoint (as (M,γ) is balanced), there are no nonzero periodic domains with boundary a linear combination of only α-curves, or only β-curves. Hence, there are no homotopy classes of disks with boundary lying entirely on $\mathbb{T}_\alpha$ or on $\mathbb{T}_\beta$, and disjoint from the hypersurfaces V_{z_i}. This ensures that in the sutured Floer chain complex $\partial^2 = 0$, as in the boundary of a 1-parameter family of pseudo-holomorphic Whitney disks, we do not have bubbles with boundary completely on one of the two Lagrangians, and only broken flow-lines appear.

Similarly to the closed case, we can assign a relative Spin^c-structure to every intersection point $\mathbf{x} \in \mathbb{T}_\alpha \cap \mathbb{T}_\beta$ by first taking a self-indexing Morse-function $f\colon M \to \mathbb{R}$ and a Riemannian metric on M such that f has no index zero or three critical points, $R_-(\gamma) = f^{-1}(0.5)$, $R_+(\gamma) = f^{-1}(2.5)$, $\Sigma = f^{-1}(1.5)$, and the gradient vector field v of f induces the diagram $(\Sigma, \boldsymbol{\alpha}, \boldsymbol{\beta})$ in the sense that the unstable manifolds of index one critical points intersect Σ in the α-curves, while the stable manifolds of the index two critical points intersect Σ in the β-curves. Corresponding to $\mathbf{x}$, there is a multi-trajectory $\gamma_{\mathbf{x}}$ of v connecting the index one and two critical points, and passing through the points of $\mathbf{x}$. If we modify v in a neighborhood of $\gamma_{\mathbf{x}}$ to a nowhere zero vector field v', the homology class of v' will be the Spin^c-structure $\mathfrak{s}(\mathbf{x})$. If there is a topological Whitney disk connecting $\mathbf{x}, \mathbf{y} \in \mathbb{T}_\alpha \cap \mathbb{T}_\beta$, then $\mathfrak{s}(\mathbf{x}) = \mathfrak{s}(\mathbf{y})$, hence the intersection points lying in a given Spin^c-structure generate a subcomplex of $CF(\overline{\Sigma}, \boldsymbol{\alpha}, \boldsymbol{\beta}, \underline{z})$. Each summand $SFH(M, \gamma, \mathfrak{s})$ carries a relative $\mathbb{Z}_{d(c_1(\mathfrak{s}))}$-grading, where $d(c_1(\mathfrak{s}))$ is the divisibility of the Chern class $c_1(\mathfrak{s}) \in H^2(M)$.

They key property of sutured Floer homology is that it behaves particularly well under sutured manifold decompositions. In particular, when performing a decomposition we get a subgroup. Let $(S, \partial S) \subset (M, \partial M)$ be a properly embedded surface. We say that a Spin^c-structure $\mathfrak{s} \in \text{Spin}^c(M, \gamma)$ is *outer* with respect to S if it can be represented by a nowhere zero vector field v on M such that $v_p \neq -(\nu_S)_p$ for every $p \in S$, where ν_S is the unit normal vector field of S with respect to some Riemannian metric on M. Let (M', γ') be the sutured manifold obtained by decomposing (M, γ) along S. Then outer Spin^c-structures are exactly the ones that arise by taking a relative Spin^c-structure on (M', γ'), and gluing S_- to S_+. Let R be a compact oriented surface with no closed components. We say that a curve $C \subset R$ is *boundary-coherent* if either $[C] \neq 0$ in $H_1(R)$, or if $[C] = 0$ and C is oriented as the boundary of its interior (i.e. the component R_1 of $R \backslash C$ that is disjoint from ∂R and satisfies $\partial R_1 = C$). We say that the decomposing surface S is *nice* if it is open, and for every component V of $R(\gamma)$, the set of closed components of $S \cap V$ consists of parallel oriented boundary-coherent simple closed curves. In particular, every well-groomed decomposing surface, in the terminology of Gabai, is nice. Now we are ready to state [35, Theorem 1.3], the *decomposition formula.*

Theorem 6.1. *Let (M, γ) be a balanced sutured manifold, and let $(M, \gamma) \rightsquigarrow^S (M', \gamma')$ be a sutured manifold decomposition such that S is nice. Then*

$$SFH(M', \gamma') \cong \bigoplus_{\mathfrak{s} \in O_S} SFH(M, \gamma, \mathfrak{s}).$$

In particular, $SFH(M', \gamma')$ is a direct summand of $SFH(M, \gamma)$.

Sketch of proof. The idea is that if S is nice, then one can find a diagram $(\Sigma, \boldsymbol{\alpha}, \boldsymbol{\beta})$ of (M, γ) that essentially contains S as a subsurface. More precisely, there is a subsurface P of Σ such that ∂P is a polygonal curve on Σ with corners being the points of $P \cap \partial\Sigma$ (recall that $\partial\Sigma = s(\gamma)$). Furthermore, we can label the edges of ∂P alternatingly A and B (in particular, $\partial P = A \cup B$) such that $A \cap \boldsymbol{\beta} = \varnothing$, $B \cap \boldsymbol{\alpha} = \varnothing$, and we obtain S after smoothing the corners of

$$(B \times [0, 1/2]) \cup (P \times \{1/2\}) \cup (A \times [1/2, 1]) \subset M.$$

We call the tuple $(\Sigma, \boldsymbol{\alpha}, \boldsymbol{\beta}, P)$ a *surface diagram.*

Given a surface diagram, we can construct a diagram $(\Sigma', \boldsymbol{\alpha}', \boldsymbol{\beta}')$ of (M', γ') as follows. To obtain Σ', we take two copies P_A and P_B of P, then glue P_A to $\Sigma \backslash P$ along A, and then we glue P_B to the resulting surface along B. Loosely speaking, we have doubled P. There is a projection map $p \colon \Sigma' \to \Sigma$ that is two-to-one over P and is one-to-one over $\Sigma \backslash P$. Then we let $\boldsymbol{\alpha}' = p^{-1}(\boldsymbol{\alpha}) \backslash P_B$ and $\boldsymbol{\beta}' = p^{-1}(\boldsymbol{\beta}) \backslash P_A$; i.e. we lift the α-curves to P_A and the β-curves to P_B.

We call a generator $\mathbf{x} \in \mathbb{T}_\alpha \cap \mathbb{T}_\beta$ outer if $\mathbf{x} \cap P = \varnothing$. These are exactly the intersection points with $\mathfrak{s}(\mathbf{x}) \in O_S$. We denote the set of outer generators O_P. The projection map p gives a bijection between $\mathbb{T}_{\alpha'} \cap \mathbb{T}_{\beta'}$ and O_P. The difficult part of the proof is showing that for some choice of $\boldsymbol{\alpha}$ and $\boldsymbol{\beta}$, this bijection is an isomorphism of chain complexes. To achieve this, we elaborate on the Sarkar–Wang algorithm [89], and wind the α- and β-curves until each component of $\Sigma \backslash (\boldsymbol{\alpha} \cup \boldsymbol{\beta} \cup A \cup B)$ disjoint from $\partial \Sigma$ becomes either a bigon or a rectangle. Then one can show that the domain of every Maslov index one domain in $(\Sigma, \boldsymbol{\alpha}, \boldsymbol{\beta})$ connecting two elements of O_P is an embedded bigon or rectangle that can be lifted to a corresponding bigon or rectangle via the projection map p. Hence, the map p establishes an isomorphism between $CF(\Sigma', \boldsymbol{\alpha}', \boldsymbol{\beta}')$ and the subcomplex of $CF(\Sigma, \boldsymbol{\alpha}, \boldsymbol{\beta})$ generated by O_P. □

This theorem has several generalizations, which provide alternative proofs to the original statement. One is a gluing formula for convex decomposition due to Honda, Kazez, and Matić [30]. Given a convex decomposition $(M, \gamma) \rightsquigarrow (M', \gamma')$, they define a gluing map $SFH(M', \gamma') \to SFH(M, \gamma)$, and show that this is an embedding for a sutured manifold decomposition. Another generalization is due to Grigsby and Wehrli [26], who prove a decomposition formula for sutured-multi-diagrams via a neck-stretching argument. A fourth proof is due to Zarev [95].

The decomposition formula has numerous nice implications. To be able to state these, we first review the definition of the hat version of knot Floer homology in terms of sutured Floer homology. Given a knot or link K in a closed oriented 3-manifold Y, let $Y(K)$ denote the knot complement, together with two oppositely oriented meridional sutures on each boundary torus. Then we let

$$\widehat{HFK}(Y, K) = SFH(Y(K)).$$

If we fix a Seifert surface S of K, the group $\widehat{HFK}(Y, K)$ carries the so-called Alexander grading, which essentially comes from the grading by $\text{Spin}^c(Y(K))$. The correspondence is via evaluating a relative version of $c_1(\mathfrak{s})$ on $[S]$: if $\langle c_1(\mathfrak{s}), [S]\rangle = 2i$, then $SFH(Y(K), \mathfrak{s})$ lies in Alexander grading i. The following is a special case of [35, Theorem 1.5].

Proposition 6.2. *Let K be a null-homologous knot in a rational homology 3-sphere Y, and let S be a Seifert surface of K. Then*

$$SFH(Y(S)) \cong \widehat{HFK}(Y, K, g(S)).$$

Sketch of proof. Recall that $Y(S)$ is the sutured manifold obtained by decomposing $Y(K)$ along S. By the decomposition formula, all we need to do is to identify the subset O_S of $\text{Spin}^c(Y(K))$ that survive the decomposition. It turns out that O_S consists of a single element, characterized by $\langle c_1(\mathfrak{s}), [S]\rangle = -2g(S)$. Hence

$$SFH(Y(S)) \cong \widehat{HFK}(Y, K, -g(S)) \cong \widehat{HFK}(Y, K, g(S)),$$

where the second isomorphism follows from a simple symmetry property of knot Floer homology. □

This allows us to translate results on sutured Floer homology to knot Floer homology. The following theorem says that sutured Floer homology detects tautness.

Theorem 6.3. *Let (M, γ) be an irreducible balanced sutured manifold. Then (M, γ) is taut if and only if $SFH(M, \gamma) \neq 0$.*

Sketch of proof. Suppose that (M, γ) is taut. Then, by the work of Gabai [17], there is a sutured manifold hierarchy

$$(M, \gamma) \rightsquigarrow^{S_1} (M_1, \gamma_1) \rightsquigarrow^{S_2} \cdots \rightsquigarrow^{S_n} (M_n, \gamma_n)$$

such that each surface S_i is well-groomed, and hence nice. Here (M_n, γ_n) is a product, so it has a diagram with no α- and β-curves, and $SFH(M_n, \gamma_n) \cong \mathbb{Z}$. By the decomposition formula,

$SFH(M_n, \gamma_n)$ is a direct summand of $SFH(M, \gamma)$, which implies that $\text{rk}\, SFH(M, \gamma) \geq 1$. When (M, γ) is not taut, then one can construct a diagram where $\mathbb{T}_\alpha \cap \mathbb{T}_\beta = \varnothing$. □

Using Proposition 6.2, the above theorem translates to the following result of Ozsváth and Szabó [72], which states that knot Floer homology detects the genus of a knot.

Theorem 6.4. *Let K be a knot in the rational homology 3-sphere Y with Seifert genus $g(K)$. Then*

$$\widehat{HFK}(K, g(K)) \neq 0;$$

moreover, $\widehat{HFK}(K, i) = 0$ for $i > g(K)$.

Proof. For every $i \geq g(K)$, we can choose a Seifert surface S for K such that $g(S) = i$. Consider the decomposition $Y(K) \rightsquigarrow^S Y(S)$. Proposition 6.2 implies that $\widehat{HFK}(K, i) \cong SFH(Y(S))$. The sutured manifold $Y(S)$ is irreducible, and it is taut if and only if $g(S) = g(K)$, so the result follows from Theorem 6.3. □

Note that the original proof of the above theorem also relied on Gabai's result on sutured manifold hierarchies, plus a theorem of Eliashberg and Thurston that every taut foliation can be perturbed into a tight contact structure, and then showing that the associated contact invariant in Heegaard Floer homology is nonzero via Stein fillings and Lefshetz pencils. The proof presented here only relies on Gabai's theorem and the decomposition formula for *SFH*.

We say that a decomposing surface S in (M, γ) is *horizontal* if it is open and incompressible, $\partial S = s(\gamma)$, $[S] = [R_+(\gamma)]$ in $H_2(M, \gamma)$, and $\chi(S) = \chi(R_+(\gamma))$. The sutured manifold (M, γ) is *horizontally prime* if every horizontal surface is parallel to either $R_+(\gamma)$ or $R_-(\gamma)$. It follows from the decomposition formula that decomposing along a horizontal surface does not change *SFH*.

A properly embedded annulus A in (M, γ) is called a *product annulus* if one component of ∂A lies in $R_-(\gamma)$, while the other component lies in $R_+(\gamma)$. The sutured manifold (M, γ) is said to be *reduced* if every incompressible product annulus A in (M, γ) is isotopic to a component of γ such that ∂A stays in $R(\gamma)$ throughout.

The author showed in [35] that if we have a sutured manifold decomposition $(M, \gamma) \rightsquigarrow^{A} (M', \gamma')$ such that A is a nice product annulus, then $SFH(M, \gamma) \cong SFH(M', \gamma')$. However, note that the Spin^c-gradings might be different on the two groups.

Now we state the following important result from [35].

Theorem 6.5. *Suppose that (M, γ) is a taut balanced sutured manifold that is not a product. Then*

$$\operatorname{rk} SFH(M, \gamma) \geq 2.$$

Sketch of proof. First, we note that if $H_2(M) \neq 0$, then $\chi(SFH(M, \gamma)) = 0$. Since (M, γ) is taut, $\operatorname{rk} SFH(M, \gamma) \geq 1$, so in fact $\operatorname{rk} SFH(M, \gamma) \geq 2$. Hence, it suffices to consider the case when $H_2(M) = 0$.

Our goal is to construct two different taut decompositions

$$(M, \gamma) \rightsquigarrow^{S_+} (M'_+, \gamma'_+) \quad \text{and} \quad (M, \gamma) \rightsquigarrow^{S_-} (M'_-, \gamma'_-)$$

such that $O_{S_+} \cap O_{S_-} = \varnothing$. If we can achieve this, then the decomposition formula implies that $SFH(M, \gamma)$ has a subgroup isomorphic with $SFH(M'_+, \gamma'_+) \oplus SFH(M'_-, \gamma'_-)$, which is of rank at least two since (M'_+, γ'_+) and (M'_-, γ'_-) are both taut.

Before we can carry out the above plan, we decompose (M, γ) along horizontal surfaces and product annuli until it becomes reduced and horizontally prime. This does not change $SFH(M, \gamma)$. Then we take an arbitrary nonzero class $\alpha \in H_2(M, \partial M)$. There are nice taut decomposing surfaces S_+ and S_- such that $[S_+] = \alpha$ and $[S_-] = -\alpha$ in $H_2(M, \partial M)$. (We can assume that S_- and S_+ have no closed components since $H_2(M) = 0$.) The rest of the argument uses cut-and-paste techniques to show that $O_{S_+} \cap O_{S_-} = \varnothing$, assuming that (M, γ) is reduced and horizontally prime. $\square$

Recall that $SFH(M, \gamma) \cong \mathbb{Z}$ for a product sutured manifold (M, γ). So, if (M, γ) is irreducible, then $SFH(M, \gamma) \cong \mathbb{Z}$ if and only if (M, γ) is a product. Indeed, if (M, γ) is not taut, then $SFH(M, \gamma) = 0$, and if it is taut but not a product, then $\operatorname{rk} SFH(M, \gamma) \geq 2$. Using Proposition 6.2, we can translate this to the following result, which states that knot Floer homology detects fibered knots. The genus one case was proved by Ghiggini [22], and the general case was proved simultaneously by Ni [60, 61] and the author [35, 38].

Theorem 6.6. *Let K be a null-homologous knot in the oriented 3-manifold Y such that $Y \setminus K$ is irreducible, and let S be a Seifert surface for K. Then*

$$\mathrm{rk}\widehat{HFK}(Y, K, [S], g(S)) = 1$$

if and only if K is fibered with fiber S.

Given a balanced sutured manifold (M, γ), let $S(M, \gamma) \subset \mathrm{Spin}^c(M, \gamma)$ be the support of $SFH(M, \gamma)$; i.e.

$$S(M, \gamma) = \{\mathfrak{s} \in \mathrm{Spin}^c(M, \gamma) : SFH(M, \gamma, \mathfrak{s}) \neq 0\}.$$

Choose an arbitrary affine isomorphism i between $\mathrm{Spin}^c(M, \gamma)$ and $H_1(M; \mathbb{Z})$, and let $j \colon H_1(M; \mathbb{Z}) \to H_1(M, \mathbb{R})$ be the map induced by $\mathbb{Z} \hookrightarrow \mathbb{R}$. We denote by $P(M, \gamma)$ the convex hull of $j \circ i(S(M, \gamma))$ in $H_1(M; \mathbb{R})$, we call this the *sutured Floer homology polytope* of (M, γ) (here, we have deviated slightly from the conventions of [38] to simplify the discussion). The main technical result of [38] is the following, which can be viewed as a generalization of Theorem 6.5.

Theorem 6.7. *Let (M, γ) be a taut balanced sutured manifold that is reduced and horizontally prime, and suppose that $H_2(M) = 0$. Then*

$$\dim P(M, \gamma) = \dim H_1(M; \mathbb{R}).$$

In particular,

$$\mathrm{rk}SFH(M, \gamma) \geq b_1(\partial M)/2 + 1.$$

The above results can be used to study Seifert surfaces of knots and links. There are several natural notions of equivalence between Seifert surfaces. We say that the Seifert surfaces R and R' of a knot K in Y are *strongly equivalent* if they are isotopic in the knot complement $X(K) = Y \setminus N(K)$. If R is a Seifert surface of a knot K and R' is a Seifert surface of a knot K', then S and S' are *weakly equivalent* if they are ambient isotopic in Y. The following is [36, Theorem 2.3].

Theorem 6.8. *Let K be a knot in S^3 of genus g, and let $n > 0$ be an integer. If*

$$\operatorname{rk} \widehat{HFK}(K, g) < 2^{n+1},$$

then K has at most n disjoint, pairwise strongly inequivalent minimal genus Seifert surfaces.

Proof. Suppose that $R_1, \ldots, R_n$ are disjoint, strongly inequivalent minimal genus Seifert surfaces for K. Then $R_2, \ldots, R_n$ are disjoint non-isotopic horizontal surfaces in the sutured manifold $S^3(R_1)$ complementary to R_1. If we decompose $S^3(R_1)$ along $R_2, \ldots, R_n$, the rank of *SFH* remains unchanged as each such surface is null-homologous. We end up with the disjoint union of n sutured manifolds, none of which is a product as the R_i are non-isotopic. Hence, by Theorem 6.5, the *SFH* of each piece has rank at least two, and $\operatorname{rk} SFH(S^3(R_1)) \geq 2^n$. □

As a special case, we obtain that if the leading coefficient of the Alexander polynomial of an alternating knot has absolute value less than four, then it has a unique minimal genus Seifert surface up to strong equivalence. This was not known before. An elementary proof of this fact was later given by Banks [5].

SFH is also useful for distinguishing Seifert surfaces up to weak equivalence. Given Seifert surfaces R and R' in S^3 such that ∂R and $\partial R'$ are ambient isotopic, Proposition 6.2 might be discouraging as it implies that

$$SFH(S^3(R)) \cong SFH(S^3(R')).$$

However, recall that these groups are graded by Spin^c-structures, which are affine spaces over $H_1(S^3 \backslash R)$ and $H_1(S^3 \backslash R')$, respectively. Often, even these graded groups are different. However, in many of these cases, if we also consider the Seifert forms on $H_1(R) \cong H_1(S^3 \backslash R)$ and on $H_1(R') \cong H_1(S^3 \backslash R')$ (where the isomorphisms are given by Alexander duality), we can distinguish R and R'. Note that the Seifert form encodes the way R_+ and R_- are glued together in $S^3(R)$. This idea was presented by Hedden, Sarkar, and the author [27]. The first example of Seifert surfaces R and R' where the graded groups themselves are different was given by Altman [2].

Sutured Floer homology currently has extensions in two different directions. One is bordered sutured Floer homology, due to Zarev [95]. In his theory, part of the boundary of the manifold is sutured, part of it is bordered. He obtains various nice gluing results in this context. The second is a generalization of the minus version of Heegaard Floer homology to sutured manifolds, due to Alishahi and Eftekhary [1]. The chain complex they define is over an algebra depending on the sutures, and is well-defined up to chain homotopy equivalence. The relations in the algebra correspond to the disks with boundary entirely in $\mathbb{T}_\alpha$ or $\mathbb{T}_\beta$, and which obstruct $\partial^2 = 0$.

Kronheimer and Mrowka [42] extended instanton and monopole Floer homology to balanced sutured manifolds. Applied to the sutured manifold complementary to a link, they obtained new link invariants in the instanton and monopole settings. This led to a new, considerably simpler proof of the Property P conjecture, and a proof of the fact that Khovanov homology detects the unknot, see [43]. We now outline the proof of the latter result. Kronheimer and Mrowka [41] defined another invariant of links using singular instantons, and showed that for knots it agrees with the sutured instanton knot invariant. They exhibited that there is a spectral sequence starting from reduced Khovanov homology and converging to the singular instanton knot invariant. The key step in the construction of the spectral sequence is showing that the singular instanton knot invariant satisfies an unoriented skein exact triangle. Finally, they proved an analogue of the decomposition formula for the sutured instanton invariant, which, using the arguments outlined earlier, implies that the sutured instanton knot invariant detects the Seifert genus, and hence in particular it detects the unknot. Another consequence of the above discussion is that if one could show that knot Floer homology is isomorphic to the sutured instanton knot invariant, then one would get a positive answer to Conjecture 2.2 for knots; i.e. that there is a spectral sequence from reduced Khovanov homology to knot Floer homology.

It is worth mentioning that bordered Floer homology, just like *SFH*, is also defined using Heegaard surfaces with boundary. But whereas in *SFH* all the α- and β-curves lie in the interior of Σ, in the bordered theory one might also have α- or β- arcs with ends on $\partial\Sigma$. Then they consider Lipshitz's cylindrical reformulation [50], and count curves in the 4-manifold $\Sigma \times I \times \mathbb{R}$. The complex structure on Σ is chosen such that the boundary becomes a puncture. The bordered algebra associated with the boundary of the manifold encodes how holomorphic curves limit to $\partial\Sigma$. The theory lends itself to nice gluing formulas, obtained by cutting an ordinary Heegaard diagram into two pieces along a curve.

Acknowledgments

I would like to thank Lino Campos Amorim, Dominic Joyce, Yanki Lekili, Ciprian Manolescu, Peter Ozsváth, Alexander Ritter, and Zoltán Szabó for helpful discussions, and Fyodor Gainullin, Cagatay Kutluhan, Marco Marengon, Goncalo Oliveira, Jacob Rasmussen, and András Stipsicz for their comments on earlier versions of this paper. I was supported by a Royal Society Research Fellowship and OTKA grant NK81203

References

[1] A. Alishahi and E. Eftekhary, A refinement of sutured Floer homology, math.GT/1112.3540.

[2] I. Altman, Sutured Floer homology distinguishes between Seifert surfaces, *Topol. Appl.* **159** (2012) 3143–3155.

[3] M. Atiyah, New Invariants of Three and Four Dimensional Manifolds, *Proc. Symp. Pure Math.*, Vol. 48 (1988).

[4] ———, Topological quantum field theories, *Inst. Hautes Études Sci. Publ. Math.* **68** (1988) 175–186.

[5] J. Banks, Homogeneous links, Seifert surfaces, digraphs and the reduced Alexander polynomial, *Geom. Dedicata* **166** (2013) 67–98.

[6] S. Boyer, C. Gordon and L. Watson, On *L*-spaces and left-orderable fundamental groups, *Math. Ann.* **356** (2013) 1213–1245.

[7] V. Colin, P. Ghiggini and K. Honda, Equivalence of Heegaard Floer homology and embedded contact homology via open book decompositions, *Proc. Natl. Acad. Sci. USA* **108** (2011) 8100–8105.

[8] ———, The equivalence of Heegaard Floer homology and embedded contact homology via open book decompositions II, arXiv:1208.1077.

[9] ———, The equivalence of Heegaard Floer homology and embedded contact homology via open book decompositions III: From hat to plus, arXiv:1208.1526.

[10] M. Culler, C. Gordon, J. Luecke and P. Shalen, Dehn surgery on knots, *Ann. Math.* **125** (1987) 237–300.

[11] S. K. Donaldson, Polynomial invariants for smooth four-manifolds, *Topology* **29** (1990) 257–315.

[12] N. Dunfield, S. Gukov and J. Rasmussen, The superpolynomial for knot homologies, *Experiment. Math.* **15** (2006) 129–159.

[13] S. Eilenberg and N. Steenrod, *Foundations of Alegbraic Topology*, Princeton Univ. Press, 1952.

[14] A. Floer, Morse theory for Lagrangian intersections, *J. Diff. Geom.* **28** (1988) 513–547.

[15] K. Froyshov, The Seiberg–Witten equations and four-manifolds with boundary, *Math. Res. Lett.* **3** (1996) 373–390.

[16] K. Fukaya, Y.-G. Oh, H. Ohta and K. Ono, *Lagrangian Intersection Floer Theory: Anomaly and Obstruction.* Part I, AMS/IP Studies in Advanced Mathematics, Vol. 46.1, Amer. Math. Soc., Providence, RI; International Press, Somerville, MA, 2009.

[17] D. Gabai, Foliations and the topology of 3-manifolds, *J. Differential Geom.* **18** (1983) 445–503.

[18] ———, Detecting fibred links in S^3, *Comment. Math. Helv.* **61** (1986) 519–555.

[19] P. Ghiggini, Strongly fillable contact 3manifolds without stein fillings, *Geom. Topol.* **9** (2005) 1677–1687.

[20] ———, Infinitely many universally tight contact manifolds with trivial Ozsváth–Szabó contact invariants, *Geom. Topol.* **10** (2006) 335–357.

[21] ———, Ozsváth–Szabó invariants and fillability of contact structures, *Math. Z.* **253** (2006) 159–175.

[22] ———, Knot Floer homology detects genus-one fibred knots, *Amer. J. Math.* **130** (2008) 1151–1169.

[23] P. Ghiggini, P. Lisca and A. Stipsicz, Tight contact structures on some small Seifert fibered 3-manifolds, *Amer. J. Math.* **129** (2007) 1403–1447.

[24] J. E. Greene, Lattices, graphs, and Conway mutation, arXiv:1103.0487.

[25] ———, The lens space realization problem, *Ann. Math.* **177** (2013) 449–511.

[26] E. Grigsby and S. Wehrli, On the naturality of the spectral sequence from Khovanov homology to Heegaard Floer homology, *Int. Math. Res. Not.* **21** (2010) 4159–4210.

[27] M. Hedden, A. Juhász and S. Sarkar, On sutured Floer homology and the equivalence of Seifert surfaces, *Algebr. Geom. Topol.* **13** (2013) 505–548.

[28] M. Hedden and A. Levine, Splicing knot complements and bordered Floer homology, arXiv:1201.7055.

[29] M. Hedden and P. Ording, The Ozsváth–Szabó and Rasmussen concordance invariants are not equal, *Amer. J. Math.* **130** (2008) 441–453.

[30] K. Honda, W. Kazez and G. Matić, Contact structures, sutured Floer homology and TQFT, math.GT/0807.2431.
[31] M. Hutchings, An index inequality for embedded pseudoholomorphic curves in symplectizations, *J. Eur. Math. Soc.* **4** (2002) 313–361.
[32] M. Hutchings and C. H. Taubes, Gluing pseudoholomorphic curves along branched covered cylinders. I, *J. Symplectic Geom.* **5** (2007) 43–137.
[33] ———, Gluing pseudoholomorphic curves along branched covered cylinders. II, *J. Symplectic Geom.* **7** (2009) 29–133.
[34] A. Juhász, Holomorphic discs and sutured manifolds, *Algebr. Geom. Topol.* **6** (2006) 1429–1457.
[35] ———, Floer homology and surface decompositions, *Geom. Topol.* **12** (2008) 299–350.
[36] ———, Knot Floer homology and Seifert surfaces, *Algebr. Geom. Topol.* **8** (2008) 603–608.
[37] ———, Cobordisms of sutured manifolds, arXiv:0910.4382.
[38] ———, The sutured Floer homology polytope, *Geom. Topol.* **14** (2010) 1303–1354.
[39] A. Juhász and D. Thurston, Naturality and mapping classs groups in Heegaard Floer homology, arXiv:1210.4996.
[40] P. Kronheimer and T. Mrowka, Gauge theory for embedded surfaces I, *Topology* **32** (1993) 773–826.
[41] ———, Knot homology groups from instantons, arXiv:0806.1053.
[42] ———, Knots, sutures and excision, *J. Differential Geom.* **84** (2010) 301–364.
[43] ———, Khovanov homology is an unknot-detector, *Publ. Math. IHÉS* **113** (2011) 97–208.
[44] P. Kronheimer, T. Mrowka, P. Ozsváth and Z. Szabó, Monopoles and lens space surgeries, *Ann. Math.* **165** (2007) 457–546.
[45] C. Kutluhan, Y.-J. Lee and C. H. Taubes, *HF = HM I*: Heegaard Floer homology and Seiberg–Witten Floer homology, arXiv:1007.1979.
[46] ———, *HF=HM II*: Reeb orbits and holomorphic curves for the ech/Heegaard-Floer correspondence, arXiv:1008.1595.
[47] ———, *HF=HM III*: Holomorphic curves and the differential for the ech/Heegaard Floer correspondence, arXiv:1010.3456.
[48] ———, *HF=HM IV*: The Seiberg–Witten Floer homology and ech correspondence, arXiv:1107.2297.
[49] Y. Lekili, Heegaard–Floer homology for broken fibrations over the circle, *Adv. Math.* **244** (2013) 268–302.
[50] R. Lipshitz, A cylindrical reformulation of Heegaard Floer homology, *Geom. Topol.* **10** (2006) 955–1097.
[51] R. Lipshitz, P. Ozsváth and D. Thurston, Bordered Heegaard Floer homology: Invariance and pairing, arXiv:0810.0687.
[52] ———, Bimodules in Heegaard Floer homology, arXiv:1003.0598.
[53] ———, Tour of bordered Floer theory, *Proc. Natl. Acad. Sci. USA* **108** (2011) 8085–8092.

[54] P. Lisca and A. Stipsicz, On the existence of tight contact structures on Seifert fibered 3-manifolds, *Duke Math. J.* **148** (2009) 175–209.
[55] C. Manolescu, Grid diagrams in Heegaard Floer theory, arXiv:1210.3845.
[56] C. Manolescu and P. Ozsváth, Heegaard Floer homology and integer surgeries on links, arXiv:1011.1317.
[57] C. Manolescu, P. Ozsváth and S. Sarkar, A combinatorial description of knot Floer homology, *Ann. Math.* **169** (2009) 633–660.
[58] C. Manolescu, P. Ozsváth and D. Thurston, Grid diagrams and Heegaard Floer invariants. arXiv:0910.0078.
[59] A. Némethi, Lattice cohomology for normal surface singularities, *Publ. RIMS Kyoto Univ.* **44** (2008) 507–543.
[60] Y. Ni, Knot Floer homology detects fibred knots, *Invent. Math.* **170** (2007) 577–608.
[61] ———, Erratum: Knot Floer homology detects fibred knots, *Invent. Math.* **170** (2009) 235–238.
[62] ———, Heegaard Floer homology and fibred 3-manifolds, *Amer. J. Math.* **131** (2009) 1047–1063.
[63] Y-G. Oh, Floer cohomology of Lagrangian intersections and pseudo-holomorphic disks I, *Comm. Pure Appl. Math.* **46** (1993) 949–994.
[64] ———, Floer cohomology of Lagrangian intersections and pseudo-holomorphic disks II, *Comm. Pure Appl. Math.* **46** (1993) 995–1012.
[65] ———, Fredholm theory of holomorphic discs under the perturbation of boundary conditions, *Math. Z.* **222** (1996) 505–520.
[66] ———, On the structure of pseudo-holomorphic discs with totally real boundary conditions, *J. Geom. Anal.* **7** (1997) 305–327.
[67] P. Ozsváth and Z. Szabó, Absolutely graded Floer homologies and intersection forms for four-manifolds with boundary, *Adv. Mat.* **173** (2003) 179–261.
[68] ———, Knot Floer homology and the four-ball genus, *Geom. Topol.* **7** (2003) 615–639.
[69] ———, On the Floer homology of plumbed three-manifolds, *Geom. Topol.* **7** (2003) 185–224.
[70] ———, Heegaard diagrams and holomorphic disks, in *Different Faces of Geometry*, eds. S. Donaldson, Ya. Eliashberg and M. Gromov, Int. Math. Ser., Vol. 3, pp. 301–348, Kluwer/Plenum, New York, 2004.
[71] ———, Holomorphic disks and 3-manifold invariants: Properties and applications, *Ann. Math.* **159** (2004) 1159–1245.
[72] ———, Holomorphic disks and genus bounds, *Geom. Topol.* **8** (2004) 311–334.
[73] ———, Holomorphic disks and knot invariants, *Adv. Math.* **186** (2004) 58–116.
[74] ———, Holomorphic disks and topological invariants for closed three-manifolds, *Ann. Math.* **159** (2004) 1027–1158.
[75] ———, Heegaard Floer homology and contact structures, *Duke Math. J.* **129** (2005) 39–61.
[76] ———, Knots with unknotting number one and Heegaard Floer homology, *Topology* **44** (2005) 705–745.

[77] ———, On knot Floer homology and lens space surgeries, *Topology* **44** (2005) 1281–1300.
[78] ———, On the Heegaard Floer homology of branched double covers, *Adv. Math.* **194** (2005) 1–33.
[79] ———, Holomorphic triangles and invariants for smooth four–manifolds, *Adv. Math.* **202** (2006) 326–400.
[80] ———, Holomorphic disks, link invariants, and the multi-variable Alexander polynomial, *Algebr. Geom. Topol.* **8** (2008) 615–692.
[81] P. Ozsváth, Z. Szabó and D. Thurston, Legendrian knots, transverse knots and combinatorial Floer homology, *Geom. Topol.* **12** (2008) 941–980.
[82] T. Perutz, Hamiltonian handleslides for Heegaard Floer homology, in *Proc. of Gökova Geometry-Topology Conference* 2007, eds. S. Akbulut, T. Önder, and R. Stern, Gökova Geometry/Topology Conference, Gökova, 2008, pp. 15–35.
[83] J. Rasmussen, Floer homology and knot complements, Ph.D. thesis, Harvard University, (2003).
[84] ———, Knot polynomials and knot homologies, in *Geometry and Topology of Manifolds*, eds. H. Boden, I. Hambleton, A. Nicas and D. Park, Fields Inst. Commun., Vol. 47, pp. 261–280. Amer. Math. Soc., Providence, RI, 2005.
[85] ———, Khovanov homology and the slice genus, *Invent. Math.* **182** (2010) 419–447.
[86] K. Reidemeister, Zur dreidimensionalen Topologie, *Abh. Math. Sem. Univ. Hamburg* **9** (1933) 189–194.
[87] S. Sarkar, Grid diagrams and the Ozsváth–Szabó tau-invariant, *Math. Res. Lett.* **18** (2011) 1239–1257.
[88] ———, Moving basepoints and the induced automorphism of link Floer homology, arXiv:1109.2168.
[89] S. Sarkar and J. Wang, An algorithm for computing some Heegaard Floer homologies, *Ann. Math.* **171** (2010) 1213–1236.
[90] P. Seidel, Fukaya categories and deformations, in *Proc. International Congress of Mathematicians*, Vol. II, ed. T. Li, pp. 351–360, Higher Ed. Press, Beijing, 2002.
[91] J. Singer, Three-dimensional manifolds and their Heegaard diagrams, *Trans. Amer. Math. Soc.* **35** (1933) 88–111.
[92] C. H. Taubes, Embedded contact homology and Seiberg–Witten Floer cohomology I–V, *Geom. Topol.* **14** (2010) 2497–3000.
[93] V. Turaev, Euler structures, nonsingular vector fields, and Reidemeister-type torsions, *Math. USSR-Izv.* **34** (1990) 627–662.
[94] ———, Torsion invariants of Spin^c-structures on 3-manifolds, *Math. Res. Lett.* **4** (1997) 679–695.
[95] R. Zarev, Bordered Floer homology for sutured manifolds, math. GT/0908.1106.

On the Framization of Knot Algebras

Jesús Juyumaya* and Sofia Lambropoulou†

*Instituto de Matemáticas, Universidad de Valparaíso,
Gran Bretaña 1091, Valparaíso, Chile
juyumaya@uvach.cl

†Department of Mathematics, National Technical University of Athens,
Zografou campus, GR–157 80 Athens, Greece
sofia@math.ntua.gr

This paper presents results on the framization of some knot algebras, defined by the authors. We explain the motivations of the concept of framization, coming from the Yokonuma–Hecke algebras, as well as recent results on the framization of the Temperley–Lieb algebra. Finally, we propose framizations for other knot algebras such as the BMW algebra, the B-type related Hecke algebras and the singular Hecke algebra.

0. Introduction

Modular framization (or simply framization) is a mechanism proposed recently by the authors and it consists in constructing a nontrivial extension of a knot algebra via the addition of framing generators. In this way we obtain a new algebra which is related to framed braids and framed knots.

By knot algebra we mean an algebra that is applied to the construction of invariants of knots and links. We are focused on the framization of those knot algebras that define invariants of knots via the Jones' construction. More precisely, such a knot algebra A is a triplet (A, π, τ), where π is a representation of a braid group in A

Table 1. Examples of knot algebras.

Knot algebra	Invariant
Temperley–Lieb algebra	Jones polynomial and bracket polynomial
Iwahori–Hecke algebra	Homflypt polynomial
BMW algebra	Kauffman polynomial
B-type and affine Hecke algebras	Lambropoulou invariants
Singular Hecke algebra	Kauffman–Vogel and Paris–Rabenda invariants
Rook algebra	Alexander polynomial

and τ a Markov trace defined on A. The invariant obtained by the knot algebra A is constructed essentially from the composition $\tau \circ \pi$ after re-scaling and normalizing τ according to the braid equivalence in the given braid category. In Table 1 we list some knot algebras with related knot invariants.

The inspiring example of framization is the so-called Yokonuma–Hecke algebra. Indeed, this algebra corresponds to a framization of the Iwahori–Hecke algebra. The Yokonuma–Hecke algebra appears in the field of group theory and was defined by Yokonuma [44] as a generalization of the Iwahori–Hecke algebra. More precisely, he considered the centralizer algebra of the permutation representation associated to any finite Chevalley group with respect to one maximal unipotent subgroup. Then, in analogy to the classical presentation of the Iwahori–Hecke algebra, he found a presentation of the Yokonuma–Hecke algebra by generators and relations [44, Theorem 1].

In [19, 21] another presentation of the Yokonuma–Hecke algebra was established (Definition 2.1). By using this new presentation, the Yokonuma–Hecke algebra $\mathrm{Y}_{d,n}(u)$ was defined for any positive integers n and d and a fixed nonzero complex number u. Then, $\mathrm{Y}_{d,n}(u)$ could be naturally viewed as a quotient of the framed braid group $\mathcal{F}_n$ or as an algebra obtained from the Iwahori–Hecke algebra $\mathrm{H}_n(u)$ by adding framing generators and by replacing the Hecke algebra quadratic relation by a quadratic relation which also involves intrinsically the framing generators (Eq. (2.3)). For $d = 1$, the algebra $\mathrm{Y}_{1,n}(u)$ coincides with the algebra $\mathrm{H}_n(u)$.

Further, a Markov trace tr was constructed by the first author on the algebras $\mathrm{Y}_{d,n}(u)$ [20], the "Juyumaya trace", with parameters $z, x_1, \ldots, x_{d-1}$, using an appropriate inductive linear basis. Parameter z takes care of the braiding, while parameters $x_1, \ldots, x_{d-1}$ take care of the framing. For $d = 1$ the trace tr coincides with the well-known Ocneanu trace τ on the algebras $\mathrm{H}_n(u)$, from which the 2-variable Jones or Homflypt polynomial for oriented links was extracted [18].

Then, in order to obtain link invariants via the trace tr, the "E-condition" needed to be imposed on the framing parameters $x_1, \ldots, x_{d-1}$ (see Eq. (3.2)) for re-scaling tr in order that it conforms with negative stabilization (cf. [26]). The trace tr is the only known Markov trace on knot algebras that does not re-scale directly. As it was shown by Gérardin (cf. Appendix [26]) solutions of the "E-system" (3.3) are parametrized by the non-empty subsets of $\mathbb{Z}/d\mathbb{Z}$.

Subsequently, for any solution of the E-system, tr yielded an invariant for framed knots [26], for classical knots [25] and for singular knots [24]. Moreover, the algebras $\mathrm{Y}_{d,n}(u)$ with the trace tr seem to relate naturally to the domain of transverse knots [12]. All these invariants are still under investigation [8, 12], especially as to how they compare with the Homflypt polynomial. In [8] it is shown that only in trivial cases the classical link invariants coincide with the Homflypt polynomial. By construction these invariants are at least as strong as the Homflypt polynomial, however computational evidence indicates that they are topologically equivalent. This is not easy to see either by algebraic or by diagrammatic methods. In [12] some conjectures are stated in this direction. Still, we believe that it is remarkable that one can obtain invariants for all these different knot categories from a single algebra.

The Yokonuma–Hecke algebras, equipped with a Markov trace, are interesting on their own right. Their representation theory has been studied thoroughly in [10, 41]. In particular, in [10] a completely combinatorial approach is taken to the subject.

All these results are presented in Secs. 2–4. The above comprise our motivation for constructing framizations of other knot algebras. In this paper we present possible framizations of most of the algebras listed in Table 1. Up to Sec. 4.4 the paper is mostly a survey of results but afterwards it continues as an announcement of new results, some by the authors, some by the authors with co-authors and some by other authors.

In Sec. 5 we present three quotients of the Yokonuma–Hecke algebra $\mathrm{Y}_{d,n}(u)$ as possible framizations of the Temperley–Lieb algebra. For each one we present necessary and sufficient conditions for the trace tr to pass through to the quotient algebra and we discuss related un-oriented knot invariants. These results comprise the doctoral thesis of Goundaroulis [16, 17]. For the first quotient, the Yokonuma–Temperley–Lieb $\mathrm{YTL}_{d,n}(u)$, the ideal is similar to the one in the classical case. Then, as it turns out, the trace tr passes to $\mathrm{YTL}_{d,n}(u)$ only if the trace parameters x_i are dth roots of unity. In this case we recover the Jones polynomial. See [16]. The second candidate, $\mathrm{FTL}_{d,n}(u)$, which we select as the "Framization of the Temperley–Lieb algebra", has the property that the conditions on the x_i's so that tr passes through to the quotient, include explicitly all solutions of the E-system mentioned above. Finally, the conditions on the trace parameters for the third candidate, the Complex Reflection Temperley–Lieb algebra $\mathrm{CTL}_{d,n}(u)$, involve only parameter z and not the framing parameters $x_1, \ldots, x_{d-1}$. So, in order to obtain knot invariants from the algebra $\mathrm{CTL}_{d,n}(u)$ we have to impose the E-condition (Eq. (3.2)). It follows that the knot invariants we obtain coincide with those from $\mathrm{FTL}_{d,n}(u)$. The main disadvantage of $\mathrm{CTL}_{d,n}(u)$ is that it is unnecessarily large for our purposes.

In Sec. 6 we propose framizations for the Hecke algebra of B-type, for the cyclotomic Hecke algebras of B-type, and for the generalized Hecke algebra of B-type, which is isomorphic to the affine Hecke algebra of A-type. These definitions were first given in [27]. All these algebras are related to the knot theory of the solid torus and to the Lambropoulou invariants [14, 33, 34]. These B-type framizations are further studied in [11], where Markov traces are also constructed and a corresponding E-condition is given.

In Sec. 7 we propose a framization of the Birman–Murakami–Wenzl or simply BMW algebra [6, 35], which is related to the Kauffman polynomial invariant of knots [28]. This framization was introduced and further studied in [27].

Finally, in Sec. 8 we propose a framization of the singular Hecke algebra [37], which is related to the invariants of Kauffman–Vogel [31] and Paris–Rabenda [37]. This was introduced in [27].

All these framization knot algebras are related to the framed braid group and they are of interest to algebraists. There are many more other knot algebras, such as other quotients of the classical braid group, quotients of the virtual braid group [29, 30], or the Rook algebra [4] which is related to the Alexander polynomial. For all these one could construct appropriate framization counterparts.

1. Notations and Background

1.1. In this paper the term algebra means a $\mathbb{C}$-associative algebra with unity denoted by 1, where as usual $\mathbb{C}$ denotes the field of complex numbers. Notice that $\mathbb{C}$ can be regarded as included in the algebra as a central subalgebra. We also denote by $\mathbb{C}G$ the group algebra of a group G.

1.2. The letters n and d denote two positive integers. We denote by S_n the symmetric group on n-symbols and by s_i the elementary transposition $(i, i+1)$. Let B_n be the *classical Artin braid group*. B_n is presented by the braiding generators $\sigma_1, \dots, \sigma_{n-1}$ and the *braid relations*:

$$\begin{aligned} \sigma_i\sigma_j &= \sigma_j\sigma_i && \text{for } |i-j| > 1, \\ \sigma_i\sigma_j\sigma_i &= \sigma_j\sigma_i\sigma_j && \text{for } |i-j| = 1. \end{aligned} \tag{1.1}$$

The *framed braid group* $\mathcal{F}_n$ is the group defined by adding to the above presentation of B_n the framing generators $t_1, \dots, t_n$ and the following relations:

$$\begin{aligned} t_it_j &= t_jt_i && \text{for } 1 \leq i, j \leq n \\ t_j\sigma_i &= \sigma_i t_{s_i(j)} && \text{for } 1 \leq i \leq n-1 \quad \text{and} \quad 1 \leq j \leq n, \end{aligned} \tag{1.2}$$

where $s_i(j)$ is the result of applying s_i to j. The *d-modular framed braid group*, denoted $\mathcal{F}_{d,n}$, is defined by adding to the above presentation of $\mathcal{F}_n$ the relations:

$$t_i^d = 1 \quad \text{for } 1 \leq i \leq n. \tag{1.3}$$

We denote C the infinite cyclic group and by C_d the cyclic group of order d. We have $C \cong \mathbb{Z}$ and $C_d \cong \mathbb{Z}/d\mathbb{Z}$. Further, if t is a generator of C, the group C_d can be presented as $C_d = \langle t; t^d = 1 \rangle$. From the above we have: $\mathcal{F}_n = C^n \rtimes B_n$ and $\mathcal{F}_{d,n} = C_d^n \rtimes B_n$. Finally, we shall denote the group:

$$C_{d,n} := C_d^n \rtimes S_n. \tag{1.4}$$

1.3. From now on we fix a nonzero complex number u and a positive integer d.

Note 1.1. One of the authors does not agree with the denomination "Juyumaya trace" and the other author does not agree with the denomination "Lambropoulou invariants".

2. The Framization of the Hecke Algebra of Type A

2.1. The *Iwahori–Hecke algebra of type* A, $\mathrm{H}_n(u)$, is the algebra with $\mathbb{C}$-linear basis $\{h_w \mid w \in S_n\}$ and the following rules of multiplication:

$$h_{s_i} h_w = \begin{cases} h_{s_i w} & \text{for } \ell(s_i w) > \ell(w) \\ u h_{s_i w} + (u-1) h_w & \text{for } \ell(s_i w) < \ell(w) \end{cases}$$

where ℓ is the usual length function on the symmetric group. Set $h_i := h_{s_i}$. As usual we consider the presentation of the algebra $\mathrm{H}_n(u)$ by "braiding" generators $h_1, \ldots, h_{n-1}$, subject to the braid relations (1.1) together with the quadratic relations:

$$h_i^2 = (u-1) h_i + u. \tag{2.1}$$

Note that $\mathrm{H}_n(1)$ coincides with the group algebra $\mathbb{C}S_n$. The above presentation of $\mathrm{H}_n(u)$ says that the mapping $\pi : \sigma_i \mapsto h_i$ defines an epimorphism from $\mathbb{C}B_n$ onto $\mathrm{H}_n(u)$ whose kernel is generated by the expressions: $\sigma_i^2 - (u-1)\sigma_i - u$.

2.2. The Hecke algebra $\mathrm{H}_n(u)$ is the knot algebra used to define the 2-variable Jones or Homflypt polynomial according to the Jones construction [18]. Namely, one uses the Markov braid equivalence on $\cup_n B_n$, comprising conjugation in the groups B_n and positive and negative stabilization and destabilization ($\alpha\sigma_n \sim \alpha \sim \alpha\sigma_n^{-1}$; $\alpha \in B_n$), the map π from B_n to $\mathrm{H}_n(u)$ and the Ocneanu trace defined on $\mathrm{H}_n(u)$:

Theorem 2.1. (Theorem 5.1 [18]) *Let ζ be an indeterminate over $\mathbb{C}$. Then, there exists a unique $\mathbb{C}$-linear map τ from the inductive limit of the family $\{\mathrm{H}_n(u)\}_n$ to $\mathbb{C}(\zeta)$, such that $\tau(1) = 1$ and satisfying the following rules for all $a, b \in \mathrm{H}_n(u)$:*

$$\begin{aligned}\tau(ab) &= \tau(ba)\\ \tau(ah_n) &= \zeta\tau(a) \quad \textit{(Markov property)}.\end{aligned}$$

Diagrammatically, the Markov property of τ is illustrated on the left–hand side of Fig. 3.

Then τ has to be *normalized*, so that the closed braids $\widehat{\alpha}$ and $\widehat{\alpha\sigma_n}$ ($\alpha \in B_n$) be assigned the same value of the invariant, and also *re-scaled*, so that the closed braids $\widehat{\alpha\sigma_n^{-1}}$ and $\widehat{\alpha\sigma_n}$ get also assigned the same value of the invariant. So, defining:

$$\lambda := \frac{1-u+\zeta}{u\zeta} \quad \text{and} \quad C := \frac{1}{\zeta\sqrt{\lambda}}$$

then the *Homflypt polynomial* $P = P(u, \lambda)$ of an oriented link L is defined as follows:

$$P(\widehat{\alpha}) = C^{n-1}(\tau \circ \pi)(\alpha),$$

where α is a braid on n strands whose closure $\widehat{\alpha}$ is isotopic to L.

2.3. In group theory the Hecke algebra has a natural generalization, the Yokonuma–Hecke algebra $\mathrm{Y}_{d,n}(u)$, see [44]. In [19, 21] a new presentation was found for this algebra. Subsequently, in [20] this new presentation was considered for the Yokonuma–Hecke algebra and was proved that it supports a Markov trace. We define $\mathrm{Y}_{d,n}(u)$ as follows.

Definition 2.1. The *Yokonuma–Hecke algebra of type A*, $\mathrm{Y}_{d,n}(u)$, is the algebra defined by the braiding generators $g_1, \ldots, g_{n-1}$, the framing generators $t_1, \ldots, t_n$ and the relations:

$$\begin{aligned}
g_i g_j &= g_j g_i && \text{for } |i-j| > 1 \\
g_i g_j g_i &= g_j g_i g_j && \text{for } |i-j| = 1 \\
t_i t_j &= t_j t_i && \text{for } 1 \leq i, j \leq n \\
t_j g_i &= g_i t_{s_i(j)} && \text{for } 1 \leq i \leq n-1 \quad \text{and} \quad 1 \leq j \leq n \\
t_i^d &= 1 && \text{for } 1 \leq i \leq n
\end{aligned} \tag{2.2}$$

where $s_i(j)$ denotes the result of applying s_i to j, together with the following quadratic relations:

$$g_i^2 = 1 + (u-1)e_i + (u-1)e_i g_i \quad \text{for all } 1 \leq i \leq n-1, \tag{2.3}$$

where

$$e_i := \frac{1}{d} \sum_{s=0}^{d-1} t_i^s t_{i+1}^{d-s}. \tag{2.4}$$

The elements e_i are easily seen to be idempotents. Using this fact it follows from the quadratic relations (2.3) that the generators g_i are invertible:

$$g_i^{-1} = g_i + (u^{-1} - 1)e_i + (u^{-1} - 1)e_i g_i. \tag{2.5}$$

From the above presentation it is clear that the algebra $\mathrm{Y}_{d,n}(u)$ is a quotient of the modular framed braid group algebra $\mathbb{C}\mathcal{F}_{d,n}$ under the quadratic relations (2.3). This observation leads to diagrammatic interpretations for the elements in $\mathrm{Y}_{d,n}(u)$. For example, the elements e_i (which are in $\mathbb{C}\mathcal{F}_{d,n}$ and in $\mathrm{Y}_{d,n}(u)$) can be represented as in Fig. 1. Also, Eq. (2.5), which is equivalent to the quadratic relation, is illustrated in Fig. 2.

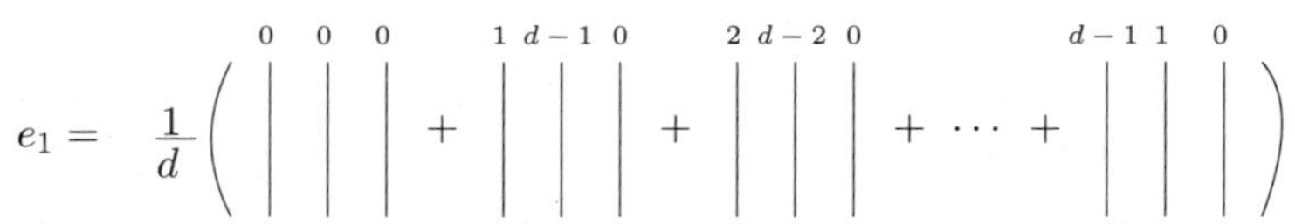

Fig. 1. The element $e_1 \in \mathbb{C}\mathcal{F}_{d,3}$.

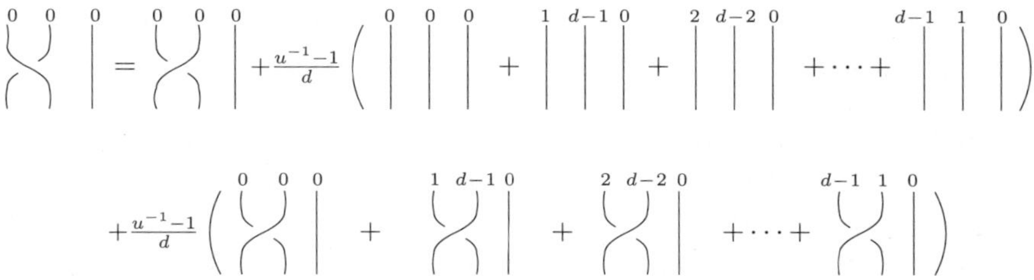

Fig. 2. The element $g_1^{-1} \in \mathrm{Y}_{d,3}(u)$.

2.4. A consequence of the above definition is that every word in the defining generators of $\mathrm{Y}_{d,n}(u)$ can be written in the *split form* $t_1^{a_1} \cdots t_n^{a_n}\, g$, where the a_i's are integers modulo d and g is a word in the g_i's. Since the g_i's satisfy the braid relations we have that, if $w = s_{i_1} \cdots s_{i_m} \in S_n$ is a reduced expression, then the following element $g_w := g_{i_1} \cdots g_{i_m}$ is well-defined. In [19] it is proved that the multiplication rules in $\mathrm{Y}_{d,n}(u)$ are governed by the group $C_{d,n}$. In fact, the multiplication rules among the framing generators and between the framing generators and the braiding generators are the same multiplication rules as in the group $C_{d,n}$. For the multiplication among the braiding generators we have:

$$g_i g_w = \begin{cases} g_{s_i w} & \text{for } \ell(s_i w) > \ell(w), \\ g_{s_i w} + (u-1)e_i g_{s_i w} + (u-1)e_i g_w & \text{for } \ell(s_i w) < \ell(w). \end{cases}$$

Notice now that $\mathrm{Y}_{d,n}(1) = \mathbb{C}C_{d,n}$. This says that $\mathrm{Y}_{d,n}(u)$ is essentially obtained from $\mathrm{H}_n(u)$ by adding framing generators, since $\mathrm{H}_n(1) = \mathbb{C}S_n$. For this reason we shall call the Yokonuma–Hecke algebra $\mathrm{Y}_{d,n}(u)$ a *framization of the Iwahori–Hecke algebra* $\mathrm{H}_n(u)$. The key point in this framization is the quadratic relation (2.3), which is considered as the framization of the Hecke algebra quadratic relation.

The representation theory of the Yokonuma–Hecke algebra has been studied in [41] and [10]. Finally, it is worth mentioning that the notions of the modular framed braid group $\mathcal{F}_{d,n}$ as well as of the Yokonuma–Hecke algebra $\mathrm{Y}_{d,n}(u)$ have been extended to constructions on the p-adic level [22, 23] and the adelic level [25].

2.5. Another crucial property of the Yokonuma–Hecke algebra is that it supports a Markov trace:

Theorem 2.2. (Theorem 12 [20]) *Let* $z, x_1, \ldots, x_{d-1}$ *be indeterminates over* $\mathbb{C}$*. Then, there exists a unique* $\mathbb{C}$*-linear map* tr *from the inductive limit of the family* $\{\mathrm{Y}_{d,n}(u)\}_n$ *to* $\mathbb{C}(z, x_1, \ldots, x_{d-1})$*, such that* $\mathrm{tr}(1) = 1$ *and satisfying the following rules for all* $a, b \in \mathrm{Y}_{d,n}(u)$ *:*

(1) $\mathrm{tr}(ab) = \mathrm{tr}(ba)$

(2) $\mathrm{tr}(ag_n) = z\,\mathrm{tr}(a)$ (*Markov property*)

(3) $\mathrm{tr}(at_{n+1}^m) = x_m\ \mathrm{tr}(a)$ $(1 \leq m \leq d-1)$.

The topological interpretations for rules (2) and (3) are given in Fig. 3.

The trace tr shall be called the *Juyumaya trace*. Note that for $d = 1$, the algebra $\mathrm{Y}_{1,n}(u)$ is the Iwahori–Hecke algebra $\mathrm{H}_n(u)$ and the Juyumaya trace coincides with the Ocneanu trace with parameter z. Note also that the trace tr lifts to the p-adic level [22, 26] and to the adelic level [25].

2.6. A Markov trace is a key ingredient for constructing knot invariants from an algebra. The first key requirement is that there is a representation of some braid category to the algebra. Another key requirement is to have a Markov-type braid equivalence for the braid category, corresponding to isotopy in the related knot category. The last key ingredient is to re-scale and normalize the trace according to the given braid equivalence.

In the case of the Yokonuma–Hecke algebras we have a natural connection with the framed braid category and with the classical

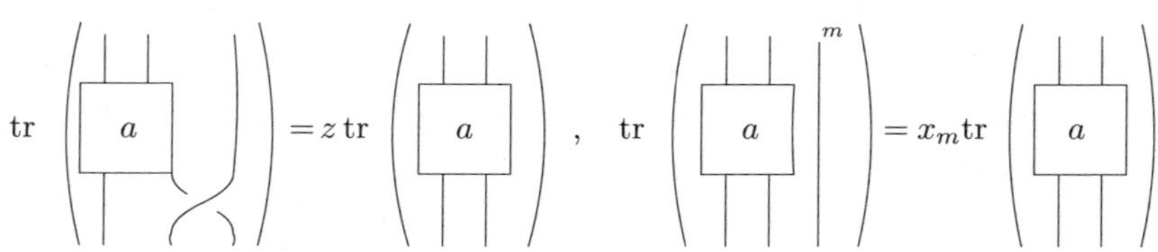

Fig. 3. Topological interpretations of the trace rules.

braid category. More precisely, the defining relations of $\mathrm{Y}_{d,n}(u)$ yield two natural representations. One of the framed braid group:

$$\begin{aligned} \gamma : \mathbb{C}\mathcal{F}_n &\to \mathrm{Y}_{d,n}(u) \\ \sigma_i &\mapsto g_i \\ t_j^s &\mapsto t_j^{s(\mathrm{mod}\, d)} \end{aligned} \tag{2.6}$$

and another of the classical braid group:

$$\begin{aligned} \delta : \mathbb{C}B_n &\to \mathrm{Y}_{d,n}(u) \\ \sigma_i &\mapsto g_i. \end{aligned} \tag{2.7}$$

The map δ can be viewed as the composition of the map γ with the natural injection of B_n into $\mathcal{F}_n$, whereby $\sigma_i \mapsto \sigma_i$, and classical braids are considered to have all framings zero.

In the sequel we will sometimes identify a braid α with its image through γ or δ.

3. The E-System

The framed braid equivalence comprises conjugation in the groups $\mathcal{F}_n$ and positive and negative stabilization and destabilization (see for example [32]). So, with the trace tr in hand, in [26] the authors tried to obtain topological invariants for framed links after the method of V. F. R. Jones [18]. This meant that tr would have to be normalized, so that the closed framed braids $\widehat{\alpha}$ and $\widehat{\alpha\,\sigma_n}$ ($\alpha \in \mathcal{F}_n$) get assigned the same value of the invariant, and also re-scaled, so that the closed framed braids $\widehat{\alpha\sigma_n^{-1}}$ and $\widehat{\alpha\sigma_n}$ ($\alpha \in \mathcal{F}_n$) also get assigned the same value of the invariant. However, as it turned out, $\mathrm{tr}(\alpha g_n^{-1})$ does not factor through $\mathrm{tr}(\alpha)$. That is, remarkably:

$$\mathrm{tr}(\alpha\, g_n^{-1}) \neq \mathrm{tr}(g_n^{-1})\ \mathrm{tr}(\alpha), \tag{3.1}$$

since, from (2.5), we deduce that

$$\mathrm{tr}(\alpha g_n^{-1}) = \mathrm{tr}(\alpha g_n) + (u^{-1} - 1)\mathrm{tr}(\alpha e_n) + (u^{-1} - 1)\,\mathrm{tr}(\alpha e_n g_n)$$

and, although $\mathrm{tr}(\alpha e_n g_n) = z\,\mathrm{tr}(\alpha)$, the term $\mathrm{tr}(\alpha e_n)$ does not factor through $\mathrm{tr}(\alpha)$. Forcing the E-*condition*:

$$\mathrm{tr}(\alpha e_n) = \mathrm{tr}(e_n)\,\mathrm{tr}(\alpha) \quad (a \in \mathrm{Y}_{d,n}(u)) \tag{3.2}$$

implies equivalently that the trace parameters $x_1, \ldots, x_{d-1}$ have to satisfy the E-*system* [26], the following nonlinear system of equations in $\mathbb{C}$:

$$E^{(m)} = x_m E \qquad (1 \leq m \leq d-1) \tag{3.3}$$

where

$$E = E^{(0)} := \mathrm{tr}(e_i) = \frac{1}{d}\sum_{s=0}^{d-1} x_s x_{d-s} \quad \text{and} \quad E^{(m)} := \frac{1}{d}\sum_{s=0}^{d-1} x_{m+s} x_{d-s},$$

where the sub-indices on the x_j's are regarded modulo d and $x_0 := 1$.

As it was shown by P. Gérardin ([26, Appendix]), the solutions of the E-system are parametrized by the non-empty subsets of $\mathbb{Z}/d\mathbb{Z}$.

It is worth noting that the solutions of the E-system can be interpreted as a generalization of the Ramanujan sum. Indeed, by taking the subset R of $\mathbb{Z}/d\mathbb{Z}$ comprising the numbers coprime to d, then the solution parametrized by R is, up to the factor $|R|$, the Ramanujan sum $c_d(k)$ (see [38]).

It is also worth mentioning that solutions of the E-system lift to solutions on the p-adic level [22, 26] and on the adelic level [25].

4. Knot Invariants from the Yokonuma–Hecke Algebras

The Yokonuma–Hecke algebras have been used for constructing invariants for framed knots [26], for classical knots [25] and for singular knots [24]. These invariants qualify the algebra $\mathrm{Y}_{d,n}(u)$ as a knot algebra and they comprise our main motivation for extending the notion of framization to other known knot algebras. We shall now recall briefly the construction of these invariants.

4.1. Let $X_D = (\mathrm{x}_1, \ldots, \mathrm{x}_{d-1})$ be a solution of the E-system parametrized by the non-empty subset D of $\mathbb{Z}/d\mathbb{Z}$.

Definition 4.1. (Definition 3 [8]) The trace map tr_D defined as the trace tr with the parameters x_i specialized to the values x_i, shall be called the *specialized Juyumaya trace* with parameter z.

Note that for $d = 1$ the trace tr_D coincides with the trace tr (and with the Ocneanu trace with parameter z). As it turns out [25, 26],

$$E_D := \mathrm{tr}_D(e_i) = \frac{1}{|D|},$$

where $|D|$ is the cardinal of the subset D.

4.2. Let $\mathcal{L}_f$ denote the set of oriented framed links. From the above, and re-scaling g_i to $\sqrt{\lambda_D}\, g_i$, so that $\mathrm{tr}_D(g_i^{-1}) = \lambda_D z$, it turns out that

$$\lambda_D := \frac{z + (1-u)E_D}{uz} = \frac{|D|z + 1 - u}{|D|uz} \tag{4.1}$$

and we have the following (mapping $\sigma_i \mapsto \sqrt{\lambda_D}\, g_i$):

Theorem 4.1. ([26]) *Given a solution X_D of the* E*-system, for any framed braid $\alpha \in \mathcal{F}_n$ we define for the framed link $\widehat{\alpha} \in \mathcal{L}_f$:*

$$\Gamma_D(\widehat{\alpha}) = P_D^{n-1}(\sqrt{\lambda_D})^{\epsilon(\alpha)}\, (\mathrm{tr}_D \circ \gamma)\, (\alpha),$$

where $P_D = \frac{1}{z\sqrt{\lambda_D}}$, $\epsilon(\alpha)$ is the algebraic sum of the exponents of the σ_i's in α and γ the epimorphism (2.6)*. Then the map $\Gamma_D(u, \lambda_D)$ is a 2-variable isotopy invariant of oriented framed links.*

Further, in [26] a skein relation has been found for the invariant $\Gamma_D(u, \lambda_D)$ involving the braiding and the framing generators:

$$\begin{aligned}\sqrt{\lambda}\,\Gamma_D(L_-) = \frac{1}{\sqrt{\lambda}}\Gamma_D(L_+)\frac{u^{-1}-1}{d}\sum_{s=0}^{d-1}\Gamma_D(L_s)\\ + \frac{u^{-1}-1}{d\sqrt{\lambda}}\sum_{s=0}^{d-1}\Gamma_D(L_{s\times}),\end{aligned} \tag{4.2}$$

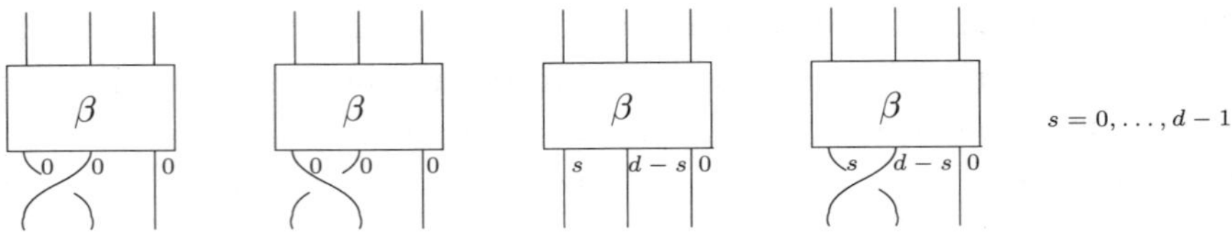

$L_+ = \widehat{\beta\sigma_1}$ $\quad L_- = \widehat{\beta\sigma_1{}^{-1}}$ $\quad L_s = \widehat{\beta t_1^s t_2^{d-s}}$ $\quad L_{s\times} = \widehat{\beta t_1^s t_2^{d-s}\sigma_1}$

Fig. 4. The framed links L_+, L_-, L_s and $L_{s\times}$.

where L_+, L_-, L_s and $L_{s\times}$, $s = 0, \ldots, d-1$, are diagrams of oriented framed links, which are all identical except near one crossing, where they differ by the ways indicated in Fig. 4.

Finally, a p-adic (respectively adelic) invariant for oriented framed links, Γ_{p^∞}, has been defined through the invariants Γ_D. For more details see [25, 26].

Note 4.1. It is logical that one should try to extract from the invariants for framed links invariants of 3-manifolds. But for this, one needs to start with invariants of *unoriented* framed links. This can be achieved from our constructions so far by taking appropriate quotients of the Yokonuma–Hecke algebras, analogous to the Temperley–Lieb algebras. This is presented in Sec. 5.

4.3. Let $\mathcal{L}$ denote the set of oriented classical links. A classical link may be viewed as a framed link with all framings zero. By the mapping (2.7) of the classical braid group B_n in $\mathrm{Y}_{d,n}(u)$, by the classical Markov braid equivalence and using the construction and notations above we obtain invariants for classical links, where the t_j's are treated as formal generators [25]. Namely:

Theorem 4.2. ([25]) *Given a solution X_D of the* E-*system, for any braid $\alpha \in B_n$ we define for the link $\widehat{\alpha} \in \mathcal{L}$.*

$$\Delta_D(\widehat{\alpha}) = P_D^{n-1}(\sqrt{\lambda_D})^{\epsilon(\alpha)}\,(\mathrm{tr}_D \circ \delta)\,(\alpha),$$

where P_D, λ_D are as defined in Sec. 4.2, $\epsilon(\alpha)$ is the algebraic sum of the exponents of the σ_i's in α and δ is the natural algebra *homomorphism* (2.7). *Then the map* $\Delta_D(u, \lambda_D)$ *is a 2-variable isotopy invariant of oriented links.*

The invariant $\Delta_D(u, \lambda_D)$ can be viewed as the restriction of $\Gamma_D(u, \lambda_D)$ on the set of framed links with all framings zero.

The invariants $\Delta_D(u, \lambda_D)$ need to be compared with known invariants of classical links, especially with the Homflypt polynomial $P(u, \lambda)$, recall Sec. 2.2. This is not easy to do on the algebraic level as there are no algebra homomorphisms connecting the algebras and the traces [8]. Further, the skein relation of the invariant $\Delta_D(u, \lambda_D)$ has no topological interpretation in the case of classical links. This also makes the comparison very difficult using diagrammatic methods. It is worth noting at this point that in $\mathrm{Y}_{d,n}(u)$ a "closed" cubic relation is satisfied [23, 25], closed in the sense of involving only the braiding generators, which is of minimal degree. Namely:

$$g_i^3 = ug_i^2 + g_i - u. \tag{4.3}$$

The cubic relation gives rise to a cubic skein relation for the invariant $\Delta_D(u, \lambda_D)$, involving only the braiding generators [25]:

$$\begin{aligned}\sqrt{\lambda_D}\,\Delta_D(L_-) &= -\frac{1}{u\,\lambda_D}\,\Delta_D(L_{++})\\ &\quad+\frac{1}{\sqrt{\lambda_D}}\,\Delta_D(L_+) + \frac{1}{u}\,\Delta_D(L_0),\end{aligned} \tag{4.4}$$

where the links L_{++}, L_+, L_0 and L_- have identical diagrams, except for a region where they differ in the manners illustrated in Fig. 5. However, the above skein relation is not sufficient for determining the invariant $\Delta_D(u, z)$ diagrammatically with a simple set of initial conditions.

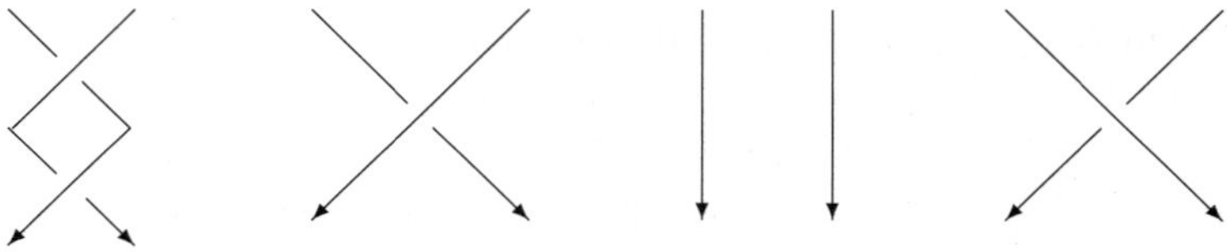

Fig. 5. The classical links L_{++}, L_+, L_0 and L_-.

As it turns out, the cubic relation factors to the quadratic relation of the Iwahori–Hecke algebra $\mathrm{H}_n(u)$:

$$g_i^3 - ug_i^2 - g_i + u = (g_i - 1)\big(g_i^2 - (u-1)g_i - u\big). \tag{4.5}$$

Unfortunately, this factoring does not give information about the comparison of the invariants.

From our construction it follows that we have, at least, recovered the Homflypt polynomial. In [8] it is shown that for generic values of the parameters u, z the invariants $\Delta_D(u, \lambda_D)$ do not coincide with the Homflypt polynomial except in the trivial cases $u = 1$ or $E = 1$. More precisely, for $E = 1$ an algebra homomorphism $h : \mathrm{Y}_{d,n}(u) \to \mathrm{H}_n(u)$ can be defined and the composition $\tau \circ h$ is a Markov trace on $\mathrm{Y}_{d,n}(u)$ which takes the same values as the specialized Juyumaya trace tr_D whereby the x_i's are specialized to dth roots of unity [26, Appendix]. In this case we also obtain $|D| = 1$. For details see [8, §3].

Yet, computational data [12] indicate that these invariants do not distinguish more knot pairs than the Homflypt polynomial, so they are probably topologically equivalent to the Homflypt polynomial. A complete answer to this question is still under investigation. In [12] some conjectures are formulated and tackled in this direction. Also, it is proved that the specialized Juyumaya trace can be computed for classical braids with rule (3) replaced by another rule involving the elements e_i.

4.4. Let $\mathcal{L}_S$ denote the set of oriented singular links. Oriented singular links are represented by singular braids, which form the singular braid monoids $\mathcal{S}B_n$ [2, 5, 39]. $\mathcal{S}B_n$ is generated by the classical braiding generators σ_i with their inverses and by the elementary singular braids τ_i which are not invertible. In [24] the following monoid homomorphism was constructed:

$$\begin{aligned} \eta : \mathcal{S}B_n &\to \mathrm{Y}_{d,n}(u) \\ \sigma_i &\mapsto g_i \\ \tau_i &\mapsto p_i = e_i(1 + g_i). \end{aligned} \tag{4.6}$$

In view of the elements p_i the quadratic relations (2.3) may be rewritten as: $g_i^2 = 1 + (u-1)p_i$. Using now the singular braid equivalence [15], the map η and the specialized Juyumaya trace tr_D we obtain isotopy invariants for oriented singular links [24]:

Theorem 4.3. ([24]) *Given a solution X_D of the* E*-system, for any oriented singular link $\widehat{\alpha} \in \mathcal{L}_{\mathcal{S}}$, where $\alpha \in \mathcal{S}B_n$, we define:*

$$H_D(\widehat{\alpha}) = P_D^{n-1}(\sqrt{\lambda_D})^{\epsilon(\alpha)}\,(\mathrm{tr}_D \circ \eta)\,(\alpha),$$

where P_D, λ_D are as defined in Sec. 4.2, η as defined in (4.6) and where $\epsilon(\alpha)$ is defined as follows: Let $\alpha = \mu_1^{\epsilon_1}\mu_2^{\epsilon_2}\cdots\mu_m^{\epsilon_m}$ where $\mu_j \in \{\sigma_i, \tau_i\,;\, 1 \leq i \leq n-1\}$. Then $\epsilon(\alpha) := \epsilon_1 + \cdots + \epsilon_m$, where $\epsilon_j = +1$ or -1 if $\mu_j = \sigma_j$ and $\epsilon_j = +1$ if $\mu_j = \tau_j$. Then the map $H_D(u, \lambda_D)$ is a 2-variable isotopy invariant of oriented singular links.

Moreover, in the image $\eta(\mathcal{S}B_n)$ the following relations hold:

$$g_i^{-1} - g_i = (u^{-1} - 1)p_i \tag{4.7}$$

which give rise to the following skein relation:

$$\sqrt{\lambda_D}\,H_D(L_-) - \frac{1}{\sqrt{\lambda_D}}\,H_D(L_+) = \frac{u^{-1}-1}{\sqrt{\lambda_D}}\,H_D(L_\times), \tag{4.8}$$

where L_+, L_- and $L_\times$ are diagrams of three oriented singular links, which are identical except near one crossing, where they are as depicted in Fig. 6.

For further details the reader is referred to [24]. We note that there are some differences from [24] in the signs in (4.6), (4.7) and (4.8); this is due to the change of sign in the quadratic relation (2.3).

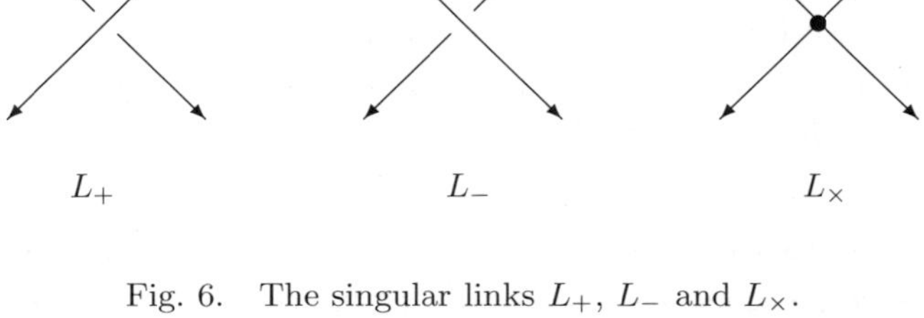

Fig. 6. The singular links L_+, L_- and $L_\times$.

4.5. Another category of knots which seems to be naturally related to the Yokonuma–Hecke algebras is the category of *transverse knots*, for the following reasons: transverse knots are naturally framed and oriented, and their equivalence is more refined than framed knot isotopy. In 1983 D. Bennequin [3] noted that the closed braid presentation of knots is convenient for describing transverse knots. Further, S. Orevkov and V. Shevchishin [36] and independently N. Wrinkle [43] gave a transverse analogue of the Markov theorem, comprising conjugation in the framed braid groups and *positive only* stabilizations and destabilizations: $\alpha \sim \alpha\sigma_n \in \mathcal{F}_{n+1}$, where $\alpha \in \mathcal{F}_n$.

Now, rule (2) of the trace tr (Theorem 2.2) tells us that it respects positive stabilizations but property (3.1) tells us that tr does not behave well under negative stabilizations. Using the above, an invariant $M(u, \lambda, x_1, \ldots, x_{d-1})$ for transverse knots has been constructed in [12], which coincides with the invariant $\Gamma_D(u, \lambda_D)$ of framed knots when $(x_1, \ldots, x_{d-1})$ specializes to a solution of the E-system. However, as it turns out, the invariants $M(u, \lambda, x_1, \ldots, x_{d-1})$ are only topological invariants of framed knots. For details see [12].

4.6. Due to the quadratic relation (2.3) it is very difficult to do computations in the Yokonuma–Hecke algebras and the trace tr. In http://www.math.ntua.gr/˜sofia/yokonuma/index.html the reader can find a program by Sergei Chmutov and Konstantinos Karvounis, together with instructions for using it.

5. Framization of the Temperley–Lieb Algebra

In this section we define possible framizations for the Temperley–Lieb algebra and we indicate one of them as our favorite. The results of this section comprise the Ph.D. thesis of Goundaroulis and they are contained in [16] and [17].

5.1. The Temperley–Lieb algebra can be defined in several ways. Here, it is convenient to define the *Temperley–Lieb algebra* $\mathrm{TL}_n(u)$

as the quotient of the Hecke algebra $\mathrm{H}_n(u)$ by the two-sided ideal generated by the *Steinberg elements* [18]:

$$h_{i,i+1} := h_i h_{i+1} h_i + h_{i+1} h_i + h_i h_{i+1} + h_i + h_{i+1} + 1,$$

that is:

$$\mathrm{TL}_n(u) = \frac{\mathrm{H}_n(u)}{\langle h_{i,i+1};\ \text{all } i\rangle}. \tag{5.1}$$

V. F. R. Jones recovered his Markov trace on the Temperley–Lieb algebra from the Ocneanu trace τ (recall Sec. 2.2) as follows: He required first that τ is zero on the Steinberg elements. This yielded two values for the parameter ζ, namely:

$$\zeta = -\frac{1}{u+1}, \qquad \text{where } u \neq -1 \quad \text{and} \quad \zeta = -1. \tag{5.2}$$

He then showed that for these values τ is zero on all elements of the defining ideal of $\mathrm{TL}_n(u)$.

Subsequently, the Jones polynomial $V(u)$ is obtained from the Homflypt polynomial (recall Sec. 2.2) by taking

$$\zeta = -\frac{1}{u+1}.$$

Hence $\lambda = u$. That is,

$$V(u) = P(u, u).$$

In order to define a framization of the Temperley–Lieb algebra, the most natural approach is to take an appropriate quotient of the Yokonuma–Hecke algebra, in analogy with the classical case. So, we try to define in $\mathrm{Y}_{d,n}(u)$ elements analogous to the $h_{i,i+1}$'s as generators of a two-sided ideal. Now, having in mind that the multiplication rules of the Hecke algebra are governed by the symmetric group and those of $\mathrm{Y}_{d,n}(u)$ are governed by the group $C_{d,n}$ (recall Eq. (1.4)), we have two obvious possibilities for generating elements, which give rise to two natural candidates for the framization of the Temperley–Lieb algebra.

5.2. For the first possibility we define for all i the elements:

$$g_{i,i+1} := g_i g_{i+1} g_i + g_{i+1} g_i + g_i g_{i+1} + g_i + g_{i+1} + 1. \tag{5.3}$$

Then, we have:

Definition 5.1. ([16]) The algebra *Yokonuma–Temperley–Lieb*, denoted $\mathrm{YTL}_{d,n}(u)$, is defined for any $n \geq 3$ as the quotient:

$$\mathrm{YTL}_{d,n}(u) := \frac{\mathrm{Y}_{d,n}(u)}{\langle g_{i,i+1}\,; \quad \text{all } i\rangle}.$$

In [9] the representations of $\mathrm{YTL}_{d,n}(u)$ are determined, as well as the dimension and an explicit basis. In [16] $\mathrm{YTL}_{d,n}(u)$ is studied as a knot algebra. A presentation with non-invertible generators is given. Also, the necessary and sufficient conditions are established for the Juyumaya trace tr on $\mathrm{Y}_{d,n}(u)$ to pass to the quotient algebra $\mathrm{YTL}_{d,n}(u)$. Indeed we have:

Theorem 5.1. (Theorem 5 [16]) *The trace* tr *passes to the quotient* $\mathrm{YTL}_{d,n}(u)$ *if and only if the* x_i*'s are solutions of the* E*-system and one of the two cases holds*:

(i) *For some* $0 \leq m_1 \leq d-1$ *the* x_ℓ*'s are expressed as*:

$$x_\ell = \exp(\ell m_1) \quad (0 \leq \ell \leq d-1).$$

In this case the x_ℓ*'s are* d*th roots of unity and* $z = -\frac{1}{u+1}$ *or* $z = -1$.

(ii) *For some* m_1, m_2 *such that* $0 \leq m_1 \neq m_2 \leq d-1$ *the* x_ℓ*'s are expressed as:*

$$x_\ell = \frac{1}{2}\left(\exp(\ell m_1) + \exp(\ell m_2)\right) \quad (0 \leq \ell \leq d-1).$$

In this case we have $z = -\frac{1}{2}$.

The cases where $z = -1$ or $z = -1/2$ are of no topological interest. The only interesting case is case (i) for $z = -\frac{1}{u+1}$. In this case the x_ℓ's are dth roots of unity, which is equivalent to $E = 1$ and $|D| = 1$ [26, Appendix]. By the discussion at the end of Sec. 4.3 the invariants $\Delta_D(u, \lambda_D)$ coincide with the Homflypt polynomial. Further, $z = -\frac{1}{u+1}$ implies $\lambda_D = u$. Hence, the invariants derived for framed, classical and singular knots are $\Gamma_D(u, u)$, $\Delta_D(u, u)$ and $H_D(u, u)$ respectively, recall Sec. 4.

Remark 5.1. In particular, the invariants:

$$V_D(u) := \Delta_D(u, u)$$

for classical knots all *coincide with the Jones polynomial.* From our point of view, this is the characteristic property of the algebra $\mathrm{YTL}_{d,n}(u)$.

To recapitulate, the conditions on the trace tr are too strong, resulting in a trivial framization of the Jones polynomial. For this reason the algebra $\mathrm{YTL}_{d,n}(u)$ is discarded as a framization of $\mathrm{TL}_n(u)$.

5.3. For the second possibility we define for all i the following elements, where the framings are now also involved.

$$c_{i,i+1} := \sum_{\alpha,\beta,\gamma\in\mathbb{Z}/d\mathbb{Z}} t_i^{\alpha} t_{i+1}^{\beta} t_{i+2}^{\gamma}\, g_{i,i+1},$$

where $g_{i,i+1}$ as defined in (5.3). Then, we obtain:

Definition 5.2. ([17]) The *Complex Reflection Temperley–Lieb algebra*, denoted $\mathrm{CTL}_{d,n}(u)$, is defined as the quotient:

$$\mathrm{CTL}_{d,n}(u) := \frac{\mathrm{Y}_{d,n}(u)}{\langle c_{i,i+1};\ \text{all } i\rangle}.$$

In [17] the necessary and sufficient conditions are determined for the trace tr to pass to $\mathrm{CTL}_{d,n}(u)$. More precisely, we have the following:

Theorem 5.2. ([17]) *The trace* tr *passes to the quotient* $\mathrm{CTL}_{d,n}(u)$ *if and only if the parameter* z *and the* x_i*'s are related through the equation:*

$$(u+1)z^2 \sum_{k\in\mathbb{Z}/d\mathbb{Z}} x_k + (u+2)z \sum_{k\in\mathbb{Z}/d\mathbb{Z}} E^{(k)} + \sum_{k\in\mathbb{Z}} \mathrm{tr}(e_1^{(k)} e_2) = 0,$$

where

$$e_1^{(k)} := \frac{1}{d}\sum_{s=0}^{d-1} t_1^{k+s} t_2^{d-s} \quad \text{and} \quad E^{(k)} := \mathrm{tr}(e_1^{(k)}) \quad (0 \le k \le d-1)$$

and where $e_1^{(0)} = e_1$ *and* $E^{(0)} = E$.

Remark 5.2. Contrary to the case of $\mathrm{YTL}_{d,n}(u)$, the conditions of Theorem 5.2 are too relaxed on the trace parameters x_i. However, in order to obtain framed, classical or singular link invariants from the algebras $\mathrm{CTL}_{d,n}(u)$ one has to impose the E-condition on the x_i's. This in turn leads to the following values for z (see [17]):

$$z = -\frac{1}{(u+1)|D|} \quad \text{or} \quad z = -\frac{1}{|D|}.$$

Again, the interesting value is $z = -1/(u+1)|D|$ and, as it turns out, the link invariants one obtains coincide with those from the algebras $\mathrm{FTL}_{d,n}(u)$ that we define next.

5.4. From the above constructions and from the knot algebras point of view it follows that it would be more sensible to consider an intermediate algebra between $\mathrm{CTL}_{d,n}(u)$ and $\mathrm{YTL}_{d,n}(u)$, for which the conditions for the trace tr to pass through include explicitly all solutions of the E-system. Indeed, for all i we define the elements:

$$r_{i,i+1} := \sum_{\substack{\alpha,\beta,\gamma\in\mathbb{Z}/d\mathbb{Z} \\ \alpha+\beta+\gamma=0}} t_i^{\alpha} t_{i+1}^{\beta} t_{i+2}^{\gamma}\, g_{i,i+1} = e_i e_{i+1}\, g_{i,i+1}. \tag{5.4}$$

We now define:

Definition 5.3. ([17]) The *framization of the Temperley–Lieb algebra*, denoted $\mathrm{FTL}_{d,n}(u)$, is defined as the quotient:

$$\mathrm{FTL}_{d,n}(u) := \frac{\mathrm{Y}_{d,n}(u)}{\langle r_{i,i+1}\,;\ \text{all } i\rangle}.$$

For the algebra $\mathrm{FTL}_{d,n}(u)$ we have determined in [17] the necessary and sufficient conditions on the trace parameters z and the x_i's for the trace tr to pass to $\mathrm{FTL}_n(u)$. In order to state our result we need to introduce the following notation: given a sequence $(x_1, \ldots, x_{d-1})$ of $d-1$ complex numbers, we denote by x the function on $\mathbb{Z}/d\mathbb{Z}$ with values in $\mathbb{C}$, such that 0 is mapped to 1 and i in mapped to x_i, for $1 \le i \le d-1$. Further, we denote $\widehat{x}$ the Fourier transform of x (cf. [17, §3]).

We then have the following:

Theorem 5.3. ([17]) *The trace* tr *passes to* $\mathrm{FTL}_{d,n}(u)$ *if and only if the trace parameters satisfy*

$$x_m = -z \left(\sum_{k \in D_1} \exp(km) + \sum_{k \in D_2} \exp(km) \right)$$

and

$$z = -\frac{1}{|D_1| + (u+1)|D_2|}$$

where the disjoint union $D_1 \cup D_2$ *is the support of the Fourier transform* $\hat{x}$ *of* x,

$$D_1 := \{k \in \mathbb{Z}/d\mathbb{Z}\,;\, y_k = -dz\},$$
$$D_2 := \{k \in \mathbb{Z}/d\mathbb{Z}\,;\, y_k = -dz(u+1)\},$$

and the y_k*'s denote the values of* $\hat{x}$.

In particular, the above theorem implies that whenever the trace parameters are solutions of the E-system, then the trace tr passes to $\mathrm{FTL}_{d,n}(u)$. Indeed, we have the following corollary.

Corollary 5.1. ([17]) *In the case where one of the sets* D_1 *or* D_2 *is the empty set we obtain that the values in the above theorem become solutions of the* E*-system. More precisely, in the case where* D_1 *is the empty set, the* x_m*'s become the solutions of the* E*-system parametrized by* D_2 *and* z *takes the value* $z = -1/(u+1)|D_2|$. *In the case where* D_2 *is the empty set we have that the* x_m*'s become the solutions of the* E*-system parametrized by* D_1 *and* z *takes the value* $z = -1/|D_1|$.

The above corollary allows one to define a nontrivial version of a framed Jones polynomial in the same manner as the Jones polynomial coincides with $P(u,u)$, the specialization of the Homflypt polynomial $P(u,\lambda)$ for $z = -1/(u+1)$. More precisely, we give the following definition, which is possible by Corollary 5.1 and Eq. (3.2).

Definition 5.4. ([17]) The *framed Jones polynomial* $\mathcal{V}_D(u)$ is defined as the specialization of the polynomial $\Gamma_D(u, \lambda_D)$ at the value:

$$z = -\frac{1}{(u+1)|D|}.$$

Hence

$$\mathcal{V}_D(u) := \Gamma_D(u,u).$$

Similarly, one derives the invariants $\Delta_D(u,u)$ and $H_D(u,u)$ for classical and singular links respectively from the invariants $\Delta_D(u,\lambda_D)$ and $H_D(u,\lambda_D)$ (recall Sec. 4) by specializing at $z = -1/(u+1)|D|$.

Remark 5.3. The invariants for classical links $\Delta_D(u,u)$ may prove to be topologically equivalent with the Jones polynomial, in analogy with $\Delta_D(u,\lambda_D)$ and the Homflypt polynomial (recall discussion at the end of Sec. 4.3). In case they are not, one could compare corresponding 3-manifold invariants with the Witten invariants from the Jones polynomial, see [42].

5.5. All three quotient algebras $\mathrm{YTL}_{d,n}(u)$, $\mathrm{FTL}_{d,n}(u)$ and $\mathrm{CTL}_{d,n}(u)$ equipped with the Markov traces are interesting on their own right. They are related via the following algebra epimorphisms:

$$\mathrm{Y}_{d,n}(u) \twoheadrightarrow \mathrm{CTL}_{d,n}(u) \twoheadrightarrow \mathrm{FTL}_{d,n}(u) \twoheadrightarrow \mathrm{YTL}_{d,n}(u),$$

which follow from inclusions of their defining ideals [17].

6. Framization of the Hecke–Related Algebras of Type B

6.1. The *Artin group of type B*, denoted $B_{1,n}$, is related to the following Dynkin diagram:

Fig. 7. The Dynkin diagram of type B.

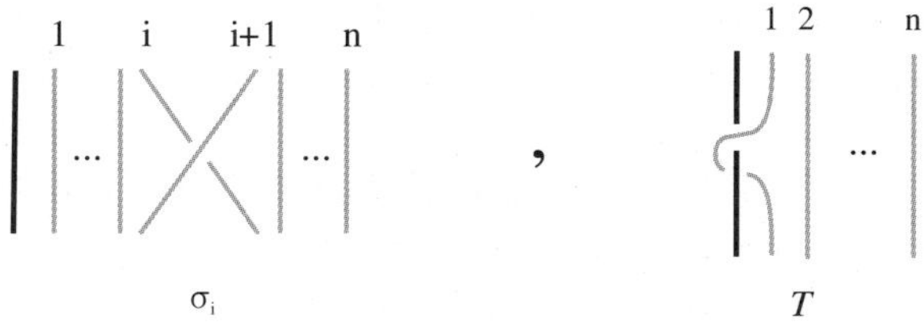

Fig. 8. The braiding generators and the loop generator of $B_{1,n}$.

Fig. 9. A braid in $B_{1,n}$ and a link in the solid torus.

That is, $B_{1,n}$ is presented by the braiding generators $\sigma_1, \ldots, \sigma_{n-1}$ and the loop generator T (see Fig. 8), satisfying the braid relations (1.1) and the relations:

$$\begin{aligned} T\sigma_1 T\sigma_1 &= \sigma_1 T\sigma_1 T \\ T\sigma_i &= \sigma_i T \qquad \text{for } 2 \leq i \leq n-1. \end{aligned}$$

Geometrically, a braid in $B_{1,n}$ has $n+1$ strands, with the first strand identically fixed and the other n strands numbered from 1 to n. See Fig. 9 for an example. Its closure is an oriented link in the solid torus, where the complement solid torus is represented by the closure of the fixed strand [33, 34].

Remark 6.1. The Artin braid group of type B, $B_{1,n}$, is isomorphic to the affine Artin braid group of type A, denoted $\widetilde{B_n}$. Thus, another geometric interpretation for elements in $B_{1,n}$ is as cylinder braids, that is, as braids in a thickened cylinder. Then, depending on how closure is defined (by simple closed arcs in the thickened cylinder or by endpoints identifications), they could give rise to oriented links in the solid torus or in the thickened torus.

6.2. For $u, U \in \mathbb{C}\backslash\{0\}$, the classical *Iwahori–Hecke algebra of type* B [13], denoted here $\mathrm{H}_{1,n}(u, U)$, can be viewed as the quotient of the group algebra $\mathbb{C}B_{1,n}$ over the expressions:

$$\sigma_i^2 - (u-1)\sigma_i - u \quad \text{and} \quad T^2 - (U-1)T - U.$$

Further, for $u, u_1, \ldots, u_r \in \mathbb{C}\backslash\{0\}$, the *cyclotomic Hecke algebra of type* B *and degree* r [1, 7], denoted here $\mathrm{H}_{1,n}(u, u_1, \ldots, u_r)$, can be defined as the quotient of the group algebra $\mathbb{C}B_{1,n}$ over the expressions:

$$\sigma_i^2 - (u-1)\sigma_i - u \quad \text{and} \quad (T-u_1)(T-u_2)\cdots(T-u_r).$$

For $r = 2$, the algebra $\mathrm{H}_{1,n}(u, u_1, u_2)$ can be proved to be isomorphic to $\mathrm{H}_{1,n}(u, U)$.

Finally, the *generalized Hecke algebra of type* B [34], denoted $\mathrm{H}_{1,n}(u)$, is defined as the quotient of the group algebra $\mathbb{C}B_{1,n}$ over the expressions:

$$\sigma_i^2 - (u-1)\sigma_i - u.$$

The algebra $\mathrm{H}_{1,n}(u)$ was observed by T. Tom Dieck [34, Remark 1] to be isomorphic to the *affine Hecke algebra of type* A.

In [14, 33] and [34] Markov traces have been constructed on all these algebras, giving rise to all possible analogues of the Homflypt polynomial for oriented links in the solid torus. The two rules of these traces, namely conjugation and Markov property, are analogous to the two rules of the Ocneanu trace (recall Theorem 2.1). Then, there is also a third inductive rule that takes care of the loopings, namely:

$$\tau\big(a(h_n \cdots h_1 T^k h_1^{-1} \cdots h_n^{-1})\big) = s_k \tau(a), \tag{6.1}$$

where σ_i corresponds to the braiding generator h_i and T corresponds to T, and where $a \in \mathrm{H}_{1,n}(u, U)$ or $\mathrm{H}_{1,n}(u, u_1, \ldots, u_r)$ or $\mathrm{H}_{1,n}(u)$, depending on the algebra we are in each time. For details see [14, 33, 34].

6.3. From the above it is natural to try to define framizations of the Hecke algebra of type B, of the cyclotomic Hecke algebra of type B and of the generalized Hecke algebra of type B, all B-type analogues of the Yokonuma–Hecke algebra, with the prospective to obtain invariants of links in the solid torus, framed, classical or singular, analogous to those defined by the Yokonuma–Hecke algebra. For this we define first:

Definition 6.1. (Definition 4 [27]) The *framed braid group of type* B, $\mathcal{F}_{1,n}$, is the group presented by the generators $T, \sigma_1, \ldots, \sigma_{n-1}$ of $B_{1,n}$ together with the framing generators $t_1, \ldots, t_n$, subject to the relations of $B_{1,n}$ together with the framing relations (1.2) and the relations:

$$t_i T = T t_i \quad \text{for all } 1 \leq i \leq n.$$

Geometrically, elements in $\mathcal{F}_{1,n}$ have framings on the n numbered strands and upon closure they represent framed links in the solid torus. The *d-modular framed braid group of type* B, denoted $\mathcal{F}_{d,1,n}$, is defined by adding to the above presentation of $\mathcal{F}_{1,n}$ the relations:

$$t_i^d = 1 \quad \text{for } 1 \leq i \leq n.$$

Remark 6.2. By Remark 6.1 $\mathcal{F}_{1,n}$ is isomorphic to the affine framed braid group of type A, denoted $\widetilde{\mathcal{F}_n}$. So, elements in $\mathcal{F}_{1,n}$ can also be interpreted as framed braids in a thickened cylinder. Then, upon different types of closures, they give rise to oriented framed links in the solid torus or in the thickened torus.

6.4. Define now the following elements in $\mathbb{C}\mathcal{F}_{d,1,n}$:

$$e_i := \frac{1}{d} \sum_{m=0}^{d-1} t_i^m t_{i+1}^{d-m} \quad \text{for all } 1 \leq i \leq n-1.$$

It is clear that the elements e_i are idempotents. The e_i's are represented geometrically as in Fig. 1, but with the addition in all terms of a first fixed strand with no framing.

We shall now define framizations of the B-type algebras above. These definitions have all been given in [27, Definition 5].

Definition 6.2. ([27]) For $u, U \in \mathbb{C}\backslash\{0\}$ we define the *Yokonuma–Hecke algebra of type B*, $\mathrm{Y}_{d,1,n}(u, U)$, as the algebra presented by the braiding generators $g_1, \ldots, g_{n-1}$, the loop generator T and the framing generators $t_1, \ldots, t_n$, subject to the relations (2.2) for the braiding and framing generators, together with the following extra relations:

$$\begin{aligned} Tg_1Tg_1 &= g_1Tg_1T \\ Tg_i &= g_iT \qquad \text{for } 2 \leq i \leq n-1 \\ Tt_j &= t_jT \qquad \text{for } 1 \leq j \leq n \end{aligned} \tag{6.2}$$

and the quadratic relations:

$$g_i^2 = 1 + (u-1)e_i + (u-1)e_ig_i \quad (1 \leq i \leq n-1) \tag{6.3}$$

and

$$T^2 = (U-1)T + U. \tag{6.4}$$

The generators g_i and T are easily seen to be invertible:

$$\begin{aligned} g_i^{-1} &= g_i + (u^{-1}-1)e_i + (u^{-1}-1)e_ig_i \quad \text{and} \\ T^{-1} &= U^{-1}T + (U^{-1}-1). \end{aligned} \tag{6.5}$$

From the above presentation it is clear that the algebra $\mathrm{Y}_{d,1,n}(u, U)$ is a quotient of the modular framed braid group algebra $\mathbb{C}\mathcal{F}_{d,1,n}$ under the quadratic relations (6.3) and (6.4). This observation leads to diagrammatic interpretations for the elements of $\mathrm{Y}_{d,1,n}(u, U)$.

Definition 6.3. ([27]) For $u, u_1, \ldots, u_r \in \mathbb{C}\backslash\{0\}$ we define the *cyclotomic Yokonuma–Hecke algebra of type B and of degree r*, $\mathrm{Y}_{d,1,n}(u, u_1, \ldots, u_r)$, as the algebra presented by the braiding generators $g_1, \ldots, g_{n-1}$, the loop generator T and the framing generators $t_1, \ldots, t_n$ subject to the relations (2.2) for the braiding and framing generators, relations (6.2) for the loop generator, the quadratic relations (6.3) for the braiding generators and the following polynomial relation for the loop generator T in place of (6.4):

$$(T-u_1)(T-u_2)\cdots(T-u_r) = 0. \tag{6.6}$$

Note that the generators g_i and T are invertible. Also, that the algebra $\mathrm{Y}_{d,1,n}(u, u_1, \ldots, u_r)$ is clearly a quotient of the modular framed braid group algebra $\mathbb{C}\mathcal{F}_{d,1,n}$ under the quadratic relations (6.3) and relation (6.6).

Definition 6.4. ([27]) For $u \in \mathbb{C}\backslash\{0\}$ we define the *generalized Yokonuma–Hecke algebra of type* B, $\mathrm{Y}_{d,1,n}(u)$, as the algebra presented by the braiding generators $g_1, \ldots, g_{n-1}$, the loop generator T and the framing generators $t_1, \ldots, t_n$ subject to the relations (2.2) for the braiding and framing generators, relations (6.2) for the loop generator and the quadratic relations (6.3) for the braiding generators.

Remark 6.3. By [34, Remark 1] and Remark 6.2, the algebra $\mathrm{Y}_{d,1,n}(u)$ can be considered isomorphically as the *affine Yokonuma–Hecke algebra of type A*. See also [10].

Note that for $d = 1$ the algebras $\mathrm{Y}_{1,1,n}(u, U)$, $\mathrm{Y}_{1,1,n}(u, u_1, \ldots, u_r)$ and $\mathrm{Y}_{1,1,n}(u)$ coincide with the corresponding algebras of B-type defined above. Note also that the algebras $\mathrm{Y}_{d,1,n}(u, U)$ and $\mathrm{Y}_{d,1,n}(u, u_1, \ldots, u_r)$ can be clearly viewed as quotients of $\mathrm{Y}_{d,1,n}(u)$ by the relations (6.4) and (6.6) respectively. Namely, we have algebra epimorphisms:

$$\mathbb{C}\mathcal{F}_{d,1,n} \twoheadrightarrow \mathrm{Y}_{d,1,n}(u) \twoheadrightarrow \mathrm{Y}_{d,1,n}(u, u_1, \ldots, u_r).$$

On all these algebras unique Markov traces can be constructed with the three rules analogous to the rules of the Markov traces discussed in Sec. 6.2, together with a fourth inductive rule that takes care of the framings, analogous to rule (3) of the Juyumaya trace tr (recall Theorem 2.2). Such traces are constructed in [11]. Further, invariants of framed links, classical links and singular links in the solid torus can be constructed, up to necessary conditions analogous to the E-condition. These conditions are given in [11], where the representation theory of all these algebras is also studied.

7. Framization of the BMW Algebra

7.1. Let $l, m \in \mathbb{C}\backslash\{0\}$. Birman and Wenzl [6] and simultaneously but independently Murakami [35] defined a unital associative algebra $C_n = C_n(l, m)$ which is known as the *Birman–Wenzl–Murakami algebra* or simply the *BMW algebra*. The algebra C_n is defined by two sets of generators: the "braiding" generators $g_1, \ldots, g_{n-1}$ and the 'tangle' generators $h_1, \ldots, h_{n-1}$, satisfying: the braid relations (1.1) among the g_i's, together with the relations:

$$\begin{aligned} g_i h_i &= l^{-1} h_i && \text{for } 1 \leq i \leq n-1 \\ h_i g_{i\pm 1} h_i &= l h_i && \text{for all } i \\ g_j h_i &= h_i g_j && \text{for } |i-j| \geq 2 \end{aligned} \tag{7.1}$$

and the quadratic relations:

$$g_i^2 = 1 - m g_i + m l^{-1} h_i \qquad (1 \leq i \leq n-1). \tag{7.2}$$

For diagrammatic interpretations of the "tangle" generators h_i the reader may view Fig. 10, where the framings 0 and 1 should be ignored and where h_i corresponds to the two horizontal arcs joining the i and $i+1$ endpoints.

From the defining relations of C_n we deduce that the g_i's are invertible:

$$g_i^{-1} = g_i - m h_i + m \tag{7.3}$$

and also the following important relations:

$$\begin{aligned} h_i g_i &= l^{-1} h_i && \text{for } 1 \leq i \leq n-1 \\ h_i h_j &= h_j h_i && \text{for } |i-j| \geq 2 \\ h_i^2 &= y h_i && \text{for } 1 \leq i \leq n-1 \end{aligned} \tag{7.4}$$

where

$$y := 1 + \frac{l^{-1} - l}{m}.$$

The algebra C_n is a quotient of the classical braid group algebra $\mathbb{C}B_n$. To see this consider C_n generated by the g_i's only, and view Eq. (7.3) as the defining relations for the h_i's. Further, the element

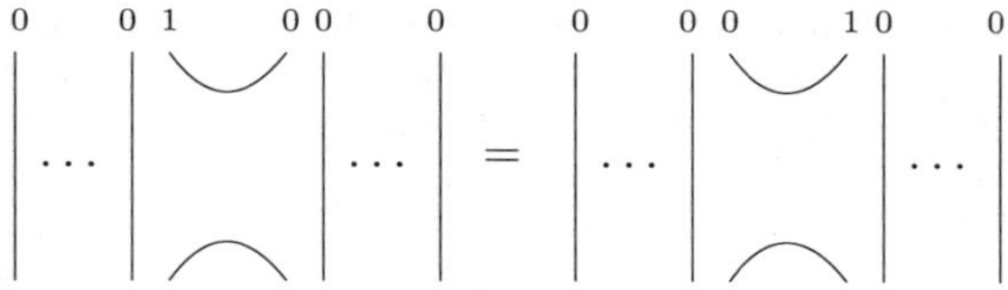

Fig. 10. The relation $t_i h_i = t_{i+1} h_i$.

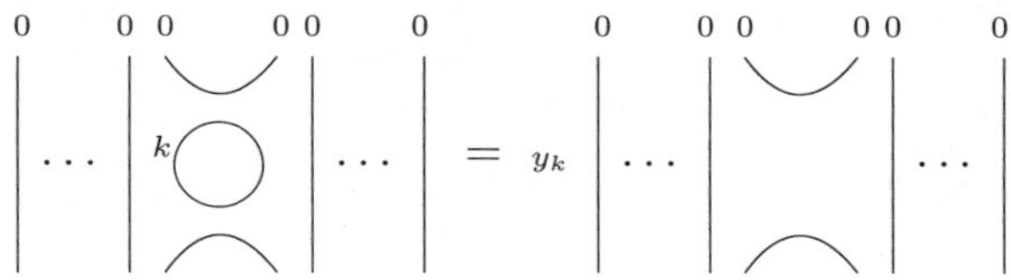

Fig. 11. The relation $h_i t_i^k h_i = y_k h_i$.

h_i can be seen represented in the category of (n, n)-tangles as the elementary tangle consisting in two curved parallel horizontal arcs joining the endpoints i and $i+1$ at the top and at the bottom of the otherwise identity tangle (see Fig. 10). The algebra C_n is related to the *Kauffman polynomial* [28].

7.2. In [27] the framization of the BMW algebra has been introduced. We shall describe it here briefly.

Definition 7.1. Let $y_0 := y$ and $y_1, \ldots, y_{d-1}$ in $\mathbb{C}\backslash\{0, 1\}$. The *framization of the BMW algebra* C_n, denoted $F_{d,n} = F_{d,n}(l, m, y_0, \ldots, y_{d-1})$, is defined through three sets of generators: the two sets of generators of the algebra C_n given above, together with the framing generators $t_1, \ldots, t_n$, satisfying all defining relations of C_n except for the quadratic relations (7.2), which are replaced by the following quadratic relations:

$$g_i^2 = (1 - m) - me_i(g_i - 1) + ml^{-1}h_i \qquad (1 \leq i \leq n-1) \qquad (7.5)$$

where e_i as defined in (2.4), together with the following relations for the framing generators:

$$\begin{array}{rlll} t_i^d = 1 \quad \text{and} \quad t_i t_j &= t_j t_i & \text{for } 1 \leq i, j \leq n \\ t_j g_i &= g_i t_{s_i(j)} & \text{for } 1 \leq i \leq n-1 \text{ and } 1 \leq j \leq n \\ t_i h_i &= t_{i+1} h_i & \text{for } 1 \leq i \leq n-1 \\ h_i t_i &= h_i t_{i+1} & \text{for } 1 \leq i \leq n-1 \\ h_i t_i^k h_i &= y_k \, h_i & \text{for } 1 \leq i \leq n-1 \text{ and } 0 \leq k \leq d-1 \\ h_i t_j &= t_j h_i & \text{for } j \neq i, i+1, \end{array} \tag{7.6}$$

where $s_i(j)$ is the effect of the transposition $s_i = (i, i+1)$ on j.

Note that for $d = 1$ we have $e_i = 1$, hence $F_{1,n}$ coincides with C_n. Also, the elements g_i are invertible [27, Proposition 1]:

$$g_i^{-1} = \frac{1}{1-m} g_i - \frac{m}{1-m} g_i e_i - m h_i + m e_i. \tag{7.7}$$

The algebra $F_{d,n}$ can be viewed as a quotient of the modular framed braid group algebra $\mathbb{C}\mathcal{F}_{d,n}$ by removing the h_i's from the set of generators and taking Eq. (7.7) as their defining relations. So, elements in the algebra $F_{d,n}$ can be viewed as framed (n, n)-tangles, with framings modulo d. In this context, Figs. 10 and 11 illustrate two of the relations of (7.6).

In [27] it is shown that $F_{d,n}$ is finite-dimensional. We also have the following important result.

Proposition 7.1. (Proposition 2 [27]) *The elements g_i satisfy the quartic relation:*

$$g_i^4 + m g_i^3 + (m-2) g_i^2 + m(m-1) g_i - (m-1) = m l^{-1} (m + l^{-2} - 1) h_i$$

and this is of minimal degree not containing the framing generators t_i. Also, they satisfy the "closed" quintic relation:

$$(x - l^{-1})(x^4 + m x^3 + (m-2) x^2 + m(m-1) x - (m-1)) = 0$$

and this is of minimal degree not containing the generators t_i *and* h_i. *Moreover,*

$$x^4+mx^3+(m-2)x^2+m(m-1)x-(m-1) = (x^2+mx-1)(x^2+m-1).$$

Finally we have the following result:

Proposition 7.2. [27, Proposition 3] *Any element in* $F_{d,n}$ *can be written as a* $\mathbb{C}$*-linear combination of monomials of the form* $\alpha f \beta$, *where* α *and* β *are monomials in* $1, g_1, \ldots, g_{n-2}, h_1, \ldots, h_{n-2}, t_1, \ldots, t_{n-1}$ *and* $f \in X_n := \{t_n^s,\ g_{n-1},\ t_{n-1}^s h_{n-1} t_{n-1}^r | 0 \leq r, s \leq d-1\}$.

This result is in the right direction for defining Markov traces on $F_{d,n}$ via inductive rules.

8. The Framization of the Singular Hecke Algebra

A definition of the *singular Hecke algebra*, denoted $S\mathrm{H}_n(u)$, was proposed by Paris and Rabenda [37]. This algebra is a complex associative unital algebra defined as the quotient of the algebra $\mathbb{C}SB_n$ of the singular braid monoid SB_n over the Hecke algebra quadratic relations (2.1).

Recall that SB_n was introduced in different contexts by Baez [2], Birman [5] and Smolin [39] and it is defined by the classical elementary braids σ_i with their inverses σ_i^{-1}, $i = 1, \ldots, n-1$, which are subject to the braid relations (1.1), and by the elementary singular braids τ_i, $i = 1, \ldots, n-1$, together with the following relations:

$$\begin{aligned} [\sigma_i, \tau_j] = [\tau_i, \tau_j] &= 0 && \text{for } |i-j| > 1 \\ [\sigma_i, \tau_i] &= 0 && \text{for all } i \\ \sigma_i \sigma_j \tau_i &= \tau_j \sigma_i \sigma_j && \text{for } |i-j| = 1. \end{aligned} \tag{8.1}$$

Paris and Rabenda [37] also constructed a universal Markov trace on these algebras, which lead to an invariant for singular knots, which is basically equivalent to the singular link invariant of Kauffman and Vogel defined in [31].

We can now proceed with the following definition.

Definition 8.1. The *framization of the algebra* $S\mathrm{H}_n(u)$, denoted $FS_{d,n}(u)$, is defined as the unital associative algebra over $\mathbb{C}$, defined through three sets of generators: $g_1, \ldots, g_{n-1}$, $\tau_1, \ldots, \tau_{n-1}$ corresponding to the two sets of generators of the algebra $S\mathrm{H}_n(u)$ given above, together with the framing generators $t_1, \ldots, t_n$, satisfying all the Yokonuma–Hecke algebra relations, (2.2) and (2.3), together with the above relations (8.1) of $SB_n(u)$, whereby σ_i corresponds to g_i and τ_i to τ_i.

For the algebra $FS_{d,n}(u)$ one needs to find appropriate inductive basis and define on it a Markov trace analogue to the one by Paris and Rabenda.

9. Concluding Note

We presented framizations of several knot algebras, starting from the example of the classical Iwahori–Hecke algebra, and discussed questions that need to be further investigated. There are many more other knot algebras. For example, other quotients of the classical braid group, quotients of the virtual braid group [29, 30], or the Rook algebra [4], which is related to the Alexander polynomial.

Acknowledgments

This research has been co-financed by the European Union (European Social Fund — ESF) and Greek national funds through the Operational Program "Education and Lifelong Learning" of the National Strategic Reference Framework (NSRF) — Research Funding Program: THALES. Moreover, the first author was partially supported by Fondecyt 1141254.

We would like to thank the Referee for the very careful reading and for the very interesting comments.

References

[1] S. Ariki and K. Koike, A Hecke algebra of $\mathbb{Z}/r\mathbb{Z} \wr S_n$ and construction of its irreducible representations, *Adv. Math.* **106** (1994) 216–243.

[2] J. C. Baez, Link invariants of finite type and perturbation theory. *Lett. Math. Phys.* **26** (1992) 43–51.

[3] D. Bennequin, Entrelacements et équations de Pfaffe, *Asterisque* **107–108** (1983) 87–161.

[4] S. Bigelow, E. Ramos and R. Yin, The Alexander and Jones Polynomials through representations of the Rook algebra, *J. Knot Theory Ramifications* **21** (2012) 1250114.

[5] J. S. Birman, New points of view in knot theory, *Bull. Amer. Math. Soc. (N.S.)* **28** (1993) 253–287.

[6] J. Birman and H. Wenzl, Braids link polynomials and a new algebra, *Trans. Amer. Math. Soc.* **313** (1989) 249–273.

[7] M. Broué and G. Malle, Zyklotomische Heckealgebren, *Astérisque* **212** (1993) 119–203.

[8] M. Chlouveraki and S. Lambropoulou, The Yokonuma–Hecke algebras and the HOMFLYPT polynomial, *J. Knot Theory Ramifications* **22**, DOI: 10.1142/S0218216513500806.

[9] M. Chlouveraki and G. Pouchin, Determination of the representations and a basis for the Yokonuma–Temperley–Lieb algebra, to appear in *Algebra and Representation Theory*, arXiv:1311.5626.

[10] M. Chlouveraki and L. Poulain D'Andecy, Representation theory of the Yokonuma–Hecke algebra, *Adv. Math.* **259** (2014) 134–172.

[11] M. Chlouveraki and L. Poulain D'Andecy, Markov traces on affine and cyclotomic Yokonuma–Hecke algebras, arXiv:1406.3207v1.

[12] S. Chmutov, S. Jablan, J. Juyumaya, K. Karvounis and S. Lambropoulou, On the knot invariants from the Yokonuma–Hecke Algebras, in preparation. See http://www.math.ntua.gr/sofia/yokonuma/index.html.

[13] R. Dipper and G. D. James, Representations of Hecke algebras of type B_n, *J. Algebra* **146** (1992) 454–481.

[14] M. Geck and S. Lambropoulou, Markov traces and knot invariants related to Iwahori–Hecke algebras of type B, *J. Reine Angew. Mathe.* **482** (1997) 191–213.

[15] B. Gemein, Singular braids and Markov's theorem, *J. Knot Theory Ramifications* **6** (1997) 441–454.

[16] D. Goundaroulis, J. Juyumaya, A. Kontogeorgis and S. Lambropoulou, *The Yokonuma–Temperley–Lieb Algebra*, to appear in *Banach Center Pub.* **103**, December 2014, arXiv:1012.1557.

[17] D. Goundaroulis, J. Juyumaya, A. Kontogeorgis, and S. Lambropoulou, Framization of the Temperley–Lieb algebra, submitted for publication, arXiv:1304.7440.

[18] V. F. R. Jones, Hecke algebra representations of braid groups and link polynomials, *Ann. Math.* **126** (1987) 335–388.

[19] J. Juyumaya, Sur les nouveaux générateurs de l algèbre de Hecke $H(G, U, 1)$, *J. Algebra* **204** (1998) 49–68.

[20] J. Juyumaya, Markov trace on the Yokonuma–Hecke algebra, *J. Knot Theory Ramifications* **13** (2004) 25–39.

[21] J. Juyumaya and S. S. Kannan, Braid relations in the Yokonuma–Hecke algebra, *J. Algebra* **239** (2001) 272–297.
[22] J. Juyumaya and S. Lambropoulou, p-adic framed braids and p-adic Markov traces, in *Proc. ITLD 2006, Hiroshima, Japan, Series on Knots and Everything* (World Scientific, 2007).
[23] J. Juyumaya and S. Lambropoulou, p-adic framed braids, *Topol. Appl.* **154** (2007) 1804–1826.
[24] J. Juyumaya and S. Lambropoulou, An invariant for singular knots, *J. Knot Theory Ramifications* **18** (2009) 825–840.
[25] J. Juyumaya and S. Lambropoulou, An adelic extension of the Jones polynomial, eds. M. Banagl and D. Vogel, *The Mathematics of Knots*, Contributions in the Mathematical and Computational Sciences, Vol. 1 (Springer, 2011).
[26] J. Juyumaya and S. Lambropoulou, p-adic framed braid II, *Adv. Math.* **234** (2013) 149–191.
[27] J. Juyumaya and S. Lambropoulou, Modular framization of the BMW algebra, arXiv:1007.0092.
[28] L. H. Kauffman, An invariant of regular isotopy, *Trans. Amer. Math. Soc.* **318** (1990) 417–471.
[29] L. H. Kauffman, *Virtual Knot Theory, Eur. J. Combin.* **20** (1999) 663–690.
[30] L. H. Kauffman and S. Lambropoulou, Virtual braids and the L-move, *J. Knot Theory Ramifications* **15** (2006) 773–811.
[31] L. H. Kauffman and P. Vogel, Link polynomials and a graphical calculus, *J. Knot Theory Ramifications* **1** (1992) 59–104.
[32] K. H. Ko and L. Smolinsky, The framed braid group and 3-manifolds, *Proc. AMS* **115** (1992) 541–551.
[33] S. Lambropoulou, Solid torus links and Hecke algebras of B-type, in *Quantum Topology*, ed. D. N. Yetter (World Scientific, 1994), pp. 225–245.
[34] S. Lambropoulou, Knot theory related to generalized and cyclotomic Hecke algebras of type B, *J. Knot Theory Ramifications* **8** (1999) 621–658.
[35] J. Murakami, The Kauffman polynomial of links and representation theory, *Osaka J. Math.* **24** (1987) 745–758.
[36] S. Yu. Orevkov and V. V. Shevchishin, Markov theorem for transversal links, *J. Knot Theory Ramifications* **12** (2003) 905–913.
[37] L. Paris and L. Rabenda, Singular Hecke algebras, Markov traces and HOMFLY-type invariants, *Ann. Inst. Fourier (Grenoble)* **58** (2008) 2413–2443.
[38] S. Ramanujan, On certain trigonometric sums and their applications in the theory of numbers, *Trans. Cambridge Philos. Soc.* **22** (1918) 297–276.
[39] L. Smolin, Knot theory, loop space and the diffeomorphism group, in *New Perspectives in Canonical Gravity* (Monogr. Textbooks Phys. Sci. Lecture Notes, Vol. 5 (Bibliopolis, 1988), pp. 245–266.
[40] H. N. V. Temperley and E. H. Lieb, Relations between the "percolation" and "colouring" problem and other graph — theoretical problems associated

with regular planar lattice: Some exact results for the "percolations problems", *Proc. Roy. Soc. London Ser. A* **322** (1971) 251–280.

[41] N. Thiem, Unipotent Hecke algebras of $GL_n(\mathbb{F}_q)$, *J. Algebra* **284** (2005) 559–577.

[42] H. Wenzl, Braids and invariants of 3-manifolds, *Invent. Math.* **114** (1993) 235–275.

[43] N. Wrinkle, The Markov theorem for transverse knots, arXiv:math/0202055v1.

[44] T. Yokonuma, Sur la structure des anneaux de Hecke d'un groupe de Chevalley fini, *C. R. Acad. Sc. Paris* **264** (1967) 344–347.

Virtual Knot Cobordism

Louis Hirsch Kauffman

Department of Mathematics, Statistics,
and Computer Science (m/c 249) 851 South Morgan Street,
University of Illinois at Chicago,
Chicago, IL 60607-7045, USA
kauffman@uic.edu

This paper defines a theory of cobordism for virtual knots and studies this theory for standard and rotational virtuals. Nontrivial examples of virtual slice knots are given and determinations of the four-ball genus of positive virtual knots are discussed in relation to [2]. Problems related to band-passing are discussed and a theory of isotopy of virtual surfaces is formulated in terms of a generalization of the Yoshikawa moves.

1. Introduction

This paper defines a theory of cobordism for virtual knots. It is organized as follows. In Sec. 2, we include a description of basics in virtual knot theory and the problems that arise from it. This section includes different interpretations of virtual knot theory including rotational virtual knot theory (where the detour moves are constrained to be regular homotopies in the plane), problems related to the bracket and Jones polynomial for the theory, a discussion of parity, the odd writhe and a description of the parity bracket polynomial formulated by Manturov [28]. All of this background material is used in the remainder of the paper where we apply these ideas and techniques to virtual knot cobordism. Section 3 gives the definitions for cobordism of virtual knots and the definition of virtual Seifert surfaces. We define the four-ball genus $g_4(K)$ for virtual knots and links, and show that every virtual link K bounds a virtual surface that is a natural generalization of the Seifert surface for a classical link. We state a new result determining the four-ball genus for postive virtual knots. This result is proved in our paper [2]. We give many properties of a key example, the virtual stevedore's knot. Then we discuss cobordism for rotational virtual knots and show that a virtual rotational link bounds a virtual surface if and only if it has an even number of virtual crossings. Section 3.4 is an exploration of band-passing for virtual knots, and the possibility to generalize classical results about the Arf invariant. In Sec. 4, we develop a theory of virtual surfaces in four-space based on a generalization of the Yoshikawa moves, and we give examples of virtual 2-spheres in four-space that are related to the virtual stevedore's knot.

This paper is intended to initiate the study of virtual knot cobodrism at the level of the usual virtual knot theory, a theory based on looking at knot diagrams without demanding planarity of the diagram. Each crossing is locally in the plane, but the connections among the crossings can leave the plane, and so arcs appear to cross over one another to make these connections. These crossings are called *virtual*, and the theory is expressed in terms of diagrams that contain both classical and virtual crossings. There are a number of different ways to interpret virtual knots and links topologically. We describe these ways in Sec. 2. One can also develop more combinatorial approaches to virtual knot theory by giving up even more structure. One can use Gauss codes or Gauss diagrams to represent virtual knots, and release certain structures related to the codes to make combinatorial theories that inform the virtual knot. Such work has been initiated by Turaev [34] and significantly by Manturov [7, 27, 29]. We intend to connect the work of this paper with the work of Manturov in a joint paper under preparation. We should also mention the following papers on virtual knot theory that are useful background, but not cited directly in this paper [3, 15, 17, 21, 23–25, 31].

It should be remarked that the work [27] can be used to prove that many virtual knots are not concordant to any slice classical knot. This may not be immediately apparent to many readers, as the paper is focused on the cobordism of free knots. Results about free knots (undecorated Gauss diagrams taken up to Reidemeister move equivalence) are often applicable to standard virtual knots by simply forgetting some of the structure. A crucial question about our formulation of virtual knot cobordism is: If two classical knots are concordant as virtual ones, are they concordant in the usual sense? At this writing, we do not know the answer to this question.

It is worth mentioning areas of low dimensional topology and classical knot theory that are related to and informed by virtual knot theory. Just as classical combinatorial knot theory can be studied up to regular isotopy, it is useful to study virtual knots up to *rotational isotopy* (see Sec. 2.3 below). Then we have:

1. Every quantum invariant of classical knots extends to an invariant of rotational virtual knots.
2. Many quantum invariants, including the Jones polynomial, extend naturally to invariants of standard virtual knots. In the case of the Jones polynomial, it is inherent in its structure that it extends in a number of ways to invariants of virtual knots and rotational virtual knots. This means that the Jones polynomial and these extensions become one of the main motivations for studying virtual knot theory. The virtual knot theory becomes a way to ask new questions about the nature of the Jones polynomial.
3. Many problems related to categorification and link homology can be formulated in this domain.
4. The generalization of the classical braid group to the virtual braid group fits naturally into relationships of knot theory with quantum link invariants. In fact, in the context of the algebraic Yang–Baxter equation, the virtual braid group arises naturally for algebraic reasons (see [18]).
5. Finally, the virtual knot theory has a number of variations. Standard virtual knot theory is equivalent to stabilized knot theory in thickened surfaces. When we add one of the forbidden moves, the scene changes and we have *welded knot theory* which is related to the Dahm braid group of circles in three-dimensional space and

to embeddings or tori in four-dimensional space. Then if we let go of structure, we get flat virtual knot theory which is equivalent to stabilized immersions of curves in surfaces, and, letting go of even more structure, we have free virtual knot theory which is the study of Gauss diagrams without any orientation up to the Reidemeister moves. All of these subjects are interrelated and the combinatorial approach, combined with geometric topology makes a rich mixture of problems and ideas.

In this paper we formulate a theory of cobordism and concordance of virtual knots and links that is a generalization of the theory of cobordism of classical knots. This includes a virtual analog of ambient isotopy for surfaces in four-space and corresponding invariants of these isotopies. We give examples showing the viability of the generalizations and the problems that ensue.

2. Virtual Knot Theory

Knot theory studies the embeddings of curves in three-dimensional space. Virtual knot theory studies the embeddings of curves in thickened surfaces of arbitrary genus, up to the addition and removal of empty handles from the surface. Virtual knots have a special diagrammatic theory, described below. Classical knot theory embeds in virtual knot theory.

In the diagrammatic theory of virtual knots one adds a *virtual crossing* (see Fig. 1) that is neither an over-crossing nor an under-crossing. A virtual crossing is represented by two crossing segments with a small circle placed around the crossing point.

Moves on virtual diagrams generalize the Reidemeister moves for classical knot and link diagrams. See Fig. 1. Classical crossings interact with one another according to the usual Reidemeister moves, while virtual crossings are artifacts of the structure in the plane. Adding the global detour move to the Reidemeister moves completes the description of moves on virtual diagrams. In Fig. 1 we illustrate a set of local moves involving virtual crossings. The global detour move is a consequence of moves (B) and (C) in Fig. 1. The detour

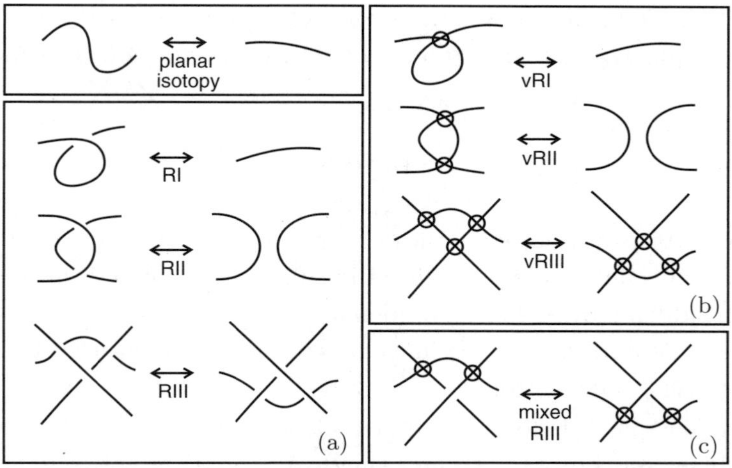

Fig. 1. Moves.

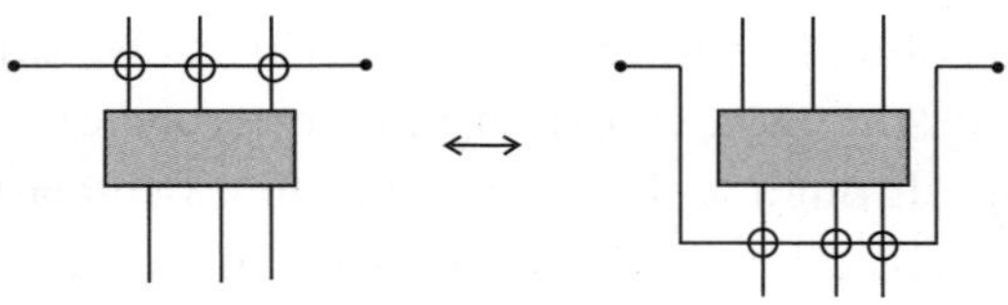

Fig. 2. Detour move.

move is illustrated in Fig. 2. Virtual knot and link diagrams that can be connected by a finite sequence of these moves are said to be *equivalent* or *virtually isotopic*.

Another way to understand virtual diagrams is to regard them as representatives for oriented Gauss codes [6, 14, 20] (Gauss diagrams). Such codes do not always have planar realizations. An attempt to embed such a code in the plane leads to the production of the virtual crossings. The detour move makes the particular choice of virtual crossings irrelevant. *Virtual isotopy is the same as the equivalence relation generated on the collection of oriented Gauss codes by abstract Reidemeister moves on these codes.*

Figure 3 illustrates the two *forbidden moves*. Neither of these follows from Reidmeister moves plus detour move, and indeed it is

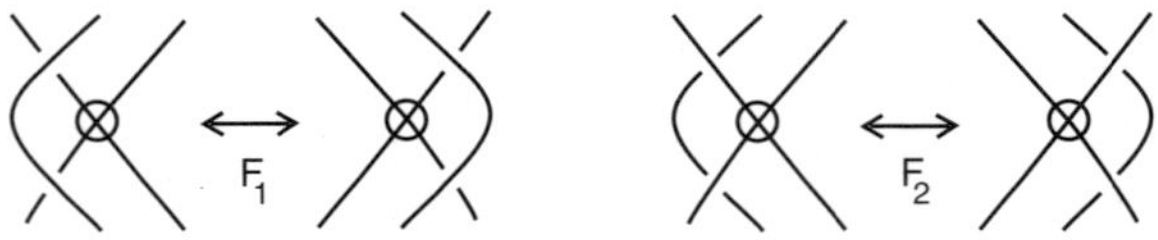

Fig. 3. Forbidden moves.

not hard to construct examples of virtual knots that are nontrivial, but will become unknotted on the application of one or both of the forbidden moves. The forbidden moves change the structure of the Gauss code and, if desired, must be considered separately from the virtual knot theory proper.

2.1. *Interpretation of virtuals links as stable classes of links in thickened surfaces*

There is a useful topological interpretation [14, 15] for this virtual theory in terms of embeddings of links in thickened surfaces. Regard each virtual crossing as a shorthand for a detour of one of the arcs in the crossing through a 1-handle that has been attached to the 2-sphere of the original diagram. By interpreting each virtual crossing in this way, we obtain an embedding of a collection of circles into a thickened surface $S_g \times R$ where g is the number of virtual crossings in the original diagram L, S_g is a compact oriented surface of genus g and R denotes the real line. We say that two such surface embeddings are *stably equivalent* if one can be obtained from another by isotopy in the thickened surfaces, homeomorphisms of the surfaces and the

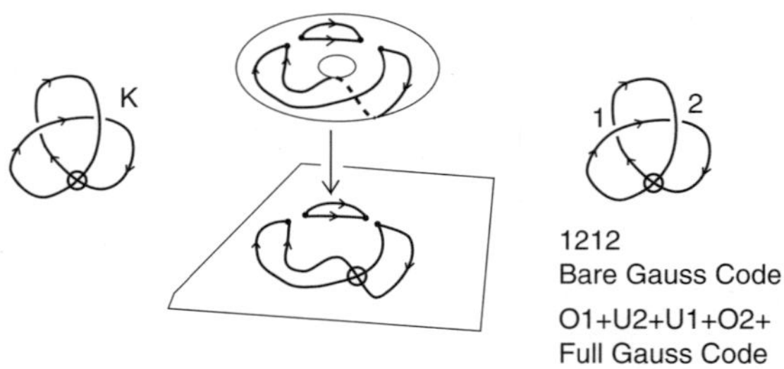

Fig. 4. Surface representation.

addition or subtraction of empty handles (i.e. the knot does not go through the handle).

We have

Theorem 2.1. ([1, 14, 15]) *Two virtual link diagrams are isotopic if and only if their corresponding surface embeddings are stably equivalent.*

The reader will find more information about this correspondence [14, 15] in other papers by the author and in the literature of virtual knot theory.

2.2. *Review of the bracket polynomial for virtual knots*

In this section we recall how the bracket state summation model [16] for the Jones polynomial is defined for virtual knots and links.

The bracket polynomial [16] model for the Jones polynomial [8–10, 35] is usually described by the expansion

$$\langle \times \rangle = A\langle \asymp \rangle + A^{-1}\langle)(\rangle \tag{1}$$

and we have

$$\langle K \bigcirc \rangle = (-A^2 - A^{-2})\langle K \rangle, \tag{2}$$

$$\langle \gamma \rangle = (-A^3)\langle \smile \rangle, \tag{3}$$

$$\langle \gamma \rangle = (-A^{-3})\langle \smile \rangle. \tag{4}$$

We call a diagram in the plane *purely virtual* if the only crossings in the diagram are virtual crossings. Each purely virtual diagram is equivalent by the virtual moves to a disjoint collection of circles in the plane.

A state S of a link diagram K is obtained by choosing a smoothing for each crossing in the diagram and labeling that smoothing with either A or A^{-1} according to the convention indicated in the bracket expansion above. Then, given a state S, one has the evaluation $\langle K|S \rangle$ equal to the product of the labels at the smoothings, and one has the evaluation $||S||$ equal to the number of loops in the state (the smoothings produce purely virtual diagrams). One then has the formula

$$\langle K \rangle = \sum_{S} \langle K|S \rangle d^{||S||-1},$$

where the summation runs over the states S of the diagram K, and $d = -A^2 - A^{-2}$. This state summation is invariant under all classical and virtual moves except the first Reidemeister move. The bracket polynomial is normalized to an invariant $f_K(A)$ of all the moves by the formula $f_K(A) = (-A^3)^{-w(K)} \langle K \rangle$ where $w(K)$ is the writhe of the (now) oriented diagram K. The writhe is the sum of the orientation signs (± 1) of the crossings of the diagram. The Jones polynomial, $V_K(t)$ is given in terms of this model by the formula

$$V_K(t) = f_K(t^{-1/4}).$$

This definition is a direct generalization to the virtual category of the state sum model for the original Jones polynomial. It is straightforward to verify the invariances stated above. In this way one has the Jones polynomial for virtual knots and links.

We have [15] the

Theorem. *To each nontrivial classical knot diagram of one component K there is a corresponding nontrivial virtual knot diagram $Virt(K)$ with unit Jones polynomial.*

The main ideas behind this theorem are indicated in Figs. 5 and 6. In Fig. 5 we indicate the virtualization operation that replaces a classical crossing by using two virtual crossings and changing the implicit orientation of the classical crossing. We also show how the bracket polynomial sees this operation as though the crossing had been switched in the classical knot. Thus, if we virtualize as set of classical crossings whose switching will unknot the knot, then the virtualized knot will have unit Jones polynonmial. On the other hand, the virtualization is invisible to the quandle, as shown in Fig. 6. This implies (by properties of the quandle) that virtual knots obtained in this way from classical nontrivial knots will themselves be nontrivial.

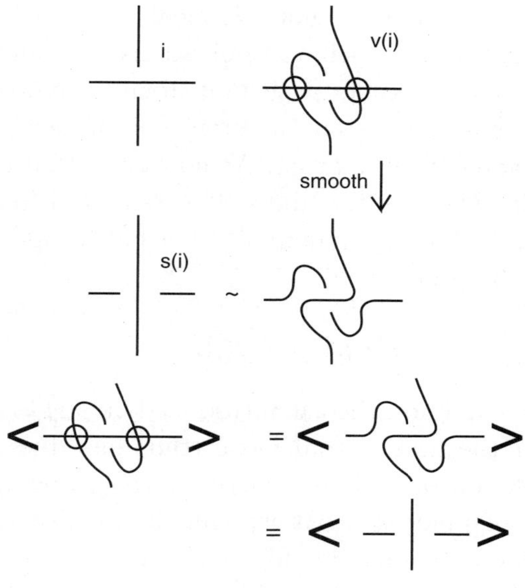

Fig. 5. Virtualizing a crossing and crossing switches.

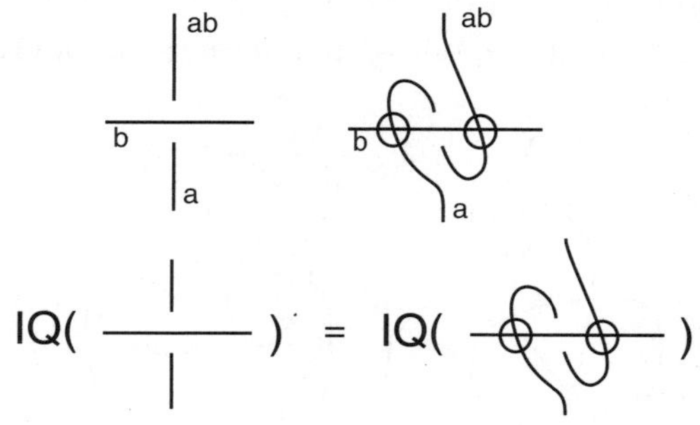

Fig. 6. Quandle invariance under virtualization.

It is an open problem whether there are classical knots (actually knotted) having unit Jones polynomial. (There are linked links whose linkedness is unseen [4] by the Jones polynomial.) If there exists a classical knot with unit Jones polynomial, then one of the knots $Virt(K)$ produced by this theorem may be isotopic to a classical knot. Such examples are guaranteed to be non-trivial, but they are usually also not classical. We do not know at this writing whether all such virtualizations of nontrivial classical knots, yielding virtual knots with unit Jones polynomial, are non-classical. This has led to an investigation of new invariants for virtual knots. Note added in proof: We now know that no virtual knot of the form $Virt(K)$ as in the above Theorem (K is a classical knot, $Virt(K)$ has unit Jones polynomial.) can be isotopic to a classical knot. We give the proof in [2].

2.3. *Rotational virtual knot theory*

Rotational virtual knot theory introduced in [14] is virtual knot theory without the first virtual move (thus one does not allow the addition or deletion of a virtual curl). All diagrams are represented in the surface of plane so that we can distinguish clockwise from counterclockwise rotations. See Fig. 7 for an example of a rotational virtual knot. The rotational version of virtual knot theory is significant because *all quantum link invariants originally defined for classical links extend to rotational virtual knot theory.* We give a quick illustration of this point in Fig. 8 where we indicate how a quantum link invariant depends on matrices or operators assigned

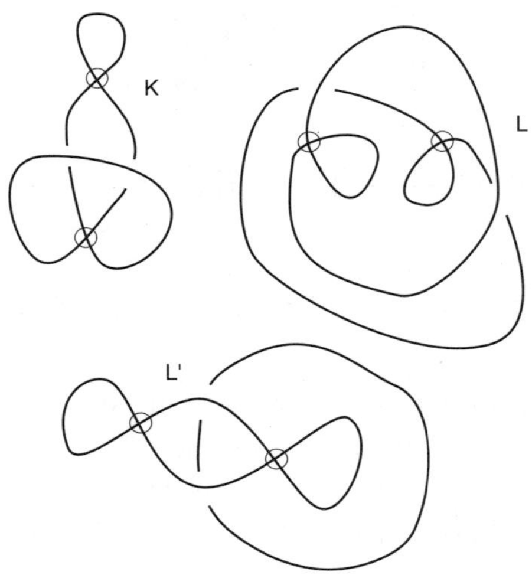

Fig. 7. A rotational virtual knot and two rotational virtual links.

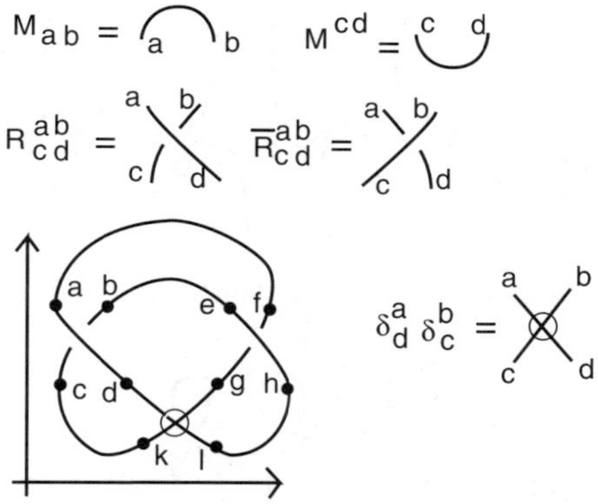

$$Z_K = M_{af} M_{be} M^{ck} M^{lh} R^{ab}_{cd} R^{ef}_{gh} \delta^d_l \delta^g_k$$

Fig. 8. Quantum link invariants.

to each crossing and each maximum and minimum of the diagram. One extends this to virtual crossings by using a crossed identity operator (a transposition) at the vritual crossings. The possibly nontrivial maxima and minima create the need to use the rotational context. This theory has been explored [14, 18] and deserves further exploration.

We formulate [14] a version of the bracket polynomial for rotational virtuals that assigns variables according to the Whitney degree of state curves. For rotational virtuals we extend the bracket just as we did for virtual knots and links except that the state curves are now disjoint unions in the plane of curves that have only virtual self-intersections, and are taken up to *regular homotopy* in the plane (we can take regular homotopy of curves to mean that the flat virtual versions of the second and third Reidemeister moves are allowed and that regular homotopy is the equivalence relation generated by these moves and planar homeomorphisms). When we expand the bracket we obtain a state sum of the form

$$[K] = \sum_S \langle K|S\rangle [S],$$

where the summation is over all states obtained by smoothing every crossing in the virtual diagram K and $\langle K|S\rangle$ is the product of the

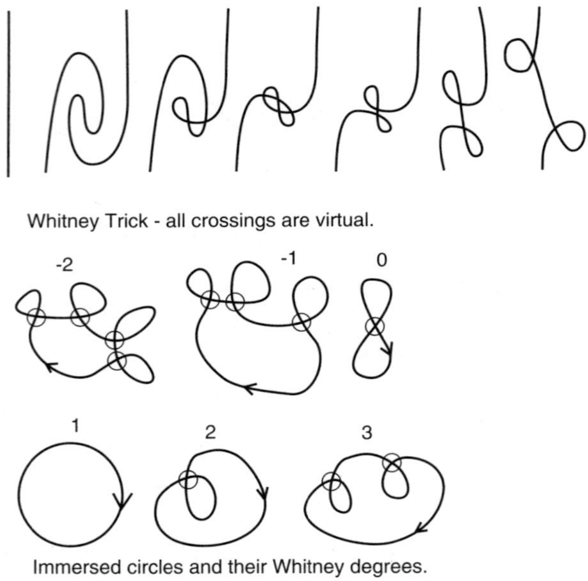

Fig. 9. Whitney trick and Whitney degrees.

weights A and A^{-1} just as before. An empty loop with no virtual crossings (in its virtual equivalence class) will be evaluated as $d = -A^2 - A^{-2}$. The symbol $[S]$ is the *planar class* of the state S. By the planar class of the state we mean its equivalence class up to virtual rotational equivalence. This means that each state loop is taken as a regular homotopy class. These individual classes are in 1-1 correspondence with the integers, as shown in Fig. 9 (via the Whitney trick and the winding degree of the plane curves), and can be handled by using combinatorial regular isotopy as in [11]. A configuration of loops (possibly nested) is equivalent to a disjoint union of adjacent loops. We can thus regard each virtual loop as a variable d_n where n is an integer and $d_1 = d_{-1} = -A^2 - A^{-2}$. Here we give an example of a computation of $[K]$ for a rotational virtual knot in Fig. 10. The reader will note that in this example, even if we let $A = -1 = B$ and $d = -2$ the invariant is still nontrivial due to the appearance of the two loops with Whitney degree zero. Thus the example in Fig. 10

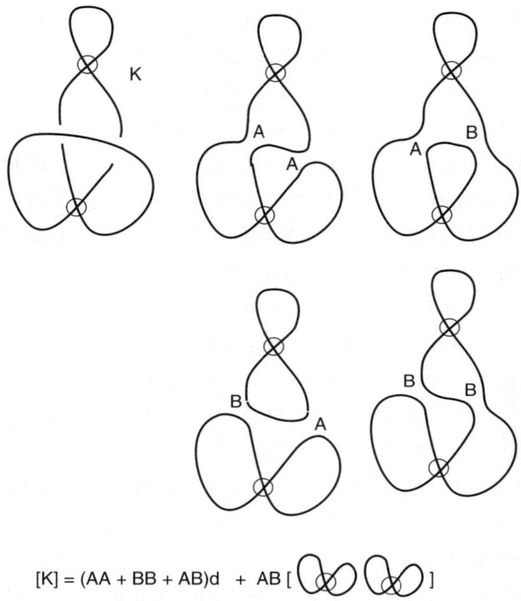

Fig. 10. Bracket expansion of a rotational virtual knot.

also gives a nontrivial flat rotational virtual knot. We shall look at cobordism of rotational virtual knots later in the paper.

2.4. *Parity, odd writhe and the parity bracket polynomial*

Parity is an important theme in virtual knot theory and figures in many investigations of this subject. In a virtual knot diagram there can be both even and odd crossings. A crossing is *odd* if it flanks an odd number of symbols in the Gauss code of the diagram. A crossing is *even* if it flanks an even number of symbols in the Gauss code of the diagram. For example, in Fig. 4 we illustrate the virtual knot K with bare Gauss code 1212. Both crossings in the diagram K are odd. In any classical knot diagram all crossings are even.

In [22] we introduced the *odd writhe* $J(K)$ for any virtual diagram K. $J(K)$ is the sum of the signs of the odd crossings. Classical diagrams have zero odd writhe. Thus if $J(K)$ is nonzero, then K is not equivalent to any classical knot. For the mirror image K^* of any diagram K, we have the formula $J(K^*) = -J(K)$. Thus, when $J(K) \neq 0$, we know that the knot K is not classical and not equivalent to its mirror image. Parity does all the work in this simple invariant. For example, if K is the virtual knot in Fig. 4, the we have $J(K) = 2$. Thus K, the simplest virtual knot, is non-classical and it is chiral (inequivalent to its mirror image).

In this section we introduce the Manturov Parity Bracket [28]. This is a form of the bracket polynomial defined for virtual knots and for free knots (unlabeled Gauss diagrams taken up to abstract Reidemeister move equivalence) that uses the parity of the crossings. To compute the parity bracket, we first make all the odd crossings into graphical vertices. Then we expand the resulting diagram on the remaining even crossings. The result is a sum of graphs with polynomial coefficients.

More precisely, let K be a virtual knot diagram. Let $E(K)$ denote the result of making all the odd crossings in K into graphical nodes as illustrated in Fig. 11. Let $SE(K)$ denote the set of all bracket states of $E(K)$ obtained by smoothing each classical crossing in $E(K)$ in one of the two possible ways. Then we define the *parity bracket*

$$\langle K \rangle_P = (1/d) \sum_{S \in SE(K)} A^{i(S)}[S],$$

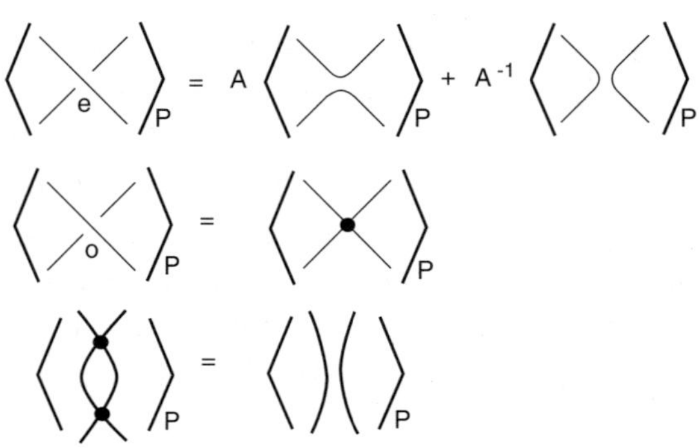

Fig. 11. Parity bracket expansion.

where $d = -A^2 - A^{-2}$, $i(S)$ denotes the product of A or A^{-1} from each smoothing site according to the conventions of Fig. 11, and $[S]$ denotes the reduced class of the virtual graph S. The graphs are subject to a reduction move that eliminates bigons as in the second Reidemeister move on a knot diagram as shown in Fig. 11. Thus $[S]$ represents the unique minimal representative for the virtual graph S under virtual graph isotopy coupled with the bigon reduction move. A graph that reduces to a circle (the circle is a graph for our purposes) is replaced by the value d above. Thus $\langle K \rangle_P$ is an element of a module generated by reduced graphs with coefficients Laurent polynomials in A.

With the usual bracket polynomial variable A, the parity bracket is an invariant of standard virtual knots. With $A = \pm 1$ it is an invariant of flat virtual knots. Even more simply, with $A = 1$ and taken modulo two, we have an invariant of flat knots with loop value zero. See Fig. 12 for an illustration of the application of the parity bracket to the Kishino diagram illustrated there. The Kishino diagram is notorious for being hard to detect by the usual polynomial invariants such as the Jones polynomial. It is a perfect example of the power of the parity bracket. All the crossings of the Kishino diagram are odd. Thus there is exactly one term in the evaluation of the Kishino diagram by the parity bracket, and this term is the Kishino diagram itself, with its crossings made into graphical nodes. The resulting graph is irreducible and so the Kishino diagram becomes its own invariant. We conclude that this diagram will be found from any isotopic version of the Kishino diagram. This allows strong conclusions about many properties of the diagram. For example, it is easy to check that the least surface on which this diagram can be represented with the given planar cyclic orders at the nodes

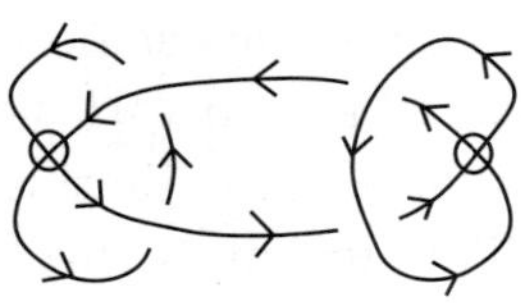

Fig. 12. Kishino diagram.

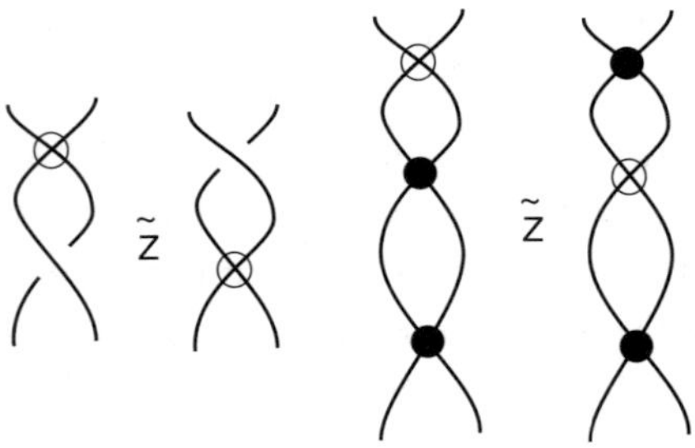

Fig. 13. Z-move and graphical Z-move.

is genus two. Thus we conclude that the least genus for a surface representation of the Kishino diagram as a flat knot or virtual knot is two.

In Fig. 13 we illustrate the *Z-move* and the *graphical Z-move.* Two virtual knots or links that are related by a Z-move have the same standard bracket polynomial. This follows directly from our discussion in the previous section. We would like to analyze the structure of Z-moves using the parity bracket. In order to do this we need a version of the parity bracket that is invariant under the Z-move. In order to accomplish this, we need to add a corresponding Z-move in the graphical reduction process for the parity bracket. This extra graphical reduction is indicated in Fig. 13 where we show a graphical Z-move. The reader will note that graphs that are irreducible without the graphical Z-move can become reducible if we allow graphical Z-moves in the reduction process. For example, the graph associated with the Kishino knot is reducible under graphical Z-moves. However, there are examples of graphs that are not reducible under graphical Z-moves and Reidemister two moves. An example of such a graph occurs in the parity bracket of the knot KS shown in Figs. 14 and 15. This knot has one even classical crossing and four odd crossings. One smoothing of the even crossing yields a state that reduces to a loop with no graphical nodes, while the other smoothing yields a state that is irreducible even when the Z-move is allowed. The upshot is that this knot KS is not Z-equivalent to any classical knot. Since one can verify that KS has unit Jones polynomial, this example is a counterexample to a conjecture

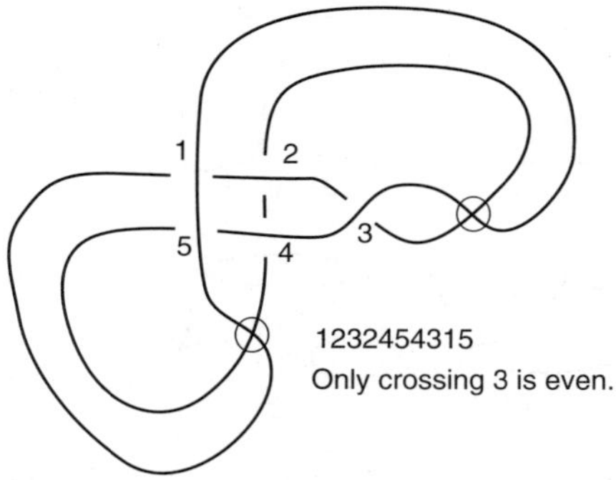

Fig. 14. A knot KS with unit Jones polynomial.

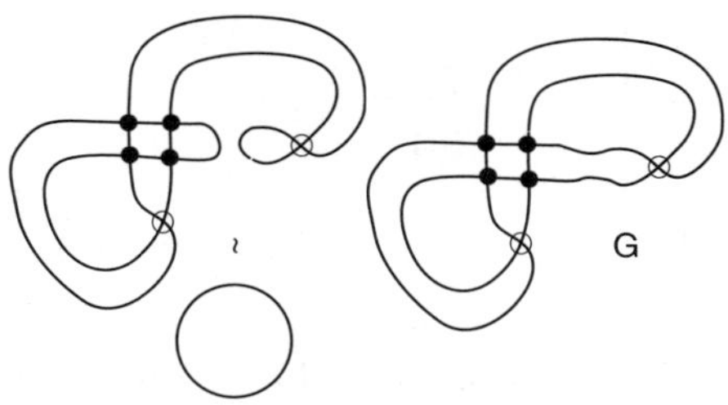

Fig. 15. Parity bracket states for the knot KS.

of Fenn, Kauffman and Maturov [5] that suggested that a knot with unit Jones polynomial should be Z-equivalent to a classical knot.

3. Virtual Knot Cobordism

Definition. Two oriented knots or links K and K' are *virtually cobordant* if one may be obtained from the other by a sequence of virtual isotopies (Reidemeister moves plus detour moves) plus births, deaths and oriented saddle points, as illustrated in Fig. 16. A *birth* is the introduction into the diagram of an isolated unknotted circle. A *death* is the removal from the diagram of an isolated unknotted circle. A saddle point move results from bringing oppositely oriented arcs into proximity and resmoothing the resulting site to obtain two new oppositely oriented arcs. See the figure for an illustration of the process. Figure 16 also illustrates the *schema* of surfaces that are

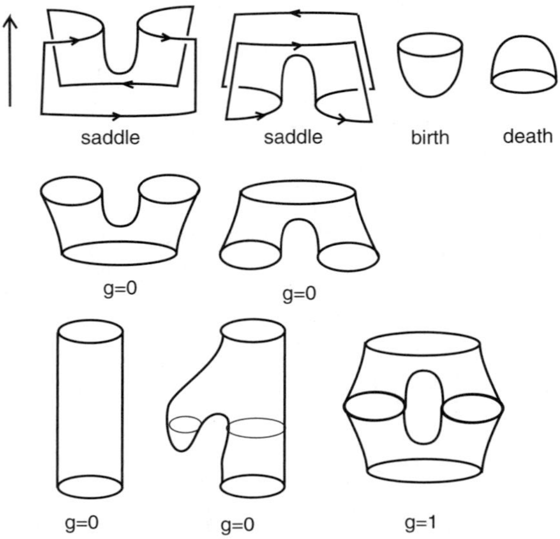

Fig. 16. Saddles, births and deaths.

generated by the cobordism process. These are abstract surfaces with well-defined genus in terms of the sequence of steps in the cobordism. In the figure we illustrate two examples of genus zero, and one example of genus 1. We say that a cobordism has genus g if its schema has that genus. Two knots are *cocordant* if there is a cobordism of genus zero connecting them. A virtual knot is said to be a *slice* knot if it is virtually concordant to the unknot, or equivalently if it is virtually concordant to the empty knot. (The unknot is concordant to the empty knot via one death.) As we shall see below, *every virtual knot or link is concordant to the unknot.* Another way to say this, is to say that there is a *virtual surface* (schema) whose boundary is the given virtual knot. The reader should note that when we speak of a virtual surface, we mean a surface schema that is generated by saddle moves, maxima and minima as described above.

Definition. We define the *four-ball genus* $g_4(K)$ of a virtual knot or link K to be the least genus among all virtual surfaces that bound K.

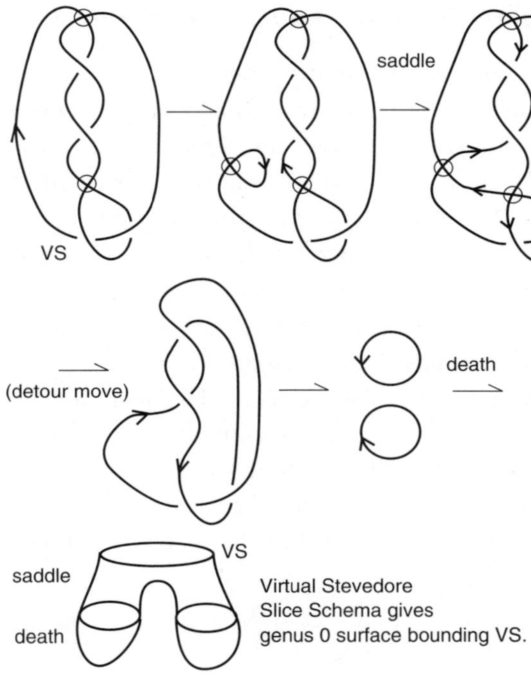

Fig. 17. Virtual stevedore is slice.

As we shall see below, there is a simple upper bound on the four-ball genus for any virtual knot or link and a definite result for the four-ball genus of positive virtual knots [2].

In Fig. 17 we illustrate the *virtual stevedore's knot, VS* and show that it is a slice knot in the sense of the above definition. This figure illustrates how the surface schema whose boundary in the virtual stevedore is evolved via the saddle point that produces two virtually unlinked curves that are isotopic to a pair of curves that can undergo deaths to produce the genus zero slicing surface. We will use this example to illustrate our theory of virtual knot cobordism, and the questions that we are investigating. Before looking at the virtual stevedore in this detail, we make a digression about spanning surfaces and the four-ball genus.

3.1. *Spanning surfaces for knots and virtual knots*

It is well known that every oriented classical knot or link bounds an embedded orientable surface in three-space. A representative surface of this kind can be obtained by the algorithm due to Seifert (see [11, 12, 19]). We have illustrated Seifert's algorithm for a trefoil diagram in Fig. 18. The algorithm proceeds as follows: At each oriented crossing in a given diagram K, smooth that crossing in the oriented manner (reconnecting the arcs locally so that the crossing disappears and the connections respect the orientation). The result of this operation is a collection of oriented simple closed curves in the plane, usually called the *Seifert circles*. To form the *Seifert surface* $F(K)$ for the diagram K, attach disjoint discs to each of the Seifert circles, and connect these discs to one another by local half-twisted bands at the sites of the smoothing of the diagram. This process is indicated in Fig. 18. In that figure we have not completed the illustration of the outer disc.

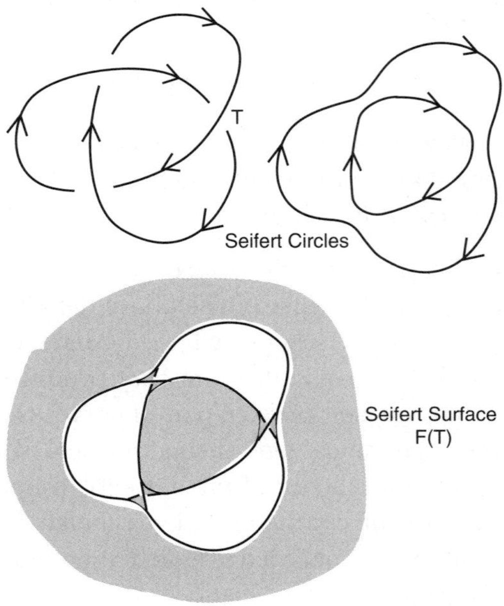

Fig. 18. Classical Seifert surface.

It is important to observe that we can calculate the genus of the resulting surface quite easily from the combinatorics of the classical knot diagram K. For purposes of simplicity, we shall assume that we are dealing with a knot diagram (one boundary component) and leave the case of links to the reader. We then have

Lemma. *Let K be a classical knot diagram with n crossings and r Seifert circles, then the genus of the Seifert surface $F(K)$ is given by the formula*

$$g(F(K)) = \frac{1}{2}(-r + n + 1).$$

Proof. The surface $F(K)$, as described prior to the statement of the lemma, retracts to a cell complex consisting of the projected graph of the knot diagram with two cells attached to each cycle in the graph that corresponds to a Seifert circle. Thus we have that the Euler characteristic of this suface is given by the formula

$$\chi(F(K)) = n - e + r,$$

where n, the number of crossings in the diagram, is the number of zero cells, e is the number of one cells (edges) in the projected diagram (from node to node), and r is the number of Seifert circles as these are in correspondence with the two cells. However, we know that $4n = 2e$ since there are four edges locally incident to each crossing. Thus,

$$\chi(F(K)) = -n + r.$$

Furthermore, we have that $\chi(F(K)) = 1 - 2g(F(K))$, since this surface has a single boundary component and is orientable. From this it follows that $1 - 2g(F(K)) = -n + r$, and hence

$$g(F(K)) = \frac{1}{2}(-r + n + 1).$$

This completes the proof. □

We now observe that *for any classical knot K, there is a surface bounding that knot in the four-ball that is homeomorphic to the Seifert surface.* One can construct this surface by pushing the Seifert surface into the four-ball keeping it fixed along the boundary. We will give here a different description of this surface as indicated in

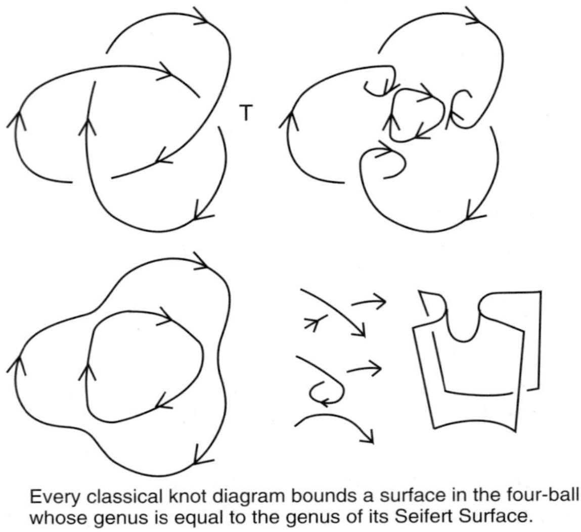

Fig. 19. Classical cobordism surface.

Fig. 19. In that figure we *perform a saddle point transformation at every crossing of the diagram.* The result is a collection of unknotted and unlinked curves. By our interpretation of surfaces in the four-ball obtained by saddle moves and isotopies, we can then bound each of these curves by discs (via deaths of circles) and obtain a surface $S(K)$ embedded in the four-ball with boundary K. As the reader can easily see, the curves produced by the saddle transformations are in one-to-one correspondence with the Seifert circles for K, and it is easy to verify that $S(K)$ is homeomorphic with the Seifert surface $F(K)$. Thus we know that $g(S(K)) = \frac{1}{2}(-r + n + 1)$. In fact the same argument that we used to analyze the genus of the Seifert surface applies directly to the construction of $S(K)$ via saddles and minima.

Now the stage is set for generalizing the Seifert surface to a surface $S(K)$ for virtual knots K. View Figs. 20 and 21. In these figures we have performed a saddle transformation at each classical crossing of a virtual knot K. The result is a collection of unknotted curves that are isotopic (by the first classical Reidemeister move) to curves with only virtual crossings. Once the first Reidemeister moves are performed,

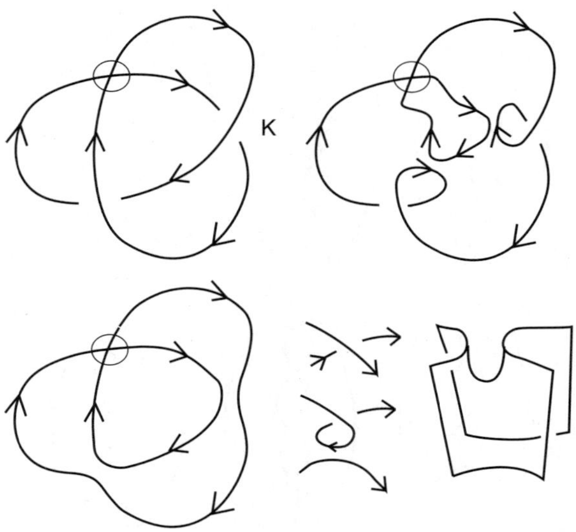

Fig. 20. Virtual cobordism Seifert surface.

these curves are identical with the *virtual Seifert circles* obtained from the diagram K by smoothing all of its classical crossings. We can then isotope these circles into a disjoint collection of circles (since they have no classical crossings) and cap them with discs in the four-ball. The result is a virtual surface $S(K)$ whose boundary is the given virtual knot K. We will use the terminology *virtual surface in the four-ball* for this surface schema. In the case of a virtual slice knot, we have that the knot bounds a virtual surface of genus zero. But with this construction we have proved the

Lemma. *Let K be a virtual knot, then the virtual Seifert surface $S(K)$ constructed above has genus given by the formula*

$$g(S(K)) = \frac{1}{2}(-r + n + 1),$$

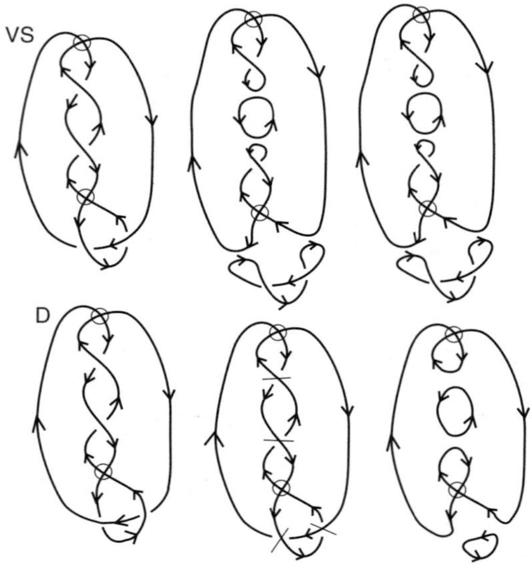

Fig. 21. Virtual stevedore cobordism Seifert surface.

where r is the number of virtual Seifert circles in the diagram K and n is the number of classical crossings in the diagram K.

Proof. The proof follows by the same argument that we already gave in the classical case. Here the projected virtual diagram gives a four-regular graph G (not necessarily planar) whose nodes are in one-to-one correspondence with the classical crossings of K. The edges of G are in one-to-one correspondence with the edges in the diagram that extend from one classical crossing to the next. We regard G as an abstract graph so the the virtual crossings disappear. The argument then goes over verbatim in the sense that G with two cells attached to the virtual Seifert circles is a retract of the surface $S(K)$ constructed by the cobordism. The counting argment for the genus is identical to the classical case. This completes the proof. □

Remark. For the virtual stevedore in Fig. 21 we have the interesting phenomenon that there is a much lower genus surface that can be produced by cobordism than the virtual Seifert surface. In that same figure we have illustrated a diagram D with the same projected diagram as the virtual stevedore, but D has all positive crossings. In this case we can prove [2] that there is no virtual surface for this diagram D of four-ball genus less than 1. In fact, we have the following result which is proved in [2]. This theorem is a generalization of a corresponding result for classcial knots due to Rasmussen [30].

Theorem 3.1. *Let K be a positive virtual knot (all classical crossings in K are positive), then the four-ball genus $g_4(K)$ is given by the formula*

$$g_4(K) = \frac{1}{2}(-r + n + 1) = g(S(K)),$$

where r is the number of virtual Seifert circles in the diagram K and n is the number of classical crossings in this diagram. In other words, that virtual Seifert surface for K represents its minimal four-ball genus.

Discussion. This theorem is proved by using a generalization of integral Khovanov homology to virtual knot theory devised by Manturov [32]. In [2] we reformulate this theory and show that it generalizes to the Lee homology theory (a variant of Khovanov homology) as well. In the Lee theory the chain complex is defined by a Frobenius algebra with the generators g and r as shown in Fig. 22. As this figure and Fig. 23 show, each link diagrm is reoriented in "source-sink" form so that every crossing has two arrows in and two arrows out that alternate cyclically around the crossing. In a virtual diagram it can be the case that these local orientations do not fit

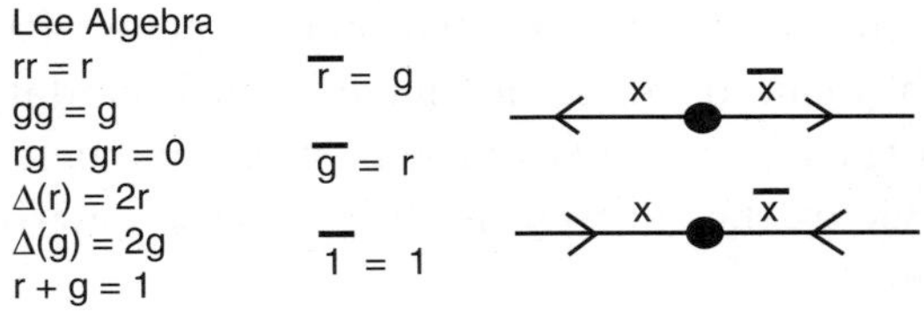

Fig. 22. Lee algebra undergoes involution at a cut locus.

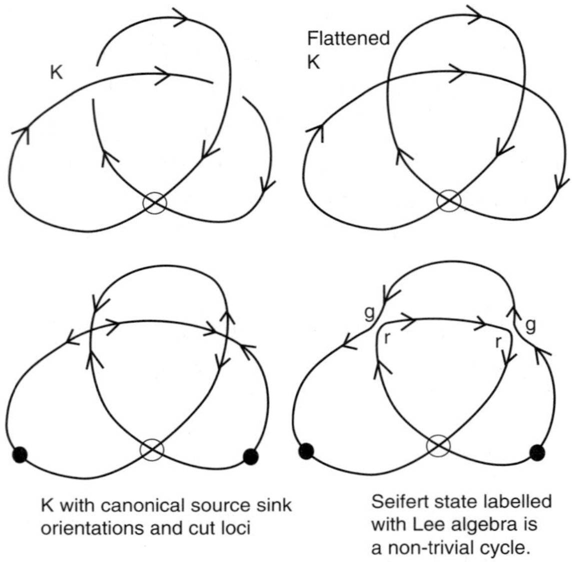

Fig. 23. Lee algebra labels seifert state.

globally. There will be some edges where a switch of local orientation occurs from one end of the edge to the other. We mark points on those edges where a switch of orientation is required, and call these points the *cut locus*. Loops in the state are labeled with algebra elements g and r so that g is changed to r and vice versa when one moves across the cut locus. Such labeled states are generators of the Lee homology chain complex. It follows from the fact that $rg = gr = 0$ that such a labeled state represents an element of the kernel of the boundary mapping for the chain complex (hence a cycle in the complex) if an r and a g are paired at each smoothing site in the state. Such labelings occur for Seifert smoothing states as illustrated in Figs. 23 and 24. Since such cycles for positive knots occur with only A-type smoothings, they cannot be in the image of the boundary map. Hence they represent the nontriviality of the positive (virtual) knot. Further analysis then yields the theorem in direct analogy with Rasmussen's original arguments.

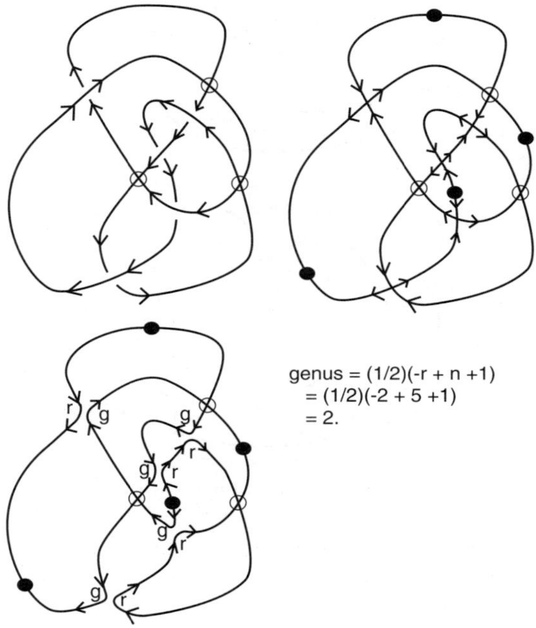

Fig. 24. Lee algebra labels Seifert state for specific knot.

3.2. *Properties of the virtual stevedore's knot*

We first point out that the virtual stevedore (VS) is an example that illustrates the viability of our theory. We prove that VS is not classical by showing that it is represented on a surface of geus one and no smaller. The reader should note the difference between representation of a virtual knot or link *on* a surface (as an embedding into the thickened surface) and the previous subsection's work on spanning surfaces.

The technique for finding this surface genus for the virtual stevedore is to use the bracket expansion on a toral representative of VS and examine the structure of the state loops on that surface. See Figs. 25 and 26. Note that in these figures the virtual crossings correspond to parts of the diagram that loop around the torus, and

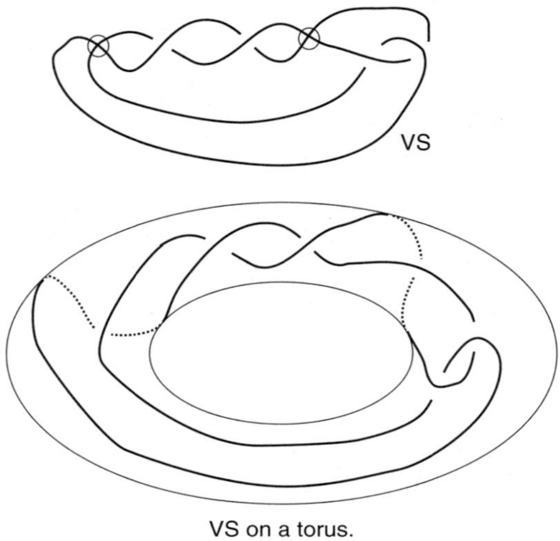

Fig. 25. Virtual stevedore on a torus.

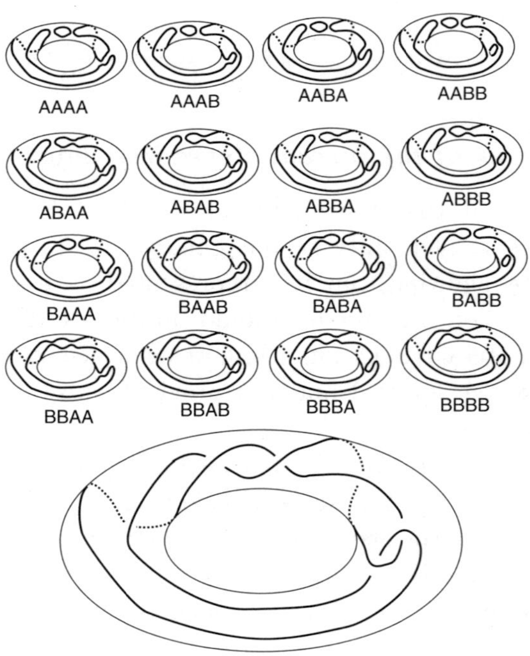

Fig. 26. Virtual stevedore is not classical.

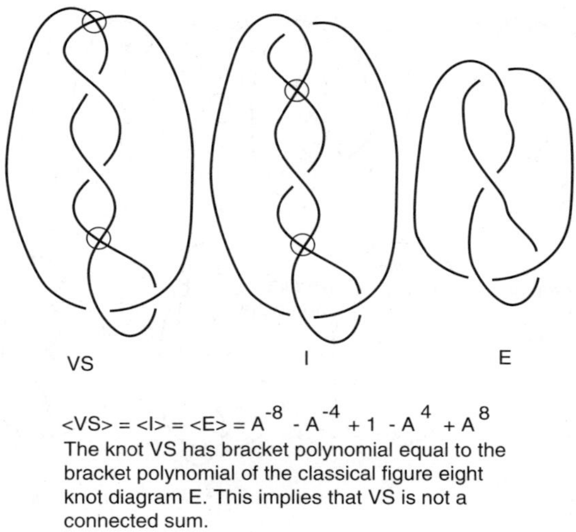

Fig. 27. Bracket polynomial of the virtual stevedore.

do not weave on the surface of the torus. An analysis of the homology classes of the state loops shows that the knot cannot be isotoped off the handle structure of the torus. See [13] for more information about using the surface bracket.

Next we examine the bracket polynomial of the virtual stevedore, and show as in Fig. 27 that it has the same bracket polynomial as the classical figure-eight knot. The technique for showing this is to use the basic bracket identity for a crossing flanked by virtual crossings as discussed in the previous section. This calculation shows that VS is not a connected sum of two virtual knots. Thus we know that VS is a nontrivial example of a virtual slice knot. We now can state the problem: *Classify virtual knots up to concordance.* We will discuss this problem in this paper, but not solve it. The reader should note that the corresponding problem for classical knots is not solved, but has more techniques available. We will need to forge new techniques for the virtual problem.

3.3. *Rotational cobordisms*

In Sec. 1 we introduced rotational virtual knots, where the detour move is restricted to regular homotopy of plane curves. This means

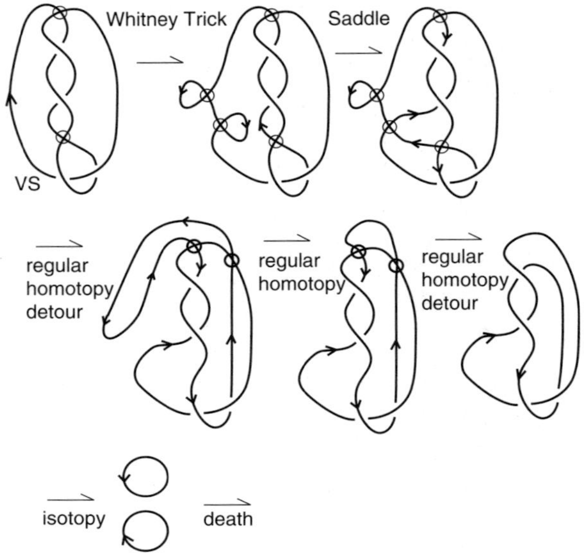

Fig. 28. The virtual stevedore is rotationally slice.

that the virtual curl of Fig. 1 cannot be directly simplified, but two opposite virtual curls can be created or destroyed by using the Whitney Trick of Fig. 9. We define *rotational virtual cobordism* by allowing saddles, deaths and births as before to extend the rotational equivalence relation. This section is a brief introduction to this theory in the form of two examples.

The first example is shown in Fig. 28. We show that the virtual stevedore diagram is rotationally slice. This is accomplished by first doing a Whitney Trick, then using one curl, so produced, to produce a saddle. In this case there is a combination of regular homotopy and virtual isotopy (just labeled isotopy in the figure) that produces two disjoint circles with no virtual curls. These die to produce the virtual disk, showing that the knot is slice in the rotational category.

In Fig. 29 we illustrate how a single curl C is (having Whitney degree zero) equivalent to its reverse (reversing orientation) and so can go through a saddle point with a copy of itself to a curve that is regularly homotopic to a trivial circle. Hence two curls can together rotationally bound a virtual surface of genus zero. This is a matter of

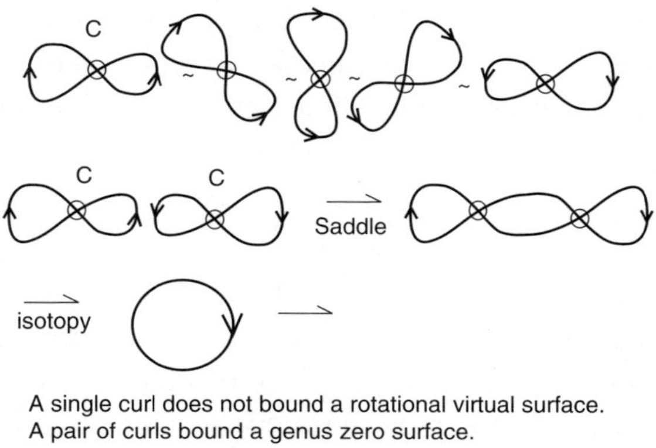

Fig. 29. Parity and cobordism of curls.

parity! Given a rotational virtual link L, define the *rotational parity of* L, $RotP(L)$ to be 0 if L has an even number of virtual crossings, and 1 if L has an odd number of virtual crossings. We have

Theorem. *If L is a rotational virtual link, then L bounds a rotational virtual surface if and only if L has even rotational parity, $RotP(L) = 0$.*

Proof. Note that the parity of the number of virtual crossings is an invariant of rotational equivalence of virtual knots and links *and* it is also an invariant of saddle moves, death and births since these moves do not change the number of virtual crossings. We only allow deaths and births for circles that have no virtual crossings — call these *free circles.* Thus $RotP$ is an invariant of rotational cobordism. Therefore if $RotP(L)$ is odd, L cannot be cobordant to a disjoint union of free circles and so cannot rotationally bound a virtual surface. Conversely, if $RotP(L) = 0$ then the same cobordism we used before, a saddle move at every crossing, combined with isotopy, produces a collection of closed curves with only virtual crossings. These curves can be made into a disjoint collection of curves by regular isotopy, and then each curve is regularly isotopic to curve with only "external curls", as illustrated in Fig. 31. Then each such curve can undergo saddle moves to transform it to a disjoint uinion of curls of the form C of Fig. 29. This is also illustrated in Fig. 31. The final number of curls of type C is even since we assumed that the parity is even. Therefore the curls cancel in pairs as explained above, and we obtain a disjoint union of one-half their number as free circles. The circles bound disks. This finishes the construction of the surface and hence finishes this proof. □

We do not expect all virtual slice knots to be rotationally slice. Consider the rotational knot K from Fig. 7. We proved in Sec. 1 that K is a nontrivial non-classical rotational virtual knot. In Fig. 30 we illustrate a cobordism of K to the disjoint union of two circles, each of which has curl. Neither of these circles can bound a rotational virtual disc by our rules. So this cobordism stops short of exhibiting K as a rotational slice knot. On the other hand, the two curls can interact through a saddle point to produce a free circle. Thus K does rotationally bound a virtual surface of genus one. We conjecture that K is not rotationally slice and that its least four-ball genus is one.

Clearly much more work needs to be done in the study of cobordisms of rotational virtual knots and links. Since there are

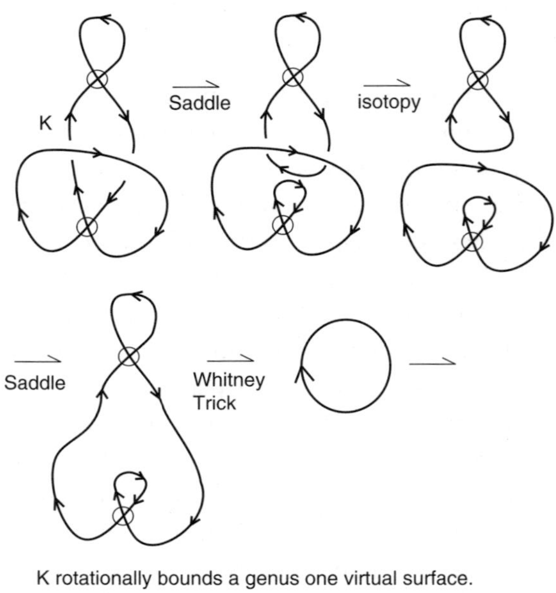

Fig. 30. A rotational cobordism.

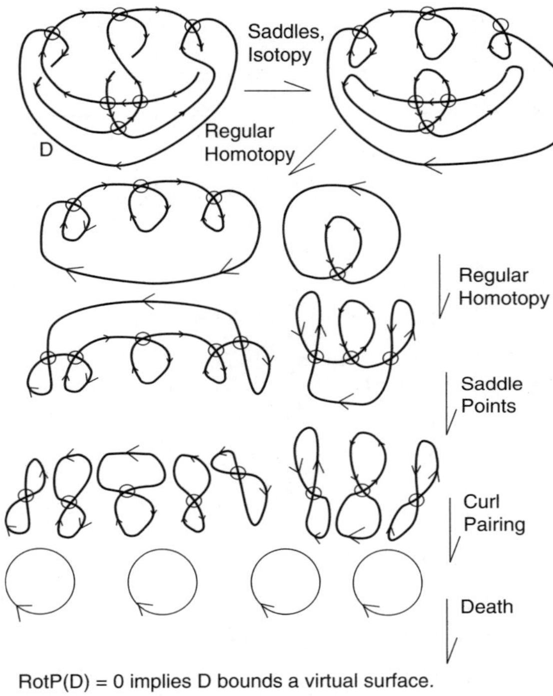

Fig. 31. Diagram D bounds rotational virtual surface if $RotP(D) = 0$.

many invariants of rotational knots and links (all the quantum link invariants), we can ask *how do quantum link invariants behave under rotational cobordism?* This will be the subject of subsequent papers.

3.4. *Band passing*

The Arf invariant of a classical knot can be interpreted as the *pass-class* of the knot, where two knots are *pass-equivalent* [12] if one can be obtained from the other by ambient isotopy combined with switching pairs of oppositely oriented pairs of parallel strands as illustrated in Fig. 32. The pass-class is a concordance invariant of classical knots and closely related to the Alexander polynomial. Any classical knot is pass-equivalent to either the trefoil knot or the unknot. The trefoil is pass-equivalent to its mirror image and is in

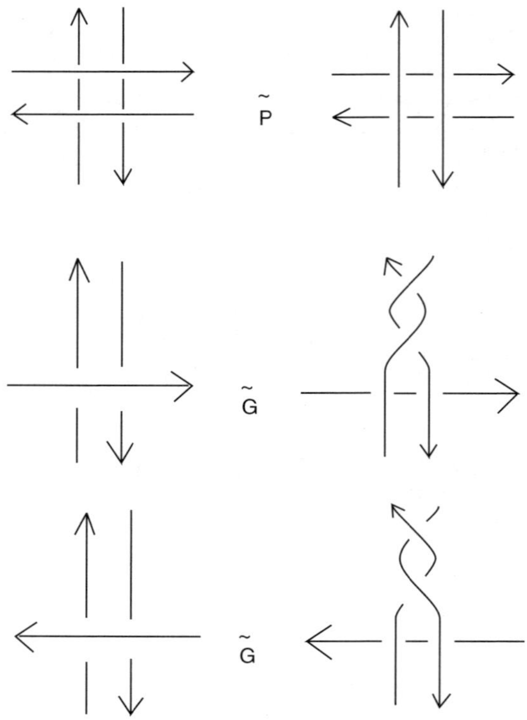

Fig. 32. Pass and gamma moves.

a distinct pass-class from the unknot. The reader can get an idea of how this works for classical knots by examining Fig. 35 where we show how a complicated surface (with boundary a clasical knot) can be simplified by band-passing. See [12] for more information about classical band passing.

We would like to determine the pass-classes of virtual knots. This problem appears difficult at this time due the lack of invariants of the passing operation. We can obtain partial results by restricting passing to only odd crossings (crossings with an odd interstice in the Gauss code) but this is only a step on the way to understanding the pass equivalence relation for virtual knots. We expect that understanding this relation will shed light on problems of knot concordance.

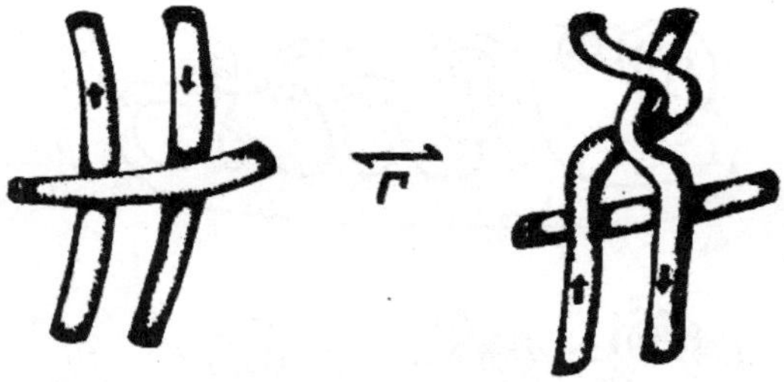

Fig. 33. Gamma move.

In Fig. 32 we ilustrate pass-equivalence and also illustrate another move denoted by "G" in that figure and we refer to this move as the *gamma move.* The gamma move, illustrated separately in Fig. 33, switches one strand past two oppositely oriented strands and places a 2π twist in these two strands. It is obvious that two gamma moves will accomplish a single pass-move, since the twist introduced by G-passing one strand is canceled by the twist introduced by G-passing a second, oppositely oriented strand. It is also the case that any gamma move can be accomplished by a combination of ambient isotopy and a pass-move, as shown in Fig. 34. Thus, pass-equivalence and gamma-equivalence are identical as equivalence relations on classical or virtual links. In the classical case, gamma-equivalence is of direct interest, as it is easy to see that a classical ribbon

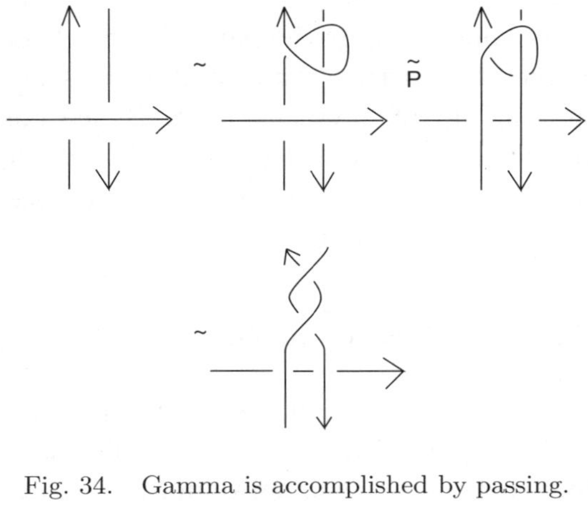

Fig. 34. Gamma is accomplished by passing.

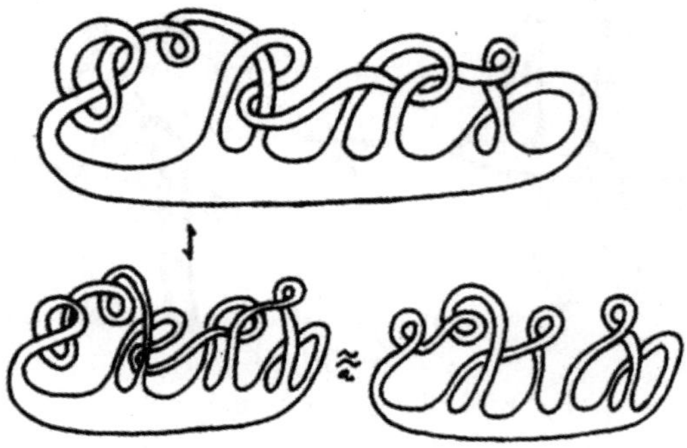

Fig. 35. Simplifying a surface by passing bands.

knot can be trivialized by gamma-moves (via the removal of ribbon singularities from an immersed disk that spans the link in three-dimensional space). This is one way to see that classical ribbon knots are pass-equivalent to unknots. Thus, a classical knot that is not pass-equivalent to the unknot cannot be ribbon.

In the virtual case, we can see in some cases (such as the virtual stevedore's knot) that the knot is gamma-equivalent (hence pass-equivalent) to an unknot. But there exist examples of virtual knots that are slice, but are *not* pass-equivalent to the unknot. For example, consider the Kishino knot of Fig. 12. The Kishino has two oppositely oriented parallel strands in its middle, and one saddle point move transforms the Kishino into two virtual unknots. Hence the Kishino is slice. However, we know (e.g. via the parity bracket) that the Kishino overlies a nontrivial flat virtual diagram. No pass-move changes the underlying flat diagram of a virtual knot. Hence pass-equivalence cannot unknot the Kishino diagram. This shows that there is a sharp difference between pass-equivalence for classical knots and pass-equivalence for virtual knots.

4. Virtual Surfaces in Four-Space

We now define a theory of virtual surfaces in four-space that is given by moves on planar diagrams. One aim is to investigate the relationships between this diagrammatic definition and more

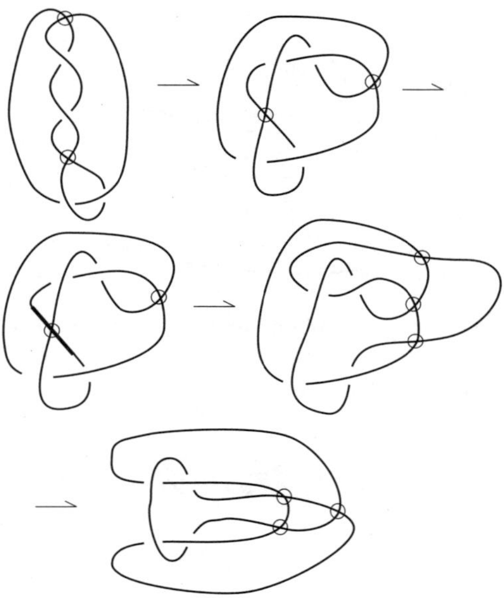

Fig. 36. Converting virtual stevedore to a "ribbon diagram".

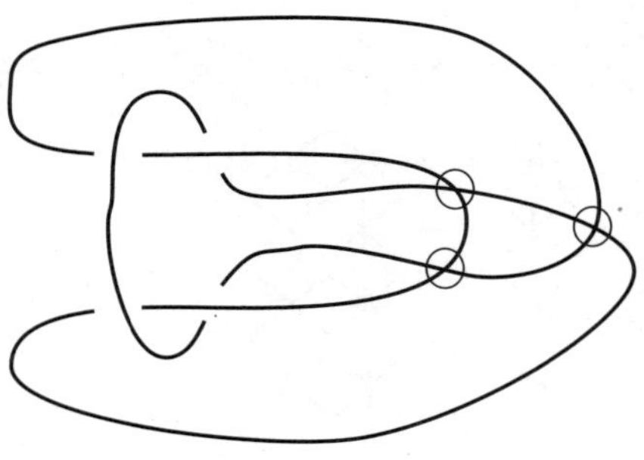

Fig. 37. Ribbon version of virtual stevedore.

geometric approaches to virtrual 2-knots due to Jonathan Schneider and to Takeda [33]. We make diagrammatic definitions as follows: We use middle level markers as indicated in Fig. 39 to encode two directions of smoothing a marked crossing in a planar diagram. The classical interpretation of such a marker is that it represents a cobordism through a saddle point at the middle level ($t = 0$ in the figure) where the forms of smoothing above ($t = 1$) and below

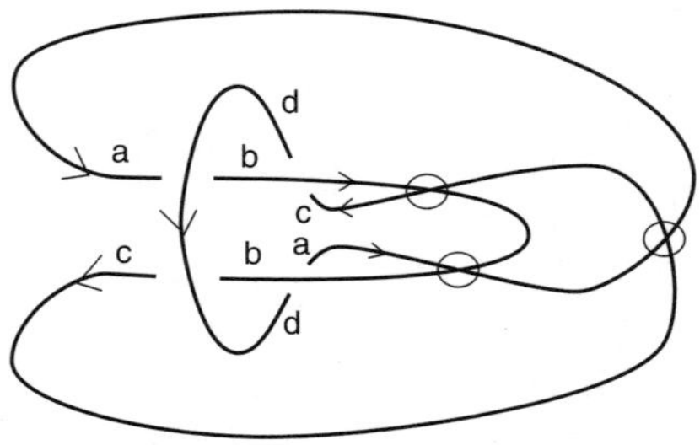

Fig. 38. Labeled ribbon version of virtual stevedore.

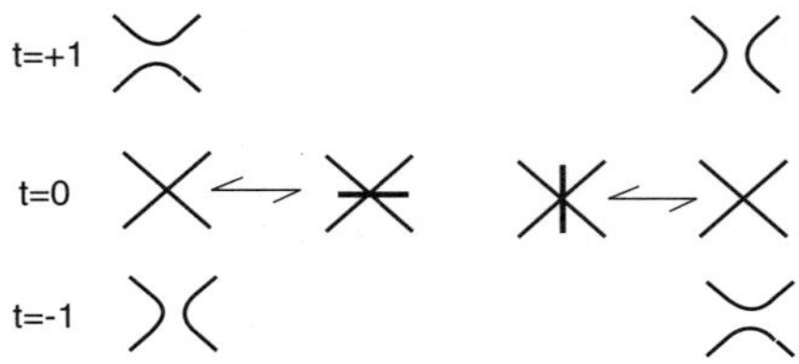

Fig. 39. Middle level markers.

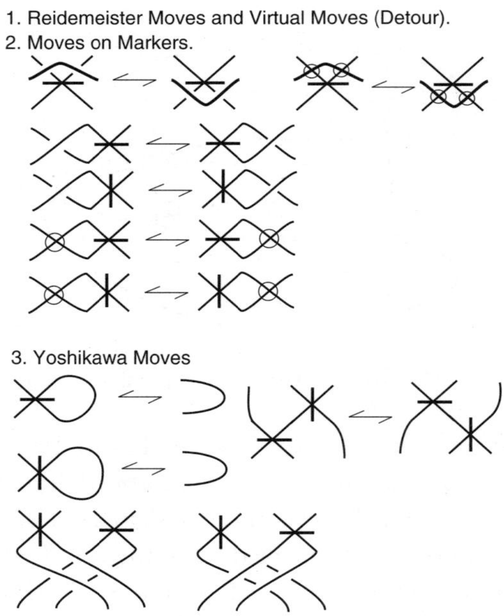

Fig. 40. Middle level moves.

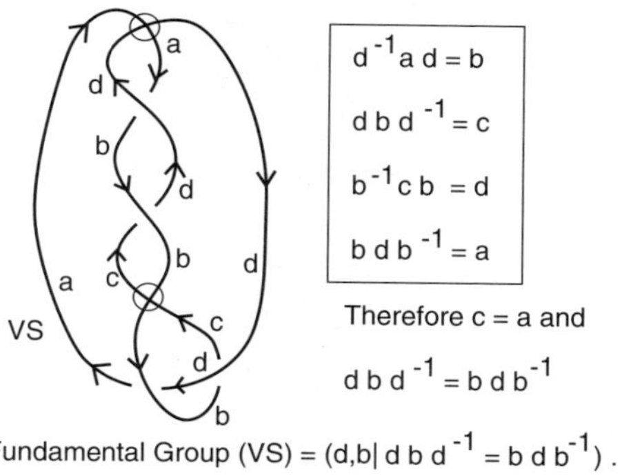

Fig. 41. Fundamental group of VS.

$(t = -1)$ are shown via the conventions in the figure. A diagram with markers can then be interpreted as two cobordisms attached at the middle. One cobordism goes downward to a collection of possibly linked and knottted loops, the other goes upward to another collection of linked and knotted loops. We will refer to these as the *up-cobordism* and the *down-cobordism.* A marked diagram is said to be *excellent* if both the up and the down cobordisms end in collections of unlinked circles that can be capped off with births (from the bottom) and deaths (at the top). The resulting schema is then a two-sphere and classically represents a two-sphere in four-space. We take exactly this definition for a *virtual two-sphere* where it is understood that the ends of the two cobordisms will be trivial virtual links.

Just as in classical theory, if a virtual knot is slice, then we can make a virtual two-sphere from it by using the same cobordism both up and down. In Fig. 42 we indicate the schema for such a surface involving one saddle point up and one saddle point down. Births for the original cobordism have to be represented directly in the middle level. The slicing example for VS, the virtual stevedore's knot, can be made into a two-sphere this way. We show the middle level diagram for this sphere, called S in Fig. 43. In this same figure, we show another middle level diagram for a virtual two-sphere S'. In this case we have used the fact (the reader can verify) that VS can be sliced from its right-hand side. The sphere S' is obtained by slicing upward from the left and downward from the right.

Fig. 42. Abstract double slice surface.

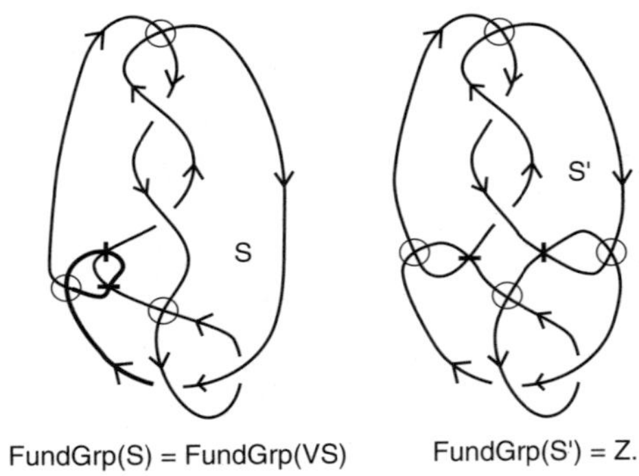

Fig. 43. Two two-spheres.

We give moves on the middle level diagrams to define isotopy of the virtual two-spheres obtained from the middle level diagrams. The moves are indicated in Fig. 40. They are a virtual generalization of the Yoshikawa moves that have been studied [26, 32] for isotopies of the classical middle level formulations. Thus we say the two two-spheres are *isotopic* if one can be obtained from the other via these *Generalized Yoshikawa moves*. In particular, the fundamental group of the two-sphere, defined by adding relations at saddle points exactly as in the classical case (but from the virtual knot theoretic fundamental group) is an isotopy invariant. For example, in Fig. 41 we calculate the fundamental group of VS and find that, in it the arcs whose elements must be identified to obtain the fundamental group of the sphere S of Fig. 43 are already identified in the fundamental group of VS. Thus we find that the sphere S is knotted since it has the same nontrivial fundamental group as VS. On the other hand, it is not hard to see that the fundamental group of the sphere S is isomorphic to the integers. At this writing I do not know if this sphere is virtually unknotted.

The generalized Yoshikawa moves present a useful first formulation for a theory of virtual surfaces. One of the advantages of this approach is that we can adapt the generalization of the bracket polyomial of Lee [26] to obtain a bracket invariant for virtual two-spheres. This will be an important subject of investigation for this proposal. We want to know how this diagrammatic formulation is related to immersions of surfaces in four-space that could represent virtual two-knots. In this case the levels (movie of a cobordism) description that we have adopted gives such an immersion, and one can begin the investigation at that point. For these reasons, we believe that this formulation of virtual cobordism and virtual surfaces will be very fruitful and lead to many new results.

References

[1] J. Scott Carter, S. Kamada and M. Saito, Stable equivalence of knots on surfaces and virtual knot cobordisms, math.GT/0008118.

[2] H. A. Dye, A. Kaestner and L. H. Kauffman, Khovanov homology, Lee homology and a Rasmussen invariant for virtual knots, submitted, math. GT/1409.5088.

[3] H. A. Dye and L. H. Kauffman, Virtual crossing number and the arrow polynomial, *J. Knot theory Ramifications* **18** (2009) 1335–1357.

[4] S. Eliahou, L. Kauffman and M. Thisttethwaite, Infinite families of links with trivial Jones polynomial, *Topology* **42** (2003) 155–169.

[5] R. A. Fenn, L. H. Kauffman and V. O. Manturov, Virtual knots: Unsolved problems, in *Fundamenta Mathematicae, Proc. of the Conf. on* "Knots in Poland-2003" **188** (2005), pp. 293–323.

[6] M. Goussarov, M. Polyak and O. Viro, Finite type invariants of classical and virtual knots, math.GT/9810073.

[7] D. P. Ilyutko and V. O. Manturov, Cobordisms of free knots, *Dokl. Akad. Nauk* **429** (2009) 439–441 [translation in *Dokl. Math.* **80** (2009) 844–846].

[8] V. F. R. Jones, A polynomial invariant for links via von Neumann algebras, *Bull. Amer. Math. Soc.* **129** (1985) 103–112.

[9] V. F. R. Jones, Hecke algebra representations of braid groups and link polynomials, *Ann. Math.* **126** (1987) 335–338.
[10] V. F. R. Jones, On knot invariants related to some statistical mechanics models, *Pacific J. Math.* **137** (1989) 311–334.
[11] L. H. Kauffman, *Formal Knot Theory*, Princeton University Press, 1983. Lecture Notes Series, Vol. 30.
[12] L. H. Kauffman, *On Knots*, Princeton University Press, 1987.
[13] L. H. Kauffman and H. A. Dye, Minimal surface representations of virtual knots and links. *Algebr. Geom. Topol.* **5** (2005) 509–535.
[14] L. H., Kauffman, Virtual knot theory, *Eur. J. Combinatorics* **20** (1999) 663–690.
[15] L. H. Kauffman, Detecting virtual knots, *Atti Sem. Mat. Fix. Univ. Modena*, Suppl. **IL** (2001) 241–282.
[16] L. H. Kauffman, State models and the Jones polynomial, *Topology* **26** (1987) 395–407.
[17] L. H. Kauffman, An extended bracket polynomial for virtual knots and links, *J. Knot Theory Ramifications* **18** (2009) 1369–1422.
[18] L. H. Kauffman and S. Lambropoulou, A categorical model for the virtual braid group, *J. Knot Theory Ramifications* **21** (2012) 1240008.
[19] L. H. Kauffman, *Knots and Physics* (World Scientific, 1991; 2nd edn., 1993; 3rd edn., 2002; 4th edn., 2012.
[20] L. H. Kauffman, A survey of virtual knot theory, in *Proceedings of Knots in Hellas '98*, World Scientific, 2000, pp. 143–202.
[21] L. H. Kauffman, Introduction to virtual knot theory, *J. Knot Theory Ramifications* **21** (2012) 1240007.
[22] L. H. Kauffman, A self-linking invariant of virtual knots, *Fund. Math.* **184** (2004) 135–158.
[23] L. H. Kauffman, math.GN/0410329, Knot diagrammatics, *Handbook of Knot Theory*, eds. W. Menasco and M. Thistlethwaite, Elsevier 2005, pp. 233–318.
[24] L. H. Kauffman, An affine index polynomial invariant of virtual knots, *J. Knot Theory Ramifications* **22** (2013) 1340007.
[25] G. Kuperberg, What is a virtual link?, arXiv:$math.GT/0208039v$.
[26] S. Y. Lee, Towards invariants of surfaces in 4-space via classical link invariants, *Trans. Amer. Math. Soc.* **361** (2009) 237–265.
[27] V. O. Manturov, Parity and cobordisms of free knots, *Mat. Sb.* **203** (2012) 45–76 [translation in *Sb. Math.* **203** (2012) 196–223].
[28] V. O. Manturov, Parity in knot theory, *Math. Sb.* **201** (2010) 693–733 [original Russian text in *Math. Sb.* **201** (2010) 65–110].
[29] V. O. Manturov, *Virtual Knot Theory — The State of the Art*, World Scientific, 2012.
[30] J. Rasmussen, Khovanov homology and the slice genus, *Invent. Math.* **182** (2010) 419–447.
[31] S. Satoh, Virtual knot presentation of ribbon torus-knots, *J. Knot Theory Ramifications* **9** (2000) 531–542.

[32] F. Swenton, On a calculus for surfaces and 2-knots in 4-space, *J. Knot Theory Ramifications* **10** (2001) 1133–1141.
[33] Y. Takeda, Introduction to virtual surface-knot theory, *J. Knot Theory Ramifications* **21** (2012) 1250131.
[34] V. Turaev, Cobordisms of words, *Commun. Contemp. Math.* **10** (2008) 927.
[35] E. Witten, Quantum field theory and the Jones polynomial, *Commun. Math. Phys.* **121** (1989) 351–399.

Mutant Knots

H. R. Morton

Department of Mathematical Sciences,
University of Liverpool, Peach Street, Liverpool L69 7ZL, UK
morton@liv.ac.uk

Mutants provide pairs of knots with many common properties. The study of invariants which can distinguish them has stimulated an interest in their use as a test-bed for dependence among knot invariants. This paper is a survey of the behaviour of a range of invariants, both recent and classical, which have been used in studying mutants and some of their restrictions and generalisations.

1. History

Remarkably little of John Conway's published work is on knot theory, considering his substantial influence on it. He had a really good feel for the geometry, particularly the diagrammatic representations, and a knack for extracting and codifying significant information. He was responsible for the terms *tangle*, *skein* and *mutant*, which have been widely used since his knot theory work dating from around 1960. Many of his ideas at that time were treated almost as a hobby and communicated to others either over coffee or in talks or seminars, only coming to be written in published form on a sporadic basis.

His substantial paper [9] is quoted widely as his source of the terms, and the comparison of his and the Kinoshita–Teresaka 11-crossing mutant pair of knots, shown in Fig. 1. While Conway certainly talks of tangles in [9], and uses methods that clearly belong with linear skein theory and mutants, there is no mention at all

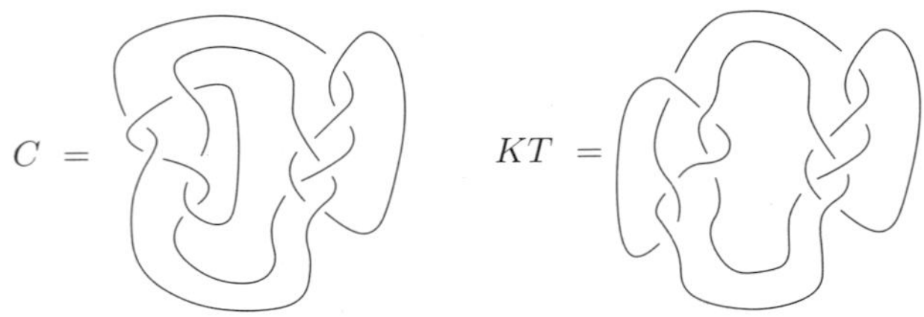

Fig. 1. The Conway and Kinoshita–Teresaka mutant pair.

of mutants, in those words or any other, in the text. Undoubtedly though he is the instigator of these terms and the paper gives one of the few tangible references to his work on knots.

In [9] Conway gives a table of 11 crossing knots, where he reckons to be confident of differences among them, although without explicit invariants in all cases to be certain of this.

The two 11-crossing knots, C and KT, found by Conway and Kinoshita–Teresaka are probably the best-known example of inequivalent mutant knots. Conway's knot is given in his table of 11-crossing knots [9], while KT appears in [19] as one of a family of knots with trivial Alexander polynomial. These two knots are shown in Fig. 1.

The first proof that the two knots in Fig. 1 are inequivalent was, I believe, given by Riley [39].

Perko [37] tidied up the tables up to 11 crossings, and used double cover techniques in places to distinguish pairs of knots. These methods, however, would not be enough to distinguish a mutant pair, by Theorem 2.3.

Gabai [15] used foliations in showing that C has genus 3 while KT has genus 2. The genus of a knot had until then been a difficult invariant to determine exactly. Gabai's work gave a much wider range of certainty, while in principle the extension of the Alexander polynomial via Heegaard Floer homology gives an exact calculation of the genus.

A recent systematic attempt to document mutant pairs among knots up to 18 crossings has been undertaken by Stoimenow [42]. He also gives comments on the history and techniques available for distinguishing mutants, and the practical limitations for calculations.

Disclaimer

While I have tried to find and credit historical work on mutants I have come across considerable difficulties in even identifying the initial sources of some of the terms, such as *Conway sphere*. I have realised that much of the work has been either in the realm of "folk-lore", or implicit in places where the authors have not felt it necessary to point up results arising from the general methods being discussed. Certainly one of the benefits of having to read older papers is the realisation of what can be deduced from an understanding of the ideas that underlie the work in question.

I would not want this paper to be taken as providing a reliable historical account, and I apologise for any omissions, both in material and in attribution, that I suspect will be found in it.

2. Definitions

The most commonly used description of mutation is combinatorial, arising directly from Conway's definition of a tangle.

In his setting a tangle is a part of a knot diagram consisting of two arcs contained in a circular region which meet the boundary circle in four diagonally placed points.

In line with current terminology I shall refer to this as a 2-*tangle*. In general a 2-tangle may contain closed curves as well as the two arcs, but since in this paper we will only be considering knots there will not be any occasion to look at 2-tangles with additional curves.

I shall also adjust the diagrams so that the containing region is a rectangle rather than a circle, with two boundary points of the arcs at the top and two at the bottom.

Suppose then that a knot K can be decomposed into two 2-tangles F and G

$K =$ 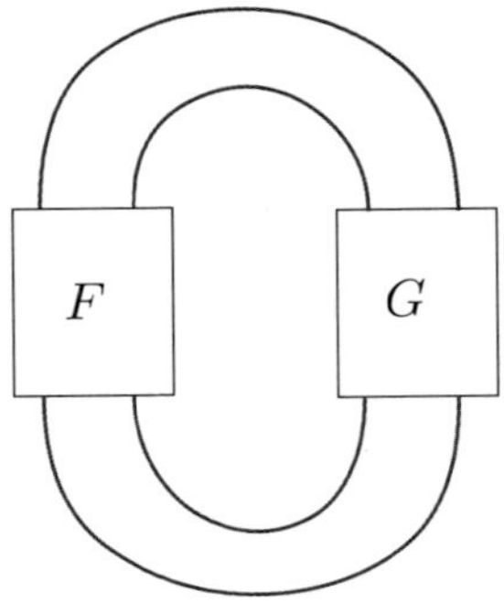

A new knot K' can be formed by replacing the tangle F with the tangle $F' = \rho_i(F)$ given by rotating F through π in one of three ways,

$\rho_1(F) =$ 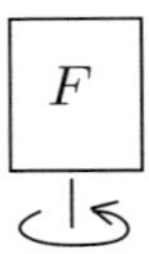

, $\rho_2(F) =$ 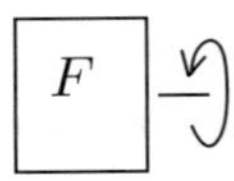

, $\rho_3(F) =$ 

.

Any of the three knots

$K' =$ 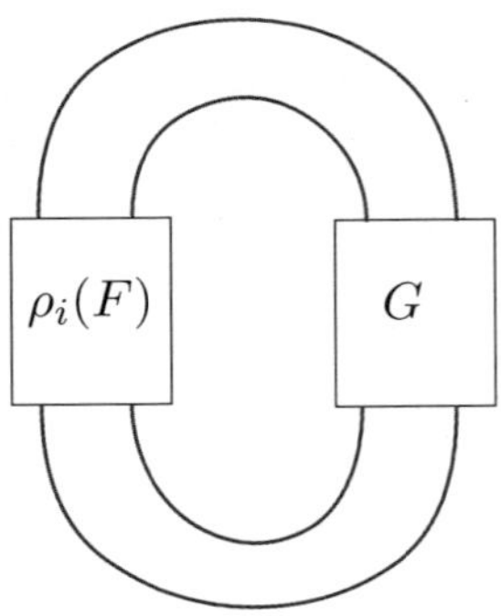

is called a *mutant* of K.

Remark. In my recent paper on mutants with symmetry [33] I have used the notation τ_3, τ_1, τ_2 respectively for these three rotations ρ_1, ρ_2, ρ_3.

2.1. *Equivalence of mutants*

As is implicit in the comments above, a knot which can be decomposed into two 2-tangles may well turn out to be equivalent to one or more of the three resulting mutants.

The simplest way in which this may happen is if the tangles F and one or more of $\rho_i(F)$ are equivalent, in other words if F is symmetric under one of the three rotations.

Since the knot K' can be redrawn as

$K' =$ 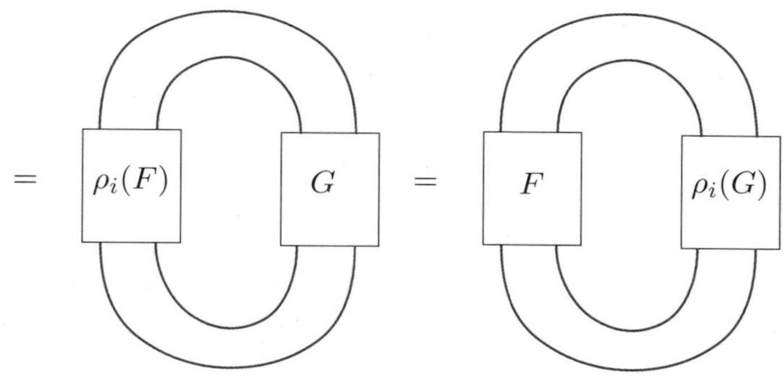

we will equally find that mutants are equivalent if the other tangle G has rotational symmetry. Indeed, if F is symmetric under one of the rotations and G is symmetric under a *different* rotation, then all three mutants will be equivalent.

If one of the tangles has all three rotational symmetries then again all the resulting mutants will be equivalent. This is of course the case when the tangle F consists simply of two non-crossing arcs. It is also true where F is a *rational tangle*, in Conway's sense.

Rational tangles arise in Conway's description from nicely arranged consequences of his notation where certain tangles appear in 1-1 correspondence with rational numbers using a continued fraction decomposition.

The 3-string braid group B_3 operates on the set of 2-tangles by braiding the strings coming from three of the boundary points. We can describe the tangles $\sigma_1 F$ and $\sigma_2 F$ by the diagrams

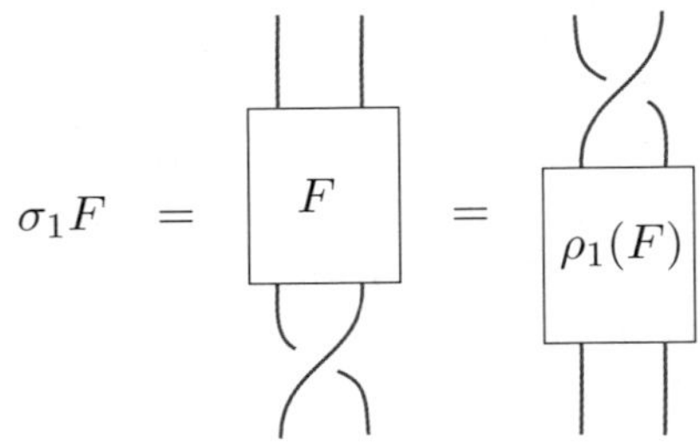

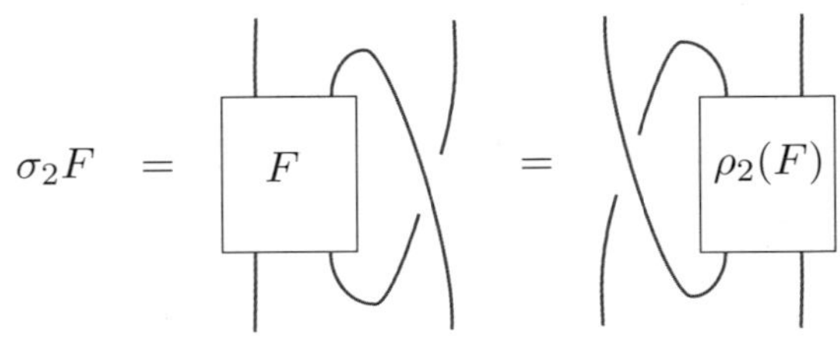

Conway's rational tangles are those which can be generated from a simple tangle without crossings by repeated operation of $\sigma_1^{\pm 1}$ and $\sigma_2^{\pm 1}$.

Theorem 2.1. *If a 2-tangle F is symmetric under all three rotations then so are the tangles $\sigma_1 F$ and $\sigma_2 F$.*

Proof. After applying each rotation to the tangles $\sigma_1 F$ we can observe that $\rho_1(\sigma_1 F) = \sigma_1 \rho_1(F)$, $\rho_2(\sigma_1 F) = \sigma_1 \rho_3(F)$ and $\rho_3(\sigma_1 F) = \sigma_1 \rho_2(F)$. Similar relations hold for the rotations of $\sigma_2 F$. $\square$

Corollary 2.2. *Rational tangles are symmetric under all three π-rotations ρ_i.*

Remark. Stoimenow makes use of this fact in his searches for mutant pairs among knots up to 18 crossings [42], as he is able to exclude decompositions in which one of the tangles is rational. By Corollary 2.2 any mutation of such a decomposition does not produce a different knot.

2.2. *A three-dimensional view*

There is a natural way of looking at mutants in a three-dimensional context, involving embedded 2-spheres in S^3 meeting a knot transversely in four points. Ruberman [41] adopted the current term *Conway sphere* for such an embedded sphere.

The two 3-balls which lie on either side of a Conway sphere then correspond to a decomposition of the knot into two 2-tangles, although there will be a choice involved in representing each of these by a diagram. In effect the diagram will be determined up to the action of the braid group B_3 on the punctures on the sphere.

If the knot in S^3 is regarded as an orbifold with cone angle π along the knot, then the Conway spheres play a natural role in the theory of orbifold decompositions, which is mirrored by their torus covers in the 2-fold cyclic cover of S^3 branched over the knot. Bonahon and Siebenmann [6] prove a uniqueness result for orbifold decompositions in a general setting. Their results apply in this case with suitable Conway spheres providing the counterpart to the tori in the Jaco–Shalen decomposition of the covering manifold.

The following result is noted by Viro [44], who uses the term *twin* rather than mutant.

Theorem 2.3. *The double covers of S^3 branched over mutant knots are homeomorphic.*

Proof. The double cover of a knot and its mutant by ρ_i are constructed from the double covers of the two constituent tangles by gluing along the torus covering the Conway sphere. The two double covers then differ by the homeomorphism of the torus which covers ρ_i. For each i this homeomorphism is isotopic to the identity. □

In his paper analysing the behaviour of Conway spheres in knots Lickorish [23] uses the term *untangled* to denote a tangle which is homeomorphic to the trivial tangle, that is, where there is a homeomorphism of the 3-ball carrying the two arcs inside the Conway sphere to a pair of parallel unknotted arcs. Such tangles are exactly the rational tangles of Conway. (Viro notes that any π-rotated untangled tangle is isotopic to the original tangle by an isotopy which fixes the boundary sphere.)

It is then easy to give a quick proof of the following result (see also the proof of Rolfsen [40]).

Theorem 2.4. *The only mutant of the unknot is the unknot.*

Proof. Suppose that we have a Conway sphere meeting the unknot in four points. The fundamental group of the four-punctured sphere is free on 3 generators. This cannot inject into both fundamental groups of the tangles on the two sides, otherwise it would inject into the fundamental group of the knot complement. Hence, by Dehn's lemma, there is a nontrivial closed curve on the Conway sphere which bounds a disc disjoint from the arcs in one of the tangles. This disc must separate the two arcs in the ball. Each arc must be unknotted, as it is then a connected summand of the unknot. Hence this tangle is untangled in the sense of Lickorish. It is then a rational tangle and is symmetric under all three rotations, and so the mutants are all equivalent. □

We have seen here that certain tangle decompositions will only give rise to equivalent mutants. When analysing a tangle decomposition a first check should then be made on the possible symmetries of the constituent tangles. On the other hand, when we suspect that two mutants may not be equivalent, there remains the question of showing that they are indeed different.

In the next section I shall give a selection of classical methods, both geometric and algebraic, which have been used to distinguish between mutant pairs, and some early limitations which were noted. In the following sections I shall give some of the known invariants which all mutants must share, and further conditions under which a greater range of invariants are shared. In this way the use of mutants contributes a means of looking at possible relations between new and existing invariants, in terms of the extent to which they may agree on various classes of mutants.

3. Classical Ways to Distinguish Mutants

Here the term "classical" refers to techniques that were in use up to the discovery of the Jones polynomial in 1984. I have loosely separated the methods used under the general headings of algebraic and geometric.

3.1. *Algebraic*

The simplest method of distinguishing knots is by means of the Alexander polynomial, which can be calculated readily from a Seifert matrix for the knot.

Viro [44] uses comparable Seifert matrices for mutant knots to observe that their Alexander polynomials are the same, along with the homology groups and forms of linking coefficients in branched covers, Minkowski units and signatures, all of which can be found from a Seifert matrix. The Seifert matrices constructed by Viro are either identical or have the form

$$S = \begin{pmatrix} A & C & 0 \\ C^T & a & D^T \\ 0 & D & B \end{pmatrix}, \quad S' = \begin{pmatrix} A & C & 0 \\ C^T & a & D^T \\ 0 & D & B^T \end{pmatrix},$$

where A and B are square matrices, C and D are column matrices.

Viro proves that the Whitehead doubles of the connected sums $K\#K$ and $K\#K^r$ are inequivalent mutants when K is a knot which is inequivalent to its reverse K^r. This follows since equivalence of the Whitehead doubles implies equivalence of the connected sums, and decomposition of connected sums is unique.

He notes also that the whole series of knots with trivial Alexander polynomial described by Kinoshita and Teresaka in [19] all have obvious mutants although he does not give a systematic way of ensuring that all of these pairs are inequivalent.

In the absence of information from the Alexander polynomial, the most basic algebraic way to show that two knots are inequivalent is to compare more directly their groups, in other words the fundamental groups of their complements.

Apart from questions of mirror images and orientation, two inequivalent mutants will have non-isomorphic groups, by the general results of Gordon and Luecke. There is then a good chance of detecting a difference by comparing homomorphisms of their groups into suitable finite groups.

Early distinctions among knots were made by this method by Riley [39], who separated the Conway and Kinoshita–Teresaka 11-crossing knots by means of homomorphisms from the knot group to $PSL(2,7)$. Riley treats this group as a subgroup of the symmetric group S_7, and considers representations in which meridians are mapped to 7-cycles. Calculation of the homology groups of the resulting 7-fold coverings branched over the knot demonstrates a difference between the two knots.

The use of homomorphisms of knot groups to finite groups is known generally as *knot colouring*. These methods include the classical 3-colouring and n-colouring, where the finite group used is the dihedral group D_n. These techniques were much used in their original form by Fox, who extended the methods in [14]. The techniques are essentially those developed, notably by Fenn and Rourke [13], under the current term of *quandle*. It is immediate that n-colouring will not distinguish mutant pairs, since the existence of an n-colouring depends on the Alexander polynomial, and Alexander polynomials are shared by mutants.

3.2. *Geometric*

Geometric methods available for distinguishing knots include comparison of related 3-dimensional manifolds covering S^3 and branched over the knot in different ways.

3.2.1. *Covers*

The simplest of such constructions is the double cover. Viro gives a nice summary of the behaviour of mutants, referred to as *twins* in [44]. He proves in Theorem 2.3 that the double covers of S^3

branched over two mutant knots are homeomorphic. So something more elaborate is needed to distinguish mutants by this type of argument.

More complicated covers can be related to homomorphisms from the knot group to a finite group, as in Riley's arguments in [39], which proved to be effective in distinguishing between the 11-crossing knots C and KT.

3.2.2. *Genus*

An early geometric invariant of a knot is its *genus*, which is the least genus among orientable surfaces spanning the knot in S^3. The simplicity of its definition has made it a popular invariant, but it is not easy to calculate in general. It can be bounded below in terms of the Alexander polynomial, but in many cases this bound is not exact. Since the Alexander polynomial agrees on mutants, this bound will not be helpful in distinguishing mutants.

It is a surprise that the genus of mutants can differ. Gabai developed techniques for calculating the genus, based on the use of foliations, which gave an early distinction between C and KT. He showed in [15] that C has genus 3 while KT has genus 2.

It is only much more recently that the Heegaard–Floer homology of a knot has provided an exact calculation of the genus in all cases.

3.2.3. *Diagrammatic invariants*

Although the genus can be different for mutants, it is not known whether invariants such as the crossing number, the braid index or the arc index, can ever differ on mutants.

The best hope for settling any of these questions would be to give a direct argument that mutants must have the same braid index (defined as the least number n of strings needed to present the knot as a closed n-braid). Attempts to show that two mutants have different braid index run up against the difficulty that one of the best ways to find a lower bound for the braid index relies on the use of knot polynomials and many of these are shared by mutants.

3.2.4. *Hyperbolic geometry*

Bonahon and Siebenmann, along with others from Orsay, analysed the structure of classes of knots with the goal of extending and systematising Conway's constructions of knots in [9]. Their original work was contained in an influential series of notes, which never itself formed a complete publication, although much is available in their ongoing draft monograph [7]. In the course of their work they made much use of Conway sphere-based decompositions. This culminated in an extensive analysis of knots from the point of view of orbifolds, where the knot is regarded as a singular subset of the 3-manifold S^3 with cone angle π. These early geometric observations for knots were used in [6] to formulate an orbifold decomposition theorem which is a counterpart to the Jaco–Shalen-Johannson decomposition for 3-manifolds.

Recent work on this, and related bibliographies can be found in work of Boileau *et al.* [5] and Paoluzzi [36], for example.

With the advent of Thurston's work on hyperbolic and other geometric structures on 3-manfolds there followed a more systematic view of knot complements from a geometric point of view.

In particular the default position for a knot complement in the absence of certain special features turns out to be that there is a complete hyperbolic structure of finite volume on the complement. The volume is an invariant of the knot. However, Ruberman [41] shows that if a knot K is hyperbolic, then any mutant is also hyperbolic, and they have the same hyperbolic volume.

Weeks [45] developed the amazingly powerful program SnapPea to calculate details of hyperbolic manifolds, including the volume and other details making up an invariant "canonical structure". While the volume on its own is not enough, it is possible to use the canonical structures to distinguish inequivalent mutants.

3.2.5. *Symmetry*

In the same paper [41] Ruberman remarks that the work of Bonahon and Siebenmann [6, 7] ensures that if two mutant knots are equivalent, then there must be some rotational symmetries in the constituent 2-tangles. In principle then mutants can be distinguished by showing that the tangles have no suitable symmetry. Subsequently Ruberman and Cochran [8] were able to find a means of ruling out symmetry in some tangles, and apply it to give examples of inequivalent mutants.

4. Polynomials and Quantum Invariants

The enormous range of invariants which followed the discovery of the Jones polynomial and its generalisations from 1984 onwards has made available many further theoretical and practical ways of comparing knots. Besides using the new invariants to compare mutants, it has also proved fruitful to regard mutants and their refinements as a tool for analysing possible dependence among invariants.

Very shortly after the discovery of the new invariants Lickorish proved, using simple skein theoretic arguments, that mutants must also have identical Homfly and Kauffman polynomials, and hence the same Jones polynomial. A good account of this can be found in his survey article [24].

Calculations of Morton and Short for a number of examples led to the conjecture [31] that two equally twisted 2-cables of a mutant pair would also share the same Homfly polynomial. This was proved by Lickorish and Lipson [25], also using skein theory. They showed further that the same result holds for reverse-string 2-cables (that is, for 2-cables of two components whose orientations run in opposite directions, giving a "reverse parallel" satellite). This holds equivalently for equally twisted Whitehead doubles. These results were also derived independently by Przytycki [38].

Although the Homfly polynomials of doubles or 2-cables were found not to distinguish mutants it was already clear that invariants of more complicated satellites of knots could provide extra information in general.

4.1. *Homfly invariants*

In 1984 V. F. R. Jones constructed a new invariant of oriented links $V_L(t) \in \mathbf{Z}[t^{\pm\frac{1}{2}}]$, which turned out to have the property that

$$t^{-1}V_{L_+} - tV_{L_-} = (\sqrt{t} - 1/\sqrt{t})V_{L_0} \qquad (1)$$

for links $L_\pm$ and L_0 related as in the Conway polynomial relation. This was quickly extended to a 2-variable invariant $P_L(v,z) \in \mathbf{Z}[v^{\pm 1}, z^{\pm 1}]$, with the property that

$$v^{-1}P_{L_+} - vP_{L_-} = zP_{L_0}. \tag{2}$$

The name "Homfly polynomial" has come to be attached to P, being the initial letters of six of the eight people involved in this further development. The name is sometimes extended to the more unwieldy 'HOMFLYPT', to make reference to all eight. The polynomial P contains both the Conway/Alexander polynomial, and Jones' invariant, and can be shown to contain more information in general than both of these taken together. It satisfies the equations

$$\begin{aligned} P(1,z) &= \nabla(z), \\ P(1, s-s^{-1}) &= \Delta(s^2), \\ P(s^2, s-s^{-1}) &= V(s^2), \\ P(s, s-s^{-1}) &= \pm 1. \end{aligned}$$

The skein relation (2) can readily be shown to determine P and V once its value on the trivial knot is given. It has been usual to take $P = 1$ on the trivial knot, although in some recent applications a different normalisation can be more appropriate.

Given the existence of V and P we can then make some calculations. For example, the unlink with two components has

$$\begin{aligned} P &= \frac{v^{-1}-v}{z}, \\ V(s^2) &= -(s+s^{-1}), \end{aligned}$$

while the Hopf link with linking number $+1$ has

$$\begin{aligned} P &= vz + (v^{-1}-v)v^2z^{-1}, \\ V(s^2) &= s^3 - s - (s+s^{-1})s^4 = -s(1+s^4). \end{aligned}$$

The Hopf link with linking number -1 has

$$\begin{aligned} P &= -v^{-1}z + (v^{-1}-v)v^{-2}z^{-1}, \\ V(s^2) &= -s^{-1}(1+s^{-4}). \end{aligned}$$

This illustrates the general feature that for the mirror image $\bar{L}$ of a link L, (where the signs of all crossings are changed), we have $P_{\bar{L}}(v,z) = P_L(v^{-1},-z)$ and so $V_{\bar{L}}(s^2) = V_L(s^{-2})$. It is thus quite possible to use V in many cases to distinguish a knot from its mirror-image, while there will be no difference in their Conway polynomials. It is worth noting that although there are still knots which cannot be distinguished from each other by P in spite of being inequivalent, no nontrivial knot has so far been found for which $P = 1$, or even $V = 1$.

4.1.1. *Framed versions*

The original Homfly polynomial is invariant under *all* Reidemeister moves, but there is a convenient version which is an invariant of a framed oriented link. A more extended discussion of the exact choice of framing normalisations can be found elsewhere, [2, 26, 27].

In its most adaptable form [17], $P_L(v,s)$, the framed invariant lies in the ring

$$\Lambda = \mathbf{Z}[v^{\pm 1}, s^{\pm 1}, (s^r - s^{-r})^{-1}], \quad r > 0.$$

Its defining characteristics are the two local skein relations.

1. 

2. 

These relate the invariants of links whose diagrams differ only locally as shown.

They are enough to allow its recursive calculation from simpler diagrams in terms of the value for the unknot.

4.2. *Satellite invariants*

Invariants such as the Homfly polynomial P of any choice of satellite of a knot K may be regarded as invariants of K itself. These provide a whole range of satellite invariants, which can be compared for mutants K and K'.

4.2.1. *Framed links*

Framed links are made from pieces of ribbon rather than rope, so that each component has a preferred annulus neighbourhood.

Combinatorially they can be modelled by diagrams in S^2 up to the Reidemeister moves R_{II} and R_{III}, excluding R_I, by use of the "blackboard framing" convention. The ribbons are determined by taking parallel curves on the diagram. Reidemeister moves R_{II} and R_{III} on a diagram give rise to isotopic ribbons. Any apparent twists in a ribbon can be flattened out using R_I.

Oriented link diagrams D have a *writhe* $w(D)$ which is the sum of the signs of all crossings. This is unchanged by moves R_{II} and R_{III}.

The unframed version of the Homfly polynomial for an oriented link L, invariant under *all* Reidemeister moves, is given from this framed version by $v^{w(D)}P_L(v,s)$ where D is a diagram for the framed link.

Remark. For a framed knot the writhe is sometimes called its "self-linking number", which is independent of the orientation of the diagram. Generally a framing of a link is determined by a choice of writhe for each component.

4.2.2. *Satellites*

A *satellite* of a framed knot K is determined by choosing a diagram Q in the standard annulus, and then drawing Q on the annular neighbourhood of K determined by the framing, to give the satellite knot $K * Q$. We refer to this construction as *decorating K with the pattern Q* (see Fig. 2).

Fig. 2. Satellite construction.

Morton and Traczyk [32] showed that the Jones polynomial V cannot be used in combination with any choice of satellite to distinguish a mutant pair, K and K'. Thus $V_{K*Q} = V_{K'*Q}$ for any choice of pattern Q, provided that the same framing of K and K' is used.

4.2.3. *A parameter space for Homfly satellite invariants*

The local nature of the Homfly skein relations allows us to make a useful simplification in studying Homfly satellite invariants P_{K*Q} as the pattern Q varies.

Compare for example three patterns $Q_\pm$ and Q_0.

$Q_+ =$ 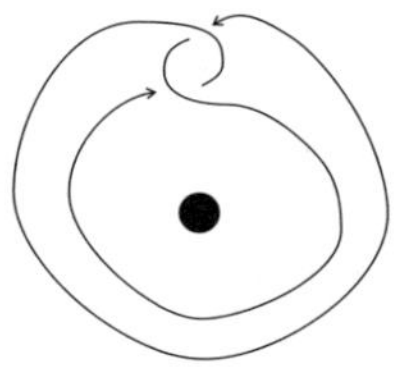 , $Q_- =$ 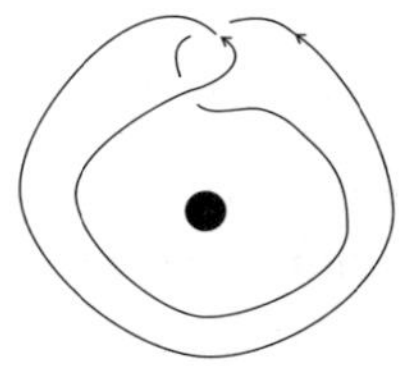 ,

$Q_0 =$ 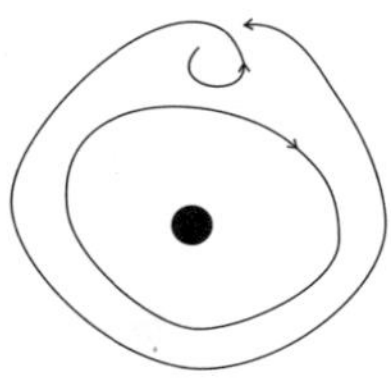 .

The framed Homfly invariants of $K * Q_\pm$ and $K * Q_0$ then satisfy

$$P_{K*Q_+} - P_{K*Q_-} = (s - s^{-1}) P_{K*Q_0}.$$

Since $K * Q_-$ is the unknot for any K, this relates the invariants of the Whitehead double $K * Q_+$ of K and those of its reverse parallel.

More generally, consider the linear space $\mathcal{C}$ of Λ-linear combinations of diagrams in the annulus (up to R_{II}, R_{III}) and impose the local relations

1. 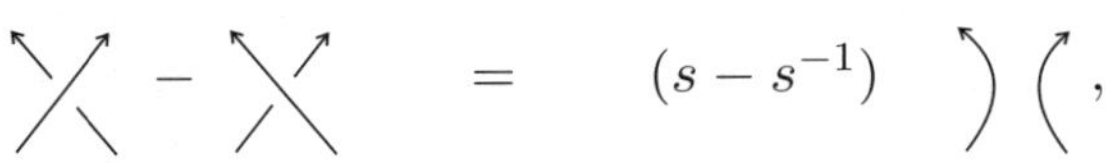

2. 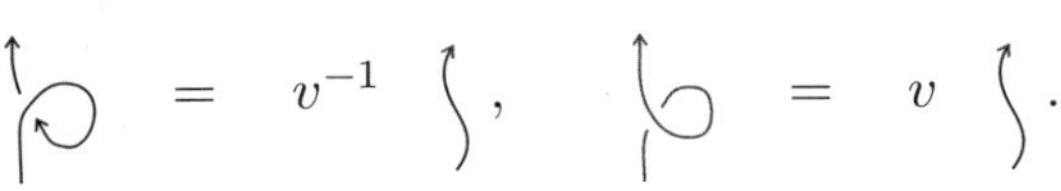

Decorating K by an element $\sum a_i Q_i$ of the linear space $\mathcal{C}$, which is known as the *framed Homfly skein of the annulus*, gives a well-defined Homfly invariant $\sum a_i P_{K*Q_i}$ since the skein relations are respected when the Homfly polynomials of the satellites are compared.

We can summarise our calculation above by saying that in the skein $\mathcal{C}$ we have

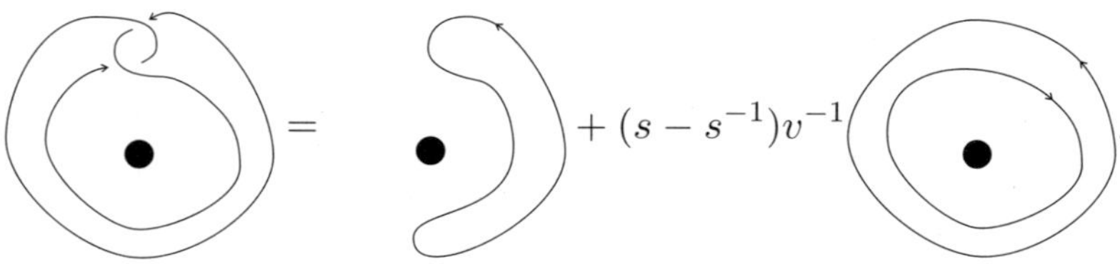

and hence

$$P_{K*Q_+} = P_{\text{unknot}} + (s - s^{-1})v^{-1} P_{\text{reverse parallel}}.$$

The space $\mathcal{C}$ then gives a more effective parameter space for satellite invariants, as we only need to know the pattern as an element of $\mathcal{C}$.

For example, any of the twist patterns

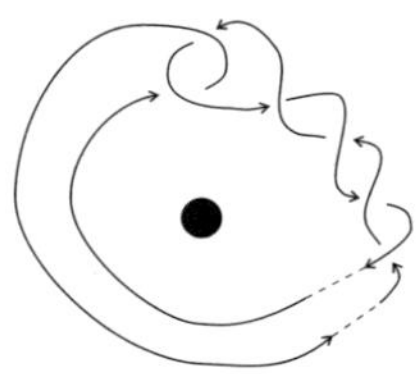

is a linear combination of the reverse parallel and the trivial pattern, so the Homfly polynomial of any twisted double can be found from the reverse parallel.

The linear subspaces $\mathcal{C}_m$ for $m > 0$ within $\mathcal{C}$ spanned by the closure of oriented m-tangles with all m ends of arcs directed in the same way are finite dimensional. The space $\mathcal{C}_m$ has a basis consisting of elements Q_λ, one for each partition λ of m. These basis elements Q_λ play an important role in relating Homfly satellite invariants to unitary quantum group invariants, as mentioned in the next section. For a more detailed account of their definition, and interpretation, see the recent article by Morton and Manchon [28], or earlier work of Aiston and Lukac [1, 2, 26].

4.2.4. *Homfly satellite invariants on mutants*

The results of Lickorish and Lipson [25] show that $P_{K*Q} = P_{K'*Q}$ for mutants K, K' when the pattern Q is the closure in the annulus of any 2-tangle, for example the twist pattern above, and P is the Homfly polynomial.

A contrasting result occurs when the pattern Q is a closed 3-tangle. Homfly invariants of 3-parallels were realised at an early stage to give possibilities for distinguishing mutants.

Calculations made in 1986 by Morton and Traczyk showed that the Homfly polynomials P_{C*Q} and P_{KT*Q} are different for the 3-parallel pattern Q. Since the computing facilities available were limited, they fixed the value of one variable and reduced the integer coefficients mod p for some small fixed value of p. Although they were able to establish that the two polynomials were different, it was not easy to appreciate the extent and nature of the difference from their calculations. Jun Murakami [34] also made calculations based on 3-parallels of other mutant pairs, and gave necessary conditions for Homfly-based satellite invariants to distinguish mutants. These involve identification of dimension 1 subspaces in representation theory.

Subsequent more sophisticated calculations by Cromwell and Morton [27] give much more detail. Their method of calculation involves a truncation which amounts to retaining only Vassiliev invariants up to a certain type, in this case type 12 is enough. Such a truncation at a fixed type is very easily implemented in terms of the calculations based on the Morton–Short algorithm for finding Homfly polynomials [31], and it gives a very satisfactory outcome when the difference of the invariants for two mutants is studied.

4.2.5. *Kauffman polynomial*

The 2-variable Kauffman polynomial, discovered shortly after the appearance of the Homfly polynomial, also has the property that it does not distinguish mutants. Nor does it distinguish the 2-parallels of mutants.

Rather less work has been done on establishing which Kauffman satellite invariants can distinguish mutants. Like the Homfly polynomial its satellite invariants are closely related to quantum group invariants [46]. More recent calculations have been made by Stoimenow [42] who has shown that the Kauffman polynomial of the 3-parallel can distinguish some mutants with symmetry, in contrast to the corresponding Homfly polynomial.

5. Unitary Quantum Group Invariants

Following closely after the discovery of the Homfly and Kauffman polynomial invariants came the work of Reshetikhin and Turaev on the development of knot invariants based on quantum groups.

Quantum groups give rise to 1-parameter invariants $J(K;W)$ of an oriented framed knot K depending on a choice of finite dimensional module W over the quantum group, following constructions of Turaev and others [43, 46]. This choice is referred to as *colouring* K by W, and can be extended for a link to allow a choice of colour for each component.

5.1. *Basic constructions of quantum invariants*

A quantum group $\mathcal{G}$ is an algebra over a formal power series ring $\mathbf{Q}[[h]]$, typically a deformed version of a classical Lie algebra. A finite dimensional module over $\mathcal{G}$ is a linear space on which $\mathcal{G}$ acts.

Crucially, $\mathcal{G}$ has a coproduct Δ which ensures that the tensor product $V \otimes W$ of two modules is also a module. It also has a *universal R-matrix* (in a completion of $\mathcal{G} \otimes \mathcal{G}$) which determines a well-behaved module isomorphism

$$R_{VW} : V \otimes W \to W \otimes V.$$

This has a diagrammatic view indicating its use in converting coloured tangles to module homomorphisms.

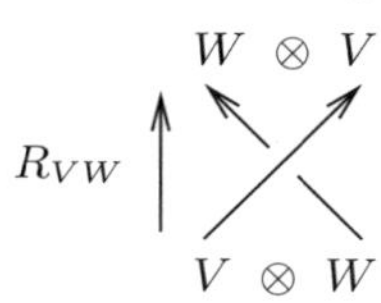

A braid β on m strings with permutation $\pi \in S_m$ and a colouring of the strings by modules $V_1, \ldots, V_m$ leads to a module homomorphism

$$J_\beta : V_1 \otimes \cdots \otimes V_m \to V_{\pi(1)} \otimes \cdots \otimes V_{\pi(m)}$$

using $R^{\pm 1}_{V_i,V_j}$ at each elementary braid crossing. The homomorphism J_β depends *only on the braid* β itself, not its decomposition into crossings, by the Yang–Baxter relation for the universal R-matrix.

When $V_i = V$ for all i we get a module homomorphism $J_\beta : W \to W$, where $W = V^{\otimes m}$. Now any module W decomposes as a direct sum $\bigoplus (W_\mu \otimes V_\mu^{(N)})$, where $W_\mu \subset W$ is a linear subspace consisting of the *highest weight vectors* of type μ associated to the module $V_\mu^{(N)}$. Highest weight subspaces of each type are preserved by module homomorphisms, and so J_β determines (and is determined by) the restrictions $J_\beta(\mu) : W_\mu \to W_\mu$ for each μ, where μ runs over partitions with at most N parts.

If a knot (or one component of a link) K is decorated by a pattern T which is the closure of an m-braid β, then its quantum invariant $J(K * T; V)$ can be found from the endomorphism J_β of $W = V^{\otimes m}$ in terms of the quantum invariants of K and the restriction maps $J_\beta(\mu) : W_\mu \to W_\mu$ by the formula

$$J(K * T; V) = \sum c_\mu J(K; V_\mu^{(N)}) \tag{3}$$

with $c_\mu = \operatorname{tr} J_\beta(\mu)$. This formula follows from Lemma II.4.4 in [43]. We set $c_\mu = 0$ when W has no highest weight vectors of type μ.

More generally the methods of Reshetikhin and Turaev allow the quantum groups $\mathcal{G} = sl(N)_q$ to be used to represent oriented tangles whose components are coloured by $\mathcal{G}$-modules as $\mathcal{G}$-module homomorphisms. One additional feature is needed, namely the use of the dual module V^* defined by means of the antipode in $\mathcal{G}$, (an anti-automorphism of $\mathcal{G}$ which is part of its structure as a Hopf algebra). When the components of the tangle are coloured by modules the tangle itself is represented by a homomorphism from the tensor product of the modules which colour the strings at the bottom to the tensor product of the modules which colour the strings at the top, provided that the string orientations are inwards at the bottom and outwards at the top. The dual module V^* comes into play in place of V when an arc of the tangle coloured by V has an output at the bottom or an input at the top.

For example, the $(4,2)$-tangle below, when coloured as shown, is represented by a homomorphism $U \otimes W^* \to U \otimes X^* \otimes X \otimes W^*$.

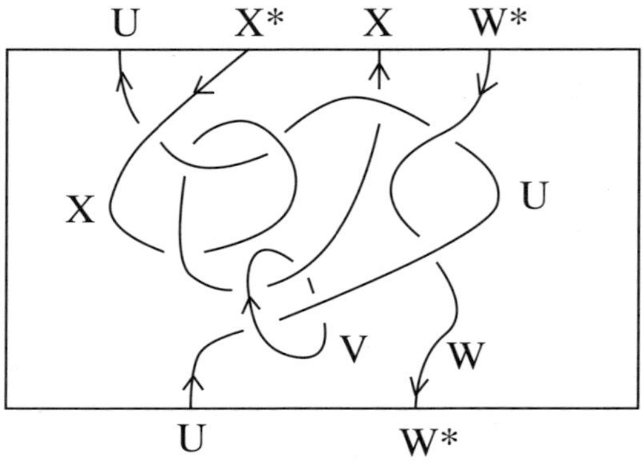

It is possible to build up the definition so that consistently coloured tangles are represented by the appropriate composite homomorphisms, starting from a definition of the homomorphisms for the elementary oriented tangles. Two cases, depending on the orientation, must be considered for both the local maximum and the local minimum, and a little care is needed here to ensure consistency. The final result is a definition of a homomorphism which is invariant when the coloured tangle is altered by R_{II} and R_{III}. When applied to an oriented k-component link diagram L regarded as an oriented $(0,0)$-tangle, it gives an element $J(L; V_1, \ldots, V_k) \in \Lambda = \mathbf{Q}[[h]]$ for each colouring of the components of L by $\mathcal{G}$-modules, which is an invariant of the framed oriented link L.

The construction is simplified in the case of $sl(2)_q$ by the fact that all modules are isomorphic to their dual, and so orientation of the strings plays no role.

5.2. *Quantum invariants of mutant knots*

We turn to the question of distinguishing mutants such as C and KT by means of quantum group invariants, especially those which use the unitary quantum groups $sl(N)_q$. These are closely related to the Homfly satellite invariants of a knot, and can provide complementary insights into their behaviour.

In the case of mutant knots K, K' the basic quantum invariants are the 1-parameter invariants $J(K; V_\lambda)$ and $J(K'; V_\lambda)$ where V_λ is an irreducible module over the quantum group.

The decomposition of oriented knots K and K' into 2-tangles

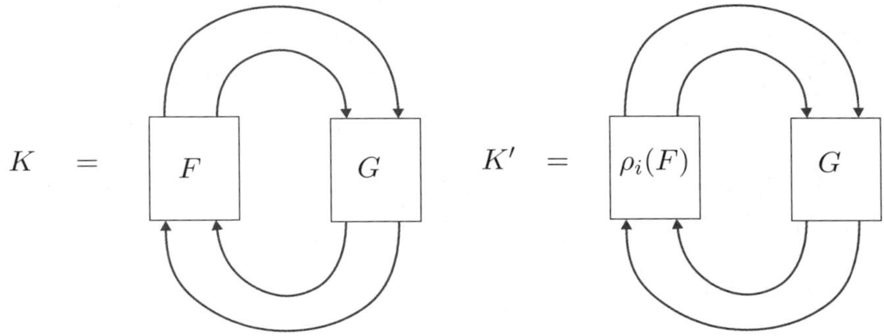

means that there are endomorphisms of $V_\lambda \otimes V_\lambda$ determined by the tangles F and G and the quantum invariant $J(K; V_\lambda)$ is a suitably weighted trace of their composite. Where F is replaced by $\rho_1(F)$ the endomorphism is replaced by the conjugate with the R-matrix for V_λ. In [27], Morton and Cromwell give conditions on V_λ which ensure that the endomorphism for F will commute with the R-matrix, so that $J(K; V_\lambda) = J(K'; V_\lambda)$, when K' is the mutant constructed using the rotation ρ_1.

Remark. This is the case known as the positive mutant in Sec. 8.2.1. They also give conditions which ensure equality for quantum invariants of the other mutants.

5.3. *Unitary quantum invariants and Homfly invariants*

When dealing with $sl(N)_q$ for any fixed natural number N it is usual to write $q = e^h$. Where the framed knot K is coloured by a finite dimensional module W over the unitary quantum group $sl(N)_q$ its invariant $J(K;W)$ depends on the variable h as a Laurent polynomial in one variable $s = e^{h/2} = \sqrt{q}$, up to an overall fractional power of q.

The invariant J is linear under direct sums of modules and all the modules over $sl(N)_q$ are semi-simple, so we can restrict our attention to the irreducible modules $V_\lambda^{(N)}$. For $sl(N)_q$ these are indexed by partitions λ with at most N parts, without distinguishing two partitions which differ in some initial columns with N cells each.

There is a close relation between Homfly satellite invariants and unitary quantum invariants of K. To help in our comparison of these invariants we write $P(K;Q)$ for P_{K*Q} and more generally

$$P(L; Q_1, Q_2, \dots, Q_k)$$

for the Homfly polynomial of a link L when its components are decorated by $Q_1, \dots, Q_k$ respectively.

Theorem 5.1. (Comparison theorem)

1. *The $sl(N)_q$ invariant for the irreducible module $V_\lambda^{(N)}$ is the Homfly invariant for the knot decorated by Q_λ with $v = s^{-N}$, suitably normalised as in [26]. Explicitly,*

   $$P(K;Q_\lambda)|_{v=s^{-N}} = x^{k|\lambda|^2} J(K; V_\lambda^{(N)}),$$

   where k is the writhe of K, and $x = s^{1/N}$.
2. *Each invariant $P(K;Q)|_{v=s^{-N}}$ is a linear combination of quantum invariants $\sum c_\alpha J(K;W_\alpha)$.*
3. *Each $J(K;W)$ is a linear combination of Homfly invariants*

   $$\sum d_j P(K;Q_j)|_{v=s^{-N}}.$$

Remark.

- In the special case when $N = 2$ we can interpret quantum invariants of K in terms of Kauffman bracket satellite invariants, using the skein of the annulus based on the Kauffman bracket relations. This simpler skein is a quotient of the algebra $\mathcal{C}$. More generally the $sl(N)_q$ invariants depend only on a quotient of the algebra $\mathcal{C}$ for each N.

- The quantum group invariants based on $sl(3)_q$ also admit a combinatorial simplification due to Kuperberg to allow an easier diagrammatic calculation of them. At the same time the quantum group itself is straightforward enough to make it possible to work directly with some of the smaller dimensional modules [29, 33].
- The 2-variable invariant $P(K;Q)$ can be recovered from the specialisations $P(K;Q)|_{v=s^{-N}}$ for sufficiently many N.
- If the pattern Q is a closed braid on m strings then we only need to use partitions $\lambda \vdash m$, since $\mathcal{C}_m$ is spanned by $\{Q_\lambda\}_{\lambda\vdash m}$. Conversely, to realise $J(K;V_\lambda^{(N)})$ with $\lambda \vdash m$ we can use closed m-braid patterns.

The basic condition on the quantum group module V_λ in [27] is that when the module $V_\lambda \otimes V_\lambda$ is decomposed as a direct sum of irreducible modules there should be no repeated summands, up to isomorphism. In this case *any* two endomorphisms of $V_\lambda \otimes V_\lambda$ will commute.

Since this is the case for all irreducible $sl(2)_q$ modules V_λ, it gives an alternative proof of the results of Morton–Traczyk about the Jones polynomial of satellites of mutants.

It is also the case for the fundamental irreducible $sl(N)_q$ module with Young diagram $\square$, which, taken together for all N, determine the Homfly polynomial, and for the irreducible $sl(N)_q$ modules with Young diagrams $\square\square$ and $\begin{smallmatrix}\square\\\square\end{smallmatrix}$, which determine the Homfly polynomial of the directed 2-cables.

It is interesting that this condition does not establish Lickorish and Lipson's result that the Homfly polynomial of *reverse* 2-parallels must agree for mutants; their result then yields a nontrivial consequence for quantum invariants. The simplest example of this is that the $sl(3)_q$ invariant of a knot when coloured by the irreducible module with Young diagram $\begin{smallmatrix}\square\square\\\square\end{smallmatrix}$ will agree on a mutant pair. I suspect that this invariant is at the heart of Stoimenow's use of the Whitehead double in showing that a pair of knots are not mutants [42]. He gives a pair of knots whose Homfly polynomials of their 2-parallels, and of the knots themselves, agree, and proves that the knots are not mutants because the Homfly polynomials of their Whitehead doubles are different.

The calculations of Cromwell and Morton [27] about the Homfly polynomials of 3-parallels show, on the other hand, that the $sl(4)_q$ invariant for the module with Young diagram  does distinguish some mutant pair, namely C and KT, as does the $sl(N)_q$ invariant with Young diagram , for every $N \geq 4$.

6. Vassiliev Invariants

The invariants, known variously as *finite type* invariants, or *Vassiliev* invariants, developed by Vassiliev and Gusarov in the late '90s, can be related readily to polynomial and quantum group invariants, originally by Birman and Lin [3]. They provide a rather transverse view of a whole collection of these invariants, and their behaviour on mutants has been a matter of continuing interest.

Chmutov, Duzhin and Lando [12] prove that all Vassiliev invariants of degree at most 8 agree on any mutant pair of knots.

This is extended to Vassiliev invariants up to degree 10 by Jun Murakami [35], where he also confirms the degree 11 invariant used by Morton and Cromwell in [27] which can be used to distinguish the knots C and KT.

Morton and Cromwell expand the difference between the Homfly polynomials of the 3-parallels of the knots C and KT to isolate a framed Vassiliev invariant of type 11 which distinguishes these two mutants, and go on to explain some features of the difference $P_{K*Q} - P_{K'*Q}$ for general K, K', where the pattern Q is the closure of a 3-braid.

Further results about Vassiliev invariants on extended and restricted classes of mutants are noted in Sec. 8.

7. Further Invariants

Among the homology invariants which have been developed in the past ten years the Heegaard–Floer homology can certainly distinguish some mutants, since it is able to calculate the genus of the knot.

On the other hand, Bloom [4] shows that odd Khovanov homology for knots is unchanged by mutation. As a corollary he notes that Khovanov homology over $\mathbf{Z}_2$ is also mutation invariant. Homfly Khovanov homology is shown by Jaeger [16] to be invariant under *positive* mutation, as defined in Sec. 8.2.1.

Kim and Livingston [20] show that the 4-ball genus of a knot can be changed by mutation, but the algebraic concordance class is invariant under mutation.

7.1. *Behaviour on mutants*

I have gathered together here a summary of the results noted about the behaviour of a selection of invariants on mutants. Where the invariants are known to be the same on mutant knots (shared) I give a reference to a proof, not necessarily the original one. Where there are mutants on which the invariant is known to differ, I give a reference to an example.

Invariant	**Shared**	**Differs**
Alexander Polynomial	[44]	
Signature	[44]	
Genus		[15]
n-colouring	By Alexander	
Homomorphism to $PSL(2,7)$		[39]
Double branched cover	[44]	
Hyperbolic volume, where available	[41]	
Crossing number	<16, [42]	Unknown if ≥ 16
Braid index	Unknown	
Arc index	Unknown	
Jones polynomial	[23]	
Homfly polynomial	[23]	
Kauffman polynomial	[23]	
Jones satellites	[32]	
Homfly 2-satellites	[25, 38]	
Kauffman 2-satellites	[25]	

Invariant	**Shared**	**Differs**
Homfly 3-parallel		[27, 34]
Kauffman 3-parallel		[42]
Vassiliev degree ≤8	[12]	
Vassiliev degree ≤10	[35]	
Vassiliev degree 11		[27, 35]
Heegaard Floer homology		By genus
Odd Khovanov homology	[4]	
4-ball genus		[18, 20]
Algebraic concordance	[20]	

For unitary quantum groups I give a similar summary. The irreducible module used in colouring is specified by giving its defining Young diagram.

Quantum group	**Colouring**	**Shared**	**Differs**
$sl(2)_q$	Any	[27]	
$sl(N)_q$	row/column	[27]	
$sl(3)_q$		[27]	
$sl(>3)_q$			[27]
$sl(3)_q$			[29]

8. Generalisations and Restrictions

8.1. *Generalisations*

Various generalisations of the original ideas of mutants have been made.

8.1.1. *Rotors*

An obvious possibility is to decompose a knot by a sphere meeting the knot in $2n$ points with $n > 2$, and then replace one side of the sphere after some transformation. In many cases the resulting knot does not have enough properties in common with the original for this to be worthwhile. However, Rolfsen [40] has used the idea of a *rotor*, based on $2n$ intersection points around the equator of a sphere with a rotation of order $2n$ on one side as the transformation. For unoriented knot diagrams this operation preserves the Jones polynomial, although Rolfsen has so far not been able to use the method in his searches for a nontrivial knot with Jones polynomial $V = 1$.

8.1.2. *Genus 2 mutants*

A more fruitful class of generalised mutants are constructed by finding an embedded genus 2 surface in the knot complement, and regluing the two sides after a suitable degree 2 transformation (a hyperelliptic involution). This construction was used by Ruberman [41] for general 3-manifolds, and by Cooper and Lickorish [10] in the context of knots in S^3.

The construction has a close relation to Conway mutation for knots, which can be realised by applying a sequence of one or two genus 2 mutations. An extensive discussion of genus 2 mutation, and properties which are known to be preserved, is given by Dunfield *et al.* [11]. Further calculations related to genus 2 mutants by Morton and Nathan Ryder appear in [30].

Here are some of the known coincidences and differences.

Invariant	**Shared**	**Differs**
Alexander Polynomial	[10]	
Signature	[10]	
Hyperbolic volume, where available	[41]	
Homfly polynomial		[11]
Kauffman polynomial		[30]
Jones satellites	[11]	
Vassiliev degree 8		[30]
Khovanov homology		[11]

8.2. *Restrictions*

On the other hand, certain more tightly defined classes of mutants are known to share more invariants than is true for a general pair of Conway mutants.

Such restricted classes of mutants are consequently more difficult to distinguish, and provide more sensitive test-beds for examining possible dependence among invariants.

8.2.1. *Positive mutants*

The simplest restriction, which splits mutants into two classes, has led to a number of helpful observations.

Recall that when a knot K can be decomposed into two 2-tangles by a Conway sphere, we construct mutants by replacing one of the tangles after one of three possible π-rotations.

$\rho_1(F) =$ 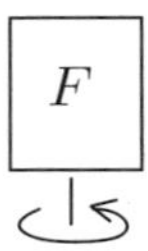

, $\rho_2(F) =$ 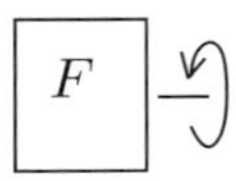

, $\rho_3(F) =$ 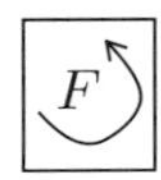

.

We may orient the original knot K, and we will find that only one of the three rotations respects the string orientation when regluing the 2-tangles.

Assume, as we may do, that the original knot K has been oriented as shown below

$K =$ 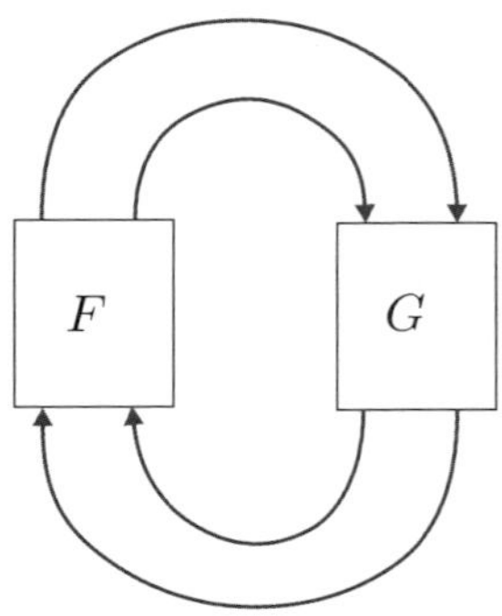

Definition. The mutant K' where F is replaced by $\rho_1(F)$ can maintain the orientation of the strings in F, and is called a *positive* mutant of K.

Remark. The classical Conway and Kinoshita–Teresaka pair of knots are positive mutants.

To orient the remaining two mutants where F is replaced by $\rho_2(F)$ or $\rho_3(F)$, the orientations of the strings in one of the two tangles must be reversed. There are then two pairs of positive mutants among the four knots consisting of the original knot K and its three mutants.

Kirk and Livingston [21] adopted the term *positive*, and showed that positive mutants are S-equivalent. This condition on Seifert matrices is necessary, but not sufficient for concordance of the knots. All the same, they prove that concordance is not preserved even for positive mutants.

Jaeger [16] shows that the Homfly Khovanov homology is preserved for positive mutants.

8.2.2. *Mutants with symmetry*

The symmetric mutants discussed by Morton in [33] also have much more restricted properties.

These knots are made up by interchanging three 2-tangles A, B and C as shown in Fig. 3.

Assume that the tangles A and B are both symmetric under the π-rotation ρ_1, so that

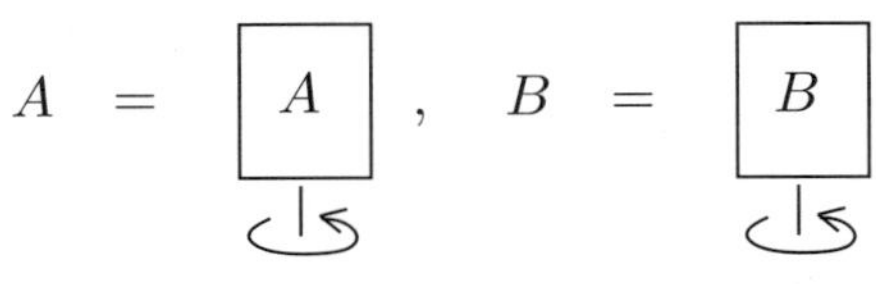

and hence A and B both commute with the half-twist.

Under these conditions the two knots K and K' are shown to share the same $sl(N)$ quantum invariants when coloured by the irreducible

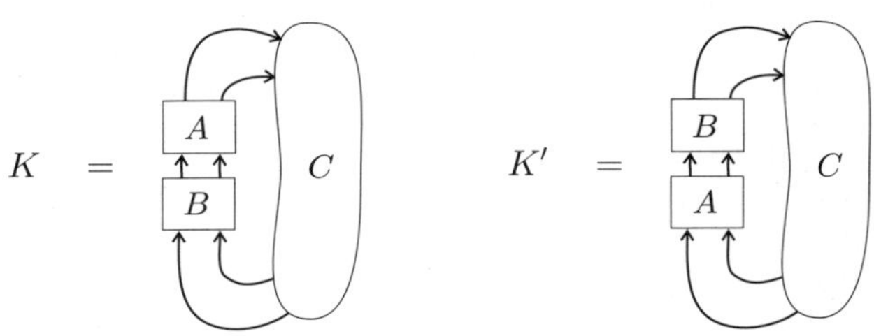

Fig. 3. Tangle interchange.

module V_λ with partition λ of $m = |\lambda|$, for any $m \leq 5$. The proof depends on showing that the module endomorphisms of $V_\lambda \otimes V_\lambda$ determined by A and by B commute, knowing by the symmetry assumption that they commute with the R-matrix.

The relations between Homfly satellite invariants and unitary quantum invariants then ensure that K and K' share the same Homfly satellite invariants for their m-parallels, and indeed any directed m-string satellite, when $m \leq 5$.

In an extension of this result it is shown that the Homfly polynomial of a genuine connected cable, based on the (m, n) torus knot pattern, with m and n coprime, does not distinguish mutants with symmetry K and K' above, for *any* number of strings, m. On the other hand, there are examples where the 6-string *parallels* have different Homfly polynomials.

In the course of these calculations [33] the pairs of mutants with symmetry could be distinguished by a Vassiliev invariant of degree 14. Interestingly, calculations of Stoimenow [42] on the 12-crossing Ochiai–Morimura mutants with symmetry 12_{1653} and 12_{1654} using the Kauffman polynomial of the 3-parallel showed that these too were distinguished by a degree 14 Vassiliev invariant. This leads to the speculation that mutants with symmetry may share their Vassiliev invariants of degree $\leq$13.

References

[1] A. K. Aiston, Skein theoretic idempotents of Hecke algebras and quantum group invariants, Ph.D. thesis, University of Liverpool, 1996.

[2] A. K. Aiston and H. R. Morton, Idempotents of Hecke algebras of type A, *J. Knot Theory Ramifications* **7** (1998) 463–487.

[3] J. S. Birman and X.-S. Lin, Knot polynomials and Vassiliev's invariants, *Invent. Math.* **111** (1993) 225–270.

[4] J. M. Bloom, Odd Khovanov homology is mutation invariant, *Math. Res. Lett.* **17** (2010) 1–10.

[5] M. Boileau, B. Leeb and J. Porti, Geometrization of 3-dimensional orbifolds, *Ann. Math.* **162** (2005) 195–290.

[6] F. Bonahon and L. C. Siebenmann, The characteristic toric splitting of irreducible compact 3-orbifolds, *Math. Ann.* **278** (1987) 441–479.

[7] F. Bonahon and L. C. Siebenmann, New geometric splittings of classical knots and the classification and symmetries of arborescent knots, preprint 2010, in Geometry and Topology Monographs, to appear.

[8] T. Cochran and D. Ruberman, Invariants of tangles, *Math. Proc. Camb. Philos. Soc.* **105** (1989) 299–306.

[9] J. H. Conway, On enumerations of knots and links, in *Computational Problems in Abstract Algebra*, ed. J. Leech, pp. 329–358, Pergamon Press, 1969.

[10] D. Cooper and W. B. R. Lickorish, Mutations of links in genus 2 handlebodies, *Proc. Amer. Math. Soc.* **127** (1999) 309–314.

[11] N. M. Dunfield, S. Garoufalidis, A. Shumakovitch and M. Thistlethwaite, Behavior of knot invariants under genus 2 mutation, arXiv:math/0607258.

[12] S. V. Chmutov, S. V. Duzhin and S. K. Lando, Vassiliev knot invariants. I. Introduction, in *Singularities and Bifurcations*, 117–126, Adv. Soviet Math., Vol. 21, Amer. Math. Soc., Providence, RI, 1994.

[13] R. Fenn and C. Rourke, Racks and links in codimension two, *J. Knot Theory Ramifications* **1** (1992) 343–406.

[14] R. H. Fox, A quick trip through knot theory, in *Topology of 3-manifolds and Related Topics*, ed. M. K. Fort, pp. 120–167, Prentice-Hall, NJ (1961).

[15] D. Gabai, Genera of the arborescent links, *Mem. Amer. Math. Soc.* Vol. 59, pp. 1–98, Amer. Math. Soc., 1986.

[16] T. C. Jaeger, Khovanov–Rozansky homology and Conway mutation, arXiv:1101.3302.

[17] L. H. Kauffman, *On Knots*, Annals Study No. 115, Princeton Univ. Press, 1987.

[18] C. Kearton, Mutation of knots, *Proc. Amer. Math. Soc.* **105** (1989) 206–208.

[19] S. Kinoshita and H. Terasaka, On unions of knots, *Osaka Math. J.* **9** (1957) 131–153.

[20] S.-G. Kim and C. Livingston, Knot mutation: 4-genus and algebraic concordance, *Pacific J. Math.* **220** (2005) 87–105.

[21] P. Kirk and C. Livingston, Twisted knot polynomials: Inversion, mutation and concordance, *Topology* **38** (1999) 663–671.

[22] P. Kirk and C. Livingston, Concordance and mutation, *Geom. Topol.* **5** (2001) 831–883.

[23] W. B. R. Lickorish, Prime knots and tangles, *Trans. Amer. Math. Soc.* **267** (1981) 321–332.

[24] W. B. R. Lickorish, Polynomials for links, *Bull. London Math. Soc.* **20** (1988) 558–588.

[25] W. B. R. Lickorish and A. S. Lipson, Polynomials of 2-cable-like links, *Proc. Amer. Math. Soc.* **100** (1987) 355–361.

[26] S. G. Lukac, Homfly skeins and the Hopf link, Ph.D. thesis, University of Liverpool, 2001.

[27] H. R. Morton and P. R. Cromwell, Distinguishing mutants by knot polynomials, *J. Knot Theory Ramifications* **5** (1996) 225–238.

[28] H. R. Morton and P. M. G. Manchón, Geometrical relations and plethysms in the Homfly skein of the annulus, *J. London Math. Soc. (2)* **78** (2008) 305–328.

[29] H. R. Morton and H. J. Ryder, Mutants and $SU(3)q$ invariants, in Geometry and Topology Monographs, Vol. 1: The Epstein Birthday Schrift, pp. 365–381 (1998).

[30] H. R. Morton and N. Ryder, Invariants of genus 2 mutants, *J. Knot Theory Ramifications* **18** (2009) 1423–1438.

[31] H. R. Morton and H. B. Short, The 2-variable polynomial of cable knots, *Math. Proc. Camb. Phil. Soc.* **101** (1987) 267–278.

[32] H. R. Morton and P. Traczyk, The Jones polynomial of satellite links around mutants, in *Braids*, eds. J. S. Birman and A. Libgober, Contemporary Mathematics, Vol. 78, pp. 587–592, Amer. Math. Soc., 1988.

[33] H. R. Morton, Mutant knots with symmetry, *Math. Proc. Camb. Philos. Soc.* **146** (2009) 95–107.

[34] J. Murakami, The parallel version of polynomial invariants of links, *Osaka J. Math.* **26** (1989) 1–55.

[35] J. Murakami, Finite type invariants detecting the mutant knots, in *Knot Theory.* A Volume Dedicated to Professor Kunio Murasugi for His 70th Birthday, eds. M. Sakuma *et al.*, pp. 258–267, Osaka University (2000).

[36] L. Paoluzzi, Hyperbolic knots and cyclic branched covers, *Publ. Mat.* **49** (2005) 257–284.

[37] K. A. Perko, Invariants of 11-crossing knots, Prépublications Math. d'Orsay 80, Université de Paris-Sud (1980).

[38] J. Przytycki, Equivalence of cables of mutants of knots, *Can. J. Math.* **XLI** (1989) 250–273.

[39] R. Riley, Homomorphisms of knot groups on finite groups, *Math. Comput.* **25** (1971) 603–617.

[40] D. Rolfsen, The quest for a knot with trivial Jones polynomial: Diagram surgery and the Temperley-Lieb algebra, in *Topics in Knot Theory*, ed. M. E. Bozhüyük, pp. 195–210, Kluwer Academic Publ., Dordrecht, 1993.

[41] D. Ruberman, Mutation and volume of knots in S^3, *Invent. Math.* **90** (1987) 189–215.

[42] A. Stoimenow, Tabulating and distinguishing mutants, *Int. J. Algebra Comput.* **20** (2010) 525–559.

[43] V. G. Turaev, Quantum invariants of knots and 3-manifolds, De Gruyter Studies in Mathematics, Vol. 18, Walter de Gruyter and Co., Berlin, 1994.

[44] O. Ya Viro, Non-projecting isotopies and knots with homeomorphic covers, *J. Sov. Math.* **12** (1979) 86–96 [This is a translation of his article *LOMI Zap. Nauchnykh Sem.* **66** (1976) 133–147].

[45] J. R. Weeks, SnapPea: A computer program for creating and studying hyperbolic 3-manifolds, available from http://thames.northnet.org/weeks/index/SnapPea.html.

[46] H. Wenzl, Quantum groups and subfactors of type B, C and D, *Comm. Math. Phys.* **133** (1990) 383–432.

Knots and Distributive Homology: From Arc Colorings to Yang–Baxter Homology*

Józef H. Przytycki

Department of Mathematics,
The George Washington University,
Washington, DC 20052, USA
University of Maryland, College Park, USA
University of Gdańsk, Poland
przytyck@gwu.edu

1. Introduction

While homology theory of associative structures, such as groups and rings, has been extensively studied in the past beginning with the work of Hurewicz, Hopf, Eilenberg, and Hochschild, the non-associative structures, such as racks or quandles, were neglected until recently. The distributive structures[a] have been studied for a long time and already C. S. Peirce (1839–1914) in 1880 [79] emphasized the importance of (right) self-distributivity in algebraic structures,

*This paper has its roots in two series of talks I gave: in Russia (Lomonosov Moscow State University, May 29–June 1, 2012), where the visualization of Fig. 8 was observed, Korea (TAPU Workshop on Knot Theory, July 23–27, 2012), and in a talk at Oberwolfach Conference (June 3–9, 2012). The short version of this paper was published in Oberwolfach Proceedings [85]. While I keep novelty of the talks (many new ideas were crystallized then), I added a lot of supporting material so the paper is mostly self-sufficient. I kept also, to some extent, the structure of talks; it may lead to some repetitions but I hope it is useful for a reader.

[a]The word *distributivity* was coined in 1814 by French mathematician Francois Servois (1767–1847).

and his friend E. Schröder [99] gave an example of a three-element magma $(X; *)$ which is not associative[b] (compare Sec. 2).

However, homology for such universal algebras was introduced only between 1990 and 1995 by Fenn, Rourke, and Sanderson [39, 42–44]. We develop theory in the historical context and propose a general framework to study homology of distributive structures. We outline potential relations to Khovanov homology and categorification, via Yang–Baxter operators. We use here the fact that Yang–Baxter equation can be thought of as a generalization of self-distributivity.

1.1. *Invariants of arc colorings*

Consider a link diagram D, say , and a finite set X. We may define a diagram invariant to be the number of colorings of arcs[c] of D by elements of X, $col_X(D)$. Even such a naive definition leads to a link invariant $col_X(L) = \min_{D \in L}\ col_X(D)$, where $D \in L$ means that D is a diagram of L.[d] More sensible approach would start with a magma $(X; *)$, that is a set with binary operation, and with the coloring convention of Fig. 1.

---

[b] The example Schröder (1841–1902) gave is

$*$	0	1	2
0	0	2	1
1	2	1	0
2	1	0	2

and we named elements of the magma by $0, 1$ and 2 as this example is the base for Fox three colorings of links (developed about 1956) [25, 26, 82], and the operation can be written as $x * y = 2y - x$ modulo 3; it happens to be self-distributive from both sides, that is $(x * y) * z = (x * z) * (y * z)$ and $x * (y * z) = (x * y) * (x * z)$ (see Example 2.3(5)).

[c] We use the term *arc* for a part of the diagram from an undercrossing to the next undercrossing (including possibility of a trivial component), and the term *semi-arc* for a part of the diagram from a crossing to the next crossing. Thus, for example, a standard trefoil knot diagram has three arcs and six semi-arcs.

[d] One can say that it is nonsense but an invariant is nontrivial: $col_X(L) = |X|^{cr(L)+t(L)}$, where $cr(L)$ is the crossing number of L and $t(L)$ the number of trivial components in L. This is the case as for a knot diagram D with at least one crossing the number of arcs equals to the number of crossings.

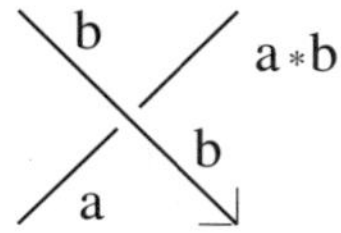

Fig. 1. Convention for a magma coloring of a crossing.

Again, let for a finite X, $col_{(X;*)}(D)$ denote the number of colorings of arcs of D by elements of X, according to the convention given in Fig. 1, at every crossing. We can define an oriented link invariant by considering $col_{(X;*)}(L) = \min_{D \in L} col_{(X;*)}(D)$; Alternatively, we can minimize $col_{(X;*)}(L)$ over minimal crossing diagrams of L only. Such an invariant would be very difficult to compute so it is better to look for properties of $(X;*)$ so that $col_{(X;*)}(D)$ is invariant under Reidemeister moves: R_1 (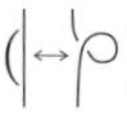) gives idempotent relation $a * a = a$.[e] R_2 (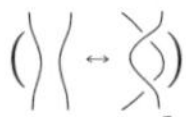) forces $*$ to be an invertible operation, and the third move illustrated in detail in Fig. 2, forces on $*$ a right self-distributivity $(a * b) * c = (a * c) * (b * c)$.

The magma $(X;*)$ satisfying all three conditions is called a quandle, the last two — a rack, and only the last condition — a shelf or RDS (right distributive system). Thus, if $(X;*)$ is a quandle then $col_{(X;*)}(D)$ (which we denote from now on succinctly $col_X(D)$) is a link invariant. We also use the notation $Col_X(D)$ for the set of X-colorings of D, thus $col_X(D) = |Col_X(D)|$. We can also try more

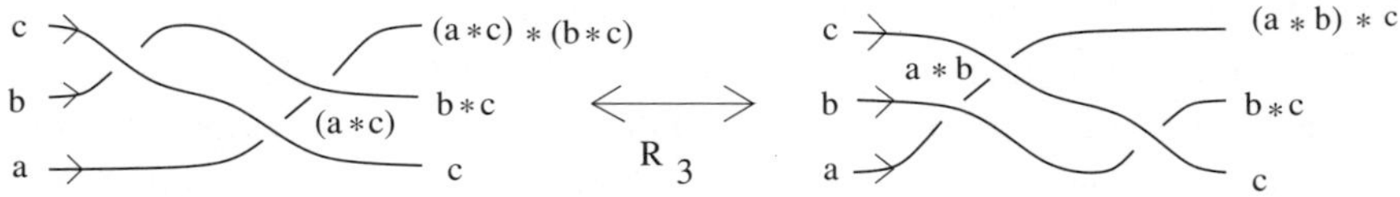

Fig. 2. Magma coloring of a third Reidemeister move and distributivity.

[e]The names *idempotent* (same power) and nilpotent (zero power) were introduced in 1870 by Benjamin Peirce [78], p. 20.

generally to color semi-arcs of a diagram by elements of X and declare for each colored crossing a weight of the crossing. This approach would lead to state sum and Yang–Baxter operators (see Sec. 12 and Fig. 15).

We can also do more with distributive magmas (after Carter, Kamada and Saito [18]; compare also Greene thesis [48]). We can sum over all crossings the pairs $\pm(a, b)$ according to the convention 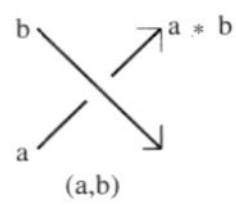

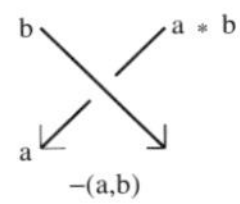

; the investigation of invariance of $\sum \pm(a, b)$ under Reidemeister moves was a hint toward construction of (co)homology of quandles.

We also encounter right distributivity by asking the following question: for a given coloring $\phi \in Col_X(D)$ by a magma $(X; *)$, and an element $x \in X$ is coloring, is $\phi * x$ also a magma coloring? If, as before, coloring is given by a, b, and $c = a * b$ then the new coloring of a crossing is given by $a * x$, $b * x$, and $c * x$. For a magma coloring we need $(a * b) * x = (a * x) * (b * x)$ which is exactly right self-distributivity. To put it on a more solid footing, we observe that the map $*_x : X \to X$ with $*_x(a) = a * x$ is a magma homomorphism: $*_x(a * b) = (a * b) * x = (a * x) * (b * x) = *_x(a) * *_x(b)$. For any magma homomorphism $g : X \to X$ if $f : arcs(D) \to X$ is a magma coloring then gf defined by $(gf)(arc) = g(f(arc))$ is a magma coloring of D. These are classical observations thus it is interesting to notice the slightly more general fact concerning the following question:

Let $(X; *_1)$ be a magma and f and g two $(X; *_1)$ magma colorings of a diagram D. Find the magma operation $*_2$ so that $f *_2 g$ is also a $(X; *_1)$ magma coloring. The question reduces to the previous one if g is a trivial coloring and $*_1 = *_2$. The nice condition which answers the question was first discussed by Niebrzydowski at his talk at Knots in Washington XXXV conference in December of 2012 [72] (compare also [23]).

Lemma 1.1. *Let $f, g : arcs(D) \to X$ be two colorings of a diagram D by $(X; *_1)$ (that is $f, g \in Col_{(X;*_1)}(D)$). Then $f *_2 g$ where $*_2$ is* *another binary operation on* X *and* $(f *_2 g)(arc) = f(arc) *_2 g(arc)$, *is an* $(X; *_1)$ *coloring of* X *if operations* $*_1$, *and* $*_2$ *are entropic one with respect to the other, that is*: $(a *_1 b) *_2 (c *_1 d) = (a *_2 c) *_1 (b *_2 d)$.

Proof. For every crossing with initial under-arc a and over-arc b we need

$$(f(a) *_2 g(a)) *_1 (f(b) *_2 g(b))$$
$$= (f(a) *_1 f(b)) *_2 (g(a) *_1 g(b))$$

which is exactly the entropic condition in Fig. 3 (compare Sec. 8.2). □

Notice that if g is a trivial coloring, say $g(arc) = x$ for any arc then any crossing forces $x *_1 x = x$ and the entropic equation reduces to $(f(a) *_2 x) *_1 (f(b) *_2 x) = (f(a) *_1 f(b)) *_2 (x *_1 x) = (f(a) *_1 f(b)) *_2 x$ (right distributivity). For use of entropic magmas in Knot Theory, see [76, 80, 81, 87, 93, 94, 101]; compare also Proposition 2.6.

We introduce now a monoid of binary operations and show that distributivity can be studied in the context of this monoid. Then we compare homology for associative structures (semigroups) with that for distributive structures (shelves). We also compare extensions in associative and distributive cases.

The paper is planned as a continuation of pioneering essay [84] and for completeness we recall parts of the essay.

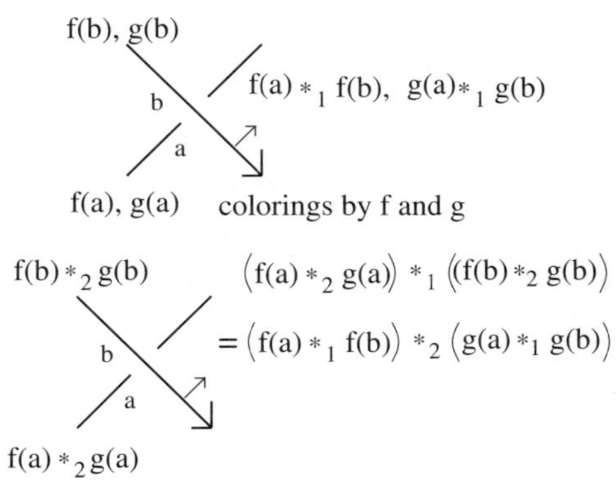

Fig. 3. Entropy condition for composition of f and g: $f *_2 g$.

2. Monoid of Binary Operations

Let X be a set and $*: X \times X \to X$ a binary operation. We call $(X;*)$ a *magma*.[f] For any $b \in X$, the adjoint map $*_b : X \to X$ is defined by $*_b(a) = a * b$. Let $Bin(X)$ be the set of all binary operations on X.

Proposition 2.1. *$Bin(X)$ is a monoid (i.e. semigroup with identity) with the composition $*_1 *_2$ given by $a *_1 *_2 b = (a *_1 b) *_2 b$ and the identity $*_0$ being the right trivial operation, that is, $a *_0 b = a$ for any $a, b \in X$.*

If $* \in Bin(X)$ is invertible, then $*^{-1}$ is usually denoted by $\bar{*}$. One should remark that $*_0$ is distributive with respect to any other operation, that is, $(a*b)*_0 c = a*b = (a*_0 c)*(b*_0 c)$, and $(a*_0 b)*c = a*c = (a*c)*_0(b*c)$.

Definition 2.2. Let $(X;*)$ be a magma, then:

(i) If $*$ is right self-distributive, that is, $(a*b)*c = (a*c)*(b*c)$, then $(X;*)$ is called an RDS (right distributive structure) or a shelf (the term coined by Alissa Crans and used in knot theory [22]).
(ii) If a shelf $(X;*)$ satisfies the idempotent condition, $a*a = a$ for any $a \in X$, then it is called an RDI structure or *right spindle*, or just a spindle (again the term coined by Crans).
(iii) If a shelf $(X;*)$ has $*$ invertible in $Bin(X)$ (equivalently $*_b$ is a bijection for any $b \in X$), then it is called a *rack* (the term wrack, like in "wrack and ruin", of J. H. Conway from 1959 [21], was modified to rack in [41]).
(iv) If a rack $(X;*)$ satisfies the idempotent condition, then it is called a *quandle* (the term coined in Joyce's Ph.D. thesis of 1979 [55, 56]). Axioms of a quandle were motivated by three Reidemeister moves (idempotent condition by the first move, invertibility by the second, and right self-distributivity by the third move).

[f] The term magma was used by J.-P. Serre [100] and Bourbaki [5], replacing the older term *groupoid* which started to mean a category with every morphism invertible.

(v) If a quandle $(X;*)$ satisfies $** = *_0$ (i.e. $(a*b)*b = a$), then it is called kei or an involutive quandle. The term kei (圭) was coined in a pioneering paper by M. Takasaki[g] in 1942 [104].

The main early example of a rack (and a quandle) was a group G with a $*$ operation given by conjugation, that is, $a*b = b^{-1}ab$ (Conway thought of this as a "wrack" of a group). Another example, considered already in Conway–Wraith correspondence [21], is defined for any group with $a*b = ba^{-1}b$ and called by Joyce (after Bruck [8]) a core quandle. This example, for a group, H, abelian was already studied by Takasaki so we call it Takasaki kei (or quandle) and denote by $T(H)$ (in abelian notation we have $a*b = 2b-a$), compare [73]. $T(Z_n)$ is often called a dihedral quandle and denoted by R_n; it can be interpreted as composed of reflections of the dihedral group D_{2n} (we can mention that rack and quandle homology of $T(Z_n)$ for prime n has been computed in [20, 74, 77]).

More general examples still starting from a group are given in Joyce paper [56]:

Example 2.3. Let G be a group and $t : G \to G$ a group homomorphism, then we have the following spindle structures on G:

(1) $a *_1 b = t(ab^{-1})b$.
(2) $a *_2 b = t(b^{-1}a)b$.
(3) If t is invertible both examples give quandles where $\bar{*}_1$ and $\bar{*}_2$ are given by the formulas:
   (i) $a \bar{*}_1 b = t^{-1}(ab^{-1})b$ thus $\bar{*}_1$ yielded by the automorphism t is equal to $*_1$ yielded by the automorphism t^{-1}.
   (ii) $a \bar{*}_2 b = bt^{-1}(ab^{-1})$, e.g. we check that $(a *_2 b)\bar{*}_2 b = bt^{-1}((t(b^{-1}a)bb^{-1})) = a$. This example is related to the fundamental group of cyclic (branched) covers of S^3 along a link. Locally, at every crossing we have relations

[g]Mituhisa Takasaki worked at Harbin Technical University, likely as an assistant to Kôshichi Toyoda. Both perished when Red army entered Harbin in August 1945.

$C = \tau^{-1}(B^{-1}A)B$ and $A = B\tau(CB^{-1})$, as illustrated in Fig. 4 [28, 82, 90].

(4) If G is an abelian group both examples lead to the same spindle called Alexander spindle (for t invertible, Alexander quandle). In abelian notation we get $a * b = ta - tb + b = (1-t)b + ta$. This two-sided distributive structure was already considered in 1929 by Burstin[h] and Mayer[i] [12].

(5) If $t = -1$ we get $a * b = 2b - a$ and this structure, as mentioned before, was the main example of Kei by Takasaki so we denote it by $T(G)$.

(6) $a *_3 b = t(ba^{-1})b$ with $t^2 = t$. It is a quandle if and only if $t = \mathrm{Id}$ in which case we get a quandle called the core quandle of G.

Definition 2.2 describes properties of an individual magma $(X; *)$. It is also useful to consider subsets or submonoids of $Bin(X)$ satisfying the related conditions (compare [31, 71, 84, 96]).

Definition 2.4. We say that a subset $S \subset Bin(X)$ is a distributive set if all pairs of elements $*_\alpha, *_\beta \in S$ are right distributive, that is, $(a *_\alpha b) *_\beta c = (a *_\beta c) *_\alpha (b *_\beta c)$ (we allow $*_\alpha = *_\beta$).

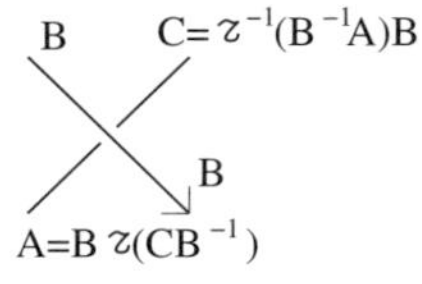

Fig. 4. Relations for cyclic covering; see Example 2.3(3).

[h]Celestyn Burstin (1888–1938) was born in Tarnopol, where he obtained "Matura" in 1907, he moved to Vienna where in 1911 he completed university. In 1929, he moved to Minsk where he was a member of the Belarusian National Academy of Sciences, and a Director of the Institute of Mathematics of the Academy. In December 1937, he was arrested on suspicion of activity as a spy for Poland and Austria. He died in October 1938, when interrogated in a prison in Minsk ("Minskaja Tjurma"); he was rehabilitated March 2, 1956 [9–11, 66, 70].

[i]Walter Mayer (1887–1948) is well known for Mayer–Vietoris sequence and for being assistant to A. Einstein at Institute for Advanced Study, Princeton [53].

(i) The pair $(X; S)$ is called a multi-shelf if S is a distributive set. If S is additionally a submonoid (respectively subgroup) of $Bin(X)$, we say that it is a distributive monoid (respectively group).
(ii) If $S \subset Bin(X)$ is a distributive set such that each $*$ in S satisfies the idempotent condition, we call $(X; S)$ a multi-spindle.
(iii) We say that $(X; S)$ is a multi-rack if S is a distributive set, and all elements of S are invertible.
(iv) We say that $(X; S)$ is a multi-quandle if S is a distributive set, and elements of S are invertible and satisfy the idempotent condition.
(v) We say that $(X; S)$ is a multi-kei if it is a multi-quandle with $** = *_0$ for any $* \in S$. Notice that if $*_1^2 = *_0$ and $*_2^2 = *_0$ then $(*_1 *_2)^2 = *_0$; more generally if $*_1^n = *_0$ and $*_2^n = *_0$ then $(*_1 *_2)^n = *_0$. This follows from the fact that elements of a multi-quandle commute pairwise (this was observed by Jabłonowski [84]).

Proposition 2.5. ([84])
(i) *If S is a distributive set and $* \in S$ is invertible, then $S \cup \{\bar{*}\}$ is also a distributive set.*
(ii) *If S is a distributive set and $M(S)$ is the monoid generated by S, then $M(S)$ is a distributive monoid.*
(iii) *If S is a distributive set of invertible operations and $G(S)$ is the group generated by S, then $G(S)$ is a distributive group.*

We show, after Mezera [68], the fact that any group can be embedded in $Bin(X)$ for some X, in particular the regular embedding of G in $Bin(G)$ is given by $g \to *_g$ with $a *_g b = ab^{-1}gb$ (compare [57] and Example X.3.15 of [32] where the operation $a *_g b = ab^{-1}gb$ is called a half-conjugacy). The expression $ab^{-1}gb$ was also discussed with respect to free rack by Fenn and Rourke (compare Remark 8.2).

Proposition 2.5 has its analogue for entropic magmas (that is magmas for which $(a*b)*(c*d) = (a*c)*(b*d)$). More precisely, we say that a subset $S \in Bin(X)$ is an entropic set if for any $*_\alpha, *_\beta \in S$

we have the entropic condition:

$(a *_\alpha b) *_\beta (c *_\alpha d) = (a *_\beta c) *_\alpha (b *_\beta d)$. Then we have:

Proposition 2.6. ([76])

(i) *If S is an entropic set and $* \in S$ is invertible, then $S \cup \{\bar{*}\}$ is also an entropic set.*
(ii) *If S is an entropic set and $M(S)$ is the monoid generated by S, then $M(S)$ is an entropic monoid.*
(iii) *If S is an entropic set of invertible operations and $G(S)$ is the group generated by S, then $G(S)$ is an entropic group.*

In the next section we consider homology theory of various magmas, it is useful here to define, for any magma $(X; *)$ a supporting structure which we call an X-set (it is an old concept for (semi)groups and for quandles it was first considered by Kamada).

Definition 2.7. Let $(X; *)$ be a magma and E a set. We say that E is an X-set (or right X-set) if there is a function (right action) $*_E : E \times X \to E$. In general we do not put any conditions on $*_E$ but if our magma satisfies some conditions (e.g. associativity or distributivity) then $*_E$ should satisfy some related conditions. In particular, we will look for a magma structure on $X \sqcup E$ having similar structure. (See examples in the next section.)

In the following few sections we discuss various homology theories for magmas (e.g. associative or distributive). In broad approach we follow [84] but we stress the use of X-sets in our definitions.

3. Homology of Magmas

We survey in this section various homology theories, starting from homology of abstract simplicial complexes, then we extract (old and new) properties to define a presimplicial module and a (weak) simplicial module. Further we give two examples of homology for associative structures (semigroups), and, an important example of homology for right self-distributive structures (RDS or shelves) in knot theory. Later we go back to very general notion of homology of a small category with coefficient in a functor to R-Mod, and recall the notion of a geometric realization in the case of a (pre)simplicial set, and (pre)cubic set. Reader interested only in distributive homology can go directly to Sec. 6.

3.1. *Homology of abstract simplicial complexes*

Our goal is to introduce homology of distributive magmas but to keep a historical perspective we start with the standard (oriented and ordered) homology of abstract simplicial complexes as they provide the framework for all homology we consider.

Definition 3.1. The abstract simplicial complex $\mathcal{K} = (V, P)$ is a pair of sets where $V = V(\mathcal{K})$ is called a set of vertices and $P(\mathcal{K}) = P \subset 2^V$, called the set of simplexes of $\mathcal{K}$ and it satisfies: elements of P are finite subsets of V, include all one-element subsets, and if $s' \subset s \in P$ then also $s' \in P$ (that is a subsimplex of a simplex is a simplex).[j] A simplex of $n+1$ vertices is called n-dimensional simplex, or succinctly, n-simplex (we write $s = \{v_{i_0}, v_{i_1}, \ldots, v_{i_n}\}$). We define $dim(\mathcal{K})$ as the maximal dimension of a simplex in $\mathcal{K}$ (may be ∞ if there is no bound).

We consider, additionally, the category of abstract simplicial complexes with a class of objects composed of abstract simplicial complexes. $Mor(K_1, K_2)$ is the set of maps from $V(K_1)$ to $V(K_2)$ which send a simplex to a simplex (that is if $f \in Mor(K_1, K_2)$, $s \in P(K_1)$, then $f(s) \in P(K_2)$).

We recall here three classical (equivalent) definitions of a homology of an abstract simplicial complex: ordered, normalized ordered, and oriented.

Definition 3.2. Recall that a chain complex $\mathcal{C}$ is a sequence of modules over a fixed ring k (here always commutative with identity),

$$\mathcal{C} : \cdots \xrightarrow{\partial_{n+2}} C_{n+1} \xrightarrow{\partial_{n+1}} C_n \xrightarrow{\partial_n} C_{n-1} \xrightarrow{\partial_{n-1}} \cdots$$

[j]Usually we do not allow $\emptyset$ as a simplex, but in some situations it is convenient to allow also an empty simplex, say of dimension -1; it will lead naturally to augmented chain complexes.

such that $\partial_n \partial_{n+1} = 0$ (succinctly $\partial^2 = 0$). Thus we have $\mathrm{im}\, \partial_{n+1} \subset \ker \partial_n$, and we define homology $H_n(\mathcal{C}) = \ker \partial_n / \mathrm{im}\, \partial_{n+1}$.

Now for an abstract simplicial complex $K = (V, P)$ one defines:

(I) (Ordered homology)
Consider a chain complex $\mathcal{C}^{ord}$ with k-modules $C_n^{ord} = C_n^{Ord}(\mathcal{C})$ a submodule of kV^{n+1} generated by all sequences $(x_0, x_1, \ldots, x_n)$, allowing repetitions, such that the set $\{x_0, x_1, \ldots, x_n\}$ is a simplex in P (possibly of dimension smaller from n). The boundary operation is given on the basis by:

$$\begin{aligned}\partial(x_0, x_1, \ldots, x_n) &= \sum_{i=0}^{n} (-1)^i d_i \\ &= \sum_{i=0}^{n} (-1)^i (x_0, \ldots, x_{i-1}, x_{i+1}, \ldots, x_n).\end{aligned}$$

The ordered homology of $\mathcal{K}$ is defined

$$H_n^{ord}(\mathcal{K}, k) = \ker \partial_n / \mathrm{im}\, \partial_{n+1}.$$

If $k = \mathbb{Z}$ we write $H_n^{ord}(\mathcal{K})$.
Notices that the maps $d_i: C_n^{ord} \to C_{n-1}^{ord}$, $(0 \leq i \leq n)$, $d_i(x_0, x_1, \ldots, x_n) = (x_0, \ldots, x_{i-1}, x_{i+1}, \ldots, x_n)$, called the face maps, satisfy:

$$(1)\ d_i d_j = d_{j-1} d_i \quad \text{for any } i < j.$$

The system (C_n, d_i) satisfying the above equality is called a presimplicial module[k] and if we limit ourselves to (V^{n+1}, d_i) it is called a presimplicial set (compare Definition 3.3). The important basic observation is that if (C_n, d_i) is a presimplicial module then (C_n, ∂_n), for $\partial_n = \sum_{i=0}^{n} (-1)^i d_i$, is a chain complex.

Motivation for the boundary operation: it is coming from the geometrical realization of an abstract simplicial complex as illustrated below (the general setting of geometric realization of

---

[k]The concept was introduced in 1950 by Eilenberg and Zilber under the name *semi-simplicial complex*, [36].

a simplicial set is discussed in Sec. 13):

$\partial(x_0,x_1,x_2) = \partial($ 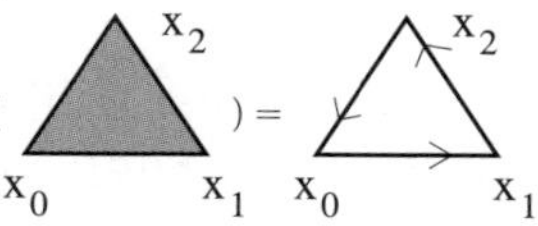 $) =$

$= (x_1, x_2) - (x_0, x_2) + (x_0, x_1).$

(II) (Normalized ordered homology).
The ordered chain complex allows degenerate simplexes (when vertices repeat, in particular neighboring vertices repeat). We define i-degeneracy maps $s_i : C_n \to C_{n+1}$ $(0 \leq i \leq n)$ by $s_i(x_0, \ldots, x_n) = (x_0, \ldots, x_{i-1}, x_i, x_i, x_{i+1}, \ldots, x_n)$. We can check here easily that:

$$(2) \quad s_i s_j = s_{j+1} s_i, \qquad 0 \leq i \leq j \leq n,$$

$$(3) \quad d_i s_j = \begin{cases} s_{j-1} d_i & \text{if } i < j \\ s_j d_{i-1} & \text{if } i > j+1 \end{cases}$$

$$(4) \quad d_i s_i = d_{i+1} s_i = \mathrm{Id}_{C_n}.$$

(C_n, d_i, s_i) satisfying properties (1)–(4) is called a simplicial module. The notion was introduced by Eilenberg and Zilber in 1950 under the name of complete semi-simplicial complex [36]. It is convenient to rephrase the definition so that it can be used for any category:

Definition 3.3. Consider a category $\mathcal{C}$, the sequence of objects X_n, $n \geq 0$ and for any n morphisms d_i, s_i, $0 \leq i \leq n$, $d_i \in Mor(X_n, X_{n-1})$, and $s_i \in Mor(X_n, X_{n+1})$. We call $(X_n; d_i, s_i)$ a simplicial category (e.g. simplicial set, simplicial module, or simplicial space) if the following four conditions hold.

(1) $d_i d_j = d_{j-1} d_i$ for $i < j$,
(2) $s_i s_j = s_{j+1} s_i$ for $i \leq j$,
(3)

$$s_i s_j = s_{j+1} s_i, \quad 0 \leq i \leq j \leq n,$$

$$d_i s_j = \begin{cases} s_{j-1} d_i & \text{if } i < j \\ s_j d_{i-1} & \text{if } i > j+1 \end{cases}$$

(4) $d_i s_i = d_{i+1} s_i = \mathrm{Id}_{X_n}$.

Eilenberg and MacLane proved in 1947 that the degenerate part of a presimplicial module is an acyclic chain complex (it has trivial homology) [34] (the proof was more specific but the method worked for all simplicial modules defined only three years later). We devote Sec. 3.2 to the proof, after [61], paying attention to which axioms of a simplicial module are used. In particular, axiom (4) cannot be ignored as the degenerate chain complex of quandles, which satisfies property (4) only partially and often has nontrivial homology.

Now back to normalized ordered homology:
Consider submodules $C_n^D(\mathcal{K})$ (named degenerated modules) and defined by

$$C_n^D = span(s_0(C_{n-1}), s_1(C_{n-1}), \ldots, s_{n-1}(C_{n-1})).$$

One checks that (C_n^D, ∂_n) is a subchain complex of $C_n^{ord}(\mathcal{K})$. Details are given in Sec. 3.2, where it is also proved that this chain complex is acyclic. We have also quotient chain complex, called normalized ordered chain complex with $C_n^N(\mathcal{K}) = C_n(\mathcal{K})/C_n^D(\mathcal{K})$. As homology of $C_n^D(\mathcal{K})$ are trivial, we have isomorphism:

$$H_n^{ord}(\mathcal{K}, k) \to H_n^N(\mathcal{K}, k).$$

(III) (Oriented homology). We can consider smaller chain complex giving the same homology of $\mathcal{K}$ by taking the quotient of $C^N(\mathcal{K})$ and considering only "oriented simplexes". Formally, let $\bar{C}_n(\mathcal{K})$ be a submodule of $C^N(\mathcal{K})$ generated by "transposition symmetrizers" $(x_0, \ldots, x_{i-1}, x_i, x_{i+1}, x_{i+2}, \ldots, x_n) + (x_0, \ldots, x_{i-1}, x_{i+1}, x_i, x_{i+2}, \ldots, x_n)$. One checks directly that $(\bar{C}_n(\mathcal{K}), \partial_n)$ is a subchain complex of $C^N(\mathcal{K})$. The oriented chain complex is the quotient: $C^{ori}(\mathcal{K}) = C^N(\mathcal{K})/\bar{C}_n(\mathcal{K})$. It requires some effort to prove that the quotient map $f : C^N(\mathcal{K}) \to C^{ori}(\mathcal{K})$ is a chain equivalence and thus $f_* : H^N(\mathcal{K}) \to H^{ori}(\mathcal{K})$ is an isomorphism of homology modules. From this we conclude that all three definitions, ordered, normalized ordered, and oriented of homology of an abstract simplicial complex give the same result.

To have more concrete view of oriented chain complex and oriented homology, we order vertices V of $\mathcal{K}$ and interpret the chain group $C_n^{ori}(\mathcal{C})$ as a subgroup of ZV^{n+1} freely generated by n-dimensional simplexes, $(x_0, x_1, \ldots, x_n)$ (we assume that $x_0 < x_1 < \cdots < x_n$ in our ordering). In essence, with ordering, we are able to choose representatives of equivalence classes in $C_n^{ori}(\mathcal{K})$ and the boundary operation

$$\partial(x_0, x_1, \ldots, x_n) = \sum_{i=0}^{n}(-1)^i d_i,$$

where

$$d_i(x_0, x_1, \ldots, x_n) = (x_0, \ldots, x_{i-1}, x_{i+1}, \ldots, x_n),$$

preserves our choice so with given ordering we have a split chain map $g : C_n^{ori}(\mathcal{K}) \to C_n^N(\mathcal{K})$. The quotient map, with our ordering can be written as $f_n((x_0, \ldots, x_n) = (-1)^{|\pi|}(x'_0, \ldots, x'_n)$ where $\pi \in S_{n+1}$ is the permutation such that $x'_i = x_{\pi(i)}$ and $x'_i < x'_{i+1}$ (if $x_i = x_j$ for some $i \neq j$ then we put $f(x_0, \ldots, x_n) = 0$). Immediately, we have $f_n g_n = \mathrm{Id}_{C_n^{ori}(\mathcal{K})}$. The proof that gf is chain homotopic to identity on $C_n^N(\mathcal{K})$ (and so f is chain equivalence) requires more effort.[1]

---

[1]The standard Eilenberg–MacLane proof uses acyclic modules method [35, 103], however in our case one can give shorter proof (the idea is still that of Eilenberg–MacLane): consider the chain map $f : \mathcal{C}^N \to \mathcal{C}^{ori}$ given by $f_n(x_0, \ldots, x_n) = (-1)^{|\pi|}(x'_0, \ldots, x'_n)$ where $\pi \in S_{n+1}$ is the permutation such that $x'_i = x_{\pi(i)}$ and $x'_i < x'_{i+1}$ (if $x_i = x_j$ for some $i \neq j$ then we put $f(x_0, \ldots, x_n) = 0$). The map $g : \mathcal{C}^{ori} \to \mathcal{C}^N$ is defined to be embedding; therefore $fg = \mathrm{Id}_{\mathcal{C}}$. We show that gf induces identity on homology of $\mathcal{C}^N$. We construct a chain homotopy between gf and the identity inductively, starting from $F_0 = 0$. The main ingredient of the proof is the fact that for a simplex $s = (x_0, \ldots, x_n)$ the subchain complex $\bar{s} = (s, 2^s)$ of $\mathcal{C}^N$ is acyclic ($H_n^N(\bar{s}) = 0$ for $n > 0$ and $H_0^N(\bar{s}) = k$).

Step n: assume that $F_{n-1}, \ldots, F_0$ are already constructed and we construct a map $F_n : C_n^N \to C_{n+1}^N$ such that $\partial_{n+1}F_n = -F_{n-1}\partial_n + \mathrm{Id} - gf$.

We compute: that

$$\begin{aligned}\partial_n(-F_{n-1}\partial_n + \mathrm{Id} - gf) &= -(\partial_n F_{n-1})\partial_n + \partial_n - \partial_n(gf)\\ &= F_{n-2}\partial_{n-1}\partial_n - \partial_n + (gf)\partial_n\partial_n - \partial_n(gf) = 0.\end{aligned}$$

3.2. *Degenerate subcomplex*

Consider a presimplicial module (C_n, d_i) with degenerate maps s_i. We define degenerate modules

$$C_n^D = span(s_0(C_{n-1}), s_1(C_{n-1}), \dots, s_{n-1}(C_{n-1})).$$

We check which conditions are needed so that (C_n^D, d_i) is a sub-presimplicial module of (C_n, d_i). We have:

$$\begin{aligned}
\partial_n s_p &= \sum_{i=0}^{n+1}(-1)^i d_i s_p \\
&= \sum_{i=0}^{p-1}(-1)^i d_i s_p + (-1)^p(d_p s_p - d_{p-1}s_p) + \sum_{i=p+2}^{n+1}(-1)^i d_i s_p \\
&\overset{(3)}{=} \sum_{i=0}^{p-1}(-1)^i s_{p-1} d_i + (-1)^p(d_p s_p - d_{p+1}s_p) + \sum_{i=p+2}^{n+1}(-1)^i s_p d_{i-1} \\
&\overset{(4')}{=} \sum_{i=0}^{p-1}(-1)^i s_{p-1} d_i + \sum_{i=p+2}^{n+1}(-1)^i s_p d_{i-1} \in C_n^D,
\end{aligned}$$

where the $(4')$ is the condition:

$$(4')\ d_p s_p = d_{p+1} s_p \quad \text{for any } p \leq n.$$

If (C_n, d_i, s_i) satisfies conditions (1), (2), (3) and $(4)'$ it is called a weak simplicial module [84]. As condition (2) was not used in calculation it is also useful to consider (C_n, d_i, s_i) satisfying conditions (1), (3) and $(4')$, we call this a weak-pseudo-simplicial module.

We strengthen the above calculation by considering the sequence of modules $F_n^i = span((s_0(C_{n-1}), s_1(C_{n-1}), \dots, s_i(C_{n-1}))$ and the

---

Because chain complex $C_n^N(\bar{s})$ is exact at place n and $-F_{n-1}\partial_n + Id - gf$ is in the kernel of this chain complex, it is also in the image, say, $\partial_{n+1}c_{n+1} = -F_{n-1}\partial_n + \mathrm{Id} - gf$. Then we declare $F_n(s) = c_{n+1}$. In fact here c_{n+1} can be obtained by putting any, fixed, vertex of s in front of $(-F_{n-1}\partial_n + \mathrm{Id} - gf)$. Such constructed F_n satisfies $\partial_{n+1}F_n + F_{n-1}\partial_n = \mathrm{Id} - gf$. Our map is well-defined as we constructed it on the basis of C_n^N.

filtration:

$$0 \subset F_n^0 \subset F_n^1 \subset \cdots \subset F_n^{n-1} = C_n^D.$$

Our calculation gives $\partial_n(s_p(C_{n-1})) \subset span(s_{p-1}(C_{n-2}), s_p(C_{n-2}))$ and consequently:

Corollary 3.4. *If (C_n, d_i, s_i) is a weak-pseudo-simplicial module, then ∂_n is filtration preserving, that is $\partial_n(F_n^p) \subset F_{n-1}^p$.*

We will prove now the Eilenberg–MacLane theorem that the degenerate complex (C_n^D, ∂_n) is acyclic, watching on the way which axioms are used. For a filtration (F_n^p) the associated graded module is defined to be $\{Gr_n^p = F_n^p/F_n^{p-1}\}$. We prove first that the chain complex $\{Gr_n^p\}$ is acyclic for any p.

Lemma 3.5. *Let (C_n, d_i, s_i) satisfies the conditions* (1), (2″), (3) *and* (4) *where*

$$(2'')\ s_{p-1}s_{p-1} = s_p s_{p-1} \quad \textit{for every } 0 < p \leq n.$$

We call such (C_n, d_i, s_i) a co-almost-simplicial module. Then the chain complex $\{Gr_n^p = F_n^p/F_n^{p-1}\}$ is acyclic, in particular $H_n(\{Gr_n^p\}) = 0$, and homologies of F_n^p and F_n^{p-1} are isomorphic.

Proof. The classical idea of Eilenberg and MacLane is to use the degenerate map s_p as a chain homotopy (we follow [61]): It suffices to show that $(\partial s_p + s_p\partial)s_p = (-1)^p s_p$ modulo $s_{p-1}C_{n-1}$, so the map s_p is a chain homotopy between $(-1)^p$ Id and the zero map on Gr_n^p. In the calculation we stress which axioms are used:

$$\begin{aligned}
&(\partial s_p + s_p\partial)s_p \\
&\quad = \left(\sum_{i=0}^{n+1}(-1)^i d_i s_p + s_p \sum_{i=0}^{n}(-1)^i d_i\right) s_p \\
&\quad = \sum_{i=0}^{p-1}(-1)^i d_i s_p s_p + (-1)^p(d_p s_p - d_{p+1}s_p)s_p + \sum_{i=p+2}^{n+1}(-1)^i d_i s_p s_p \\
&\qquad + \sum_{i=0}^{p-1}(-1)^i s_p d_i s_p + (-1)^p s_p(d_p s_p - d_{p+1}s_p) + \sum_{i=p+2}^{n}(-1)^i s_p d_i s_p
\end{aligned}$$

$$\stackrel{(3)}{=} \sum_{i=0}^{p-1}(-1)^i s_{p-1}s_{p-1}d_i + (-1)^p(d_p s_p - d_{p+1}s_p)s_p$$

$$+ \sum_{i=p+2}^{n+1}(-1)^i s_p d_{i-1}s_p + \sum_{i=0}^{p-1}(-1)^i s_p s_{p-1}d_i$$

$$+ (-1)^p s_p(d_p s_p - d_{p+1}s_p) + \sum_{i=p+2}^{n}(-1)^i s_p d_i s_p$$

$$= \sum_{i=0}^{p-1}(-1)^i s_{p-1}s_{p-1}d_i + (-1)^p(d_p s_p - d_{p+1}s_p)s_p + (-1)^p s_p d_{p+1}s_p$$

$$+ \sum_{i=0}^{p-1}(-1)^i s_p s_{p-1}d_i + (-1)^p s_p(d_p s_p - d_{p+1}s_p)$$

$$\stackrel{d_p s_p = d_{p+1}s_p}{=} \sum_{i=0}^{p-1}(-1)^i s_{p-1}s_{p-1}d_i$$

$$+ \sum_{i=0}^{p-1}(-1)^i s_p s_{p-1}d_i + (-1)^p s_p d_{p+1}s_p$$

$$\stackrel{(2)}{=} 2\sum_{i=0}^{p-1}(-1)^i s_{p-1}s_{p-1}d_i + (-1)^p s_p d_{p+1}s_p$$

$$\stackrel{\text{mod } 2s_{p-1}M_{n-1}}{=} (-1)^p s_p d_{p+1}s_p \stackrel{d_{p+1}s_p = Id}{=} (-1)^p s_p.$$

Now consider the short exact sequence $0 \to F^{p-1} \to F^p \to F^p/F^{p-1} \to 0$ and the corresponding long exact sequence of homology:

$$\cdots \to H_{n+1}(F^p/F^{p-1})$$
$$\to H_n(F^{p-1}) \to H_n(F^p) \to H_n(F^p/F^{p-1}) \to \cdots.$$

Thus because homology of F^p/F^{p-1} is trivial we obtain isomorphism $H_n(F^{p-1}) \to H_n(F^p)$. In conclusion, by induction on p we get the Eilenberg–MacLane result: $H_n^D(C) = 0$. □

From our proof follows that working modulo $2s_{p-1}M_{n-1}$, e.g. modulo 2, gives directly $H_n(F^p) = 0$. Also from axiom (2) we took only $s_p s_{p-1} = s_{p-1}s_{p-1}$ that is axiom (2″).

The above consideration does not work for a distributive case (axiom (4) usually does not hold as explained in Sec. 6 (see [24, 88, 92])). We proved however that the degenerate part of quandle homology can be obtained from the normalized one via Künneth type formula, see [89]).

3.3. *Bicomplex for a degenerate subcomplex* C_n^D

One more important observation follows from our calculations. If (C_n, d_i, s_i) is a weak simplicial module[m] (i.e. conditions (1)–(3), (4′) hold then the formula

$$\partial_n s_p = \sum_{i=0}^{p-1}(-1)^i s_{p-1}d_i + \sum_{i=p+2}^{n}(-1)^i s_p d_{i-1}$$

allows us to define a bicomplex with entries $E^0_{p,q} = Gr_{n,p} = F_n^p/F_n^{p-1}$, $n = p+q$, with horizontal and vertical boundary operation: $\partial^h_{p,q} = \sum_{i=0}^{p-1}(-1)^i$ and $\partial^v_{p,q} = \sum_{i=p+2}^{n}(-1)^i s_p d_{i-1}$ with $\partial^h_{p,q-1}\partial^v_{p,q} = -\partial^v_{p-1,q}\partial^h_{p,q} : E^0_{p,q} \to E^0_{p-1,q-1}$; see Fig. 5.

The bicomplex $(E^0_{p,q}, \partial^v, \partial^h)$ yields a spectral sequence, in fact two spectral sequences: starting from columns, that is ${}^cE^1_{pq} = \frac{\ker(\partial^v : E^0_{pq} \to E^0_{p,q-1})}{\mathrm{im}(\partial^v : E^0_{p,q+1} \to E^0_{pq})}$, and the spectral sequence starting from

$$\begin{array}{ccccccc}
 & \downarrow \partial^v & & \downarrow \partial^v & & \downarrow \partial^v & \\
\cdots \overset{d^h}{\leftarrow} & E^0_{p-1,q+1} & \overset{\partial^h}{\leftarrow} & E^0_{p,q+1} & \overset{\partial^h}{\leftarrow} & E^0_{p+1,q+1} & \overset{\partial^h}{\leftarrow} \cdots \\
 & \downarrow \partial^v & & \downarrow \partial^v & & \downarrow \partial^v & \\
\cdots \overset{\partial^h}{\leftarrow} & E^0_{p-1,q} & \overset{\partial^h}{\leftarrow} & E^0_{p,q} & \overset{\partial^h}{\leftarrow} & E^0_{p+1,q} & \overset{\partial^h}{\leftarrow} \cdots \\
 & \downarrow \partial^v & & \downarrow \partial^v & & \downarrow \partial^v & \\
\cdots \overset{\partial^h}{\leftarrow} & E^0_{p-1,q-1} & \overset{\partial^h}{\leftarrow} & E^0_{p,q-1} & \overset{\partial^h}{\leftarrow} & E^0_{p+1,q-1} & \overset{\partial^h}{\leftarrow} \cdots \\
 & \downarrow \partial^v & & \downarrow \partial^v & & \downarrow \partial^v &
\end{array}$$

Fig. 5. Bicomplex $(E^0_{p,q}, \partial^v, \partial^h)$.

[m]In fact a pseudo weak simplicial module suffices, i.e. conditions (1), (3) and (4′).

rows: ${}^{r}E^{1}_{pq} = \frac{\ker(\partial^h : E^0_{pq} \to E^0_{p-1,q})}{\mathrm{im}(\partial^h : E^0_{p+1,q} \to E^0_{pq})}$ which can be used to analyze homology of $Gr(C^D)$ and (C_n), see [89] for an application in the distributive case.

3.4. *Homology with coefficients in a k-Mod functor*

Each individual abstract simplicial complex $K = (V, \mathcal{P})$ is a small category[n] with simplexes as objects and inclusions of simplexes, $s \subset s'$, as morphisms. As usual K^{op} will denote the opposite category so restrictions, $s \supset s'$, are morphisms, more precisely $Mor_{K^{op}}(s, s')$ is empty if s does not contain s' and otherwise $Mor_{K^{op}}(s, s')$ has one morphism denoted by $(s \supset s')$. Now for any (covariant) functor $\mathcal{F} : K^{op} \to k$-Mod, where k-Mod is a category of modules over a commutative ring k, we can define oriented homology $H^{ori}_n(K, \mathcal{F})$ of an abstract simplicial complex K with coefficients in $\mathcal{F}$, as follows:

Definition 3.6. Let $K = (V, P)$ be an abstract simplicial complex with ordered vertices[o] and $\mathcal{F} : K^{op} \to k$-Mod a functor. We define the presimplicial module $(C^{ori}_n(K, \mathcal{F}), d_i)$ as follows:

$$C^{ori}_n(K, \mathcal{F}) = \bigoplus_{dim(s)=n} \mathcal{F}(s)$$

the face map $d_i : C^{ori}_n(K, \mathcal{F}) \to C^{ori}_{n-1}(K, \mathcal{F})$ is defined by

$$d_i = \mathcal{F}(s \supset (s - x_i))$$

where

$$s = (x_0, \ldots, x_n) \quad \text{and} \quad x_i < x_{i+1}$$

[n]Category is called small if objects form a set.

[o]It suffices to have V partially ordered as long as for any simplex $s = (x_0, \ldots, x_n)$ the partial order on V restricts to linear order on vertices of s. Even better we do not need a partial order, it suffices that vertices of every simplex are ordered in such a way that if $s_1 \subset s_2$ then the ordering of vertices of s_1 is a restriction of the ordering of vertices of s_2.

as usually for presimplicial modules $\partial_n = \sum_{i=1}^{n}(-1)^i d_i$ and $(C_n^{ori}(K,\mathcal{F}),\partial_i)$ is a chain complex whose homology is denoted by $H_n^{ori}(K,\mathcal{F})$.

The above definition can be thought of as a twisted version of an oriented homology of abstract simplicial complexes. Similarly we can define ordered homology of $(K,\mathcal{F})$ but oriented and ordered homology with coefficient in a functor are not necessarily isomorphic.

Definition 3.6 is related to more general Definition 3.7 on homology of a small category with a functor coefficient, usually thought to be first given by [108], who in turn refers to the earlier paper [30] in the case of the category of posets.

Definition 3.7. Let $\mathcal{P}$ be as small category (i.e. objects, $P = Ob(\mathcal{P})$ form a set), and let $\mathcal{F} : \mathcal{P} \to k$-Mod be a functor from $\mathcal{P}$ to the category of modules over a commutative ring k. We call the sequence of objects and functors, $x_0 \overset{f_0}{\to} x_1 \overset{f_1}{\to} \dots \overset{f_{n-1}}{\to} x_n$ an n-chain (more formally n-chain in the nerve of the category). We define the chain complex $C_*(\mathcal{P},\mathcal{F})$ as follows:

$$C_n = \bigoplus_{x_0 \overset{f_0}{\to} x_1 \overset{f_1}{\to} \dots \overset{f_{n-1}}{\to} x_n} \mathcal{F}(x_0),$$

where the sum is taken over all n-chains.

The boundary operation $\partial_n : C_n(\mathcal{P},\mathcal{F}) \to C_{n+1}(\mathcal{P},\mathcal{F})$ is an alternative sum of face maps, $\partial_n = \sum_{i=0}^{n}(-1)^i d_i$, where d_i are given by:

$$\begin{aligned} &d_0(\lambda; x_0 \overset{f_0}{\to} x_1 \overset{f_1}{\to} \cdots \overset{f_{n-1}}{\to} x_n) \\ &\quad = (\mathcal{F}(x_0 \overset{f_0}{\to} x_1)(\lambda); x_1 \overset{f_1}{\to} \cdots \overset{f_{n-1}}{\to} x_n), \\ &d_i(\lambda; x_0 \overset{f_0}{\to} x_1 \overset{f_1}{\to} \cdots \overset{f_{n-1}}{\to} x_n) \\ &\quad = (\lambda; x_0 \overset{f_0}{\to} x_1 \overset{f_1}{\to} \cdots \to x_{i-1} \overset{f_i f_{i-1}}{\to} x_{i+1} \to \cdots \overset{f_{n-1}}{\to} x_n) \end{aligned}$$

for $0 < i < n$, and

$$\begin{aligned} &d_n(\lambda; x_0 \overset{f_0}{\to} x_1 \overset{f_1}{\to} \cdots \overset{f_{n-1}}{\to} x_n) \\ &\quad = (\lambda; x_0 \overset{f_0}{\to} x_1 \overset{f_1}{\to} \cdots \overset{f_{n-2}}{\to} x_{n-1}). \end{aligned}$$

We denote by $H_n(\mathcal{P}, \mathcal{F})$ the homology yielded by the above chain complex, and call this the homology of a small category $\mathcal{P}$ with coefficients in a functor $\mathcal{F}$.

Similarly, if $\mathcal{F}' : \mathcal{P} \to k$-Mod is a contravariant functor we may define a homology $H_n(\mathcal{P}; \mathcal{F}')$, starting from

$$C_n(\mathcal{P}; \mathcal{F}') = \bigoplus_{x_0 \xrightarrow{f_0} x_1 \xrightarrow{f_1} \cdots \xrightarrow{f_{n-1}} x_n} \mathcal{F}'(x_n).$$

In particular, $d_n(x_0 \xrightarrow{f_0} x_1 \xrightarrow{f_1} \cdots \xrightarrow{f_{n-1}} x_n; \lambda) = (x_0 \xrightarrow{f_0} x_1 \xrightarrow{f_1} \cdots \xrightarrow{f_{n-2}} x_{n-1}); \mathcal{F}'(x_{n-1} \xrightarrow{f_{n-1}} x_n)(\lambda))$, where $\lambda \in \mathcal{F}'(x_n)$. One can also consider both functors, $\mathcal{F}$ and $\mathcal{F}'$ in the definition starting from

$$C_n(\mathcal{P}; \mathcal{F}, \mathcal{F}') = \bigoplus_{x_0 \xrightarrow{f_0} x_1 \xrightarrow{f_1} \cdots \xrightarrow{f_{n-1}} x_n} \mathcal{F}'(x_n) \otimes \mathcal{F}(x_0);$$

compare Definition 4.6. We can also start from a bifunctor $\mathcal{D} : \mathcal{P}^{op} \times \mathcal{P} \to k$-Mod and mimic the definition of the Hochschild homology (Sec. 5) [61].

Remark 3.8. Any subcategory $\mathcal{P}'$ of $\mathcal{P}$ has its chain complex, and homology (we use the functor $\mathcal{F}' = \mathcal{F}/\mathcal{P}'$, that is, the restriction of $\mathcal{F}$ to $\mathcal{P}'$). $C_*(\mathcal{P}', \mathcal{F}')$ is a subchain complex of $C_*(\mathcal{P}, \mathcal{F})$ so we can consider the short exact sequence of chain complexes:

$$0 \to C_n(\mathcal{P}, \mathcal{F}) \to C_n(\mathcal{P}', \mathcal{F}') \to C_n(\mathcal{P}, \mathcal{F})/C_n(\mathcal{P}', \mathcal{F}') \to 0$$

and yielded by it the long exact sequence of homology.

The pair (C_n, d_i) forms a presimplicial module by associativity of morphisms of a category and properties of a functor. More generally we have:

Proposition 3.9. *Let $s_i : C_n \to C_{n+1}$ be a map inserting identity morphism on the ith place in the nth chain of the nerve of the category, that is*

$$s_i((\lambda; x_0 \xrightarrow{f_0} \cdots \xrightarrow{f_{n-1}} x_n) = (\lambda; x_0 \xrightarrow{f_0} x_i \xrightarrow{\mathrm{Id}_{x_i}} x_i \cdots \xrightarrow{f_{n-1}} x_n).$$

Then (C_n, d_i, s_i) is a simplicial module.

The classical example is the homology of a simplicial complex with constant coefficients, that is $\mathcal{F}(s) = k$ and $\mathcal{F}(f) = \mathrm{Id}_k$; in that case we write $H_n(K, \mathcal{F}) = H_n(K, k)$ or just $H_n(K)$ if $k = \mathbb{Z}$. Related to this example is homology of posets: if $\mathcal{P}$ is a small category and for any objects x and y, $Mor(x, y)$ has at most one element and additionally if $Mor(x, y) \neq \emptyset$ and $Mor(y, x) \neq \emptyset$ then $x = y$, then $\mathcal{P}$ is a poset with $x \leq y$ iff $Mor(x, y) \neq \emptyset$.

Another classical example concerns homology of groups, where the category has one object and G morphisms (interpreted as multiplication by elements of G [6], that is the morphism $g : G \to G$ is given by $g(h) = hg$); compare Sec. 4.

More recent example is motivated by Khovanov homology so we call a related functor $\mathcal{F}_{(D,A,M)}$, a Khovanov functor. The functor depends on a choice of a k-Frobenius algebra A, A-Frobenius bimodule M and a link diagram (possibly virtual link, or a link diagram on a surface, L (equivalently we can work with graphs on a surface). Here for simplicity we assume that $M = A$ is an abelian Frobenius k-algebra[p] and D is a classical link diagram.

Definition 3.10. For a link diagram D, let V be the set of its crossings (in some order), and $P = 2^V$. Thus $K = (V, P)$ is a simplex (we allow also the empty, -1-dimensional simplex). Let A be a Frobenius algebra with a multiplication μ and a co-multiplication Δ (e.g. $A = Z[x]/(x^m)$, $\Delta(1) = \sum_{i+j=m-1} x^i \otimes x^j$). We define a functor $\mathcal{F}_{D,A} : K \to k$-Mod as follows. For any $s \in P$ we identify s with a Kauffman state, where $s(v) = 1$ (i.e. ) iff $v \in s$. We denote by D_s the collection of circles obtained from D by smoothing it according to s, and by $|D_s|$ the number of circles in D_s. Then we define $\mathcal{F}(s) = A^{\otimes |D_s|}$. To define $\mathcal{F}(s \supset (s - v_i))$ we first decorate circles of D_s by algebra A, (that is each circle by one copy of A); then we have two cases: (μ) $|D_{s-v_i}| = |D_s| - 1$, thus two circles are glued together when we switch the state at v_i. In this case we

[p] A is a k module with associative and commutative multiplication, μ, with co-associative and co-commutative co-multiplication, Δ, satisfying the Frobenius condition, that is $\Delta\mu = (\mu \times \mathrm{Id})(\mathrm{Id} \times \Delta)$; graphically: 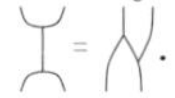. There is no need for unit and counit.

multiply the element associated with glued circles (we use the fact that A is commutative). (Δ) $|D_{s-v_i}| = |D_s| + 1$, thus a circle of D_s is split so we apply to the element of A associated to this circle a co-multiplication (we use the fact that Δ is co-commutative). $\mathcal{F}$ is a functor as A is a commutative Frobenius algebra.

This approach to Khovanov homology was first sketched in [83], where Khovanov homology was connected to Hochschild homology.

It is a classical result that homology of a baricentric subdivision of an abstract simplicial complex is isomorphic to the homology of the complex. This was an ingredient of the original proof of topological invariance of homology. The generalization of the result also holds for a homology of a simplicial complex $\mathcal{K}$ with a coefficient in a functor and the homology of $\mathcal{K}$ treated as a small category. I was informed by S. Betley and Jolanta Słomińska about at least three proofs of the fact, compare [102]. We are writing, with my student Jing Wang detailed survey with the proof following closely the classical proof with constant coefficients (in essence it is another case of acyclic model theorem of Eilenberg and Zilber [36]).

In the next few sections we discuss homology related to various magmas (e.g. associative and right distributive) and look for the common traits, for example presimplicial or simplicial module structure, geometric realization etc.

4. Group Homology of a Semigroup

In the homology of abstract simplicial complexes, the set of vertices, X, has no algebraic structure or, as we will see later, we can treat X as a magma with the trivial operation $*_0$, $x *_0 y = x$. We will now discuss homology of magma $(X, *)$ equipped with some specific structure, e.g. associativity, Jacobi identity, or distributivity.

According to [6]: *The cohomology theory of groups arose from both topological and algebraic sources. The starting point for the topological aspect of the theory was the work of Hurewicz* ([52], *1936*) *on "aspherical spaces". About a year earlier, Hurewicz had* *introduced the higher homotopy groups $\pi_n X$ of a space X ($n \geq 2$). He now singled out for study those path-connected spaces X whose higher homotopy groups are all trivial, but whose fundamental group $\pi = \pi_1 X$ need not be trivial. Such spaces are called aspherical. Hurewicz proved, among other things, that the homotopy type of an aspherical pace X is completely determined by its fundamental group π. ... Hopf* ([51], 1942) ... *expressed $H_2\pi$ in purely algebraic terms ....*

Let $(X, *)$ be a semigroup that is a set with associative binary operation. We associate with $(X, *)$ a presimplicial set, presimplicial module, chain complex, group homology and geometric realization as follows:

Definition 4.1. (i) Let $X_n = X^n$ and $d_i : X_n \to X_{n-1}$ for $0 \leq i \leq n$ is given by:

$$d_0(x_1, x_2, \ldots, x_n) = (x_2, , \ldots, x_n),$$
$$d_i(x_1, \ldots, x_n) = (x_1, \ldots, x_{i-1}, x_i * x_{i+1}, x_{i+2}, \ldots, x_n) \quad \text{for } 0 < i < n,$$
$$d_0(x_1, \ldots, x_{n-1}, x_n) = (x_1, \ldots, x_{n-1}).$$

Then (X_n, d_i) is a presimplicial set.

(ii) If we choose a commutative ring k and consider $C_n = kX^n$ and $d_i : C_n \to C_{n-1}$ the unique extension of the map d_i from (i) then (C_n, d_i) is a presimplicial module.

(iii) If $\partial_n = \sum_{i=0}^{n} (-1)^i d_i$, then (C_n, ∂_n) is a chain complex; its homologies are called group homologies of a semigroup X and denoted by $H_n(X; k)$ or just $H_n(X)$ if $k = \mathbb{Z}$.

(iv) A presimplicial set has a standard geometric realization, BX (as a CW-complex[q]). Thus the semigroup homology has a natural interpretation as a homology of a CW-complex [6].

---

[q] BX can be made into geometric simplicial complex by second baricentric subdivision because BX by the construction is glued from simplexes (such a space is called a Δ-complex in [49]), see Sec. 13, e.g. Definition 13.1.

Definition 4.1 has a classical generalization, when a semigroup $(X, *)$ is augmented by an X-right-semigroup-set E, that is a set with the right action (also denoted by $*$) of X on E such that $(e * a) * b = e * (a * b)$.

Definition 4.2. (i) Let $X_n = E \times X^n$ and $d_i : X_n \to X_{n-1}$ for $0 \leq i \leq n$ is given by:

$$d_0(e, x_1, x_2, \ldots, x_n) = (e * x_1, x_2, \ldots, x_n),$$

$$d_i(e, x_1, \ldots, x_n) = (e, x_1, \ldots, x_{i-1}, x_i * x_{i+1}, x_{i+2}, \ldots, x_n) \quad \text{for } 0 < i < n,$$

$$d_n(e, x_1, \ldots, x_{n-1}, x_n) = (e, x_1, , \ldots, x_{n-1}).$$

Then (X_n, d_i) is a presimplicial set.

(ii) For $C_n = k(E \times X^n)$, (C_n, d_i) is a presimplicial module and (C_n, ∂_n) is a chain complex with homology denoted by $H_n(X, E)$ and geometric realization $B(X, E)$.

If E has one element then we get the case of Definition 4.1.

Definition 4.2 has further generalization if, in addition to a semigroup $(X; *)$ we have the right X-set E_0 and the left X set E_w (here we need $(a * (b * e) = (a * b) * e)$, compare [13], Chap. X.

Definition 4.3. (i) Let $X_n = E_0 \times X^n \times E_w$ and $d_i : X_n \to X_{n-1}$ for $0 \leq i \leq n$ is given by:

$$d_0(e_0, x_1, x_2, \ldots, x_n, e_{n+1}) = (e * x_0, x_2, \ldots, x_n, e_{n+1}),$$

$$d_i(e_0, x_1, \ldots, x_n, e_{n+1}) = (e_0, x_1, \ldots, x_{i-1}, x_i * x_{i+1}, x_{i+2}, \ldots, x_n, e_{n+1}) \quad \text{for } 0 < i < n,$$

$$d_n(e_0, x_1, \ldots, x_{n-1}, x_n, e_{n+1}) = (e_0, x_1, \ldots, x_{n-1}, x_n * e_{n+1}).$$

Then (X_n, d_i) is a presimplicial set. We call this presimplicial set a "two walls" presimplicial set due to visualization of Fig. 6.

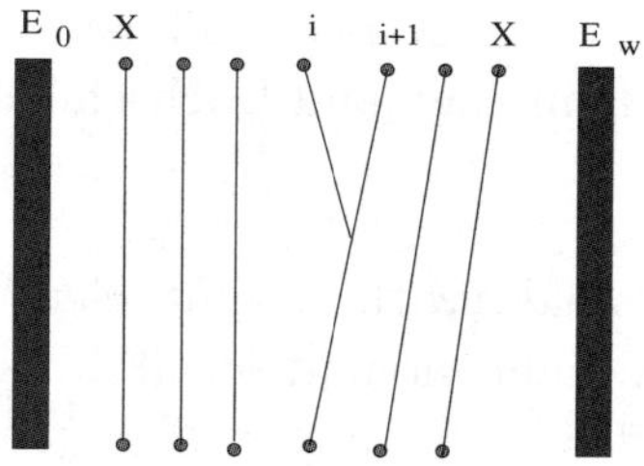

Fig. 6. ith face map in "two walls" presimplicial set for a semigroup.

(ii) For $C_n = k(E_0 \times X^n \times E_w)$, (C_n, d_i) is a presimplicial module and (C_n, ∂_n) is a chain complex with homology denoted by $H_n(X, E_0, E_w)$, and geometric realization $B(X, E_0, E_w)$.
If E_w has one element then we get the case of Definition 4.2.

A version of Definition 4.3 when we assume that $E_0 = E_w$ and $(e_1 * x) * e_2 = e_1 * (x * e_2)$ that is E_0 is an X-biset leads to the Hochschild homology. In particular, for a semigroup $(X; *)$ we have:

Definition 4.4. Let $(X; *)$ be a semigroup and E an X-biset then:

(i) Let $X_n = E \times X^n$ and $d_i : X_n \to X_{n-1}$ for $0 \leq i \leq n$ is given by:

$$d_0(e, x_1, x_2, \ldots, x_n) = (e * x_1, x_2, \ldots, x_n),$$
$$d_i(e, x_1, \ldots, x_n) = (e, x_1, \ldots, x_{i-1}, x_i * x_{i+1}, x_{i+2}, \ldots, x_n) \quad \text{for } 0 < i < n,$$
$$d_n(e, x_1, \ldots, x_{n-1}, x_n) = (x_n * e, x_1, \ldots, x_{n-1}).$$

Then (X_n, d_i) is a presimplicial set.

(ii) For $C_n = k(E \times X^n)$, (C_n, d_i) is a presimplicial module and (C_n, ∂_n) is a chain complex with (Hochschild) homology denoted by $HH_n(X, E)$ and geometric realization $BH(X, E)$.

(iii) If E_0 is an X-right-semigroup-set and E_w is an X-right-semigroup-set then we can take $E = E_w \times E_0$ and E has a natural structure of an X-biset. Thus the concepts of a "two-wall" semigroup homology and Hochschild semigroup homology are equivalent.

If $(X;*)$ is a monoid (with a unit element 1), then we say that the set E is X-right-monoid-set if it is X-right-monoid-set and additionally $e*1=e$ for any $e\in E$ (that is 1 acts trivially on E from the right). Similarly we define X-left-monoid-set (e.g. $1*e=e$). For a monoid the presimplicial sets (modules) described in Definitions 4.2–4.4 are in fact simplicial sets (modules) with the degeneracy maps s_i placing 1 between x_i and x_{i+1}, for example in the case of Definition 4.3:

$$s_0(e_0,x_1,x_2,\ldots,x_n,e_{n+1})=(e_0,1,e_0,x_1,x_2,\ldots,x_n,e_{n+1}),$$
$$s_i(e_0,x_1,x_2,\ldots,x_n,e_{n+1})=(e_0,x_1,\ldots,x_i,1,x_{i+1},\ldots,x_n,e_{n+1})\quad\text{for } 0<i<n,$$
$$s_n(e_0,x_1,x_2,\ldots,x_n,e_{n+1})=(e_0,x_1,x_2,\ldots,x_n,1,e_{n+1}).$$

Example 4.5. We can check that $d_id_{i+1}=d_id_i$ $(0<i<n)$ if and only if $*$ is associative. Furthermore, $d_0d_1=d_0d_0$ iff $(e_0*x_1)*x_2=e_0*(x_1*x_2)$ that is E_0 is an X-right-semigroup-set. Similarly $d_{n-1}d_n=d_{n-1}d_{n-1}$ iff E_w is an X-left-semigroup-set.

Let $\partial^{(\ell)}$ be a boundary map obtained from the group homology boundary operation by dropping the first term from the sum. Analogously, let $\partial^{(r)}$ be a boundary map obtained from the group homology boundary operation by dropping the last term from the sum. It is a classical observation that $(C_n,\partial^{(\ell)})$ and $(C_n,\partial^{(r)})$ are acyclic for a group (or a monoid). We show this in a slightly more general context of weak simplicial modules (used later in the distributive case) in Sec. 6. It would be of interest to analyze homology of $(C_n,\partial^{(\ell)})$ for a semigroup without identity. Can it have a torsion?

Our definition (in the present form) can be generalized to any k-algebra V not only $V = kX$. Below we give the definition for "two wall" k-algebra, and in the next section we describe the main-stream Hochschild homology of k-algebra closely related to group homology.

Definition 4.6. Let A be a k-algebra which acts from the right on a k-module M_0 ($(m*x)*y = m*(x*y)$) and from the left on a k-module M_w ($x*(y*m) = (x*y)*m$), that is M_0 is a right A-module and M_w a left A-module. We define chain groups $C_n = M_0 \otimes A^{\otimes n} \otimes M_w$ and face maps $d_i(x_0, x_1, \ldots, x_n, x_{n+1}) = (x_0, \ldots, x_i * x_{i+1}, \ldots, x_{n+1})$, $0 \leq i \leq n$, $x_0 \in M_0$, $x_{n+1} \in M_w$, and $x_i \in A$ for $0 < i \leq n$. Then (C_n, d_i) is a presimplicial module and (C_n, ∂_n), with $\partial_n = \sum_{i=0}^{n}(-1)^i d_i$ is a chain complex, whose homology is denoted by $H_n(A, M_0, M_w)$. If A is a unitary algebra (with unit 1) then we define degenerate maps $s_i(x_0, x_1, \ldots, x_n, x_{n+1}) = (x_0, x_1, \ldots, x_i, 1, x_{i+1}, \ldots, x_n, x_{n+1})$, $0 \leq i \leq n$, and one checks directly that (C_n, d_i, s_i) is a simplicial module.

If we glue together M_0 and M_w to get two-sided module (A-bimodule) $M = M_w \otimes M_0$ we obtain Hochschild homology $H_n(A, M)$, [50, 61]; see Sec. 5.

5. Hochschild Homology of a Semigroup and an Algebra

Hochschild homology was created to have a homology theory of algebras, as before homology was defined only for (semi)groups, G, and (semi)group algebras kG (Definition 4.6 is only afterthought with Hochschild homology in mind). The history of discovering homology for algebra is described in MacLane autobiography [64]: *Given his topological background and enthusiasm, Eilenberg was perhaps the first person to see this clearly. He was in active contact with Gerhard Hochschild, who was then a student of Chevalley at Princeton. Eilenberg suggested that there ought to be a cohomology (and a homology) for algebras. This turned out to be the case, and the complex used to describe the cohomology of groups (i.e. the bar resolution) was adapted to define the Hochschild cohomology of algebras.*

Nevertheless, we start from Hochschild homology of semigroups as it leads to a presimplicial set, while the general Hochschild homology gives a presimplicial module.

Let $(X;*)$ be a semigroup and E a two-sided X-semigroup-set that is $(e*a)*b = e*(a*b)$, $(a*e)*b = a*(e*b)$, and $(a*b)*e = a*(b*e)$ We define a Hochschild presimplicial module $\{C_n(X,E), d_i\}$ as follows [50, 61]: $C_n(X) = k(E\times X^n)$ and the Hochschild face map is given by $d_i : k(E\times X^n) \to k(E\times X^{n-1})$ where $d_0(e_0,x_1,\ldots,x_n) = (e_0*x_1,x_2,\ldots,x_n)$, $d_i(e_0,x_1,\ldots,x_n) = (e_0,x_1,\ldots,x_{i-1},x_i*x_{i+1},\ldots,x_n)$ for $0<i<n$, and $d_n(e_0,x_1,\ldots,x_n) = (x_n*e_0,x_1,\ldots,x_{n-1})$.
$\partial_n : \mathbb{Z}X^n \to \mathbb{Z}X^{n-1}$ is defined by:

$$\partial(x_0,x_1,\ldots,x_n) = \sum_{i=0}^{n-1}(-1)^i(x_0,\ldots,x_{i-1},x_i*x_{i+1},x_{i+2},\ldots,x_n) + (-1)^n(x_n*x_0,x_1,\ldots,x_{n-1}).$$

The resulting homology is called the Hochschild homology of a semigroup $(X,*)$ and denoted by $HH_n(X)$ (introduced by Hochschild in 1945 [50]). It is useful to define $C_{-1}=\mathbb{Z}$ and define $\partial_0(x)=1$ to obtain the augmented Hochschild chain complex and augmented Hochschild homology.

Again if $(X,*)$ is a monoid then dropping the last term gives an acyclic chain complex.

Notice that $\partial_n = \sum_{i=0}^{n}(-1)^i d_i$, where $d_i(x_0,\ldots,x_n) = (x_0,\ldots,x_{i-1},x_i*x_{i+1},x_{i+2},\ldots,x_n)$, for $0\leq i<n$ and $d_n(x_0,\ldots,x_n) = (x_n*x_0,\ldots,x_{n-1})$.

Again, (C_n,d_i) is a presimplicial module. If $(X,*)$ is a monoid, one can define $n+1$ homomorphisms $s_i : C_n \to C_{n+1}$, called degeneracy maps, by $s_i(x_0,\ldots,x_n) = (x_0,\ldots,x_i,1,x_{i+1},\ldots,x_n)$

(similarly, in the case of group homology of a semigroup, we put, $s_i(x_1,\dots,x_n) = (x_1,\dots,x_i,1,x_{i+1},\dots,x_n))$. We check that in both cases the following conditions hold:

$$(1)\quad d_i d_j = d_{j-1}d_i \quad \text{for } i < j.$$

$$(2)\quad s_i s_j = s_{j+1}s_i,\ 0 \le i \le j \le n,$$

$$(3)\quad d_i s_j = \begin{cases} s_{j-1}d_i & \text{if } i < j \\ s_j d_{i-1} & \text{if } i > j+1 \end{cases}$$

$$(4)\quad d_i s_i = d_{i+1}s_i = \mathrm{Id}_{C_n}.$$

(C_n, d_i, s_i) satisfying conditions (1)–(4) above is called a simplicial module[r] (e.g. Z-module/abelian group). If we replace (4) by a weaker condition $d_i s_i = d_{i+1}s_i$ we deal with a weak simplicial module, the concept useful in the theory of homology of distributive structures (spindles or quandles).

As we already mentioned before Hochschild homology (and presimplicial module) can be defined for any algebra A and two-sided A-module M. We put $C_n(A;M) = M \otimes A^{\otimes n}$ and $d_i(m,x_1,\dots,x_n)$ is given by:

$$d_0(m,x_1,\dots,x_n) = (mx_1,x_2,\dots,x_n).$$
$$d_i(m,x_1,\dots,x_n) = (m,x_1,\dots,x_{i-1},x_ix_{i+1},x_{i+2},\dots,x_n) \quad \text{for } 0 < i < n,$$

and

$$d_n(m,x_1,\dots,x_n) = (x_nm,x_1,\dots,x_{n-1}).$$

From associativity of A and our action of A on M follows that (C_n, d_i) is a presimplicial module. Furthermore, if A is unitary we can define a simplicial module structure (C_n, d_i, s_i), by putting $s_i(m,x_1,\dots,x_n) = (m,x_1,\dots,x_{i-1},1,x_i,\dots,x_n)$.

[r]The concept of a simplicial set was introduced by Eilenberg and Zilber who called it *complete semi-simplicial complex*; their semi-simplicial complex is now usually called *presimplicial set* [36, 67].

6. Homology of Distributive Structures

Recall that a shelf (or right distributive system (RDS)) $(X, *)$ is a set X with a right self-distributive binary operation $* : X \times X \to X$ (i.e. $(a * b) * c = (a * c) * (b * c)$).

We work, for simplicity, with chain complexes and homology over $\mathbb{Z}$ but we can replace $\mathbb{Z}$ by any commutative ring k in our considerations.

We start from atomic definition, one term distributive homology, introduced in 2010 just before Knots in Poland III conference [84].

6.1. *One-term distributive homology*

Definition 6.1. We define a (one-term) distributive chain complex $\mathcal{C}^{(*)}$ as follows: $C_n = \mathbb{Z}X^{n+1}$ and the boundary operation $\partial_n^{(*)} : C_n \to C_{n-1}$ is given by:

$$\partial_n^{(*)}(x_0, \ldots, x_n)$$
$$= (x_1, \ldots, x_n) + \sum_{i=1}^{n} (-1)^i (x_0 * x_i, \ldots, x_{i-1} * x_i, x_{i+1}, \ldots, x_n).$$

The homology of this chain complex is called a one-term distributive homology of $(X, *)$ (denoted by $H_n^{(*)}(X)$).

We directly check that $\partial^{(*)}\partial^{(*)} = 0$.

We can put $C_{-1} = \mathbb{Z}$ and $\partial_0(x) = 1$. We have $\partial_0\partial_1^{(*)} = 0$, so we obtain an augmented distributive chain complex and an augmented (one-term) distributive homology, $\tilde{H}_n^{(*)}$. As in the classical case we get:

Proposition 6.2.

$$H_n^{(*)}(X) = \begin{cases} \mathbb{Z} \oplus \tilde{H}_n^{(*)}(X) & n = 0, \\ \tilde{H}_n^{(*)}(X) & \textit{otherwise}. \end{cases}$$

If $(X; *)$ is a rack then the complex $(C_n^{(*)}, \partial^{(*)})$ is acyclic, but in the general case of a shelf or spindle homology can be nontrivial with nontrivial free and torsion parts (joint work with Crans, Putyra and Sikora [24, 88, 92]).

If we define $d_i : C_n \to C_{n-1}$, $0 \le i \le n$, by $d_i(x_0, \ldots, x_n) = (x_0 * x_i, \ldots, x_{i-1} * x_i, x_{i+1}, \ldots, x_n)$, then (C_n, d_i) is a presimplicial module and (X^{n+1}, d_i) is a presimplicial set. If we define degeneracy maps $s_i(x_0, \ldots, x_n) = (x_0, \ldots, x_{i-1}, x_i, x_i, x_{i+1}, \ldots, x_n)$ then one checks that (C_n, d_i, s_i) is a very weak simplicial module. If we assume idempotency, that is $(X; *)$ is a spindle, then (C_n, d_i, s_i) is a weak simplicial module and the degenerate part (C_n^D, ∂_n) is a subchain complex which splits from (C_n, ∂_n) (see [84]). This split is analogous to the one conjectured in [17] and proved in [60] for classical quandle homology (for a history of quandle homology, see [14]). In [74] we gave very short, easy to visualize and generalize, proof using the split map $C_n^N \to C_n$ given by $(x_0, x_1, \ldots, x_n) \to (x_0, x_1 - x_0, \ldots, x_n - x_{n-1})$.

We can repeat our definitions if $(X; *)$ is a shelf and Y is a shelf-set ($* : Y \times X \to Y$ with $(y * x_1) * x_2 = (y * x_2) * (x_1 * x_2)$ see Fig. 7 for visualization). The presimplicial set $(Y \times X^{n+1}, d_i)$ has face maps d_i defined by $d_i(y, x_0, \ldots, x_n) = (y * x_i, x_0 * x_i, \ldots, x_{i-1} * x_i, x_{i+1}, \ldots, x_n)$. The face map d_i is visualized in Fig. 8; this

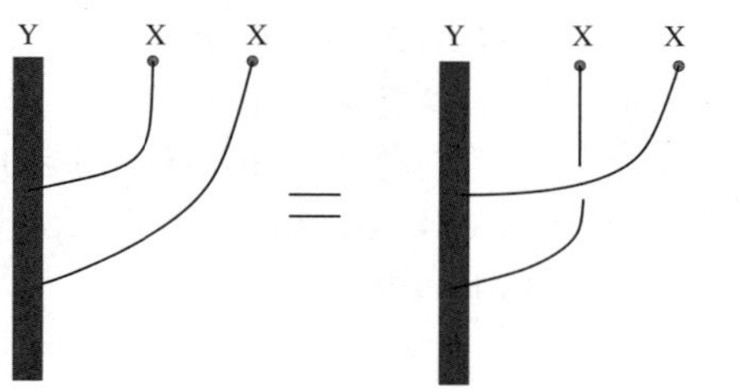

Fig. 7. Graphical interpretation of the axiom for X-shelf-set Y $(y * x_1) * x_2 = (y * x_2) * (x_1 * x_2)$.

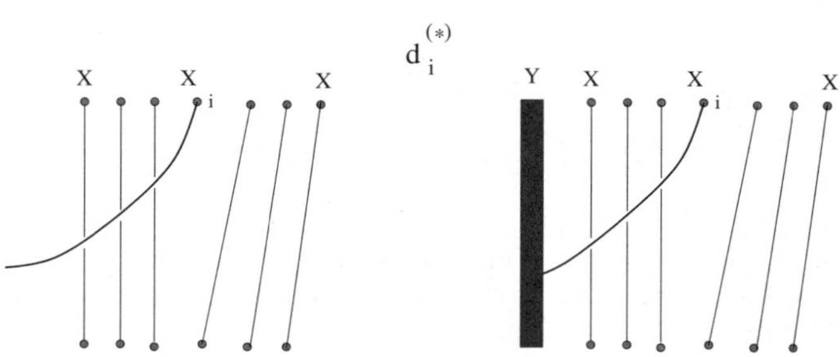

Fig. 8. Graphical interpretation of the face map $d_i^{(*)}$.

visualization will play an important role when distributive homology will be generalized to Yang–Baxter homology.

6.2. *Multi-term distributive homology*

The first homology theory related to a self-distributive structure was constructed in early '90s by Fenn, Rourke, and Sanderson [43] and motivated by (higher dimensional) knot theory.[s] For a rack $(X,*)$, they defined rack homology $H_n^R(X)$ by taking $C_n^R = \mathbb{Z}X^n$ and $\partial_n^R : C_n \to C_{n-1}$ is given by $\partial_n^R = \partial_{n-1}^{(*)} - \partial_{n-1}^{(*_0)}$. Our notation has grading shifted by 1, that is, $C_n(X) = C_{n+1}^R = \mathbb{Z}X^{n+1}$. It is routine to check that $\partial_{n-1}^R \partial_n^R = 0$. However, it is an interesting question what properties of $*_0$ and $*$ are really used. With relation to the paper [75] we noticed that it is distributivity again which makes $(C^R(X), \partial_n^R)$ a chain complex. More generally we observed that if $*_1$ and $*_2$ are right self-distributive and distributive with respect to each other, then $\partial^{(a_1,a_2)} = a_1\partial^{(*_1)} + a_2\partial^{(*_2)}$ leads to a chain complex (i.e. $\partial^{(a_1,a_2)}\partial^{(a_1,a_2)} = 0$). Below I answer a more general question: for a finite set $\{*_1, \ldots, *_k\} \subset Bin(X)$ and integers $a_1, \ldots, a_k \in \mathbb{Z}$, when is $(C_n, \partial^{(a_1,\ldots,a_k)})$ with $\partial^{(a_1,\ldots,a_k)} = a_1\partial^{(*_1)} + \cdots + a_k\partial^{(*_k)}$ a chain complex? When is $(C_n, d_i^{(a_1,\ldots,a_k)})$ a presimplicial set? We answer these questions in Lemma 6.3. In particular, for a distributive set $\{*_1, \ldots, *_k\}$ the answer is affirmative.

Lemma 6.3. (i) *If $*_1$ and $*_2$ are right self-distributive operations, then $(C_n, \partial^{(a_1,a_2)})$ is a chain complex if and only if the operations $*_1$ and $*_2$ satisfy:*

$$\begin{aligned}(a *_1 b) *_2 c &+ (a *_2 b) *_1 c \\ &= (a *_2 c) *_1 (b *_2 c) + (a *_1 c) *_2 (b *_1 c) \text{ in } \mathbb{Z}X. \qquad (6.4)\end{aligned}$$

[s]The recent paper by Roger Fenn, [39] states: "Unusually in the history of mathematics, the discovery of the homology and classifying space of a rack can be precisely dated to 2 April 1990."

We call this condition weak distributivity. If this condition (6.4) *does not hold we can take* $C_0(X)$ *to be the quotient by Eq.* (6.4): $C_0(X) = \mathbb{Z}X/(6.4)$ *and then we take* $C_n = C_0^{\otimes n+1}$.

(ii) *We say that a set* $\{*_1, \ldots, *_k\} \subset Bin(X)$ *is weakly distributive if each operation is right self-distributive and each pair of operations is weakly distributive (with two main cases: distributivity* $(a *_1 b) *_2 c = (a *_2 c) *_1 (b *_2 c)$ *and chronological distributivity*[t] $(a *_1 b) *_2 c = (a *_1 c) *_2 (b *_1 c)$). *We have:* $(C_n, d_i^{(a_1,\ldots,a_k)})$ *is a presimplicial set if and only if the set* $\{*_1, \ldots, *_k\} \subset Bin(X)$ *is weakly distributive.*

(iii) $(C_n, \partial_n^{(a_1,\ldots,a_k)})$ *is a chain complex if and only if the set* $\{*_1, \ldots, *_k\} \subset Bin(X)$ *is weakly distributive.*

We complete this section by showing that for a rack homology of a quandle or spindle, H_n^R embeds in H_{n+1}^R; we construct monomorphic "homology operation" of degree one.[u] We place it in a more general context of weak simplicial modules.

Lemma 6.5. *Let* (C_n, d_i, s_i) *be a weak simplicial module then* $\partial s_0 + s_0\partial = s_0 d_0$; *in effect* $s_0 d_0$ *induces a trivial map on homology. In particular:*

(i) *if the map* $s_0 d_0$ *is the identity then* $s_0 d_0 s_0 = s_0$ *(as in the case of a simplicial module) and then the chain complex* $(s_0(C_{n-1}), \partial_n)$ *is acyclic,*

(ii) *if* $d_0 = 0$ *then* s_0 *is a chain map (e.g. this hold for two-term rack homology),*

(iii) *in the case of one-term distributive homology, we conclude that the map replacing* x_0 *by* x_1 *in* $(x_0, x_1, \ldots, x_n)$ *is a chain map, chain homotopic to zero map.*

[t]I did not see this concept considered in literature, but it seems to be important in Putyra's work on odd Khovanov homology [95].

[u]For a quandle it is a well-known fact that rack homology in dimension n is isomorphic to "early degenerate" homology in dimension $n + 1$.

Proof. We have

$$\begin{aligned}
\partial s_0 + s_0\partial &= d_0s_0 - d_1s_0 + \sum_{i=2}^{n+1}(-1)^i d_i s_0 + s_0d_0 + \sum_{i=1}^{n}(-1)^i s_0 d_i \\
&= d_0s_0 - d_1s_0 + \sum_{i=2}^{n+1}(-1)^i s_0 d_{i-1} + s_0d_0 + \sum_{i=1}^{n}(-1)^i s_0 d_i \\
&= s_0d_0.
\end{aligned}$$

□

If s_0 has a left inverse map, say $p_n : C_{n+1} \to C_n$, $p_n s_0 = \mathrm{Id}_{C_n}$, as is the case for a weak simplicial module in (multi) spindle case, we can say more.

Lemma 6.6. *Let (C_n, d_i, s_i) be a weak simplicial module $p_n : C_{n+1} \to C_n$ is a left inverse of s_0 and additionally $p_n d_i = d_{i-1}p_n$ for $i > 0$ then $p\partial + \partial p = pd_0$; in effect pd_0 induces a trivial map on homology. In particular:*

(i) *if we deal with (multi)term distributive homology, p may be taken to be the map deleting the first coordinate of $(x_0, x_1, \ldots, x_n)$ (in one term distributive homology $p = d_0$, so d_0d_0 is a chain map trivial on homology),*
(ii) *if $d_0 = 0$ then p is a chain map (e.g. this hold for two-term rack homology),*
(iii) *if $d_0 = 0$ then p induces an epimorphism on homology and s_0 induces a monomorphism on homology; in particular s_0 induces monomorphic "homology operation" of degree one ($s_0(C_{n-1}, \partial_n)$ is called an early degenerate chain complex).*

Proof. We have

$$p\partial + \partial p = pd_0 + \sum_{i=1}^{n}(-1)^i d_{i-1}p + \sum_{i=0}^{n-1}(-1)^i d_i p = pd_0.$$

Part (iii) follows from the fact that $p_n s_0 = \mathrm{Id}_{C_n}$ and p and s_0 are chain maps. □

Thus we proved that rack homology of quandles (or spindles) cannot decrease with n ($H_n^R \subset H_{n+1}^R$).

We computed with Putyra [88] various multi-term homology, including that for finite distributive lattices (including Boolean algebras).

7. Bloh–Leibniz–Loday Algebra

Lie algebra was probably the first nonassociative structure for which homology was defined [19]. The idea of Chevalley and Eilenberg was to translate homology of a (Lie) group to homology of its Lie algebra.[v] We should stress, in particular the role of conjugacy in Lie algebra, as conjugacy was the motivation for wracks (racks) and quandles.

We discuss here homology theory of Bloh–Loday–Leibniz algebras introduced by Bloh and Loday [3, 4, 62], and which can be informally thought to be a linearization of distributive homology.[w] We follow Loday and Lebed here [58, 59, 61, 62]. BLL (Bloh–Leibniz–Loday)

---

[v]The paper starts from: *The present paper lays no claim to deep originality. Its main purpose is to give a systematic treatment of the methods by which topological questions concerning compact Lie groups maybe reduced to algebraic questions concerning Lie algebras.*

[w]BLL algebras are often call Leibniz algebras as the version of Jacobi identity they satisfy can be treated as a Leibniz rule. The history of the discovery is described by Loday as follows [61]:

"In the definition of the Chevalley–Eilenberg complex of a Lie algebra $\mathcal{G}$ the module of chains is the exterior module. The non-commutative analog of the exterior module $\Lambda\mathcal{G}$ is the tensor module $T\mathcal{G}$. If one replaces Λ by $\bigotimes$ in the classical formula for the boundary map d of the CE-complex, then one gets a well-defined map $T\mathcal{G}$ but the relation d^2 is not valid anymore. However, I discovered that, if one writes d so as to put the commutator $[x_i, x_j]$ at the place i when $i < j, \ldots$, then the relation $d^2 = 0$ is satisfied in the tensor (i.e. non-commutative) context. So, this gives rise to a new complex $T\mathcal{G}, d)$ for the Lie algebra $\mathcal{G}$. The homology groups of this complex are denoted by $HL_*(\mathcal{G})$ and called the non-commutative homology groups of $\mathcal{G}$. In the proof of the relation $d^2 = 0$ in the tensor module case, I noticed that the only property of the Lie bracket, which is needed, is the Leibniz relation $[x, [y, z]] = [[x, y], z] - [[x, z], y]$. So the complex $(T\mathcal{G}, d)$ and its homology are defined for more general objects than Lie algebras, for the *Leibniz algebras*."

algebra is a generalization of a Lie algebra we use a bracket $[-,-]$ for a bilinear map:

Definition 7.1. (1) Let V be a k-module equipped with a bilinear map $[-,-] : V \times V \to V$ satisfying the relation (Leibniz version of the Jacobi identity):

$$[x,[y,z]] = [[x,y],z] - [[x,z],y], \quad \text{for all } x,y,z \in V.$$

I see the linearization of distributivity as

$$(x * y) * z = (x * z) * (y * z) \Rightarrow (x * y) * z$$
$$= (x * z) * y + x * (y * z) \text{ BLL condition.}$$

V. Lebed formalized this "linearization" in the case V has a central element 1 that is $[x,1] = 0[1,x]$ we color crossing as follows:

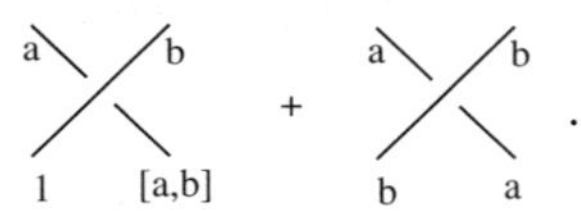

This led Lebed to describe face map for chain complex of BLL algebras as in Fig. 9; compare the figure with Definition 7.2.

(2) A BLL module M over V is a k-module with a bilinear action (still denoted $[-,-] : M \times V \to M$, satisfying the formula from (1) for any $x \in M$ and $y,z \in V$.

In a special case of $M = k$ (k a ring with identity), the map $[-,-] : M \times V \to M$ is replaced by map $\epsilon : V \to k$ which is zero on commutators (Lie character), [59]).

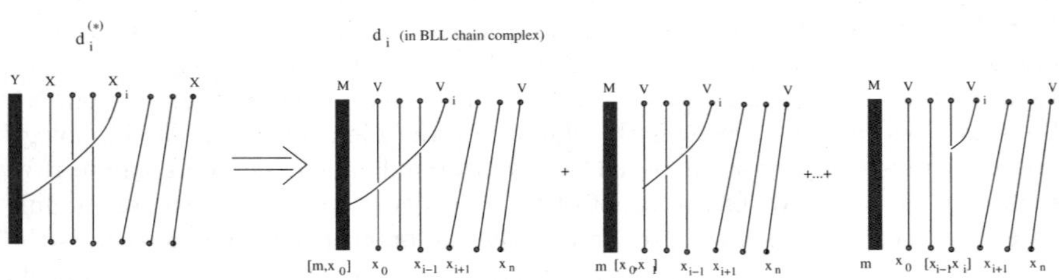

Fig. 9. Comparing face map d_i in distributive and unital BLL algebra.

Homology of (V, M) was constructed by Loday based partially on the work of Cuvier [27, 61], they were unaware of the earlier work by Bloh [4].

Definition 7.2. Let V be a BLL algebra and M a BLL module over V.

(1) We define $C_n = M \otimes V^{\otimes n+1}$, that is $C_* = TV$ (tensor algebra). For $0 \leq i \leq n$, $d_i : C_n \to C_{n-1}$ is given by:

$$d_i(x_{-1}, x_0, \ldots, x_n) = \sum_{j:-1\leq j<i} (x_{-1}, x_0, \ldots, x_{j-1}, [x_j, x_i], x_{j+1}, x_{i-1}, x_{i+1}, \ldots, x_n).$$

If we take ΛV, the exterior algebra in place of TV. We will be in the classical case of homology developed for Lie algebras by Chevalley and Eilenberg [19].

(2) Assume that V is a split unital BLL algebra (that is $V = V' \oplus k1$, and $M = k$, $[1, x] = \epsilon(x)$ $(1 \in k\ x \in V)$, then the "primitive" degenerate maps $s_i^p : C_n \to C_{n+1}$ is defined by

$$s_i^p(x_{-1}, x_0, \ldots, x_n) = \begin{cases} (x_{-1}, x_0, \ldots, x_{i-1}, (1, x_i) \\ \quad + (x_i, 1), x_{i+1}, \ldots, x_n) & \text{if } x_i \in V' \\ (x_{-1}, x_0, \ldots, x_{i-1}, 1, 1, x_{i+1}, \ldots, x_n) & \text{if } x_i = 1. \end{cases}$$

(3) Assume V is a free k-module with basis X and define degenerate maps $s_i^g : C_n \to C_{n+1}$ on X by doubling ith coordinate:

$$s_i^g(m, x_0, \ldots, x_n) = (m, x_0, \ldots, x_{i-1}, x_i, x_i, x_{i+1}, \ldots, x_n).$$

Lemma 7.3. (1) $(C_n(V, M), d_i)$ *is a presimplicial module, that is* $d_i d_j = d_{j-1} d_i$ *for* $0 \leq i < j \leq n$.

(2) $(C_n(X, M), d_i, s_i^p)$ *is a weak simplicial module, and its homology and degenerated homology are annihilated by* $\epsilon(1)$.

(3) $(C_n(X, M), d_i, s_i^g)$ *satisfies conditions* (1), (2) *and partially* (3) $(d_i s_j = s_{j-1} d_i$ *for* $i < j)$ *of a simplicial module. Condition* $(4'), d_i s_i - d_{i+1} s_i = 0$ *holds iff* $[x, x]$ *for all* $x \in X$.

Condition (3) *of a simplicial module for* $i > j+1$ *requires the following equality*:

$$[x_j, x_i] \otimes [x_j, x_i] = [x_j, x_i] \otimes x_j + x_j \otimes [x_j, x_i]$$

for basic elements $(x_j, x_i) \in X^2$ *for* $i > j+1$. *Thus if we take* $C_* = VT/\mathcal{I}$ *divided by an ideal containing the above equation, then* $(C_n(X, M), d_i, s_i^g)$ *is a very weak simplicial module. It is a weak simplicial module iff additionally* $[x, x] = 0$, *for all* $x \in X$.[x]

We leave the proof as an exercise for the reader however we make the calculation in two small but typical cases which show that our axioms are needed:

(i) Comparison of $d_0 d_1$ with $d_0 d_0$ (they should be equal):

$$\begin{aligned} d_0 d_1(m; x_0, x_1) &= d_0(([m, x_1]; x_0) + (m; [x_0, x_1])) \\ &= ([m, x_1], x_0) + [m, [x_0, x_1]], \\ d_0 d_0(m; x_0, x_1) &= d_0([m, x_0]; x_1) = [[m, x_0), x_1]. \end{aligned}$$

Thus $d_0 d_1 = d_0 d_0$ if and only if $([m, x_1], x_0] + [m, [x_0, x_1]] = [[m, x_0], x_1]$ which is the axiom of BLL-module.

(ii) Comparison of $d_1 d_2$ with $d_1 d_1$ (they should be equal):

$$\begin{aligned} &d_1 d_2(m; x_0, x_1, x_2) \\ &\quad = d_1(([m, x_2]; x_0, x_1) + (m; [x_0, x_2], x_1) + (m; x_0, [x_1, x_2])) \\ &\quad = ([[m, x_2], x_1]; x_0) + ([m, x_2]; [x_0, x_1]) \\ &\qquad + ([m, x_1]; [x_0, x_2]) + (m; [[x_0, x_2], x_1]) \\ &\qquad + ([m, [x_1, x_2]]; x_0) + (m; [x_0, [x_1, x_2]]), \end{aligned}$$

and

$$\begin{aligned} &d_1 d_1(m; x_0, x_1, x_2) \\ &\quad = d_1(([m, x_1]; x_0, x_2) + (m; [x_0, x_1], x_2)) \\ &\quad = ([[m, x_1], x_2]; x_0) + ([m, x_1]; [x_0, x_2]) \\ &\qquad + ([m, x_2]; [x_0, x_1]) + (m; [[x_0, x_1], x_2]). \end{aligned}$$

---

[x]Notice that in exterior algebra ΛV, the equation $[x_j, x_i] \otimes [x_j, x_i] = [x_j, x_i] \otimes x_j + x_j \otimes [x_j, x_i]$ holds $(0 = 0)$, but in that case our degenerate map s_i^g would be a zero map.

Thus $d_1d_2 = d_1d_1$ if and only if the following sum is equal to zero:

$$([[m, x_2], x_1]; x_0) + ([m, [x_1, x_2]]; x_0) - ([[m, x_1], x_2]; x_0) \\ + (m; [[x_0, x_2], x_1]) + (m; [x_0, [x_1, x_2]]) - (m; [[x_0, x_1], x_2]).$$

The first part is equal to zero iff M is BLL-module and the second part is equal to zero iff V is BLL-algebra.

Remark 7.4. Lie algebra homology, as proved by Cartan and Eilenberg [13] can be obtained from homology of the universal enveloping algebra $UV = TV/(a\otimes b - b\otimes a = [a, b])$ of the Lie algebra V. One hopes for a similar connection between distributive homology and homology of the group associated to a wrack or quandle. One hint in this direction is that in every group the following "distributivity" holds:

$$[[x, y^{-1}], z]^y [[y, z^{-1}], x]^z [[z, x^{-1}], y]^x = 1,$$

where $[x, y] = x^{-1}y^{-1}xy$ and $x^y = y^{-1}xy$. This leads to the graded Lie algebra associated to the group, via lower central series of the group [107].

8. Semigroup Extensions and Shelf Extensions

The theory of extension of structures and related cocycles started from two important examples from group theory:

(i) The extension of $PSL_n(C)$ by $SL_n(C)$ by Schur (1904), with related short exact sequence of groups [2]

$$0 \to \mathbb{Z}_2 \to SL_n(C) \to PSL_n(C) \to 1$$

and the study of crystallographic groups Γ where we consider a short exact sequence

$$0 \to \mathbb{Z}^n \to \Gamma \to \Gamma/\mathbb{Z}^n \to 1.$$

An extension of a group X by a group N is a short exact sequence of groups

$$1 \to N \xrightarrow{i} E \xrightarrow{\pi} X \to 1$$

(some people call this an extension of N by X [2, 6, 65]).

Consider a set-theoretic section $s : X \to E$ (that is $\pi s = \mathrm{Id}_X$). Every element of E is a unique product $as(x)$ for $a \in N$ and $x \in X$ (coset decomposition), thus we have $E = N \times X$ as sets (here $e \to (es(\pi(e^{-1})), \pi(e))$ and $es(\pi(e^{-1}))(s\pi(e)) = e$ as needed. The inverse map is $(a, x) \to as(x)$.

This motivates study of extension of magmas as study of projections $\pi : A \times X \to X$ with various structures preserved. In particular, we compare semigroup extension of a semigroup by an abelian group with the shelf extension of a shelf by an Alexander quandle. We start from a general concept of a dynamic cocycle in a magma case and then in associative and distributive cases and in both we relate to the (co)homology of our structures. Extension of modules, groups and Lie algebras is described in the classical book by Cartan and Eilenberg [13], distributive case was developed in [1, 16, 18].

Definition 8.1. Let $(X; *)$ be a magma, A a set, and $\pi : A \times X \to X$ the projection to the second coordinate. Any magma structure on $A \times X$ for which π is an epimorphism, can be given by a system of functions $\phi_{a_1,a_2}(x_1, x_2) : X \times X \to A$ by:

$$(a_1, x_1) * (a_2, x_2) = (\phi_{a_1,a_2}(x_1, x_2), x_1 * x_2).$$

Functions $\phi_{a_1,a_2}(x_1, x_2)$ are uniquely defined by the multiplication on $A \times X$, thus binary operations on $A \times X$ agreeing with π are in bijection with choices of functions ϕ_{a_1,a_2}. If we require some special structure on $(X; *)$ (e.g. associativity or right-distributivity) we obtain some property of $\phi_{a_1,a_2}(x_1, x_2)$ which we call a dynamical cocycle property for the structure.

(1) Let $(X; *)$ be a semigroup; in order that an action on $A \times X$ is associative we need:

$$\begin{aligned}
&((a_1, x_1) * (a_2, x_2)) * (a_3, x_3) \\
&\quad = (\phi_{a_1,a_2}(x_1, x_2), x_1 * x_2) * (a_3, x_3) \\
&\quad = (\phi_{\phi_{a_1,a_2}(x_1,x_2),a_3}(x_1 * x_2, x_3), (x_1 * x_2) * x_3)
\end{aligned}$$

to be equal to

$$\begin{aligned}(a_1, x_1) * ((a_2, x_2) * (a_2, x_3)) \\ &= (a_1, x_1) * (\phi_{a_2,a_3}(x_2, x_3), x_2 * x_3) \\ &= (\phi_{a_1,\phi_{a_2,a_3}}(x_2, x_3))(x_1, x_2 * x_3), x_1 * (x_2 * x_3).\end{aligned}$$

Thus the dynamical cocycle condition in the associative case has the form:

$$\phi_{\phi_{a_1,a_2}(x_1,x_2),a_3}(x_1 * x_2, x_3) = \phi_{a_1,\phi_{a_2,a_3}(x_2,x_3)}(x_1, x_2 * x_3).$$

(2) Let $(X; *)$ be a shelf; in order that an action on $A \times \dot{X}$ is right self-distributive we need:

$$\begin{aligned}((a_1, x_1) * (a_2, x_2)) * (a_2, x_3) \\ &= (\phi_{a_1,a_2}(x_1, x_2), x_1 * x_2) * (a_3, x_3) \\ &= (\phi_{\phi_{a_1,a_2}(x_1,x_2),a_3})(x_1 * x_2, x_3), (x_1 * x_2) * x_3)\end{aligned}$$

to be equal to

$$\begin{aligned}((a_1, x_1) * (a_3, x_3)) * ((a_2, x_2) * (a_3, x_3)) \\ &= (\phi_{a_1,a_3}(x_1, x_3), x_1 * x_3) * (\phi_{a_2,a_3}(x_2, x_3), x_2 * x_3) \\ &= (\phi_{\phi_{a_1,a_3}(x_1,x_3),\phi_{a_2,a_3}(x_2,x_3)}(x_1 * x_3, x_2 * x_3), \\ &\quad (x_1 * x_3) * (x_2 * x_3)).\end{aligned}$$

Thus the dynamical cocycle condition in right-distributive case has a form:

$$\begin{aligned}\phi_{\phi_{a_1,a_2}(x_1,x_2),a_3}(x_1 * x_2, x_3) \\ &= \phi_{\phi_{a_1,a_3}(x_1,x_3),\phi_{a_2,a_3}(x_2,x_3)}(x_1 * x_3, x_2 * x_3).\end{aligned}$$

(3) We assume now that $(X; *)$ is an entropic (inner turn) magma, that is $(a * b) * (c * d) = (a * c) * (b * d)$ for any $a, b, c, d \in X$. We look for condition on the dynamical cocycle so that $A \times X$

is entropic. We need

$$
\begin{aligned}
&((a_1, x_1) * (a_2, x_2)) * ((a_3, x_3) * (a_4, x_4)) \\
&\quad = (\phi_{a_1,a_2}(x_1, x_2), x_1 * x_2) * (\phi_{a_3,a_4}(x_3, x_4), x_3 * x_4) \\
&\quad = (\phi_{\phi_{a_1,a_2}(x_1,x_2),\phi_{a_3,a_4}(x_3,x_4)}(x_1 * x_2, x_3 * x_4), \\
&\qquad (x_1 * x_2) * (x_3 * x_4))
\end{aligned}
$$

to be equal to

$$
\begin{aligned}
&((a_1, x_1) * (a_3, x_3)) * ((a_2, x_2) * (a_3, x_3)) \\
&\quad = (\phi_{a_1,a_3}(x_1, x_3), x_1 * x_3) * (\phi_{a_2,a_4}(x_2, x_4), x_2 * x_4) \\
&\quad = (\phi_{\phi_{a_1,a_3}(x_1,x_3),\phi_{a_2,a_4}(x_2,x_4)}(x_1 * x_3, x_2 * x_4), \\
&\qquad (x_1 * x_3) * (x_2 * x_4)).
\end{aligned}
$$

Thus the dynamic cocycle condition in entropic case has the form:

$$
\begin{aligned}
&\phi_{\phi_{a_1,a_2}(x_1,x_2),\phi_{a_3,a_4}(x_3,x_4)}(x_1 * x_2, x_3 * x_4) \\
&\quad = \phi_{\phi_{a_1,a_3}(x_1,x_3),\phi_{a_2,a_4}(x_2,x_4)}(x_1 * x_3, x_2 * x_4).
\end{aligned}
$$

We illustrate the above by several examples, starting from a classical group extension by an abelian group. Consider the extension E of a group X by an abelian group A; this is described by a short exact sequence of groups:

$$0 \to A \to E \xrightarrow{\pi} X \to 1.$$

As noted before, $E = A \times X$ as a set and bijection depends on a section $s : X \to E$. Furthermore, X acts on A (we have $X \times A \to A$) and the action is given by conjugation: $x(a) = s(x)a(s(x)^{-1}$ and does not depend on the choice of s as A is commutative.

For a semigroup this is the starting point.

Let X be a semigroup, $\pi : A \times X$ a projection and a semigroup X acts on a set A. We define a product on $A \times X$ by the formula:

$$(a_1, x_1)(a_2, x_2) = (a_1 + x_1(a_2) + f(x_1, x_2), x_1x_2).$$

The function $f : X \times X \to A$, as in the group case, arises by comparing section of a multiplication with multiplication of sections, that is $s(x_1x_2) = f(x_1,x_2)s(x_1)s(x_2)$. We assume that the action $x : A \to A$ is a group homomorphism for any x and it is associative $(x_1(x_2(a)) = (x_1x_2)(a))$. The associativity of the product on $A \times X$ is equivalent to the condition on $f : X \times X \to A$ of the form[y] $x_1(f(x_2,x_3)) - f(x_1x_2,x_3) + f(x_1,x_2x_3) - f(x_1,x_2) = 0$ which we call a second cocycle condition (relation to homology of groups defined before, will be explained). Thus $\phi_{a_1,a_2}(x_1,x_2) = a_1 + x_1(a_2) + f(x_1,x_2)$, is an example of a dynamical cocycle for an associative structure. We should stress that for a semigroup there may be choice for a dynamical cocycle but for a group it is unique (see e.g. [6]). $f : X \times X \to A$ is a cocycle for a chain complex introduced in Definition 4.1 for the trivial action (and generally Definition 4.2); we have:

$$\begin{aligned}
&\partial^2(f)(x_1,x_2,x_3)\\
&\quad = f(\partial_2(x_1,x_2,x_3))\\
&\quad = f((x_2,x_3) - (x_1x_2,x_3) + (x_1,x_2x_3) - (x_1,x_2))\\
&\quad = f(x_2,x_3) - (x_1x_2,x_3) + f(x_1,x_2x_3) - f(x_1,x_2) = 0.
\end{aligned}$$

---

[y] Calculation is as follows: Associativity,

$$((a_1,x_1)(a_2,x_2))(a_3,x_3) = (a_1,x_1)((a_2,x_2)(a_3,x_3))$$

gives, after expanding each side:

$$\begin{aligned}
&((a_1,x_1)(a_2,x_2))(a_3,x_3)\\
&\quad = (a_1 + x_1(a_2) + f(x_1,x_2), x_1x_2)(a_3,x_3)\\
&\quad = (a_1 + x_1(a_2) + f(x_1,x_2) + (x_1x_2)(a_3) + f(x_1x_2,x_3), (x_1x_2)x_3)
\end{aligned}$$

and

$$\begin{aligned}
(a_1,x_1)((a_2,x_2)(a_3,x_3)) &= (a_1,x_1)(a_2 + x_2(a_3) + f(x_2,x_3), x_2x_3)\\
&= (a_1 + x_1(a_2 + x_2(a_3) + f(x_2,x_3) + f(x_1,x_2x_3), x_1(x_2x_3))
\end{aligned}$$

thus the associativity reduces to:

$$f(x_1,x_2) + f(x_1x_2,x_3) = x_1(f(x_2,x_3)) + f(x_1,x_2x_3)$$

which is our 2-cocycle condition.

If action of X on A is not necessarily trivial, we define cohomology $H^n(G,C)$ with a cochain complex $C_n = Hom(ZG^n \to A)$ and $\partial^n : C^n \to C^{n+1}$ is given by

$$\begin{aligned}&\partial^n(f)(x_1,\ldots,x_n,x_{n+1})\\&\quad = x_1 f(x_2,\ldots,x_{n+1}) + \sum_{i=1}^{n}(-1)^i f(x_1,\ldots,x_i x_{i+1},\ldots,x_{n+1})\\&\qquad + (-1)^{n+1}(x_1,\ldots,x_n).\end{aligned}$$

8.1. *Extensions in right distributive case*

We give here two examples of extension in right distributive case:

I. Let $(X;*)$ be a shelf and A an abelian group with a given homomorphism $t : A \to A$ (equivalently, A is a $\mathbb{Z}[t]$ module). We define a shelf structure (called Alexander extension [16, 18]) on $A\times X$ by the formula:

$$(a_1,x_1)(a_2,x_2) = (ta_1 + (1-t)a_2 + f(x_1,x_2), x_1 * x_2)$$

and right distributivity is equivalent to the condition on $f : X\times X \to A$ which satisfies twisted cocycle condition:

$$\begin{aligned}&t(f(x_2,x_3) - f(x_1,x_3) + f(x_1,x_2)) - f(x_2,x_3)\\&\quad + f(x_1 * x_2, x_3) - f(x_1 * x_3, x_2 * x_3) = 0.\end{aligned}$$

The calculation is as follows: Right self-distributivity

$$\begin{aligned}&((a_1,x_1)*(a_2,x_2))*(a_3,x_3)\\&\quad = ((a_1,x_1)*(a_3,x_3))*((a_2,x_2)*(a_3,x_3))\end{aligned}$$

gives, after expanding each side:

$$\begin{aligned}&((a_1,x_1)*(a_2,x_2))*(a_3,x_3)\\&\quad = (ta_1 + (1-t)a_2 + f(x_1,x_2), x_1 * x_2)*(a_3,x_3)\\&\quad = (t(ta_1 + (1-t)a_2 + f(x_1,x_2))\\&\qquad + (1-t)a_3 + f(x_1 * x_2, x_3), (x_1 * x_2) * x_3)\end{aligned}$$

and

$$\begin{aligned}
&((a_1,x_1)*(a_3,x_3))*((a_2,x_2)*(a_3,x_3))\\
&\quad = (ta_1+(1-t)a_3+f(x_1,x_3),x_1*x_3)\\
&\qquad *(ta_2+(1-t)a_3+f(x_2,x_3),x_2*x_3)\\
&\quad = (t(ta_1+(1-t)a_3+f(x_1,x_3))+(1-t)(ta_2+(1-t)a_3\\
&\qquad +f(x_2,x_3))+f(x_1*x_3,x_2*x_3),(x_1*x_3)*(x_2*x_3)).
\end{aligned}$$

This is equivalent to:

$$\begin{aligned}
&tf(x_1,x_2)+f(x_1*x_2,x_3)\\
&\quad = tf(x_1,x_3)+(1-t)f(x_2,x_3)+f(x_1*x_3,x_2*x_3)
\end{aligned}$$

and further to a cocycle in a (twisted) rack homology:

$$\begin{aligned}
(\partial^R f)(x_1,x_2,x_3) &= -t(f(x_2,x_3)-f(x_1,x_3))+f(x_1,x_2)+f(x_2,x_3)\\
&\quad -f(x_1*x_2,x_3)+f(x_1*x_3,x_2*x_3)=0.
\end{aligned}$$

If there are two right self-distributive binary operations, $*_1$ and $*_2$ on $A\times X$ represented by f_1 and f_2 respectively (that is $(a_1,x_1)*_i(a_2,x_2)) = (a_1*a_2+f_i(x_1,x_2),x_1*x_2,\ i=1,2)$, and there is a homomorphism $H: A\times X\to A\times X$ given by $H(a,x)=(a+c(x),x)$ for some $c: X\to A$, then the homomorphism condition $H((a_1,x_1)*_1(a_2,x_2)) = H(a_1,x_1)*_2 H(a_2,x_2)$ is equivalent to

$$\begin{aligned}
&(a_1*a_2+f_1(x_1,x_2)+c(x_1*x_2),x_1*x_2)\\
&\quad = ((a_1+c(x_1))*(a_2+c(x_2))+f_2(x_1,x_2),x_1*x_2)
\end{aligned}$$

thus $ta_1+(1-t)a_2+c(x_1*x_2)+f_1(x_1,x_2) = t(a_1+c(x_1))+(1-t)(a_2+c(x_2))+f_2(x_1,x_2)$ so $f_1(x_1,x_2)-f_2(x_1,x_2) = tc(x_1)+(1-t)c(x_2)-c(x_1x_2) = (\partial c)(x_1,x_2)$. We can say that the second cohomology (here (twisted) rack cohomology) $H^2(X,A)$ describes (shelf) extensions of X by A of type described above, modulo described above equivalence, for an abelian group A.

The dynamical cocycle is given by $\phi_{a_1,a_2}(x_1,x_2) = ta_1+(1-t)a_2+f(x_1,x_2)$, [18].

II. Another family of extensions is given by the hull construction for a multi-shelf (multi-RD-system) of Patrick Dehornoy and David

Larue [32, 57], and its "G-group" generalization (which we call a *twisted hull*) by Ishii, Iwakiri, Jang, and Oshiro [54].

For a hull construction we need a distributive set of binary operations on A indexed by elements of X, that is $(a *_x b) *_y c = (a *_y c) *_x (b *_y c)$, and the "hull" shelf structure on $A \times X$ is given by:

$$(a_1, x_1) * (a_2, x_2) = (a_1 *_{x_2} a_2, x_1).$$

To see our construction as obtained from a dynamical cocycle we put trivial operation on X ($x * y = x$), and the dynamical cocycle is given by $\phi_{a_1,a_2}(x_1, x_2) = a_1 *_{x_2} a_2$.

Remark 8.2. If $A = F(X)$ is a fee group on free generators X, then the hull $A \times X$ related to the distributive set of operations $*_x$ on A given by $a_1 *_x a_2 = a_1 a_2^{-1} x a_2$ is a free rack generated by X (denoted by $FR(X)$) as defined by Fenn and Rourke [41] (see also [18]). To summarize, we have then $FR(X) = F(X) \times X$ with $(a_1, x_1) * (a_2, x_2) = (a_1 *_{x_2} a_2, x_1) = (a_1 a_2^{-1} x_2 a_2, x_1)$.

The G-group generalization of hull to twisted hull, relaxes condition that X is indexing distributive set of operation and we allow "twisted distributivity". That is: $(a *_x b) *_y c = (a *_y c) *_{x*y} (b *_y c)$. Thus X indexes operations on A satisfying "twisted distributivity". In this case the shelf structure on $A \times X$ is given by:

$$(a_1, x_1) * (a_2, x_2) = (a_1 *_{x_2} a_2, x_1 * x_2).$$

The fundamental example leading to "twisted distributivity" was already given by Joyce: Let G be a group and X be a subgroup of $hom(G, G)$. Then we define $g_1 *_x g_2 = x(g_1 g_2^{-1}) g_2$ and we get:

$$(g_1 *_{x_2} g_2) *_{x_3} g_3 = (g_1 *_{x_3} g_3) *_{x_2 * x_3} (g_2 *_{x_3} g_3),$$

where $x_2 * x_3 = x_3 x_2 x_3^{-1}$.

Twisted distributivity is illustrated in Fig. 10.

8.2. *Extensions in entropic case*

Let $(X; *)$ be an entropic magma, that is $*$ satisfies the entropic identity: $(a * b) * (c * d) = (a * c) * (b * d)$. Let also A be an abelian

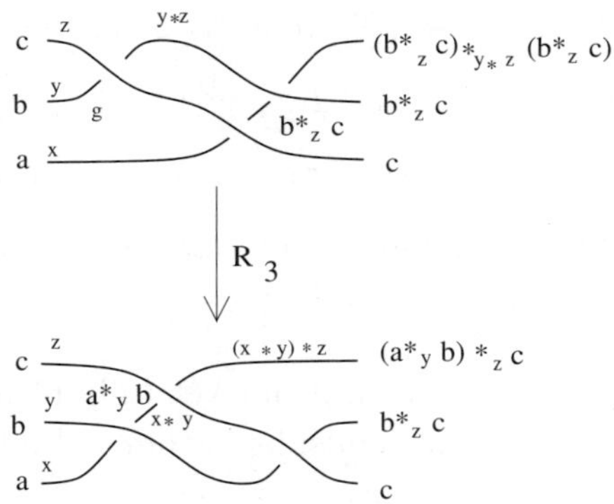

Fig. 10. Twisted distributivity.

group with a given pair of commuting homomorphisms $t, s : A \to A$ and a constant $a_0 \in A$; we consider $(A; *)$ as an entropic magma with an affine action $a * b = ta + sb + a_0$. Then we define a binary operation on $A \times X$ by $(a_1 * x_1) * (a_2 * x_2) = (a_1 * a_2 + f(x_1, x_2), x_1 * x_2)$. In order for $A \times X$ to be entropic magma we need entropic condition, or equivalently $\phi_{a_1,a_2}(x_1, x_2) = a_1 * a_2 + f(x_1, x_2)$ should be an entropic dynamic cocycle. This leads to entropic cocycle condition:

$$\begin{aligned} &tf(x_1, x_2) - tf(x_1, x_3) + sf(x_3, x_4) \\ &\quad - sf(x_2, x_4) + f(x_1 * x_2, x_3 * x_4) - f(x_1 * x_3, x_2 * x_4) = 0. \end{aligned}$$[z]

The above formula may serve as a hint how to define (co)homology in entropic case [76].

---

[z]Calculation is as follows: Entropic condition, $((a_1, x_1) * (a_2, x_2)) * ((a_3, x_3) * (a_4, x_4)) = ((a_1, x_1) * (a_3, x_3)) * ((a_2, x_2) * (a_4, x_4))$ gives, after expanding each side: $((a_1, x_1) * (a_2, x_2)) * ((a_3, x_3) * (a_4, x_4)) = (a_1 * a_2 + f(x_1, x_2), x_1 * x_2) * (a_3 * a_4 + f(x_3, x_4), x_3 * x_4) = ((a_1 * a_2) * (a_3 * a_4) + tf(x_1, x_2) + sf(x_3, x_4) + f(x_1 * x_2, x_3 * x_4), (x_1 * x_2) * (x_3 * x_4))$

Similarly $((a_1, x_1) * (a_3, x_3)) * ((a_2, x_2) * (a_4, x_4)) = ((a_1 * a_3) * (a_2 * a_4) + tf(x_1, x_3) + sf(x_2, x_4) + f(x_1 * x_3, x_2 * x_4), (x_1 * x_3) * (x_2 * x_4))$, which reduces to (entropic) cocycle condition $tf(x_1, x_2) + sf(x_3, x_4) + f(x_1 * x_2, x_3 * x_4) = tf(x_1, x_3) + sf(x_2, x_4) + (x_1 * x_3, x_2 * x_4)$.

In particular $\partial : RX^4 \to RX^2$ may be given by:

$$\partial(x_1, x_2, x_3, x_4) = t(x_1, x_2) - t(x_1, x_3) + s(x_3, x_4) - s(x_2, x_4) \\ + (x_1 * x_2, x_3 * x_4) - (x_1 * x_3, x_2 * x_4)$$

and $\partial : RX^2 \to RX$ may be given by: $\partial(x_1, x_2) = tx_1 - x_1 * x_2 + sx_2$ (here it agrees with the rack case for $s = 1 - t$).

Remark 8.3. An important, but not yet fully utilized, observation from [76], is that we can consider atomic boundary functions $\partial^{(*)}((x_1, x_2, x_3, x_4) = (x_1 * x_2, x_3 * x_4) - (x_1 * x_3, x_2 * x_4)$ and $\partial^{(*)}(x_1, x_2) = -x_1 * x_2$, and consider also the left trivial binary operations $x *_0 y = x$, and the right trivial binary operations $x *_{\sim} y = y$ and then to recover ∂ as a three-term entropic boundary function, for the multi-entropic system $(*, *_0, *_{\sim})$, by the formula $\partial = \partial^{(*)} - t\partial^{(*_0)} - s\partial^{(*_{\sim})}$.

9. Degeneracy for a Weak and Very Weak Simplicial Module

We expand here Secs. 3.2 and 3.3 and discuss degenerate part of distributive homology in the general context of weak and very weak simplicial modules.

Quandle homology is built in analogy to group homology or Hochschild homology of associate structures. In the unital associative case we deal with simplicial sets (or modules) and it is a classical result of Eilenberg and MacLane that the degenerate part of a chain complex is acyclic so homology and normalized homology are isomorphic (see Sec. 3.2). It is not the case for distributive structures, e.g. for quandles or spindles. Quandle homology or even one-term distributive homology of spindles may have nontrivial degenerate part. The underlining homological algebra structure is a weak simplicial module and in this case the degenerate part is not necessarily acyclic and the best one can say is that the degenerate part has a natural filtration so yields a spectral sequence which can be used to study degenerate homology. In the concrete case of quandle homology (motivated by an applicable to knot theory), it is proven that the homology (called the rack homology) splits into degenerate and normalized (called the quandle homology) parts [60]. Otherwise no clear general connection between degenerate and quandle part were observed. We prove in the joint paper with Krzysztof Putyra [89] that the degenerated homology of a quandle is fully determined by quandle homology via a Künneth type formula.

10. Degeneracy for a Weak Simplicial Module

Here we give a few general observations about degenerate part of a weak simplicial module. They are related to concrete work in the distributive case done in [89].

Consider a weak simplicial module (C_n, d_i, s_i) (see Sec. 3.2 and [84]). As checked in Corollary 3.4, the filtration by degenerate elements $F_n^p = span(s_0(C_{n-1}), \ldots, s_p(C_{n-1}))$ is preserved by the boundary operation $\partial_n = \sum_{i=0}^{n}(-1)^i d_i$. In Sec. 3.3 we constructed a degenerate bicomplex $(E_{i,j}^0, d^v, d^h)$.

We discuss here the fact that a weak simplicial complex also has dual filtration (or better to say it has left and right filtrations). We define the dual (or opposite) filtration $\hat{F}_n^p = span(s_{n-1}(C_{n-1}), \ldots, s_{n-p}(C_{n-1}))$.

We start our dual description from a presimplicial module:

If $(C_n; d_i)$ is a presimplicial module then we define $\hat{d}_i = d_{n-i}$ and notice that $(C_n; \hat{d}_i)$ is also a presimplicial module with $\hat{\partial}_n = (-1)^n \partial_n$ and unchanged homology. More generally we have:

Proposition 10.1. (i) *If $(C_n; d_i)$ is a presimplicial module, then $(C_n; \hat{d}_i)$ is also a presimplicial module.*

(ii) *If $s_i : C_n \to C_{n+1}$, $0 \leq i \leq n$ are degenerate map, define $\hat{s}_i = s_{n-i}$. Then if $(C_n; d_i, s_i)$ is a (weak, or very weak) simplicial module then $(C_n; \hat{d}_i, \hat{s}_i)$ is also a (weak or very weak) simplicial module.*

Proof. (i) For $i < j$ we have $n - j < n - i$, so:

$$\hat{d}_i \hat{d}_j = \hat{d}_i d_{n-j} = d_{n-1-i} d_{n-j} \overset{(1)}{=} d_{n-j} d_{n-i} = \hat{d}_{j-1} \hat{d}_i.$$

(ii) For a better presentation let us list conditions of a simplicial module for $(C_n, \hat{d}_i, \hat{s}_i)$, one by one:

$$(\hat{1}) \qquad \hat{d}_i\hat{d}_j = \hat{d}_{j-1}\hat{d}_i \quad \text{for } i < j.$$

$$(\hat{2}) \qquad \hat{s}_i\hat{s}_j = \hat{s}_{j+1}\hat{s}_i, \quad 0 \leq i \leq j \leq n,$$

$$(\hat{3}) \qquad \hat{d}_i\hat{s}_j = \begin{cases} \hat{s}_{j-1}\hat{d}_i & \text{if } i < j \\ \hat{s}_j\hat{d}_{i-1} & \text{if } i > j+1. \end{cases}$$

$$(\hat{4}') \qquad \hat{d}_i\hat{s}_i = \hat{d}_{i+1}\hat{s}_i.$$

$$(\hat{4}) \qquad \hat{d}_i\hat{s}_i = \hat{d}_{i+1}\hat{s}_i = \mathrm{Id}_{M_n}.$$

Proposition 10.1 follows from the following lemma. □

Lemma 10.2. *Consider* $(C_n; d_i, s_i)$ *and its complementary (dual)* $(C_n; \hat{d}_i, \hat{s}_i)$, *then conditions* (x) *and* $(\hat{x})$ *are equivalent.*

Proof. Equivalence of (1) and $(\hat{1})$ was already established. Other parts are equally simple but we prove them for completeness:

(2) ⇔ $(\hat{2})$ (we assume $i \leq j$ or equivalently $n - j \leq n - i$):

$$\hat{s}_i\hat{s}_j = \hat{s}_i s_{n-j} = s_{n+1-i}s_{n-j} \overset{(2)}{=} s_{n-j}s_{n-i} = \hat{s}_{j+1}\hat{s}_i.$$

(3) ⇔ $(\hat{3})$ First assume that $i < j$ (i.e. $n + 1 - i > n - j + 1$), then

$$\hat{d}_i\hat{s}_j = d_{n+1-i}s_{n-j} \overset{(3)}{=} s_{n-j}d_{n-i} = \hat{s}_{j-1}\hat{d}_i.$$

Second assume that $i > j + 1$ (i.e. $n + 1 - i < n - j$), then

$$\hat{d}_i\hat{s}_j = d_{n+1-i}s_{n-j} \overset{(3)}{=} s_{n-j-1}d_{n-i+1} = \hat{s}_j\hat{d}_{i-1}.$$

$(4')$ ⇔ $(\hat{4}')$. We have:

$$\hat{d}_i\hat{s}_i = d_{n+1-i}s_{n-i} \overset{(4')}{=} d_{n-i}s_{n-i} = \hat{d}_{i+1}\hat{s}_i.$$

(4) ⇔ $(\hat{4})$. We have:

$$\hat{d}_i\hat{s}_i = d_{n+1-i}s_{n-i} \overset{(4)}{=} \mathrm{Id}.$$

□

Remark 10.3. If $C_n = ZX^{n+1}$ then we can consider the map $\hat{I}$: $(C_n; d_i) \to (C_n; \hat{d}_i)$ given by $\hat{I}(x_0, x_1, \ldots, x_n) = (x_n, \ldots, x_1, x_0)$ (or succinctly $\hat{I}(\mathbf{x}) = \hat{\mathbf{x}})$).

Our results (Proposition 10.1 and Lemma 10.2) hold for very weak simplicial modules, weak simplicial modules, and simplicial modules. In particular, for a weak simplicial module the dual filtration of C_n^D, $\hat{F}_n^p = span(\hat{s}_0(C_{n-1}), \hat{s}_1(C_{n-1}), \ldots, \hat{s}_p(C_{n-1}))$ leads to a spectral sequence and a bicomplex. Here we give a few general remarks to summarize basic facts:

A weak simplicial module yields two filtrations: F_n^p and the dual (complementary) one $\hat{F}_n^p$. By the definition we have

$$\partial_n = \sum_{i=0}^{n}(-1)^i d_i = (-1)^n \sum_{i=0}^{n}(-1)^i \hat{d}_i = (-1)^n \hat{\partial}_n.$$

Furthermore, on $s_p(C_{n-1})$, we have

$$\begin{aligned}
\partial_n s_p &= \sum_{i=0}^{n}(-1)^i d_i s_p \\
&= \sum_{i=0}^{p-1}(-1)^i d_i s_p + (-1)^p d_p s_p + (-1)^{p+1} d_{p+1} s_p \\
&\quad + \sum_{i=p+2}^{n}(-1)^i d_i s_p \\
&\overset{(4')}{=} \sum_{i=0}^{p-1}(-1)^i d_i s_p + \sum_{i=p+2}^{n}(-1)^i d_i s_p \\
&\overset{(3)}{=} \sum_{i=0}^{p-1}(-1)^i s_{p-1} d_i + \sum_{i=p+2}^{n}(-1)^i s_p d_{i-1}.
\end{aligned}$$

Clearly, $\sum_{i=0}^{p-1}(-1)^i s_{p-1} d_i(C_{n-1})$ belongs to F_n^{p-1}.

The formulas above lead to the bicomplex with $E_{p,q}^0 = M_{p,q} = F_n^p/F_n^{p-1}$, where $n = p+q$, $d^h = \sum_{i=0}^{p-1}(-1)^i d_i s_p$ and $d^v = \sum_{i=p+2}^{n}(-1)^i d_i s_p$. The equality $d^h d^h = 0 = d^v d^v$ and $d^h d^v = -d^v d^h$ follows directly from the weak simplicial module structure.

If we replace the filtration F_n^p by $\hat{F}_n^p$ we see that the spectral sequence is modified; it is analogous, but not the same, as when d^h is replaced by d^v.

Remark 10.4. An acute observer will notice immediately[aa] that we deal not only with a bicomplex but also with pre-bisimplicial category (set or module). For completeness I recall definitions after [61], p. 459:

We define a bisimplicial object but in a same vain we can define pre-bisimplicial category, and weak bisimplicial category: "By definition a bisimplicial object in a category $\mathcal{C}$ is a functor"

$$\mathcal{X} : \Delta^{op} \times \Delta^{op} \to \mathcal{C}.$$

Such a bisimplicial object can be described equivalently by a family of objects $M_{p,q}$, $p \geq 0, q \geq 0$, together with horizontal and vertical faces and degeneracies:

$$d_i^h : M_{p,q} \to M_{p-1,q}, \quad s_i^h : M_{p,q} \to M_{p+1,q}, \quad \text{where } 0 \leq i \leq p$$
$$d_i^v : M_{p,q} \to M_{p,q-1}, \quad s_i^v : M_{p,q} \to M_{p,q+1}, \quad \text{where } 0 \leq i \leq q$$

which satisfy the classical simplicial relations horizontally and vertically and such that horizontal and vertical operations commute. For any bisimplicial set X there are three (homeomorphic) natural ways to make geometric realization, $|X|$ of X. Loday notes that any bisimplicial set X gives rise to the bisimplicial module RX and $H_*(|X|, R) = H_*(Tot(RX))$, [61].

Example 10.5. A natural example of a (pre)-bisimplicial set or module is obtained by a Cartesian (or tensor) product of (pre)-simplicial sets (or modules). Namely:

(×) Let $M_{p,q} = C_p \times C'_q$ where (C_n, d_i) and (C'_n, d'_i) are (pre)-simplicial sets. We define $d_i^h = d_i \times \mathrm{Id}_{C'_q}$ and $d_i^v = \mathrm{Id}_{C_p} \times d'_i$. In the case (C_n, d_i, s_i) and (C'_n, d'_i, s'_i) are (weak)-simplicial sets we get $M_{p,q}$ a (weak)-simplicial set with $s_i^h = s_i \times \mathrm{Id}_{C'_q}$ and $s_i^v = \mathrm{Id}_{C_p} \times s'_i$.

[aa] Victoria Lebed studied this before me in context of her prebraided category.

($\otimes$) Let $M_{p,q} = C_p \otimes C'_q$ where (C_n, d_i) and $(C'_n, d_{,i})$ are (pre)simplicial modules. Then $M_{p,q}$ is a (pre)simplicial modules with $d_i^h = d_i \otimes \mathrm{Id}_{C'_q}$ and $d_i^v = \mathrm{Id}_{C_p} \otimes d'_i$. Similarly in the case (C_n, d_i, s_i) and (C'_n, d'_i, s'_i) are (weak)-simplicial modules.

10.1. *Right filtration of degenerate distributive elements*

We restrict ourselves here to a weak simplicial module yielded by a distributive structures.

Let $(X; *)$ be a spindle that is a magma which is right distributive ($(a * b) * c = (a * c) * (b * c)$) and idempotent ($a * a = a$).

Definition 10.6. (i) Let $\hat{s}_i = s_{n-i} : C_n \to C_{n+1}$ is given by

$$\hat{s}_i(x_0, x_1, \ldots, x_n) = (x_0, \ldots, x_{n-i-1}, x_{n-i}, x_{n-i}, x_{n-i+1}, \ldots, x_n),$$

that is we double the letter on the position $n - i$ (or i from the end) if we count from zero.

(ii) We define $\hat{F}_n^p = span(\hat{s}_0(C_{n-1}), \hat{s}_1(C_{n-1}), \ldots, \hat{s}_{n-1}(C_{n-1}))$ in $C_n(X)$. $\hat{F}_n^p$ form a boundary coherent filtration of $C_n(X)$:

$$0 \subset \hat{F}_n^0 \subset \hat{F}_n^1 \subset \hat{F}_n^{n-1} = C_n^D.$$

(iii) Let $\hat{Gr}_p^n$ be the associated graded group: $\hat{Gr}_p^n = \hat{F}_n^p / \hat{F}_{p-1}^n$.

If, as before, we define face maps $\hat{d}_i = d_{n-i}$ then $(C_n, \hat{d}_i, \hat{s}_i)$ is a weak simplicial module. Thus $\hat{F}_n^p$ is a graded filtration ripe for the spectral sequence. (We already noticed that $\hat{\partial}_n = \sum_{i=0}^n (-1)^i \hat{d}_i = (-1)^n \partial_n$.)

We consider the spectral sequence of the filtration starting from the initial page $\hat{E}_{p,q}^0 = \hat{Gr}_{p+q}^p$. It is the first main observation of [89] that the spectral sequence $(\hat{E}_{p,q}^r, \hat{\partial}_{p,q}^r)$ stabilizes on the first page, and eventually one term spindle homology can be computed easily from the normalized part.

10.2. *Integration maps* $\hat{u}_i : \hat{F}_n^p \to \hat{F}_n^{p-1}$

The main tools to show that the spectral sequence $\hat{E}^r_{p,q}$ stabilizes on the first page are the maps (which we can call integration) $\hat{u}_i : \hat{F}_n^p \to \hat{F}_n^{p-1}$, illustrated in Fig. 11,[bb] they serve to show that the right degenerated filtration spectral sequence has all d^r $(r > 0)$ trivial and that homology splits.

We check that the maps $\hat{u}_i$ satisfy:

(1) for $i < p$: $\hat{d}_i(y) = \hat{d}_{p+1}\hat{u}_i(y)$, where $y \in \hat{s}_p(C_{n-1})$.
(2) For $p > i_2 > i_1$: $\hat{d}_{p+1}\hat{u}_{i_2-1}\hat{u}_{i_1} = \hat{d}_{i_1}\hat{u}_{i_2}$.

From this follows

(2′) For $p > i_2 > i_1$: $\hat{d}_p\hat{u}_{i_2-1}\hat{u}_{i_1} = \hat{d}_{i_2-1}\hat{u}_{i_1}$.
(2) and (2′) are illustrated in Figs. 12 and 13.
(j) For $p > i_j > i_{j-1} > \cdots > i_1 \geq 0$ one has $\hat{d}_{p+1}\hat{u}_{i_j-j+1}\cdots\hat{u}_{i_2-1}\hat{u}_{i_1} = \hat{d}_{i_1}\hat{u}_{i_j-j+2}\cdots\hat{u}_{i_3-1}\hat{u}_{i_2}$.

From this follows:

$$(j \geq k) \qquad \hat{d}_{p+1-k+1}\hat{u}_{i_j-j+1}\cdots\hat{u}_{i_k-k+1}\cdots\hat{u}_{i_2-1}\hat{u}_{i_1}$$
$$= \hat{d}_{i_k-k+1}\hat{u}_{i_j-j+2}\cdots\hat{u}_{i_{k+1}-(k+1)+2}\hat{u}_{i_{k-1}-(k-1)+1}\cdots\hat{u}_{i_2-1}\hat{u}_{i_1}.$$

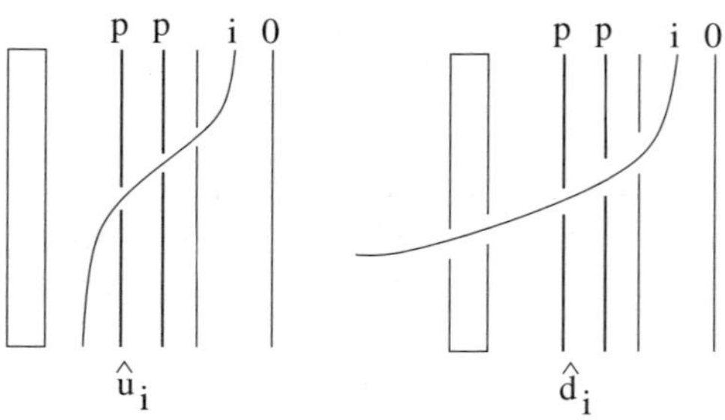

Fig. 11. The maps $\hat{u}_i$ and $\hat{d}_i$.

[bb]In braid notation, $\hat{u}_i$ can be expressed as $\sigma_p\sigma_{p-1}\cdots\sigma_{i+1}\sigma_i$.

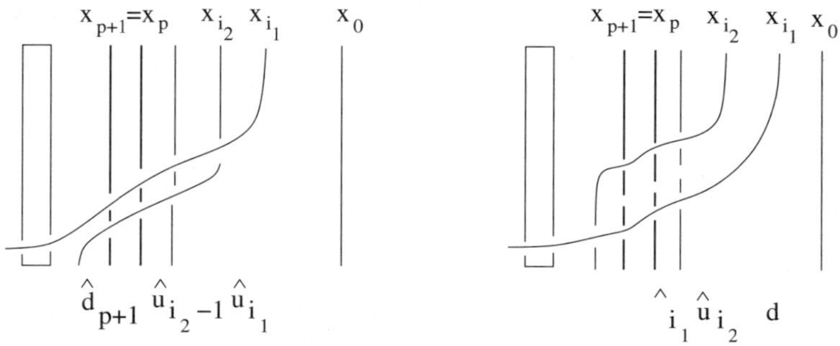

Fig. 12. Property (2): $\hat{d}_{p+1}\hat{u}_{i_2-1}\hat{u}_{i_1} = \hat{d}_{i_1}\hat{u}_{i_2}$.

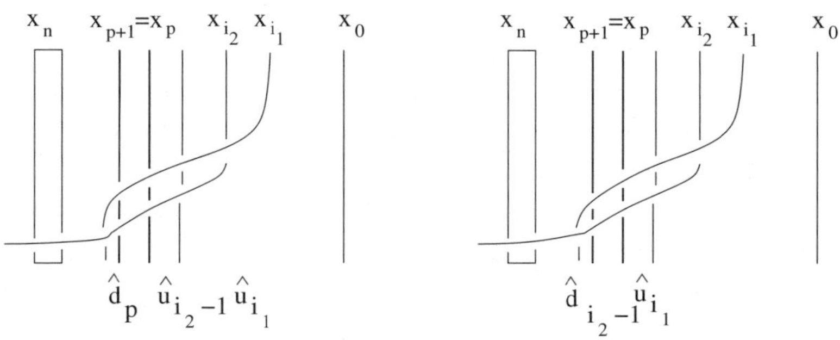

Fig. 13. Property (2): $\hat{d}_p\hat{u}_{i_2-1}\hat{u}_{i_1} = \hat{d}_{i_2-1}\hat{u}_{i_1}$.

Remark 10.7. We wrote formulas for $\hat{d}_s\hat{u}_{i_k-k+1}\cdots\hat{u}_{i_1-1}\hat{u}_{i_1}$ only in the case of $p+1-k \leq s \leq p+1$ as it is needed to compute the degenerate part of one-term distributive homology of a spindle.

10.3. *Weak simplicial modules with integration*

Here we formalize the above equations to define *weak simplicial module* with integration for which spectral sequence stabilizes on $E^1_{p,q}$ and homology splits.

We consider a weak simplicial module (C_n, d_i, s_i) with an additional structure, $u_i : s_p(C_{n-1}) \to s_{p-1}(C_{n-1})$ for $i < p$, where maps u_i satisfy condition (j) for any j. This additional structure allows us to split degenerate part.

Definition 10.8. We say that (C_n, d_i, s_i, u_i) is a weak simplicial module with integration if (C_n, d_i, s_i) is a weak simplicial module, $u_i : F_n^p \to F_n^{p-1}$ $(0 \leq i < p)$ and the following hold:

(1) $d_{p+1}u_i = d_i$ $(i < p)$.
(j) $d_{p+1}u_{i_j-j+1} \cdots u_{i_2-1}u_{i_1} = d_{i_1}u_{i_j-j+2} \cdots u_{i_3-1}u_{i_2}$, where $p > i_j > \cdots > i_1 \geq 0$.

A weak simplicial module with integration leads to bicomplex which stabilizes on $E^1_{p,q}$ and eventually splits using the maps $f_n^p : F_n^p/F_n^{p-1} \to \bigoplus_{i=0}^p F_n^i/F_n^{i-1}$.

11. Degeneracy for a Very Weak Simplicial Module

We have considered, previously, the degenerate subcomplex in the case of a weak simplicial module, however if (C_n, d_i, s_i) is only a very weak simplicial module, that is $d_i s_i$ is not necessarily equal to $d_{i+1}s_i$, we can still construct the analogue of a degenerate subcomplex (and degenerate filtration (compare Remark 3.4 in [84]).

Let $\mathcal{C} = (C_n, d_i, s_i)$ be a very weak simplicial module, that is axioms (1)–(3) of Definition 3.3 hold. We do not necessarily have the condition $d_i s_i - d_{i+1}s_i$ equal to zero so it is of interest to study an obstruction to zero: $t_i = d_i s_i - d_{i+1}s_i$. We have:

Lemma 11.1. *Let* $t_i : C_n \to C_n$ *where* $t_i = d_i s_i - d_{i+1}s_i$ *in a very weak simplicial module, then*

(i)

$$d_i t_j = \begin{cases} t_{j-1}d_i & \textit{if } i < j \\ 0 & \textit{if } i = j \\ t_j d_i & \textit{if } i > j \end{cases}$$

(ii) $t_i t_j = t_j t_i$.
(iii) *It follows from* (i) *that we have boundary preserving filtrations*:

$$F_0^t = t_0(C_n) \subset F_1^t = span(t_0(C_n), t_1(C_n)) \subset \cdots \subset F_n^t = span(t_0(C_n), \ldots, t_n(C_n)) = F^t.$$

(iv) *We also have boundary preserving filtrations:*

$$\begin{aligned}F_0^{tD} &= span(t_0(C_n), s_0(C_{n-1}) \subset \cdots \subset \\ F_{n-1}^{tD}C_n &= span(t_1(C_n), s_1(C_{n-1}), \ldots, t_{n-1}(C_n), s_{n-1}(C_{n-1}) \subset \\ F_n^{tD}C_n &= span(t_1(C_n), s_1(C_{n-1}), \ldots, t_{n-1}(C_n), s_{n-1}(C_{n-1}), t_n(C_n)) \\ &= F^{tD}.\end{aligned}$$

Or the filtration

$$0 \subset (F_0^D + F^t)/F^t \subset \cdots \subset (F_{n-1}^D + F^t)/F^t = F^{tD}/F^t.$$

(v) $(\partial_n t_p - t_p \partial_n)(t_p(C_n)) \subset t_{p-1}(C_n)$. *In particular,* t_p *is a chain map on* $F_p^t(C_n^t)/F_{p-1}^t(C_n^t)$.

F^{tD} is likely the best proxy of degenerated subchain complex so I call it the generalized degenerated subchain complex of a very weak simplicial module. The quotient $C_n/F_n^{tD}C_n$ is an analogue of a normalized chain complex (in quandle theory the quandle chain complex for any distributive structure, not necessarily spindle or quandle).

Proof. (i) ($i < j$ case): we have here, $d_i t_j = d_i d_j s_j - d_i d_{j+1} s_j = d_{j-1} d_i s_j - d_j d_i s_j = d_{j-1} s_{j-1} d_i - d_j s_{j-1} d_i = t_{j-1} d_i$,

($i = j$ case) we have, $d_j t_j = d_j d_j s_j - d_j d_{j+1} s_j = 0$,

($i > j$ case): we have, $d_i t_j = d_i d_j s_j - d_i d_{j+1} s_j =$ DO.

(ii) First we show that:

(ii′) $t_i s_j = s_j t_i$ for $i < j$ we have, stressing which property is used:

$$\begin{aligned}t_i s_j &= (d_i s_i - d_{i+1} s_i) s_j \\ &= (d_i - d_{i+1}) s_i s_j \\ &\overset{(2) s_i s_j = s_{j+1} s_i}{=} (d_i - d_{i+1}) s_{j+1} s_i \\ &= d_i s_{j+1} s_i - d_{i+1} s_{j+1} s_i \\ &\overset{(3)}{=} s_j d_i s_i - s_j d_{i+1} s_i = s_j t_i.\end{aligned}$$

(Similarly we prove that $t_i s_j = s_j t_i$ for $i > j$.) Now we complete the proof that $t_i t_j = t_j t_i$. We have, assuming

$i < j$:

$$\begin{aligned} t_i t_j &= t_i(d_j s_j - d_{j+1} s_j) \\ &\overset{(i)}{=} d_j t_i s_j - d_{j+1} t_i s_j \\ &\overset{(ii')}{=} d_j s_j t_i - d_{j+1} s_j t_i = t_j t_i \text{ as needed.} \end{aligned}$$

(iii) We start from computing $\partial_n t_p$ stressing each time which property is used:

$$\begin{aligned} \partial_n t_p &= \sum_{i=1}^{n} (-1)^i d_i t_p \\ &= \sum_{i=0}^{p-1} (-1)^i d_i t_p + 0 + \sum_{i=p+1}^{n} (-1)^i d_i t_p \\ &= \sum_{i=0}^{p-1} (-1)^i t_{p-1} d_i + \sum_{i=p+1}^{n} (-1)^i t_p d_i. \end{aligned}$$

This is to prove that filtration F_p^n is boundary preserving and also that we deal with bicomplex. Thus we can construct the spectral sequence from the bicomplex. □

Again if we work with a shelf and filtration from the right then it seems to stabilize at $E^1_{p,q}$ and homology splits, like in the case of right degenerate filtration of a spindle, $\hat{F}^p_n$. See Fig. 14 for graphical interpretation of $\hat{t}_i$. $\hat{u}^t_i$ will be defined by analogy to $\hat{u}_i$.

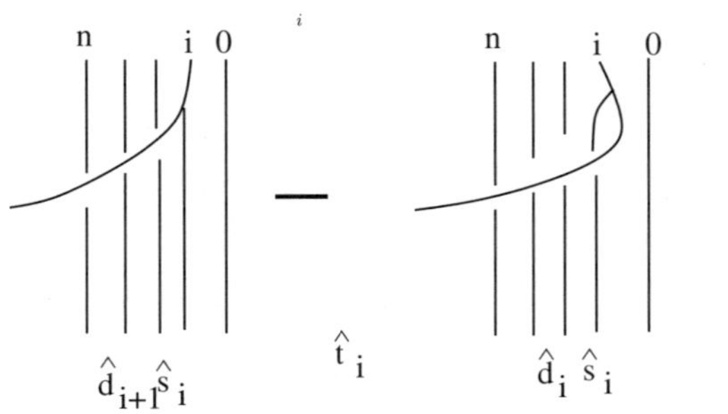

Fig. 14. The maps $\hat{t}_i$ as the difference of $\hat{d}_{i+1}\hat{s}_i$ and $\hat{d}_i\hat{s}_i$.

We leave to the reader the development of these ideas, especially in the case of multi-term distributive homology (compare [89]).

11.1. *Introduction to t-simplicial objects*

We can extract properties of maps t_i to obtain a new version of a simplicial object which we will call t-simplicial object.

Definition 11.2. Let $\mathcal{C}$ be a category and (X_n, d_i, t_i) the sequence of objects X_n, $n \geq 0$, and morphisms $d_i : X_n \to X_{n-1}$, $t_i : X_n \to X_n$, $0 \leq i \leq n$. We say that (X_n, d_i, t_i) is a t-simplicial object if the following four conditions hold (the first is the condition of a presimplicial object):

(1^t) $d_i d_j = d_{j-1} d_i$ for $i < j$.
(2^t) $t_i t_j = t_j t_i$.
(3^t)

$$d_i t_j = \begin{cases} t_{j-1} d_i & \text{if } i < j \\ t_j d_i & \text{if } i > j. \end{cases}$$

(4^t) $d_i t_i = 0$.

Let (C_n, d_i, t_i) be a t-simplicial module, with $\partial_n = \sum_{i=0}^{n} (-1)^i d_i$, then (C_n, d_i, t_i) leads to a bicomplex $(C_{p,q}, d^h, d^v)$ where d^h and d^v are defined up sign/shift by:

$$d^h = \sum_{i=0}^{p-1} (-1)^i d_i \quad \text{and} \quad d^v = \sum_{i=p+1}^{n} (-1)^i d_i.$$

We can also define a t analogue of a bisimplicial object, which we call a t-bisimplicial object, $(C_{p,q}, d_i^h, d_j^v)$, $0 \leq i \leq p$, $0 \leq j \leq q$, with $d_i^h = d_i$ and $d_j^v = d_{p+1+i}$ (however some adjustment is needed to have $p + q = n$ (or just $p + q = n - 1$).

Problem 11.3. The relation $d_i t_i = 0$ is crucial. Can one find for it general setting (say, t_i as the marker for horizontal and vertical parts of a bicomplex)? Other applications?

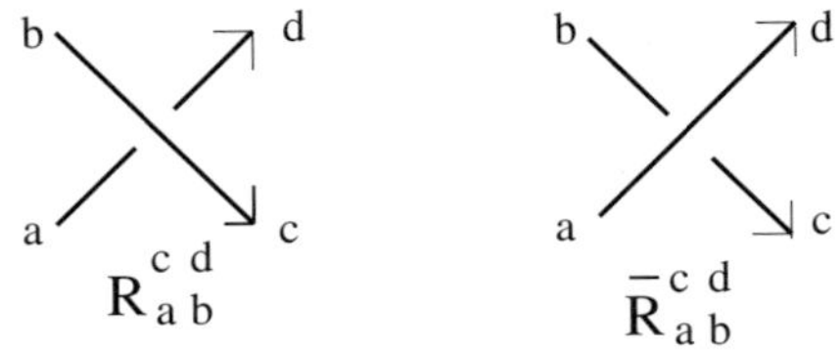

Fig. 15. Boltzmann weights $R^{c,d}_{a,b}$ and $\bar{R}^{c,d}_{a,b}$ for positive and negative crossings.

12. From Distributive Homology to Yang–Baxter Homology

We can extend the basic construction from the introduction, still using very naive point of view, as follows: Fix a finite set X and color semi-arcs of D (parts of D from a crossing to a crossing) by elements of X allowing different weights from some ring k for every crossing (following statistical mechanics terminology we call these weights Boltzmann weights). We allow also differentiating between a negative and a positive crossing; see Fig. 15.

We can now generalize the number of colorings to state sum (basic notion of statistical physics) by multiplying Boltzmann weight over all crossings and adding over all colorings:

$$col_{(X;BW)}(X) = \sum_{\phi \in col_X(D)} \prod_{p \in \{\text{crossings}\}} \hat{R}^{c,d}_{a,b}(p),$$

where $\hat{R}^{c,d}_{a,b}$ is $R^{c,d}_{a,b}$ or $\bar{R}^{c,d}_{a,b}$ depending on whether p is a positive or negative crossing. Our state sum is an invariant of a diagram but to get a link invariant we should test it on Reidemeister moves. To get analogue of a shelf invariant we start from the third Reidemeister move with all positive crossings. Here we notice that, in analogy to distributivity, where passing through a positive crossing was coded by a map $R: X\times X \to X\times X$ with $R(a,b) = (b, a*b)$. Thus in the general case passing through a positive crossing is coded by a linear map $R: kX \otimes kX \to kX \otimes kX$ and in basis X the map R is given by the $|X|^2 \times |X|^2$ matrix with entries $(R^{c,d}_{a,b})$. The third Reidemeister move leads to the equality of the following maps $V\otimes V\otimes V \to V\otimes V\otimes V$

where $V = kX$:

$$(R \otimes \mathrm{Id})(\mathrm{Id} \otimes R)(R \otimes \mathrm{Id}) = (\mathrm{Id} \otimes R)(R \otimes \mathrm{Id})\mathrm{Id} \otimes R).$$

This is called the Yang–Baxter equation and R is called a pre-Yang–Baxter operator. If R is additionally invertible it is called a Yang–Baxter operator. If entries of R^{-1} are equal to $\bar{R}^{c,d}_{a,b}$ then the state sum is invariant under "parallel" (directly oriented) second Reidemeister move.[cc]

For a given pre-Yang–Baxter operator we attempt to find presimplicial module, from which homology will be derived.

The figure below illustrate various graphical interpretation of the generating morphism d_i of the presimplicial category Δ^{op}_{pre}. They are related to homology of a set-theoretic Yang–Baxter equation of Carter–Kamada–Saito [16] and Fenn [40], and to homology of Yang–Baxter equation of Eisermann [37, 38]. We should also acknowledge stimulating observations by Ivan Dynnikov.

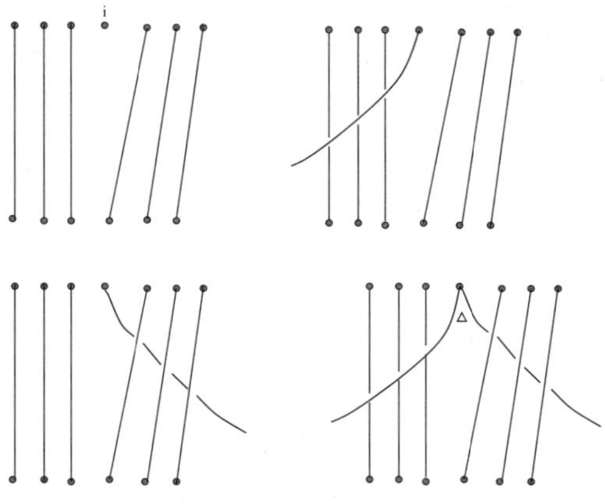

Fig. 16. Various interpretation of the graphical face map d_i.

[cc]We should stress that to find link invariants it suffices to use directly oriented second and third Reidemeister moves in addition to both first Reidemeister moves, as we can restrict ourselves to braids and use the Markov theorem. This point of view was used in [105].

12.1. *Graphical visualization of Yang–Baxter face maps*

The presimplicial set corresponding to (two-term) Yang–Baxter homology has the following visualization. In the case of a set-theoretic Yang–Baxter equation we recover the homology of [16].

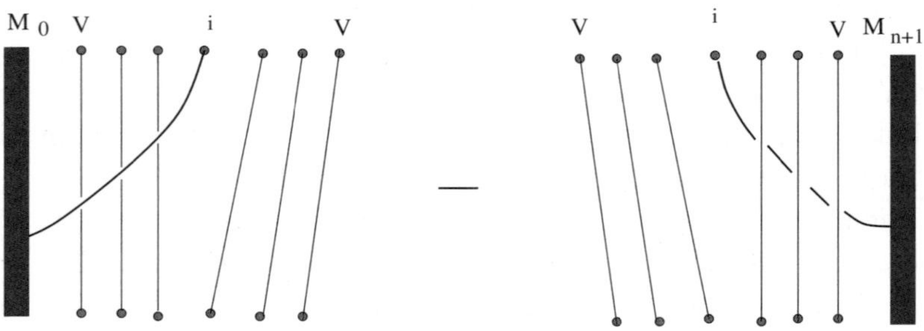

Fig. 17. Graphical interpretation of the face map d_i.

Our graphical model allows easy calculation:

Example 12.1. Assume $R : X \times X \to X \times X$ generates set-theoretic Yang–Baxter operator with $R(x, y) = (R_1(x, y), R_2(x, y))$. Then

$$\begin{aligned}
\partial^{YB}(x_1, x_2, x_3, x_4) &= \partial^{\ell} - \partial^{r}, \\
\partial^{\ell}(x_1, x_2, x_3, x_4) &= ((x_2, x_3, x_4) - (R_2(x_1, x_2), x_3, x_4) \\
&\quad + (R_2(x_1, R_1(x_2, x_3), R_2(x_2, x_3), x_4) \\
&\quad - (R_2(x_1, R_1(x_2, R_1(x_3, x_4), \\
&\quad R_2(x_2, R_1(x_3, x_4), R_2(x_3, x_4)), \\
\partial^{r}(x_1, x_2, x_3, x_4) &= (R_1(x_1, x_2), R_1(R_2(x_1, x_2), x_3), \\
&\quad R_1(R_2(R_2(x_1, x_2), x_3), x_4) \\
&\quad - (x_1, R_1(x_2, x_3), R_1(R_2(x_2, x_3), x_4) \\
&\quad + (x_1, x_2, R_1(x_3, x_4)) - (x_1, x_2, x_3).
\end{aligned}$$

We have generally for any n:

$$\partial_n^{\ell} = \sum_{i=1}^{n} (-1)^{i-1} d_i^{\ell}$$

with

$$d_i^\ell(x_1,\ldots,x_n) = (R_2(x_1,R_1(x_2,R_1(x_3,\ldots,R_1(x_{i-1},x_i))),\ldots,$$
$$R_1(x_{i-1},x_i)),x_{i+1},\ldots,x_n).$$

Similarly we have, directly from Fig. 17, for any n:

$$\partial_n^r = \sum_{i=1}^{n}(-1)^{i-1}d_i^r$$

with

$$d_i^r(x_1,\ldots,x_n) = (x_1,\ldots,x_{i-1},R_1(x_i,x_{i+1}),\ldots,$$
$$R_1(R_2(R_2(\ldots(R_2(x_i,x_{i+1}),x_{i+2}),\ldots,x_{n-1})x_n))).$$

13. Geometric Realization of Simplicial and Cubic Sets

A simplicial (or presimplicial) set (or space, more generally) can be a treated as an instruction of how to glue a topological space from pieces (simplexes in the most natural case). The result is CW complex or more precisely a Δ-complex in the terminology used in [49]. That is an object in X_n is a name/label for an n-dimensional simplex, and maps d_i (and s_i in a weak simplicial case) are given gluing instruction. Precise description (following [61]) is given below. The similar construction for a cubic (or pre-cubic) set is described at the end of the section.[dd] We speculate also what should be a natural generalization of (pre)simplicial and (pre)cubic categories.

Let $\mathcal{X}$ be a simplicial space (e.g. simplicial set with discrete topology), and $\mathcal{Y}$ a cosimplicial space (e.g. $\blacktriangle$). We define their product over Δ similar to the tensor product as follows:

$$\mathcal{X} \times_\Delta \mathcal{Y} = \bigcup_{n\geq 0}(X_n \times Y_n)/\sim_{rel},$$

where $\sim_{rel}$ is an equivalence relation generated by $(x, f_Y(y)) \sim_{rel} (f_X(x), y)$, $f : [m] \to [n]$, f_X is the image of f under contravariant functor $\Delta \to \mathcal{X}$ and f_Y is the image of f under the covariant functor $\Delta \to \mathcal{Y}$.

[dd] The definition comes under the general scheme of a *co-end*, p. 371 of [7].

Definition 13.1. The geometric realization of a simplicial space $\mathcal{X}$ is, by definition, the space

$$|\mathcal{X}| = \mathcal{X} \times_\Delta \blacktriangle = \bigcup_{n \geq 0} (X_n \times \Delta^n)/ \sim_{rel} .$$

We restrict our topological spaces in order to have $|\mathcal{X} \times \mathcal{Z}| = |\mathcal{X}| \times |\mathcal{Z}|$.

We can perform our construction also for a presimplicial set (or space). The gluing maps are then limited to that induced by d_i.

If $\mathcal{X}$ is a simplicial set we may consider only non-degenerate elements in X_n (that is elements which are not images under degeneracy maps) and build $|\mathcal{X}|$ as a CW complex: We start from the union of n-cells Δ^n indexed by non-degenerate elements in X_n; the face operations tell us how these cells are glued together to form $|\mathcal{X}|$.

Example 13.2. Let $\mathcal{X}$ be an abstract simplicial complex $X = (V, P)$. If we order its vertices then $\mathcal{X}$ is a presimplicial set with X_n being the set of n simplexes of $\mathcal{X}$ and faces are defined on each simplex in a standard way $d_i(x_0, \dots, x_n) = (x_0, \dots, x_{i-1}, x_{i+1}, \dots, x_n)$. A copresimplicial space is here the category $\blacktriangle$ with objects $Y_n = \Delta^n = \{(y_0, \dots, y_n) \in \mathbb{R}^{n+1} \mid \sum_{i=0}^n y_i = 1, y_i \geq 0\}$, one object for any $n \geq 0$. The co-face maps $d^i : Y_n \to Y_{n+1}$ are given by $d^i(y_0, \dots, y_n) = (y_0, \dots, y_{i-1}, 0, y_i, \dots, y_n)$ (of course $d^j d^i = d^i d^{j-1}$ for $i < j$). Then the topological realization $|\mathcal{X}|$ is a standard geometric simplicial complex associated to $\mathcal{X}$, that is

$$\begin{aligned} |X| &= \bigcup_{n \geq 0} (X_n \times \Delta^n)/(x, d^i(y) \\ &= (d_i(x), y) \quad \text{for } x \in X_n \text{ and } y \in \Delta^{n-1}, \end{aligned}$$

X_n with discrete topology and $|X|$ with quotient topology.

13.1. *Geometric realization of a (pre)cubic set*

First we define a precubic category $\square_{pre}$.

Definition 13.3. The precubic category $\square_{pre}$ has as objects non-negative integers $[n]$ interpreted as n points ($[0]$ has the empty

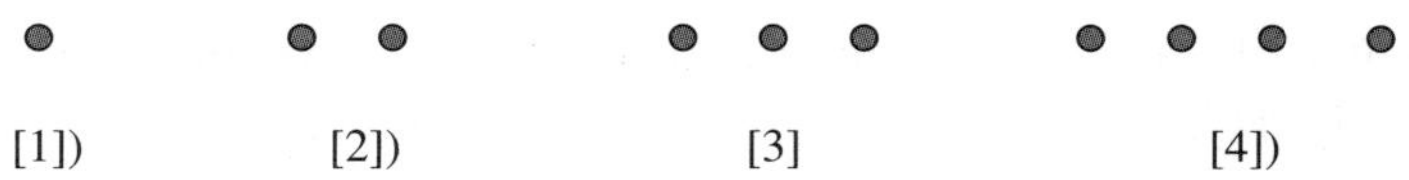

Fig. 18. Interpreting objects [1], [2], [3] and [4] in $\square_{pre}$ category.

object), see Fig. 18. Thus the objects are the same as in presimplicial Δ_{pre} category except the grading shift (now $[n]$ is in grading n).

Morphisms are strictly increasing maps that is $f \in Mor([m],[n])$ if $f : (1,2,\ldots,m) \to (1,2,\ldots,n)$ and $i < j$ implies $f(i) < f(j)$ with an additional data that points which are not in the image of f have 0 or 1 associated to them. Morphisms are generated by maps $d^i_\epsilon(1,2,\ldots,n) = (1,\ldots,i-1,i+1,\ldots,n)$ and the point i has marker ϵ (on the picture marker 0 is denoted by $\leftarrow$) (here $1 \le i \le n+1$, $\epsilon = 0$ or 1); see Fig. 19.

The presimplicial category Δ_{pre} is the quotient of the pre-cubic category $\square_{pre}$, but this functor is not that interesting in applications. The proper functor is related with triangulation of a cube [20].

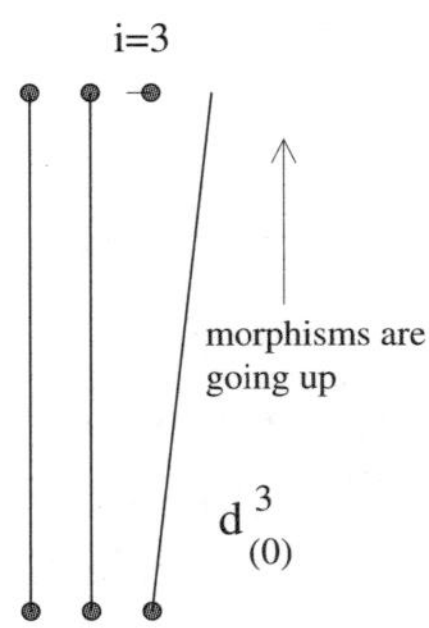

Fig. 19. Morphism (co-face map) $d^3_{(0)}$ from [3] to [4].

Definition 13.4. (i) A pre-cubic category is a contravariant functor from a pre-cubic category $\square_{pre}$ to a given category, $\mathcal{C}$, or, equivalently, a covariant functor $\mathcal{F} : \square^{op}_{pre} \to \mathcal{C}$. We can also say that a pre-cubic category is a sequence of objects X_n in $Ob(\mathcal{C})$, and morphisms $d^\epsilon_i : X_n \to X_{n-1}$ satisfying $d^\alpha_i d^\beta_j = d^\beta_{j-1} d^\alpha_i$ for $i < j$. The category

$\square_{pre}^{op}$ is visualized by the same diagrams as the category $\square_{pre}$ except morphisms will be read from the top to the bottom.

(ii) A co-pre-cubic category is a covariant functor from a pre-cubic category $\square_{pre}$ to a given category, $\mathcal{C}$. A basic example of a co-pre-cubic space ($\mathcal{C} = TOP$) is given by choosing $X^n = I^n = \{(x_1, \ldots, x_n) \in \mathbb{R}^n \mid 0 \leq x_i \leq 1\}$ and morphisms: $d_\epsilon^i(x_1, \ldots, x_n) = (x_1, \ldots, x_{i-1}, \epsilon, x_{i+1}, \ldots, x_n)$. Let us denote this co-pre-cubic space by $\blacksquare_{pre}$.

The basic co-pre-cubic space, $\blacksquare_{pre}$, can be enriched by degenerate (projection) morphisms $s^i : [n] \to [n-1]$ for $1 \leq i \leq n$, given by $s^i(x_1, \ldots, x_n) = (x_1, \ldots, x_{i-1}, x_{i+1}, \ldots, x_n)$. We denote the new category by $\blacksquare$. Then the morphisms d_ϵ^i and s^i (with the domain $X_n = I^n$) satisfy:

(1) $d_\beta^j d_\alpha^i = d_\alpha^i d_\beta^{j-1}$ for $i < j$,

(2) $s^j s^i = s^i s^{j+1}$ for $i \leq j$,

(3) $s^j d_\epsilon^i = \begin{cases} d_\epsilon^i s^{j-1} & \text{if } i < j \\ d_\epsilon^{i-1} s^j & \text{if } i > j \end{cases}$,

(4) $s^i d_\epsilon^i = \mathrm{Id}_{I^n}$.

The axioms of the cubic category are modeled on axioms (1)–(4) more precisely:

Definition 13.5. (i) The cubic category $\square$ is composed of objects $[n]$, face maps d_ϵ^i (as in pre-cubic category $\square_{pre}$) and degeneracy maps $s_i : [n-1] \to [n]$ $(1 \leq i \leq n)$ and relations between morphisms in the category are given by axioms (1)–(4).

(ii) The cubic category is a functor $\mathcal{F} : \square^{op} \to \mathcal{C}$. The classical example (of a cubic space) is giving by an approach to singular cubic homology. Here, for a topological space T, $\mathcal{F}([n])$ is the set of all continuous maps $f : I^n \to T$. $d_i^\epsilon : \mathcal{F}([n]) \to \mathcal{F}([n-1])$ is given by $d_i^\epsilon(f) = f d_\epsilon^i : I^{n-1} \xrightarrow{d_\epsilon^i} I^n \xrightarrow{f} T$.

(iii) The co-cubic category is a functor $\mathcal{F} : \square \to \mathcal{C}$. The classical example (I^n, d_ϵ^i, s^i) was described before as a motivation for a cubic category, $\blacksquare$. This example is used as a building block of a geometric realization of a pre-cubic and cubic set (Definitions 13.6 and 13.8).

As we observed already, the pre-cubic category leads to two presimplicial categories $(X_n, d_i^{(0)})$ and (X_n, d_i^1) (with shifted grading). Conversely, if we have two presimplicial categories so that $d_i^\alpha d_j^\beta = d_{j-1}^\beta d_i^\alpha$ then we combine them to in pre-cubic category (with grade shift).

With degenerate maps situation is not that clear as a conditions (3) and (4) of a simplicial category only partially agree with the analogous conditions of a cubic category.

The geometrical realization of a cubical and precubical set (or space) is analogous to that for simplicial or presimplicial sets (space). We write below a formal definition only in the case of a pre-cubic set (space) as degeneracies of a cubic set are not necessarily the one used in knot theory.

Definition 13.6. The geometric realization of a precubic set is a CW complex defined as follows (notice that X_n is indexing cubes and pre-cubic structure gives an instruction how to glue the cubes together):

$$|\mathcal{X}| = \mathcal{X} \times_{\square_{pre}} \blacksquare = \bigcup_{n \geq 0} (X_n \times I^n)/\sim_{rel},$$

where $\sim_{rel}$ is an equivalence relation generated by $(x, d_\epsilon^i(y)) \sim_{rel} (d_i^\epsilon(x), y)$, and, as before $d_\epsilon^i : I^{n-1} \to I^n$ and $d_i^\epsilon : X_i \to X_{i-1}$, $x \in X_n$, and $y \in I^{n-1}$.

More generally:

Definition 13.7. Let $\mathcal{X}$ be a pre-cubic space (e.g. pre-cubic set with discrete topology), and $\mathcal{Y}$ a co-pre-cubic space (e.g. $\blacksquare_{pre}$). The we define

$$\mathcal{X} \times_{\square_{pre}} \mathcal{Y} = \bigcup_{n \geq 0} (X_n \times Y_n)/\sim_{rel},$$

where $\sim_{rel}$ is an equivalence relation generated by $(x, d_\epsilon^i(y)) \sim_{rel} (d_i^\epsilon(x), y)$.

If (X_n, d_i^ϵ, s_i) is a cubic set (or space) we define its geometric realization similarly to Definitions 13.6 and 13.7, but including also s_i, s^i as gluing morphisms (effectively dividing by degenerate part, which is not necessary acyclic).

Definition 13.8. Let $\mathcal{X}$ be a cubic set and $\mathcal{Y}$ the co-cubic space (I^n, d_ϵ^i, s^i) then we define the geometric realization $|\mathcal{X}|$ of $\mathcal{X}$ as

$$|\mathcal{X}| = \mathcal{X} \times_\square \blacksquare = \bigcup_{n\geq 0} (X_n \times I^n)/\sim_{rel},$$

where $\sim_{rel}$ is an equivalence relation generated by $(x, d_\epsilon^i(y)) \sim_{rel} (d_i^\epsilon(x), y)$, and analogously with s_i and s^i.

14. Higher Dimensional Knot Theory $M^n \to \mathbb{R}^{n+2}$

Many ideas described in this paper can be applied to higher dimensional knot theory, where we study embedding of n-dimensional (mostly orientable) manifolds in $\mathbb{R}^{n+2}$. For this we direct reader to [18] and [91].

Acknowledgments

J. H. Przytycki was partially supported by the NSA-AMS 091111 grant, by the GWU REF grant, and Simons Collaboration Grant-316446.

References

[1] M. Andruskiewitsch and M. Grana, From racks to pointed Hopf algebras, *Adv. Math.* **178** (2002) 177–243, arxiv:0202.5084.
[2] F. R. Beyl and J. Tappe, *Group Extensions, Representations, and the Schur Multiplicator* (Springer, 1982).
[3] A. Bloh, On a generalization of the concept of Lie algebra, *Dokl. Akad. Nauk SSSR* **165** (1965) 471–473.
[4] A. Bloh, Cartan–Eilenberg homology theory for a generalized class of Lie algebras, *Dokl. Akad. Nauk SSSR* **175** (1967) 824–826.
[5] N. Bourbaki, *Elements of Mathematics*, Algebra (Algebraic Structures) (Addison-Wesley, 1974).

[6] K. S. Brown, Cohomology of Groups, Graduate Texts in Mathematics, Vol. 87 (Springer-Verlag, 1982).
[7] R. Brown, P. J. Higgins and R. Silvera, *Nonabelian Algebraic Topology*, Tracts in Mathematics, Vol. 15 (EMS, 2011).
[8] R. H. Bruck, *A Survey of Binary Systems* (Springer, 1958).
[9] C. Burstin, *Belorusskij "Memorial"* hhtp://nekropole.info/en/Celestin-Burstin, accessed November 29, 2013.
[10] C. Burstin, Letter to A. Einstein (Minsk, 1933).
[11] B. Matura, *Nowa Reforma* (afternoon edition), 285 Krakow, Monday, July 24, 1907. (Tarnopol, 22 czerwca. Egzamin dojrzałości w gimnazyum I w Tarnopolu, ...Burstin Celestyn (ekst.))
[12] C. Burstin and W. Mayer, Distributive Gruppen von endlicher Ordnung, *J. Reine Angew. Math.* **160** (1929) 111–130; Translation "Finite Distributive Groups" by Ansgar Wenzel, arXiv:1403.6326
[13] H. Cartan and S. Eilenberg, *Homological Algebra* (Princeton Univ. Press, 1956).
[14] J. S. Carter, A survey of quandle ideas, in *Introductory Lectures on Knot Theory: Selected Lectures presented at the Advanced School and Conference on Knot Theory and Its Applications to Physics and Biology, ICTP*, Trieste, Italy, 11–29 May 2009, Series on Knots and Everything, Vol. 46 (World Scientific, 2012), pp. 22–53, arXiv:1002.4429.
[15] J. S. Carter, M. Elhamdadi and M. Saito, Twisted quandle homology theory and cocycle knot invariants, *Algebr. and Geom. Topol.* **2** (2002) 95–135.
[16] J. S. Carter, M. Elhamdadi and M. Saito, Homology theory for the set-theoretic Yang–Baxter equation and knot invariants from generalizations of quandles, *Fund. Math.* **184** (2004) 31–54; arXiv:0206.5255.
[17] S. Carter, D. Jelsovsky, S. Kamada and M. Saito, Quandle homology groups, their Betti numbers, and virtual knots, *J. Pure Appl. Algebra* **157** (2001) 135–155; arXiv:math.GT9909161.
[18] S. Carter, S. Kamada and M. Saito, Surfaces in 4-space, in *Encyclopaedia of Mathematical Sciences*, Low-Dimensional Topology III, eds. R. V. Gamkrelidze and V. A. Vassiliev (Springer-Verlag, 2004).
[19] C. Chevalley and S. Eilenberg, Cohomology theory of Lie groups and Lie algebras, *Trans. Amer. Math. Soc.* **63** (1948) 85–124.
[20] F. J. B. J. Clauwens, The algebra of rack and quandle cohomology, *J. Knot Theory Ramifications* **20** (2011) 1487–1535; arXiv:1004.4423.
[21] J. H. Conway and G. Wraith, correspondence 1959.
[22] A. S. Crans, Lie 2-algebras, Ph.D. dissertation, 2004, UC Riverside, arXiv:math.QA/0409602.
[23] A. Crans and S. Nelson, Hom Quandles; arXiv:1310.0852.
[24] A. Crans, J. H. Przytycki and K. Putyra, Torsion in one term distributive homology, *Fund. Math.*, **225** (2014) 75–94; arXiv:1306.1506.
[25] R. H. Crowell, *Knots and Wheels*, National Council of Teachers of Mathematics (N.C.T.M.) Yearbook, 1961.

[26] R. H. Crowell and R. H. Fox, *An Introduction to Knot Theory* (Ginn and Co., 1963), 2nd edn. (Springer-Verlag, 1977); 3rd edn. (Dover, 2008).
[27] C. Cuvier, Homologie de Leibniz et homologie de Hochschild, *C. R. Acad. Sci. Paris Sr. I Math.* **313** (1991) 569–572. (Christian Cuvier was a student of Loday, who gave up mathematics soon after his thesis; he was the first to publish a detailed study of Leibniz algebras and their homologies, in his thesis and in the above paper; but compare [3, 4].)
[28] M. K. Dąbkowski, J. H. Przytycki and A. A. Togha, Non-left-orderable 3-manifold groups, *Canad. Math. Bull.* **48** (2005) 32–40.
[29] J. F. Davis and P. Kirk, *Lectures Notes in Algebraic Topology*, Graduate Studies in Mathematics, Vol. 35 (Amer. Math. Soc., 2001).
[30] R. Deheuvels, Homologies des ensembles ordonnés et des espaces topologiques, *Bull. Soc. Math. de France* **90** (1962) 261–321.
[31] P. Dehornoy, Multiple left distributive systems, *Comm. Math. Univ. Carol.* **38** (1997) 615–625.
[32] P. Dehornoy, *Braids and Self-Distributivity*, Progress in Mathematics, Vol. 192 (Birkhäuser Verlag, 2000).
[33] P. Dehornoy, Laver's results and low-dimensional topology, For Lavers's memorial book, to appear in Archive for Math Logic.
[34] S. Eilenberg and S. MacLane, Cohomology theory in abstract groups. I, *Ann. Math.* **48** (1947) 51–78.
[35] S. Eilenberg and S. MacLane, Acyclic models, *Am. J. Math.* **75** (1953) 189–199.
[36] S. Eilenberg and J. Zilber, Semi-simplicial complexes and singular homology, *Ann. Math. (2)* **51** (1950) 499–513.
[37] M. Eisermann, Yang–Baxter deformations of quandles and racks, *Algebr. Geom. Topol.* **5** (2005) 537–562; arXiv:0409.5202.
[38] M. Eisermann, Yang–Baxter deformations and rack cohomology, *Trans. Amer. Math. Soc.*, in press, 2014; arXiv:0808.0108.
[39] R. Fenn, Tackling the Trefoils, *J. Knot Theory Ramifications* **21** (2012) 1240004; arXiv:1110.0582v1.
[40] R. Fenn, D. P. Ilyutko, L. H. Kauffman and V. Manturow, Unsolved problems in virtual knot theory and combinatorial knot theory, *Banach Center Publications*, 103, 2014, to appear; arXiv:1409.2823.
[41] R. Fenn and C. Rourke, Racks and links in codimension two, *J. Knot Theory Ramifications* **1** (1992) 343–406; e-print: http://www.maths.sussex.ac.uk/Staff/RAF/Maths/racks.ps.
[42] R. Fenn, C. Rourke and B. Sanderson, An introduction to species and the rack space, in *Topics in Knot Theory, Proc. of the Topology Conference*, Erzurum, NATO Adv. Sci. Inst. Ser. C. Math. Phys. Sci., Vol. 399, ed. M. E. Buzhuyuk (Kluwer, 1993), pp. 33–35.
[43] R. Fenn, C. Rourke and B. Sanderson, Trunks and classifying spaces, *Appl. Categorical Structures* **3** (1995) 321–356.
[44] R. Fenn, C. Rourke and B. J. Sanderson, James bundles and applications, *Proc. London Math. Soc.* **89** (2004) 217–240.

[45] R. Fenn, C. Rourke and B. J. Sanderson, The rack space, *Trans. Amer. Math. Soc.* **359**(2) (2007) 701–740.

[46] A. Frabetti, Dialgebra (co)homology with coefficients, in J.-L. Loday, A. Frabetti, F. Chapoton and F. Goichot, eds. *Dialgebras and Related Operads*, Lectures Notes in Mathematics, Vol. 1763, (Springer-Verlag, 2001), pp. 67–103.

[47] A. Frabetti, Dialgebra homology of associative algebras, *C. R. Acad. Sci. Paris* **325** (1997) 135–140.

[48] M. Greene, *Some results in geometric topology and geometry*, Ph.D. thesis, University of Warwick, advisor: Brian Sanderson, 1997.

[49] A. Hatcher, *Algebraic Topology* (Cambridge Univ. Press, 2002); http://www.math.cornell.edu/hatcher/AT/ATch3.pdf.

[50] G. Hochschild, On the cohomology groups of an associative algebra, *Ann. Math.* **46** (1945) 58–67.

[51] H. Hopf, Fundamentalgruppe und zweite Bettische Gruppe, *Comment. Math. Helv.* **14** (1942) 257–309.

[52] W. Hurewicz, Beiträge, zur Topologie der deformationen. IV Asphärische Räume, *Nederl. Akad. Wetensch. Proc.* **39** (1936) 215–224.

[53] W. Isaacson, *Einstein: His Life and Universe* (Simon & Schuster, 2007).

[54] A. Ishii, M. Iwakiri, Y. Jang and K. Oshiro, A *G*-family of quandles and handlebody-knots, arXiv:1205.1855.

[55] D. Joyce, Ph.D. thesis 1979, adviser D. Freyd.

[56] D. Joyce, A classifying invariant of knots: the knot quandle, *J. Pure Appl. Algebra* **23** (1982) 37–65.

[57] D. Larue, Left-distributive idempotent algebras, *Comm. Algebra* **27** (1999) 2003–2029.

[58] V. Lebed, Braided objects: Unifying algebraic structures and categorifying virtual braids, December 2012, Ph.D. thesis, Université Paris 7.

[59] V. Lebed, Homologies of algebraic structures via braidings and quantum shuffles, *J. Algebra* **391** (2013) 152–192.

[60] R. A. Litherland and S. Nelson, The Betti numbers of some finite racks, *J. Pure Appl. Algebra* **178** (2003) 187–202.

[61] J.-L. Loday, Cyclic Homology, Grund. Math. Wissen., Band 301 (Springer-Verlag, 1992) (2nd edn., 1998).

[62] J.-L. Loday, Une version non commutative des algbres de Lie: les algbres de Leibniz, 1993 *Enseign. Math.* **39** (1993) 269–293.

[63] J.-L. Loday, Pirashvili Teimuraz, Universal enveloping algebras of Leibniz algebras and (co)homology, *Math. Ann.* **296** (1993) 139–158.

[64] S. MacLane, *A Mathematical Autobiography*, by Saunders MacLane (AK Peters, 2005).

[65] S. MacLane and G. Birkhoff, *Algebra* (Macmillan, 1967) (3rd edn., AMS Chelsea).

[66] L. Maligranda, Antoni Łomnicki (1881–1941), *Wiadomosci Matematyczne*, **XLIV** (2008) 61–112.

[67] J. P. May, *Simplicial Objects in Algebraic Topology* (The Univ. of Chicago Press, 1967).

[68] G. T. Mezera, Embedding groups into distributive subsets of the monoid of binary operations, arXiv:1211.0031.

[69] J. Milnor, The geometric realization of a semi-simplicial complex, *Ann. Math.* **65** (1957) 357–362.

[70] J. Mioduszewski, Celestyn Burstin (1888–1938) — członek Polskiego Towarzystwa Matematycznego, in Matematycy polskiego pochodzenia na obczyznie, XI Szkoła Historii Matematyki (Kołobrzeg, May 1997), red. S. Fudali, Wydawnictwo Naukowe Uniwersytetu Szczecińskiego, Szczecin 1998, pp. 161–165.

[71] Yu. M. Movsisyan, The multiplicative group of field and hyperidentities, *Izv. Akad. Nauk SSSR Ser. Mat.* **53** (1989) 1040–1055 [English transl. in *Math. USSR Izvestiya* **35** (1990) 337–391].

[72] M. Niebrzydowski, Knots, categories, and dynamics, presentation at *Knots in Washington* XXXV in December of 2012; http://at.yorku.ca/cgi-bin/abstract/cbfw-04.

[73] M. Niebrzydowski, J. H. Przytycki and Burnside Kei, *Fund. Math.* **190** (2006) 211–229; arXiv:math:GT0601004.

[74] M. Niebrzydowski and J. H. Przytycki, Homology of dihedral quandles, *J. Pure Appl. Algebra* **213** (2009) 742–755; arXiv:math.GT0611803.

[75] M. Niebrzydowski and J. H. Przytycki, Homology operations on homology of quandles, *J. Algebra* **324** (2010) 1529–1548; arXiv:0907.4732.

[76] M. Niebrzydowski and J. H. Przytycki, Entropic magmas, their homology, and related invariants of links and graphs, *Algebr. Geom. Topol.* **13** (2013) 3223–3243; arXiv:1211.2951.

[77] T. Nosaka, On quandle homology groups of Alexander quandles of prime order, *Trans. Amer. Math. Soc.* **365** (2013) 3413–3436.

[78] B. Peirce, Linear Associative Algebra, Read before the National Academy of Sciences, Washington City, 1870; http://www.math.harvard.edu/history/peirce_algebra/index.html.

[79] C. S. Peirce, On the algebra of logic, *Amer. J. Math.* **3** (1880) 15–57.

[80] J. H. Przytycki, Survey on recent invariants in classical knot theory, arXiv:0810.4191.

[81] J. H. Przytycki, *Teoria węzłów: podejście kombinatoryczne*, (Knots: combinatorial approach to the knot theory), Script, Warsaw, August 1995.

[82] J. H. Przytycki, 3-coloring and other elementary invariants of knots, *Banach Center Publications*, Vol. 42, *Knot Theory*, 1998, pp. 275–295; arXiv:math:GT0608172.

[83] J. H. Przytycki, When the theories meet: Khovanov homology as Hochschild homology of links, *Quantum Topol.* **1** (2010) 93–109; arXiv:math.GT0509334.

[84] J. H. Przytycki, Distributivity versus associativity in the homology theory of algebraic structures, *Demonstratio Math.* **44** (2011) 821–867; arXiv:1109.4850.

[85] J. H. Przytycki, Distributivity versus associativity in the homology theory of algebraic structures, in *Mathematisches Forschungsinstitut Oberwolfach* — Invariants in Low-Dimensional Topology and Knot Theory, Report No. 28/2012, pp. 33–36.

[86] J. H. Przytycki, From Goeritz matrices to quasi-alternating links, in *The Mathematics of Knots: Theory and Applications*, eds. M. Banagl and D. Vogel (Springer-Verlag, 2011), pp. 257–315, arXiv:0909.1118.

[87] J. H. Przytycki, Knots: From combinatorics of knot diagrams to the combinatorial topology based on knots (Cambridge Univ. Press), accepted for publication, to appear 2016.
Chap. II, arXiv.math/0703096;
Chap. III: arXiv:1209.1592v1;
Chap. IV, arXiv:0909.1118v1;
Chap. V, arXiv:math.GT/0601227;
Chap. VI, arXiv:1105.2238;
Chap. IX, arXiv:0602264;
Chap. X, arXiv:math.GT/0512630.

[88] J. H. Przytycki and K. Putyra, Homology of distributive lattices, *J. Homotopy Related Structures* **8** (2013) 35–65; arXiv:1111.4772.

[89] J. H. Przytycki and K. Putyra, Degenerate distributive complex is degenerate, arXiv:1411.5905.

[90] J. H. Przytycki and W. Rosicki, The topological interpretation of the core group of a surface in S^4, *Canad. Math. Bull.* **45** (2002) 131–137; arXiv:math.GT/0403475.

[91] J. H. Przytycki and W. Rosicki, Cocycle invariants of codimension 2-embeddings of manifolds, *Banach Center Publications*, 103, 2014, to appear, arXiv:1310.3030.

[92] J. H. Przytycki and A. S. Sikora, Distributive products and their homology, *Comm. Algebra* **42** (2014) 1258–1269; arXiv:1105.3700.

[93] J. H. Przytycki and P. Traczyk, Invariants of links of Conway type, *Kobe J. Math.* **4** (1987) 115–139.

[94] J. H. Przytycki and P. Traczyk, Conway algebras and skein equivalence of links, *Proc. Amer. Math. Soc.* **100** (1987) 744–748.

[95] K. Putyra, Cobordisms with chronologies and a generalisation of the Khovanov complex, Master degree Thesis. Jagiellonian University, Cracow, Poland, 2008, arXiv:1004.0889; (see also 2-category of chronological cobordism and odd Khovanov homology, *Banach Center Publications*, 103, 2014, to appear).

[96] A. B. Romanowska and J. D. H. Smith, *Modal Theory* (Heldermann, 1985).

[97] C. Rourke and B. Sanderson, Δ-sets. I. Homotopy theory, *Quart. J. Math. Oxford Ser. (2)* **22** (1971) 321–338.

[98] C. Rourke and B. Sanderson, A new classification of links and some calculations using it; arXiv:math/0006062.

[99] E. Schröder, Über Algorithmen und Calculn, *Arch. Math. Phys.* **5** (1887) 225–278.

[100] J.-P. Serre, *Lie Algebras and Lie Groups*, Lectures given at Harvard University, 1964 (W. A. Benjamin, 1965).

[101] A. S. Sikora, On Conway algebras and the Homflypt polynomial, *J. Knot Theory Ramifications* **6** (1997) 879–893.

[102] J. Słomińska, Coloured posets, graphs and homological algebra in functor categories, *Bull. Pol. Acad. Sci.. Math.* **60** (2012) 219–239.

[103] E. H. Spanier, *Algebraic Topology* (Springer-Verlag, 1966).

[104] M. Takasaki, Abstraction of symmetric transformation, (in Japanese) *Tohoku Math. J.* **49** (1942) 145–207; [English translation is being prepared by S. Kamada].

[105] V. G. Turaev, The Yang–Baxter equation and invariants of links, *Invent. Math.* **92** (1988) 527–553.

[106] P. Turner and E. Wagner, The homology of digraphs as a generalisation of Hochschild homology, *J. Algebra Appl.* **11** (2011) 1250031.

[107] M. Vaughan-Lee, *The Restricted Burnside Problem*, 2nd edn., London Mathematical Society Monographs. New Series, 8 (The Clarendon Press, Oxford University Press, 1993).

[108] C. E. Watts, A homology theory for small categories, in *Proc. of the Conf. on Categorical Algebra*, La Jolla, Canada 1965 (possibly the first paper defining generally homology of a small category with coefficients in a functor).

[109] C. A. Weibel, *An Introduction to Homological Algebra*, Cambridge Studies in Advanced Mathematics, Vol. 38 (Cambridge Univ. Press, 1994).

Ordering Knot Groups

Dale Rolfsen

Department of Mathematics,
University of British Columbia,
Vancouver, BC Canada V6T 1Z2
rolfsen@math.ubc.ca

In this survey we discuss the remarkable fact that classical knot groups are locally indicable and consequently they can be totally ordered in a left-invariant manner. Some knot groups, such as that of the figure-of-eight knot, have orderings which are two-sided invariant, while others (such as torus knot groups) cannot be bi-ordered. Orderability and indicability imply that the groups have special algebraic properties. Much remains to be discovered about orderability properties of knot and link groups.

1. Knot Groups and Orderings

If K is a classical knot, i.e. a smooth simple closed curve in the three-sphere S^3, then the fundamental group of its complement $\pi_1(S^3 \setminus K)$ is the *knot group* of K. A knot $K \subset S^3$ is *fibred* if there is a (locally trivial) fibre bundle map from its complement to the circle with fibre a surface.

A strict total ordering $<$ of the elements of a group G is a *left-ordering* if $g < h$ implies $fg < fh$ for all f, g, h in G; in this case G is said to be *left-orderable.* It is easy to see that a group is left-orderable if and only if it is right-orderable, by a possibly different ordering. If an ordering is simultaneously left- and right-invariant, we will call it a *bi-ordering* and say the group is *bi-orderable.*

A group is *indicable* if it has the integers as a quotient, and *locally indicable* if every nontrivial finitely-generated subgroup is indicable. Examples of locally indicable groups are torsion-free Abelian groups, free non-Abelian groups, but there are many other examples, as we will see. By a theorem of Burns and Hale [3], locally indicable groups are left-orderable.

The focus of our discussion will be the following three results, which have already appeared in the literature [5, 7, 11].

Theorem 1.1. *Every knot group is locally indicable, and hence left-orderable.*

Theorem 1.2. *If K is a fibred knot whose Alexander polynomial $\Delta_K(t)$ has all roots real and positive, then its knot group is bi-orderable.*

Theorem 1.3. *If K is a nontrivial fibred knot whose knot group is bi-orderable, then $\Delta_K(t)$ has at least two real positive roots.*

Before discussing their proofs, we will discuss why orderability is interesting, and then consider some examples.

1.1. *The importance of being orderable*

Orderable groups have some special algebraic properties, which we will discuss in this section. If $<$ is a left-ordering of the group G, we define the positive cone $P = P_<$ to be all elements of G which are greater than the identity $1 \in G$. One easily checks that

(1) P is closed under multiplication,
(2) for every $g \in G$ exactly one of $g \in P$, $g^{-1} \in P$ or $g = 1$ holds.

Conversely, if G is a group which admits a subset P satisfying (1) and (2), then there is a left-ordering $<$ of G defined by $g < h \Leftrightarrow g^{-1}h \in P$. Moreover, the ordering is a bi-ordering if and only if P is closed under conjugation by elements of G.

A subset of an ordered set is called *convex* if for every pair of elements of the subset, all members of the set between them in the ordering also belong to the subset. The collection of convex subgroups of an ordered group is totally ordered by inclusion and is an important tool in studying orderings.

Proposition 1.4. *If H is a convex normal subgroup of the left (respectively bi-) ordered group G, then G/H is left (respectively bi-) orderable.*

To see this, one just orders cosets according to the ordering of representatives; convexity insures this is well-defined. The following is also routine to check.

Proposition 1.5. *Suppose $1 \to K \to G \to H \to 1$ is an exact sequence of groups, with K regarded as a subgroup of G.*

1. *If K and H are left-orderable, then so is G, with positive cone the union of the positive cone of K and the pullback of the positive cone of H.*
2. *If K and H are bi-orderable, then so is G, if and only if there is a bi-ordering of K which is invariant under conjugation by elements of G. In this case the ordering described above is a bi-ordering of G.*
3. *If K and H are locally indicable, then so is G.*

Note that in the ordering of G described in this proposition, the subgroup K is convex.

Proposition 1.6. *Bi-orderable groups are locally indicable.*

I will just sketch a proof, based on a classic theorem of Hölder [6] that bi-ordered groups which are Archimedian actually are isomorphic, simultaneously in the algebraic and order senses, to a subgroup of the additive reals $\mathbb{R}$. This implies that such groups (if nontrivial) are indicable. *Archimedian* means that powers of each non-identity element are cofinal in the ordering. We need to show that finitely-generated nontrivial bi-ordered groups are indicable. If G is a bi-ordered group with minimal generating set $g_1, \ldots, g_n$ with g_n the largest in the ordering, one considers the union H of all *convex* subgroups of G which do not contain g_n. Then one argues that

H is a proper convex normal subgroup, G/H has an Archimedian bi-ordering and there are surjections $G \to G/H \to \mathbb{Z}$.

In summary we have implications:

$$\text{Bi-orderable} \Rightarrow \text{Locally indicable} \Rightarrow \text{Left-orderable.}$$

Neither of these implications can be reversed. Bergman [1] noted that there are finitely generated groups — in fact fundamental groups of compact 3-manifolds — which are left-orderable but not locally indicable or even indicable, because they are perfect, equal to their own commutator subgroups. To see that the first arrow cannot be reversed, we will observe that there are many knot groups which are not bi-orderable, although they are locally indicable.

A well-known algebraic open problem is the zero divisor conjecture: "if a group G is torsion-free, then the group ring $\mathbb{Z}G$ has no zero divisors." Ordering a group gives a handle on this.

Proposition 1.7. *If G is left-orderable and R a ring with no zero divisors, then the group ring RG has no zero divisors. Moreover, the only invertible elements of RG are the standard units: monomials of the form rg where $g \in G$ and r is an invertible element of R.*

Another subtle property of left-orderable groups, due to La Grange and Rhemtulla [8], is that they are determined by their integral group rings.

Proposition 1.8. *Suppose G is a left-orderable group and H is another group such that $\mathbb{Z}G \cong \mathbb{Z}H$ as rings, then G and H are isomorphic as groups.*

In a bi-orderable group (though typically not in a left-ordered group) one can multiply inequalities: $f < g$ and $f' < g'$ implies $ff' < gg'$. In particular, $f < g$ iff $f^n < g^n$ for some (hence all) positive n. The reader is invited to use this to prove the following propositions.

Proposition 1.9. *Bi-orderable groups have unique roots: if $g^n = h^n$ for nonzero exponent n, then $g = h$.*

Proposition 1.10. *In a bi-orderable group, if g commutes with h^n for some nonzero n, then g commutes with h.*

Proofs and further information about orderable groups may be found in [2].

1.2. *Examples*

First, we note that Theorem 1.1 is rather easy for the case of fibred knots. If K is a fibred knot with fibre the surface Σ, the homotopy sequence yields the short exact sequence

$$1 \to \pi_1(\Sigma) \to \pi_1(S^3 \backslash K) \to \mathbb{Z} \to 1.$$

Now Σ is a connected surface whose closure has boundary K, so $\pi_1(\Sigma)$ is a free group. Since the knot group $\pi_1(S^3 \backslash K)$ is flanked by locally indicable groups, it follows that it is also locally indicable by Proposition 1.5.

- *The trefoil* This simplest of all (nontrivial) knots is fibred, so the trefoil group is locally indicable by the above discussion. However, it cannot be bi-ordered. One way to see this is to consider its presentation $\langle x, y | xyx = yxy \rangle$ and note that this implies $xyxyxy = yxyxyx$, that is $(xy)^3 = (yx)^3$ whereas $xy \neq yx$, so we do not have unique roots.
- *Torus knots* More generally, the p, q torus knots have knot group

$$\langle a, b | a^p = b^q \rangle.$$

Note that a commutes with b^q but not with b (unless the group is Abelian, and the knot unknotted). By Proposition 1.10 we conclude:

Proposition 1.11. *Nontrivial torus knot groups are not bi-orderable.*

This could also be proved using Theorem 1.3 and the fact that torus knots are fibred. As a typical example, consider the knot 8_{19}, which is the 4, 3 torus knot. It has Alexander polynomial

$$\begin{aligned}\Delta_{8_{19}} &= 1 - t + t^3 - t^5 + t^6 \\ &= (t^2 + \sqrt{3}t + 1)(t^2 - \sqrt{3}t + 1)(t^2 - t + 1).\end{aligned}$$

Its six roots are $(\sqrt{3} \pm i)/2, (-\sqrt{3} \pm i)/2$ and $(1 \pm i\sqrt{3})/2$. More generally, the torus knot $T(p,q)$ has Alexander polynomial

$$\Delta_{T(p,q)} = \frac{(t^{pq} - 1)(t - 1)}{(t^p - 1)(t^q - 1)}$$

whose roots are on the unit circle and not real.

Yet another argument that torus knots are not bi-orderable can be made using the fact that the monodromy map associated with a fibration of the complement of a torus knot is periodic.

- *The knot* 4_1 This knot, sometimes called the figure-eight knot and pictured in the table below, is also a fibred knot. It has Alexander polynomial $1 - 3t + t^2$, whose roots are $(3 \pm \sqrt{5})/2$, both real and positive. Theorem 1.2 implies the following.

Proposition 1.12. *The group of the knot* 4_1 *is bi-orderable.*

- *Other bi-orderable knot groups* The criterion of having all roots of $\Delta_K(t)$ real and positive is not really very common. The table below contains all nontrivial prime knots with 12 or fewer crossings whose groups are known to be bi-orderable because the knots are fibred and all roots of the Alexander polynomial are in $\mathbb{R}^+$. The diagrams were produced using Rob Scharein's program Knotplot [12]. Some of the data on the knots are from *knotinfo* and its database [4], kindly provided by Chuck Livingston.

Knot		Alexander polynomial
	4_1	$1 - 3t + t^2$
	8_{12}	$1 - 7t + 13t^2 - 7t^3 + t^4$
	10_{137}	$1 - 6t + 11t^2 - 6t^3 + t^4$
	$11a_5$	$1 - 9t + 30t^2 - 45t^3 + 30t^4 - 9t^5 + t^6$
	$11n_{142}$	$1 - 8t + 15t^2 - 8t^3 + t^4$
	$12a_{0125}$	$1 - 12t + 44t^2 - 67t^3 + 44t^4 - 12t^5 + t^6$

(*Continued*)

(*Continued*)

Knot		Alexander polynomial
	$12a_{0181}$	$1 - 11t + 40t^2 - 61t^3 + 40t^4 - 11t^5 + t^6$
	$12a_{1124}$	$1 - 13t + 50t^2 - 77t^3 + 50t^4 - 13t^5 + t^6$
	$12n_{0013}$	$1 - 7t + 13t^2 - 7t^3 + t^4$
	$12n_{0145}$	$1 - 6t + 11t^2 - 6t^3 + t^4$
	$12n_{0462}$	$1 - 6t + 11t^2 - 6t^3 + t^4$
	$12n_{0838}$	$1 - 6t + 11t^2 - 6t^3 + t^4$

I think these are the only knots of at most 12 crossings known to have bi-orderable group. Added in proof: One fibred knot with 12 crossings is missing from the above table: the knot $12a_{0477}$. Its Alexander polynomial is $1 - 11t + 41t^2 - 63^3 + 41t^4 - 11t^5 + t^6$, which has all positive real roots. Several examples of non-fibred knots with bi-orderable groups have recently been discovered [16] including $6_1, 8_1, 10_1, 10_{13}, 12a_{803}$.

- *Non-bi-orderable knot groups* There are many more knot groups which are known *not* to be bi-orderable by applying Theorem 1.3. According to [4], among the knots of 12 or fewer crossings, 1246 of them are fibred and among those knots 487 have Alexander polynomials with *no* roots in $\mathbb{R}^+$, so they cannot be bi-orderable. A complete list of them can be found in [5] (see also [16]); the examples with up to ten crossings are: 3_1, 5_1, 6_3, 7_1, 7_7, 8_7, 8_{10}, 8_{16}, 8_{19}, 8_{20}, 9_1, 9_{17}, 9_{22}, 9_{26}, 9_{28}, 9_{29}, 9_{31}, 9_{32}, 9_{44}, 9_{47}, 10_5, 10_{17}, 10_{44}, 10_{47}, 10_{48}, 10_{62}, 10_{69}, 10_{73}, 10_{79}, 10_{85}, 10_{89}, 10_{91}, 10_{99}, 10_{100}, 10_{104}, 10_{109}, 10_{118}, 10_{124}, 10_{125}, 10_{126}, 10_{132}, 10_{139}, 10_{140}, 10_{143}, 10_{145}, 10_{148}, 10_{151}, 10_{152}, 10_{153}, 10_{154}, 10_{156}, 10_{159}, 10_{161}, 10_{163}.
- *Orderability unknown* There are many knot groups for which the question of bi-orderability is not known (at least not to me!). The first such knot in the usual knot tables is 5_2, which is not a fibred knot. Its Alexander polynomial is $2t^2 - 3t + 2$, which does not have real roots. Added in proof: Since the writing of this chapter, it has been discovered that the group 5_2 is NOT bi-orderable. See [15] and [17]. Many further examples also appear in [16].

2. Knot Groups are Locally Indicable

In this section we will prove Theorem 1.1: knot groups are locally indicable. First of all, we note that knot groups are indicable; if $X = S^3 \backslash K$ the Hurewicz homomorphism $h : \pi_1(X) \to H_1(X)$ is surjective, and $H_1(X)$ is infinite cyclic by Alexander duality. To prove that $\pi_1(X)$ is *locally* indicable, we need to consider an arbitrary nontrivial finitely generated subgroup G of $\pi(X)$ and argue that it admits a nontrivial homomorphism to $\mathbb{Z}$. This argument is due essentially to Howie and Short [7].

Case 1. G has finite index. Then the restriction $h|_G$ of the Hurewicz homomorphism is nontrivial and we are done.

Case 2. G has infinite index. There is a covering space $p : \tilde{X} \to X$ such that, for suitably chosen basepoint, $p_*(\pi_1 \tilde{X}) = G$. Although $\tilde{X}$ must be non-compact, its fundamental group is finitely generated, by assumption. By a theorem of P. Scott [13], $\tilde{X}$ has a compact "core" — that is a compact connected submanifold C of $\tilde{X}$ such that inclusion induces an isomorphism $i_* : \pi_1(C) \to \pi_1(\tilde{X})$.

Noting that C must have nonempty boundary, we first argue that we can assume there are no 2-sphere components of ∂C. For suppose $\Sigma \subset \partial C$ is a 2-sphere. Knot complements are irreducible, that is every tame 2-sphere bounds a ball, and it is known that irreducibility is inherited by coverings, so $\tilde{X}$ is irreducible. Therefore there is a 3-ball B in $\tilde{X}$ with $\partial B = \Sigma$. It is easy to see that C is either a subset of B or else disjoint from the interior of B. But $C \subset B \subset \tilde{X}$ would contradict the fact that i_* is a nontrivial homomorphism, so we conclude that B is disjoint from the interior of C. If we now define $C' := C \cup B$, we see that C' will also serve as a compact core for $\tilde{X}$. After repeating this a finite number of times we obtain a compact core, which we will again call C, such that $\partial C \neq \emptyset$ and every component of ∂C is a surface of positive genus.

Lemma 2.1. *The integral homology group $H_1(C)$ is infinite.*

This completes Case 2, because then one easily constructs a surjection of the Abelian group $H_1(C)$ onto $\mathbb{Z}$ and combined with the Hurewicz map we have a surjection $G \cong \pi_1(C) \to H_1(C) \to \mathbb{Z}$.

Proof of the Lemma. This is a standard argument, repeated here for the reader's convenience. We will show, equivalently, that the *rational* homology $H_1(C; \mathbb{Q})$ has positive rank. Recall that C is a compact orientable 3-manifold with nonempty boundary containing no 2-spheres. Consider the *closed* manifold $2C$ obtained from two copies of C, with their boundaries glued together by the identity map. The Euler characteristic of a closed 3-manifold is always zero, so we have $0 = \chi(2C) = 2\chi(C) - \chi(\partial C)$. Our assumption on ∂C implies that its Euler characteristic is less than or equal to zero.

Therefore $\chi(C) \leq 0$. But by definition of $\chi(C)$ as alternating sum of ranks, we conclude

$$1 - \text{rank } H_1(C;\mathbb{Q}) + \text{rank } H_2(C,\mathbb{Q}) - 0 \leq 0$$

which implies rank $H_1(C;\mathbb{Q}) \geq 1$. □

3. Fibred Knots, Bi-Ordering and Eigenvalues

A fibration $X \to S^1$, with fibre Σ can be regarded as a product of Σ with an interval $I = [0,1]$ with the ends identified via some homeomorphism h of Σ called the *monodromy* associated with the fibration:

$$X = \Sigma \times I/(x,1) \sim (h(x),0).$$

The fundamental group of X is an HNN extension of that of Σ. In the special case that X is the complement of a fibred knot or link, Σ is a connected surface with boundary and its fundamental group is a free group. We may write

$$\pi_1(X) = \langle g_1,\ldots,g_k,t | t^{-1}g_i t = h_*(g_i), i = 1,\ldots,k\rangle,$$

where $\{g_i\}$ is a set of free generators of $\pi_1(\Sigma)$ and $h_* : \pi_1(\Sigma) \to \pi_1(\Sigma)$ is the homotopy monodromy map induced by h.

We know that free groups are bi-orderable [14]. Although left-orderability is preserved under taking HNN extensions, bi-orderability may not be. In fact it is not difficult to verify that an HNN extension of a bi-orderable group G is again bi-orderable if and only if there is a bi-ordering of G which is invariant under the automorphism $\varphi : G \to G$ associated with the extension.

Again specializing to X being the complement of a fibred knot, its fundamental group is bi-orderable if and only if there exists a bi-ordering of the free group $\pi_1(\Sigma)$ which is invariant under the homotopy monodromy $h_* : \pi_1(\Sigma) \to \pi_1(\Sigma)$. We also note that for a fibred knot K the Alexander polynomial $\Delta_K(t)$ is precisely the characteristic polynomial for the *homology* monodromy $h_* : H_1(\Sigma) \to H_1(\Sigma)$. The homology groups may be considered with rational (or even real or complex) coefficients, so that h_* is a linear map of vector spaces. (Note that we are using the same symbol for the homotopy monodromy and the homology monodromy — the induced map upon abelianization. The context should make it clear which map is under discussion.)

3.1. *Digression on linear algebra*

Suppose we have an invertible linear map $L : \mathbb{R}^n \to \mathbb{R}^n$ and we wish to find a bi-ordering $<$ of $\mathbb{R}^n$ as an additive group such that L preserves the order: $v < w \Leftrightarrow L(v) < L(w)$.

It may be impossible — for example if L has finite order or even a finite orbit. Indeed suppose that $L(v) \neq v$ but $L^k(v) = v$ and $<$ is an invariant ordering. If $L(v) < v$ we have $L^2(v) < L(v) < v$ and inductively $L^k(v) < v$, a contradiction. If $L(v) > v$ a similar contradiction ensues.

On the other hand, suppose L is represented by an upper triangular matrix, as in the following equation ($n = 3$ for simpicity)

$$\begin{pmatrix} \lambda_1 & * & * \\ 0 & \lambda_2 & * \\ 0 & 0 & \lambda_3 \end{pmatrix} \begin{pmatrix} x_1 \\ x_2 \\ x_3 \end{pmatrix} = \begin{pmatrix} \lambda_1 x_1 + *x_2 + *x_3 \\ \lambda_2 x_2 + *x_3 \\ \lambda_3 x_3 \end{pmatrix}.$$

Further suppose the eigenvalues λ_i are all positive. Then we can order vectors by taking the positive cone to be all vectors whose last nonzero coordinate is greater than zero. In other words reverse lexicographic order. Then one sees from the above equation that L preserves that positive cone and hence respects the ordering. We have sketched a proof of the following.

Proposition 3.1. *If all the eigenvalues of a linear map $L : \mathbb{R}^n \to \mathbb{R}^n$ are real and positive, then there is a bi-ordering of $\mathbb{R}^n$ which is preserved by L. Similarly for $\mathbb{Q}^n$ in place of $\mathbb{R}^n$.*

There is a partial converse.

Proposition 3.2. *Suppose there is a bi-ordering of $\mathbb{R}^n$ which is preserved by the non-singular linear map $L : \mathbb{R}^n \to \mathbb{R}^n$. Then L has at least one positive real eigenvalue. Similarly for $\mathbb{Q}^n$ in place of $\mathbb{R}^n$.*

This may be proved topologically. Consider the set H of all points in $\mathbb{R}^n$ for which every neighbourhood contains points greater than zero and also points less than zero in the ordering. One argues that H is a codimension-one subspace of R^n separating the positive points from the negative points in the ordering. Let $D \subset S^{n-1}$ be the (closed) half of the unit sphere on the positive side of H. Then D is homeomorphic with an $(n-1)$-dimensional ball. Since L preserves the ordering, the map $v \to L(v)/|L(v)|$ takes D to itself. By the Brouwer fixed-point theorem, that map has a fixed point. Finally, one observes that such a fixed point is an eigenvector of L with positive eigenvalue. To argue for $\mathbb{Q}^n$ just repeat this argument, considering $\mathbb{Q}^n$ inside $\mathbb{R}^n$ in the usual way.

3.2. *Proof of Theorem 1.2*

Since roots of the Alexander polynomial are exactly the eigenvalues of the homology monodromy associated with a fibred knot, our problem reduces to showing:

Proposition 3.3. *Suppose $h : F \to F$ is an automorphism of a finitely-generated free group. If all the eigenvalues of the induced map $h_* : H_1(F;\mathbb{Q}) \to H_1(F;\mathbb{Q})$ are real and positive, then there is a bi-ordering of F preserved by h.*

Proof. One way to order a free group F is to use the lower central series $F_1 \supset F_2 \supset \cdots$ defined by

$$F_1 = F, \quad F_{i+1} = [F, F_i].$$

This has the properties that $\bigcap F_i = \{1\}$ and F_i/F_{i+1} is free Abelian.

Choose an arbitrary bi-ordering of F_i/F_{i+1}, and define a positive cone of F by declaring $1 \neq x \in F$ positive if its class in F_i/F_{i+1} is positive in the chosen ordering, where i is the last subscript such that $x \in F_i$. One can check that this is a bi-ordering of F.

If $h : F \to F$ is an automorphism it preserves the lower central series and induces maps of the lower central quotients: $h_i : F_i/F_{i+1} \to F_i/F_{i+1}$. With this notation, h_1 is just the abelianization h_{ab}. In a sense, all the h_i are determined by h_1. That is, there is an embedding of F_i/F_{i+1} in the tensor power $F_{ab}^{\otimes k}$, and the map h_i is just the restriction of $h_{ab}^{\otimes k}$. The reader is referred to [11] for details.

The assumption that all eigenvalues of h_{ab} are real and positive implies that the same is true of all its tensor powers. This allows us to find bi-orderings of the free Abelian groups F_i/F_{i+1} which are invariant under h_i for all i. Using these to bi-order F, we get invariance under h, which proves the proposition and Theorem 1.2.

□

4. Proof of Theorem 1.3

Let us turn to the proof of the third main theorem: If $K \subset S^3$ is a nontrivial fibred knot whose knot group is bi-orderable, then $\Delta_K(t)$ has at least two real positive roots. First of all, since the Alexander polynomial satisfies $\Delta_K(t) = t^{2g}\Delta_K(1/t)$ and $\Delta_K(1) = \pm 1$, any positive real root r will produce another, namely $1/r$. So we need only to find one positive real root.

Our third theorem will follow from a more general result. Suppose G is an arbitrary finitely generated group. If $\phi : G \to G$ is an automorphism, we can define its *eigenvalues* to be the eigenvalues of its induced map on the rational vector space $H_1(G, \mathbb{Q}) \cong G_{ab} \otimes \mathbb{Q}$.

Theorem 4.1. *Suppose G is a nontrivial finitely generated bi-orderable group and that the automorphism $\phi : G \to G$ preserves a bi-ordering of G. Then ϕ has a positive eigenvalue.*

To prove this, assume $\phi : G \to G$ preserving a bi-ordering of G. Then ϕ induces an automorphism $\phi_* : G/G' \to G/G'$, but (unless the commutator subgroup G' is convex) we do not know that G/G' inherits a ϕ_*-invariant ordering. However G, being finitely generated, does have a maximal proper convex subgroup C. Since ϕ respects the ordering, C is ϕ-invariant. Moreover, G/C is Abelian so $G' \subset C$ and we have the commutative diagram with exact rows:

$$\begin{array}{ccccccccc} 0 & \longrightarrow & C/G' & \longrightarrow & G/G' & \longrightarrow & G/C & \longrightarrow & 0 \\ & & \downarrow & & \phi_* \downarrow & & \phi_C \downarrow & & \\ 0 & \longrightarrow & C/G' & \longrightarrow & G/G' & \longrightarrow & G/C & \longrightarrow & 0. \end{array}$$

Moreover, since C is convex, G/C inherits an order from G which is invariant under ϕ_C.

Writing $U = C/G' \otimes \mathbb{Q}$, $V = G/G' \otimes \mathbb{Q}$, and $W = G/C \otimes \mathbb{Q}$, tensoring with $\mathbb{Q}$ yields the commutative diagram of finite-dimensional vector spaces over $\mathbb{Q}$ with exact rows:

$$\begin{array}{ccccccccc} 0 & \longrightarrow & U & \longrightarrow & V & \longrightarrow & W & \longrightarrow & 0 \\ & & \downarrow & & \phi_V \downarrow & & \phi_W \downarrow & & \\ 0 & \longrightarrow & U & \longrightarrow & V & \longrightarrow & W & \longrightarrow & 0, \end{array}$$

where $\phi_W = \phi_C \otimes \mathrm{id}$ and $\phi_V = \phi_* \otimes \mathrm{id}$.

Since ϕ_W preserves the induced ordering of W, it has a positive real eigenvalue.

Letting ϕ_U be ϕ_V restricted to U, we conclude that $\phi_V = \phi_U \oplus \phi_W$.

Therefore the characteristic polynomial of ϕ_V factors as

$$\chi_{\phi_V}(\lambda) = \chi_{\phi_U}(\lambda) \cdot \chi_{\phi_W}(\lambda).$$

The positive eigenvalue of ϕ_W is also an eigenvalue of ϕ_V, concluding the proof.

5. Surgery

We conclude with some applications to *surgery* on a knot K in Σ^3, as discussed in more detail in [5]. Recall that in surgery one removes a tubular neighbourhood of K and attaches a solid torus $\Sigma^1 \times D^2$ to the remaining exterior $X = S^3 \setminus \nu(K)$ so that the meridian $\{*\} \times S^1$ is attached to a specified "framing" curve on the boundary of the neighbourhood $\nu(K)$.

Theorem 5.1. *Suppose K is a fibred knot in S^3 and nontrivial surgery on K produces a 3-manifold M whose fundamental group is bi-orderable. Then the surgery must be longitudinal (that is, 0-framed) and $\Delta_K(t)$ has a positive real root. Moreover, M fibres over S^1.*

This is actually an easy application of Theorem 1.3. First, note that the surgery must be longitudinal for homological reasons. If X is the complement of the tubular neighbourhood of the knot K, the knot group $\pi_1(X)$ has preferred elements which generate the fundamental group of ∂X: meridian μ, represented by a loop bounding a disk transverse to the knot, and longitude λ, which is parallel to the knot and homologically trivial in X. A surgery framing curve J is then represented by a pair of relatively prime integers p and q, where

$$[J] = \mu^p \lambda^q$$

and the surgery manifold M has fundamental group obtained from $\pi_1(X)$ by killing $\mu^p\lambda^q$. Similarly, a Mayer–Vietoris argument shows that $H_1(M)$ can be calculated from $H_1(X)$, which is infinite cyclic and generated by the meridian, by killing $p\mu$ (in additive notation). So $H_1(M)$ is a *finite* cyclic group unless $p = 0$.

Now suppose $\pi_1(M)$ is bi-orderable. It is a nontrivial group by the so-called property P theorem. Recalling Proposition 1.6 and that $\pi_1(M)$ itself is finitely generated, there is a surjection $\pi_1(M) \to \mathbb{Z}$. This cannot happen if $H_1(M)$ is finite. It follows that p must be zero, or in other words $[J] = \lambda$. Since the preimages of points under the fibration map $X \to S^1$ are longitudinal curves on ∂Z, they bound meridian disks in $S^1 \times D^2$. The fibration map to S^1 thus extends to these disks and we see that M fibres over S^1.

Since the fibration of M is essentially that of the knot exterior X: the fibres $\hat{\Sigma}$ of the former are the fibres Σ of the latter with disks sewn to the boundary. The first homology of the fibres coincides in the two cases, and the same is true of the homology monodromy. By Theorem 4.1, applied to the group $G = \pi_1(\hat{\Sigma})$ we conclude that the homology monodromy has a positive eigenvalue. But this is a root of the Alexander polynomial of the knot.

Ozsváth and Szabó [10] define an *L-space* to be a closed 3-manifold M such that $H_1(M;\mathbb{Q}) = 0$ and its Heegaard–Floer homology $\widehat{HF}(M)$ is a free Abelian group of rank equal to $|H_1(M;\mathbb{Z})|$, the simplest possible. Lens spaces, and more generally 3-manifolds with finite fundamental group are examples of L-spaces, but there are also many with inifinite fundamental group.

Theorem 5.2. *If surgery on a knot K in S^3 results in an L-space, then the knot group $\pi_1(S^3\backslash K)$ is not bi-orderable.*

Of course, one may rephrase this by saying that the group of a knot is bi-orderable, then surgery on that knot never results in an L-space.

I will just outline the proof, referring the reader to [5] for details. By Yi Ni [9] if surgery on K yields an L-space, K must be fibred. Moreover, Ozsváth and Szabó show that the Alexander polynomial of K must have a special form. Then one argues that a polynomial of this form has no positive real roots, so the knot group cannot be bi-ordered.

References

[1] G. M. Bergman, Right orderable groups that are not locally indicable, *Pacific J. Math.* **147** (1991) 243–248.

[2] R. B. Mura and A. Rhemtulla, *Orderable Groups*, Marcel Dekker, New York, 1977. Lecture Notes in Pure and Applied Mathematics, Vol. 27.

[3] R. G. Burns and V. W. D. Hale, A note on group rings of certain torsion-free groups, *Canad. Math. Bull.* **15** (1972) 441–445.

[4] J. C. Cha and C. Livingston, Knotinfo: Table of knot invariants. http://www.indiana.edu/knotinfo.

[5] A. Clay and D. Rolfsen, Ordered groups, eigenvalues, knots, surgery and L-spaces, *Math. Proc. Cambridge Philos. Soc.* **152** (2012) 115–129.

[6] O. Hölder, Die Axiome der quantität und die Lehre vom Mass, *Math.-Phys. Kl* **53** (1901) 1–64.

[7] J. Howie and H. Short, The band-sum problem, *J. London Math. Soc. (2)* **31** (1985) 571–576.

[8] R. H. La Grange and A. H. Rhemtulla, A remark on the group rings of order preserving permutation groups, *Canad. Math. Bull.* **11** (1968) 679–680.

[9] Yi Ni, Knot Floer homology detects fibred knots, *Invent. Math.* **170** (2007) 577–608.

[10] P. Ozsváth and Z. Szabó, On knot Floer homology and lens space surgeries, *Topology* **44** (2005) 1281–1300.
[11] B. Perron and D. Rolfsen, On orderability of fibred knot groups, *Math. Proc. Cambridge Philos. Soc.* **135** (2003) 147–163.
[12] R. Scharein, Knotplot, www.knotplot.com.
[13] G. P. Scott, Compact submanifolds of 3-manifolds, *J. London Math. Soc. (2)* **7** (1973) 246–250.
[14] A. A. Vinogradov, On the free product of ordered groups, *Math. Sbornik N.S.* **25** (1949) 163–168.
[15] I. M. Chiswell, A. M. W. Glass and J. S. Wilson, Residual nilpotence and ordering in one-relator groups and knot groups, preprint, arXiv:1405.0994.
[16] A. Clay, C. Desmarais and P. Naylor, Testing bi-orderability of knot groups, preprint, arXiv:1410.5774.
[17] G. Naylor and D. Rolfsen, Generalized torsion in knot groups, preprint, arXiv:1409.5730.

Casson-Type Invariants from the Seiberg–Witten Equations

Daniel Ruberman

Department of Mathematics,
Brandeis University, Waltham, MA 02454, USA
ruberman@brandeis.edu

Nikolai Saveliev

Department of Mathematics,
University of Miami, Coral Gables, FL 33124, USA
saveliev@math.miami.edu

This is a survey of our recent work [16, 17, 23] with Tom Mrowka on Seiberg–Witten gauge theory and index theory for manifolds with periodic ends. We explain how this work leads to a new invariant, which is related to the classical Rohlin invariant of homology 3-spheres and to the Furuta–Ohta invariant originating in Yang–Mills gauge theory. We give some new calculations of our invariant for 4-dimensional mapping tori.

1. Introduction

Since its inception in the mid-'90s, Seiberg–Witten gauge theory has had numerous applications in topology. The natural domain of this theory comprises simply connected closed oriented smooth 4-manifolds with $b_+ > 1$, where a straightforward count of irreducible solutions to the Seiberg–Witten equations produces a diffeomorphism invariant of the manifold. Here, b_+ stands for the number of positive eigenvalues of the intersection form on the second homology of the manifold. The theory has also been extended to manifolds with

$b_+ = 1$ using wall-crossing formulas. The project surveyed in this paper deals with the Seiberg–Witten theory for a class of manifolds having $b_+ = 0$, including manifolds with integral homology of $S^1 \times S^3$. In this case, the usual count of irreducible solutions to the Seiberg–Witten equations depends heavily on metric and perturbation but in a joint project with Tom Mrowka, we succeeded in [16] in defining a diffeomorphism invariant by countering this dependence by a correction term.

The correction term is of great interest in its own right. Its definition was inspired by the work of Weimin Chen [6] and Yuhan Lim [13], who counted irreducible solutions to the Seiberg–Witten equations on a 3-dimensional homology sphere. This count is not well-defined because of potential contributions from the reducible solutions to the equations as one varies the metric and perturbation in a 1-parameter family; this issue is analogous to the one we encounter in the 4-dimensional case when $b_+ = 0$. To get a diffeomorphism invariant, Chen and Lim (independently, following a suggestion of Kronheimer) added a correction term which is a combination of η-invariants of Atiyah–Patodi–Singer [3]. Equivalently, their correction term can be expressed as a combination of the signature of a compact spin 4-manifold with boundary the homology sphere, and the index of the spin Dirac operator on said 4-manifold with the Atiyah–Patodi–Singer boundary conditions.

Our correction term is similar to that of Chen and Lim but requires a considerably more complicated analytical setting because our definition involves the spin Dirac operator on a non-compact manifold with a periodic end. A fundamental analytical issue is therefore ensuring that this Dirac operator is Fredholm and hence has a well-defined index. In dealing with this issue, we relied on the study of the Fredholm properties of elliptic operators on manifolds with periodic ends initiated by Taubes [25]. We further developed this theory, which allowed us to prove the well-definedness of our invariant in [16]. It also led us in [17] to a general index theorem for end-periodic Dirac operators in the spirit of the Atiyah–Patodi–Singer theorem [3], complete with a new η-invariant. A special case of this theorem is described in Sec. 5.

Our interest in Seiberg–Witten invariants of manifolds with the homology of $S^1 \times S^3$ is mainly explained by the prominent role these manifolds play in low-dimensional topology: several outstanding problems concerning homology cobordisms and the Rohlin invariant can be translated using the doubling construction to problems about a homology $S^1 \times S^3$. Moreover, the classification of smooth manifolds homotopy equivalent to $S^1 \times S^3$ is a basic problem for the theory of non-simply-connected 4-manifolds. We refer the reader to our paper [20] and to Sec. 4 of current paper for details.

It should be pointed out that in [20] we studied another set of gauge theoretic invariants of a homology $S^1 \times S^3$ arising from Donaldson gauge theory. As in Seiberg–Witten theory, the study of solutions to the Yang–Mills equations on a manifold with $b_+ = 0$ has some subtleties that are not present in the usual theory of Donaldson invariants. Among the invariants studied in [20] is the one due to Furuta and Ohta [8] of manifolds with the $\mathbb{Z}[\mathbb{Z}]$ homology of $S^1 \times S^3$. We conjecture that this Furuta–Ohta invariant matches the Seiberg–Witten invariants described in this paper. Our conjecture can be viewed as an extension of the Witten conjecture [28] comparing Donaldson and Seiberg–Witten invariants to manifolds with $b_+ = 0$. It is straightforward to verify that, for manifolds of the type $S^1 \times \Sigma$, where Σ is an integral homology sphere, the Furuta–Ohta invariant reproduces the Casson invariant of Σ. Moreover, our invariant in this product case is equal to the Seiberg–Witten invariant of Chen and Lim. Thus the conjecture in the product case follows from these observations together with the theorem proved by Lim [14] that Chen and Lim's invariant is the Casson invariant. In this paper, we verify the conjecture for the more general case of mapping tori of orientation preserving finite order diffeomorphisms $\tau : \Sigma \to \Sigma$ without fixed points; this is the only original result of this paper, and we provide its complete proof. While we have not been able to handle the situation when τ has fixed points in full generality, we verified the conjecture in some special cases in [23].

2. Seiberg–Witten Moduli Spaces

A homology $S^1 \times S^3$ is a smooth closed oriented 4-manifold X such that $H_*(X;\mathbb{Z}) = H_*(S^1 \times S^3;\mathbb{Z})$. One way to obtain such a manifold is to furl up a smooth homology cobordism W from an integral homology 3-sphere Σ to itself by gluing the two boundary components of W together via the identity map. If W is the product cobordism, this construction will result in $X = S^1 \times \Sigma$, and if W is the mapping cylinder of $\tau : \Sigma \to \Sigma$ the manifold X will be the mapping torus of τ.

The Seiberg–Witten invariant of X that we wish to define will depend on a choice of generator $1 \in H^1(X;\mathbb{Z}) = \mathbb{Z}$, called a *homology orientation.* The invariant will prove to be independent of several other choices, which are however necessary to just write the Seiberg–Witten equations. These are the choices of spin structure, Riemannian metric g, and perturbation $\beta \in \Omega^1(X, i\mathbb{R})$. The manifold X has two different spin structures, corresponding to the fact that $H^1(X;\mathbb{Z}/2) = \mathbb{Z}/2$. Since these spin structures are the same when viewed as spin^c structures, our invariant will be independent of this choice. The independence of g and β is much less obvious, and proving it is a major part of this project.

The Seiberg–Witten equations [11] are a system of nonlinear partial differential equations on triples (A, s, φ), where A is a U(1) connection on the determinant bundle of the spin bundle, φ is a positive chiral spinor of L^2 norm one, and $s \geq 0$ is a real number. The equations read

$$F_A^+ - s^2 \cdot \tau(\varphi) = d^+\beta, \quad D_A^+(X, g)(\varphi) = 0,$$

where $F_A^+ \in \Omega^2_+(X; i\mathbb{R})$ is the anti-self-dual part of the curvature, $D_A^+(X, g)$ is the chiral Dirac operator on X, and $\tau(\varphi)$ is a quadratic form in φ whose exact nature is immaterial for this paper. The gauge group, which consists of the maps $u : X \to S^1$, acts on the set of solutions of this system by the rule $(A, s, \varphi) \to (A - u^{-1}du, s, u \cdot \varphi)$. The gauge equivalence classes of solutions (A, s, φ) form the Seiberg–Witten moduli space $\mathcal{M}(X, g, \beta)$. Solutions are called *reducible* if $s = 0$, and *irreducible* otherwise.

Note that these are the blown up equations of Kronheimer and Mrowka [11]. For $s > 0$, the map $(A, s, \varphi) \to (A, s \cdot \varphi)$ would take us back to the original Seiberg–Witten equations, but the reducibles now appear as the boundary points of the moduli space rather than as singularities. This apparently modest change in perspective turns out to be crucial for the analysis of the change in the moduli space in a path of perturbations and metrics that arises in the proof of Theorem 4.1. Also note that for X a homology $S^1 \times S^3$, any $\omega \in \Omega^2_+(X, i\mathbb{R})$ normally used as a perturbation is of the form $\omega = d^+\beta$ because $H^2_+(X;\mathbb{Z}) = 0$.

Theorem 2.1. *Let g be a metric on X. For a generic β, the moduli space $\mathcal{M}(X, g, \beta)$ is a compact oriented 0-dimensional manifold with no reducibles.*

For a proof, see [16, Proposition 2.2]. Any pair (g, β) as in Theorem 2.1 will be called *regular*. Given a regular pair, denote by $\#\mathcal{M}(X, g, \beta)$ the signed count of points in the moduli space $\mathcal{M}(X, g, \beta)$. In general, this count will depend of the choice of (g, β).

To quantify this dependence, take two regular pairs (g_0, β_0) and (g_1, β_1) and connect them by a path (g_t, β_t). This path can be chosen so that it goes through at most finitely many non-regular pairs (g_t, β_t). The moduli spaces $\mathcal{M}(X, g_t, \beta_t)$ at such pairs will have reducibles which will prompt jumps in the count, see Fig. 1.

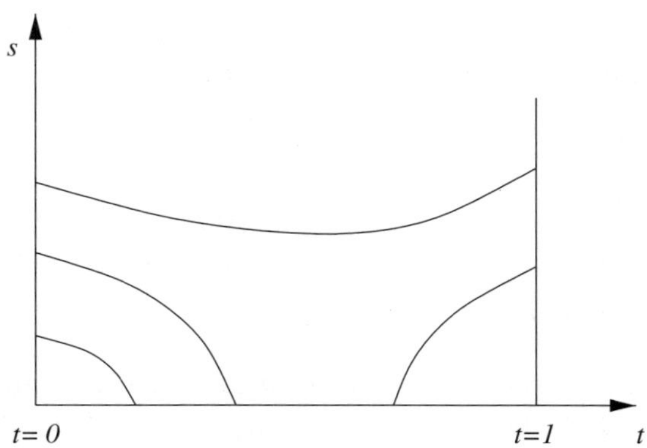

Fig. 1. The parametrized moduli space.

3. The Correction Term

Let X be a homology $S^1 \times S^3$ and $p : \tilde{X} \to X$ the infinite cyclic cover corresponding to the generator $1 \in H^1(X; \mathbb{Z})$ provided by the choice of homology orientation. Given a submanifold $Y \subset X$ dual to this generator, cut X open along Y to obtain a cobordism W from Y to itself and write

$$\tilde{X} = \bigcup_{n \in \mathbb{Z}} W_n \quad \text{with} \quad W_n = W.$$

Define

$$\tilde{X}_+ = \bigcup_{n \geq 0} W_n \quad \text{and} \quad Z_+ = Z \cup \tilde{X}_+$$

for any smooth compact spin 4-manifold Z with boundary Y. The manifold Z_+ is a manifold with periodic end in the sense of Taubes [25]. Our choices of metric g, perturbation β and spin structure on X lift to $\tilde{X}_+$ and extend to a metric g, perturbation β and spin structure on Z_+. With respect to Sobolev L^2-completions, the spin Dirac operator $D^+(Z_+, g) : L^2_1(Z_+, S^+) \to L^2(Z_+, S^-)$ is a bounded operator on the non-compact manifold Z_+. The following is proved in [16, Theorem 3.1].

Theorem 3.1. *The perturbed Dirac operator* $D^+(Z_+, g) + \beta$ *is Fredholm for any regular pair* (g, β).

Note that the operator $D^+(Z_+, g) + \beta$ is complex linear and, for any regular pair (g, β), define the *correction term*

$$w(X, g, \beta) = \operatorname{ind}_{\mathbb{C}}(D^+(Z_+, g) + \beta) + \frac{1}{8} \operatorname{sign} Z.$$

Theorem 3.2. *The correction term* $w(X, g, \beta)$ *is independent of the choices of manifolds* Z *and* $Y \subset X$*, and the way* g*,* β *and the spin structure are extended to* Z*.*

For a proof, see [16, Proposition 3.2]. Since the manifold Z_+ is not compact, the index $\operatorname{ind}_{\mathbb{C}}(D^+(Z_+, g) + \beta)$ is sensitive to changes in metric g and perturbation β. This makes the correction term $w(X, g, \beta)$ depend on the choice of regular pair (g, β). Like with the count $\#\mathcal{M}(X, g, \beta)$ in the previous section, quantifying this dependence involves tracing the jumps in $\operatorname{ind}_{\mathbb{C}}(D^+(Z_+, g_t)+\beta_t)$ along a generic path (g_t, β_t) between two regular pairs. This requires a good understanding of the index theory on manifolds with periodic ends. This theory was initiated by Taubes [25]. In our paper [16] we developed it far enough to calculate the jumps in the index and to match them with the jumps in the count $\#\mathcal{M}(X, g, \beta)$. This has led to the invariant described in the next section.

4. The Invariant

Let X be a smooth oriented 4-manifold with the integral homology of $S^1 \times S^3$ and with a fixed homology orientation $1 \in H^1(X; \mathbb{Z})$. Given a regular pair (g, β), define

$$\lambda_{\mathrm{SW}}(X) = \#\mathcal{M}(X, g, \beta) - w(X, g, \beta).$$

Theorem 4.1. *The invariant $\lambda_{\mathrm{SW}}(X)$ is well defined, that is, independent of the choice of regular pair (g, β). Moreover, the reduction of $\lambda_{\mathrm{SW}}(X)$ modulo 2 equals the Rohlin invariant of X.*

This theorem is proved in [16]. Recall that the Rohlin invariant of X is defined as $\operatorname{sign} Z/8 \pmod 2$, where Z is a smooth compact spin manifold with boundary $Y \subset X$ dual to the generator $1 \in H^1(X; \mathbb{Z})$. The proof that $\lambda_{\mathrm{SW}}(X)$ reduces to the Rohlin invariant modulo 2 requires a stronger version of Theorem 3.1 stating that the *unperturbed* Dirac operator $D^+(Z_+, g)$ is Fredholm for a generic metric g on X. This was proved in our paper [21].

The fact that $\lambda_{\mathrm{SW}}(X)$ reduces modulo 2 to the Rohlin invariant opens the door to potential applications of $\lambda_{\mathrm{SW}}(X)$ to several old problems concerning homology cobordisms. These problems are described in [20], where we attempted to address them using a different gauge theoretic invariant $\lambda_{\mathrm{FO}}(X)$ called the *Furuta–Ohta invariant.* The latter is defined using Donaldson theory as roughly one quarter times a count of irreducible instantons in the trivial $\mathrm{SU}(2)$ bundle on X. Here, X must be a $\mathbb{Z}[\mathbb{Z}]$ homology $S^1 \times S^3$ meaning that, in addition to its having the integral homology of

$S^1 \times S^3$, its infinite cyclic cover has the integral homology of S^3. This additional condition is satisfied, for instance, when a generator of $H_3(X;\mathbb{Z})$ is carried by an integral homology sphere $Y \subset X$. Later in the paper, we will present evidence for the following conjecture.

Conjecture. *Let X be a $\mathbb{Z}[\mathbb{Z}]$ homology $S^1 \times S^3$ with a fixed orientation and homology orientation. Then*

$$\lambda_{\mathrm{FO}}(X) = -\lambda_{\mathrm{SW}}(X). \tag{1}$$

Let us briefly explain how this conjecture is relevant to the study of manifolds with the homotopy type of $S^1 \times S^3$. High-dimensional surgery theory predicts a calculation of the structure set $\mathcal{S}_{\mathrm{Diff}}(S^1 \times S^3)$, consisting of homotopy equivalences

$$f : X \to S^1 \times S^3$$

with X a smooth manifold. The surgery exact sequence ([27]; see [9] for the calculations cited below) would predict that the cokernel of the map

$$N_{\mathrm{Diff}}(S^1 \times S^3 \times I, S^1 \times S^3 \times \partial I) \xrightarrow{\sigma} L_5(\mathbb{Z}[\mathbb{Z}])$$

between the normal maps on $S^1 \times S^3 \times I$ and the surgery group $L_5(\mathbb{Z}[\mathbb{Z}])$ acts freely on $\mathcal{S}_{\mathrm{Diff}}(S^1 \times S^3)$. Both of those groups are isomorphic to $\mathbb{Z}$, where the isomorphism is given by the signature of a codimension-one submanifold dual to a generator of $H^1(S^1 \times S^3)$. A computation involving Rohlin's theorem implies that the map σ is actually multiplication by 2; the upshot is that one might expect a smooth manifold homotopy equivalent to $S^1 \times S^3$, detected by the Rohlin invariant.

The conjecture above would imply that such a manifold does not exist. For it is automatic from the definition that $\lambda_{\mathrm{FO}}(X) = 0$ for any X homotopy equivalent to $S^1 \times S^3$ and so the conjecture would imply that $\lambda_{\mathrm{SW}}(X)$ vanishes as well. But then the last part of Theorem 4.1 would give the vanishing of the Rohlin invariant. We remark that although there are exotic homotopy equivalences produced by the action of $L_5(\mathbb{Z}[\pi])$ on the structure set [2, 24] there does not seem to be any known example of an exotic smooth structure on a 4-manifold produced by this action; the case of $S^1 \times S^3$ is the most basic test case.

Remark 4.2. We wish to point out that both $\lambda_{\mathrm{FO}}(X)$ and $\lambda_{\mathrm{SW}}(X)$ change sign when one reverses homology orientation of X while keeping its orientation. This is immediate for $\lambda_{\mathrm{FO}}(X)$ and $\#\mathcal{M}(X,g,\beta)$ because the above operation simply reverses the signs with which the points are counted in both instanton and monopole moduli spaces. That the correction term $w(X,g,\beta)$ reverses sign follows from Proposition 5.4 of [16] (which implies that $w(X,g,\beta)$ does not change when both orientation and homology orientation of X are reversed) and the fact that $w(X,g,\beta)$ reverses sign when one reverses orientation while keeping homology orientation. Unfortunately, we do not know what happens to either $\lambda_{\mathrm{FO}}(X)$ or $\lambda_{\mathrm{SW}}(X)$ when the orientation on X is reversed.

To conclude this section, we will mention that the invariant $\lambda_{\mathrm{SW}}(X)$ was extended in [16] to a wider class of negative-definite 4-manifolds X, those with $H_1(X;\mathbb{Z}) = \mathbb{Z}$ but not necessarily vanishing $H_2(X;\mathbb{Z})$. Such manifolds are encountered, for example, in the study of non-Kähler complex surfaces, see [19].

5. A Formula for λ_{SW}

In this section we express the invariant $\lambda_{\mathrm{SW}}(X)$ solely in terms of the manifold X, without referring to auxiliary end-periodic manifolds. This formula will follow from the general index theorem for end-periodic operators proved in [17].

Before we state the theorem we need a few definitions. Let X be a homology $S^1 \times S^3$ with a fixed orientation and a fixed homology orientation and choose a smooth function $f : X \to S^1$ so that $[df] = 1 \in H^1(X;\mathbb{Z})$. For any choice of metric g, consider the holomorphic family

$$D_z^{\pm} = D^{\pm}(X,g) - \ln z \cdot df, \quad z \in \mathbb{C}^*,$$

of twisted Dirac operators on X. All of these operators have index zero. It follows from [21] that, for a generic metric g, the operators

$D_z^\pm$ are invertible away from a discrete set $\mathcal{S} \subset \mathbb{C}^*$ and moreover, the set $\mathcal{S}$ can be chosen to be disjoint from the unit circle $|z| = 1$. In particular, all of the operators $D_z^\pm$ with $|z| = 1$ are invertible. The set $\mathcal{S}$ is called the *spectral set*; one can show that it is independent of the choice of f. The *η-invariant* is defined in [17] by the formula

$$\eta(X, g) = \frac{1}{\pi i} \int_0^\infty \oint_{|z|=1} \operatorname{Tr}(df \cdot D_z^+ e^{-tD_z^- D_z^+}) \frac{dz}{z} dt.$$

To get a better grip on $\eta(X, g)$ consider the special case of $X = S^1 \times Y$ with a product metric and spin structure so that $D^+(X, g) = \partial/\partial t - D$, where D is the self-adjoint Dirac operator on Y. Choose $f : X \to S^1$ to be the projection onto the circle factor, then the above formula will simplify to

$$\eta(S^1 \times Y, g) = \frac{1}{\sqrt{\pi}} \int_0^\infty t^{-1/2} \operatorname{Tr}(De^{-tD^2}) dt.$$

The right-hand side of this formula matches the η-invariant $\eta_{\mathrm{Dir}}(Y)$ of Atiyah, Patodi and Singer [3] hence we conclude that $\eta(S^1 \times Y, g) = \eta_{\mathrm{Dir}}(Y)$. According to [3], the η-invariant $\eta_{\mathrm{Dir}}(Y)$ also equals the value at $s = 0$ of the meromorphic extension of the function

$$\sum_{\lambda \neq 0} \operatorname{sign} \lambda \, |\lambda|^{-s}$$

defined for sufficiently large $\Re(s)$ by summing over the spectrum of D. Thus one can say that $\eta_{\mathrm{Dir}}(Y)$ measures the asymmetry of the spectrum of D. Similarly, we show in [17] that $\eta(X, g)$ measures the asymmetry of the spectral set $\mathcal{S}$ with respect to the unit circle: the integral defining $\eta(X, g)$ can be viewed as a regularization of the difference between the number of spectral points outside of the circle $|z| = 1$ and the number of those inside.

The following is a special case of the end-periodic index theorem proved in [17] for Dirac-type operators in all dimensions divisible by four.

Theorem 5.1. *Let X be a homology $S^1 \times S^3$ and Z_+ an end-periodic manifold whose end is modeled on the infinite cyclic cover of X. For* *a generic metric g on X making $D^+(Z_+, g)$ Fredholm, we have*

$$\operatorname{ind}_{\mathbb{C}} D^+(Z_+, g) = \int_Z \widehat{A} - \int_Y \omega + \int_X df \wedge \omega - \frac{1}{2}\eta(X, g),$$

where $Y \subset X$ is a submanifold dual to $1 \in H^1(X;\mathbb{Z})$ and ω is a transgressed $\widehat{A}$-class given by $d\omega = \widehat{A}$.

Under the assumption that $Y \subset X$ is chosen to have a normal neighborhood $N(Y) \subset X$ with product metric and $\operatorname{supp} f \subset N(Y)$, the index formula of this theorem can be simplified to

$$\operatorname{ind}_{\mathbb{C}} D^+(Z_+, g) = \int_Z \widehat{A} - \frac{1}{2}\eta(X, g). \tag{2}$$

In the special case of $X = S^1 \times Y$, this formula reduces to that of Atiyah, Patodi and Singer [3] for manifolds with product ends.

Corollary 5.2. *Let X be a homology $S^1 \times S^3$ with a metric g such that the pair (g, β) with $\beta = 0$ is regular, and suppose that $Y \subset X$ can be chosen to have a normal neighborhood $N(Y) \subset X$ with product metric and $\operatorname{supp} f \subset N(Y)$. Then*

$$\lambda_{\mathrm{SW}}(X) = \#\mathcal{M}(X, g, \beta) + \frac{1}{8}\eta_{\mathrm{Sign}}(Y) + \frac{1}{2}\eta(X, g),$$

where $\eta_{\mathrm{Sign}}(Y)$ is the Atiyah–Patodi–Singer η-invariant of the odd signature operator on Y.

Proof. Using formula (2) together with the signature theorem of Atiyah, Patodi and Singer [3],

$$\operatorname{sign} Z = \int_Z L - \eta_{\mathrm{Sign}}(Y),$$

and keeping in mind that $\widehat{A} = -p_1/24$ and $L = p_1/3$ in degree four, we obtain

$$\begin{aligned} w(X, g, \beta) &= \operatorname{ind}_{\mathbb{C}} D^+(Z_+, g) + \frac{1}{8}\operatorname{sign} Z \\ &= -\frac{1}{24}\int_Z p_1 - \frac{1}{2}\eta(X, g) + \frac{1}{24}\int_Z p_1 - \frac{1}{8}\eta_{\mathrm{Sign}}(Y), \end{aligned}$$

and the statement obviously follows. ∎

Finally, we mention that the requirement that (g, β) be a regular pair for $\beta = 0$ is not essential. For an arbitrary regular pair (g, β), Theorem 5.1 will hold for the perturbed Dirac operator $D^+(Z_+, g) + \beta$ once the family $D_z^\pm$ used to define $\eta(X, g)$ is replaced with the perturbed family $D_z^\pm + \beta$.

6. The Product Case

Let Σ be an oriented integral homology sphere and $X = S^1 \times \Sigma$. We will work with product metrics g on X and with perturbations β which are constant in the direction of S^1, and from now on we will suppress both in our notations. It is a well-known fact that $\mathcal{M}(S^1 \times \Sigma)$ equals $\mathcal{M}(\Sigma)$, the Seiberg–Witten moduli space in dimension 3, see for instance [11]. Together with the above discussion of the η-invariants, this implies that

$$\lambda_{\mathrm{SW}}(S^1 \times \Sigma) = \#\mathcal{M}(\Sigma) + \frac{1}{8}\eta_{\mathrm{Sign}}(\Sigma) + \frac{1}{2}\eta_{\mathrm{Dir}}(\Sigma).$$

The right-hand side of this equality was studied by Weimin Chen [6] and Yuhan Lim [14]. Lim showed that it equals, up to an overall sign, the Casson invariant $\lambda(\Sigma)$. Recall that $\lambda(\Sigma)$ is defined as one half times a signed count of the conjugacy classes of irreducible SU(2) representations of $\pi_1(\Sigma)$, see [1]. A quick calculation with the Poincaré homology sphere $\Sigma(2, 3, 5)$ fixes the overall sign to be negative, hence we conclude that

$$\lambda_{\mathrm{SW}}(S^1 \times \Sigma) = -\lambda(\Sigma).$$

This confirms our conjecture (1) in the product case since we showed in [22] that $\lambda_{\mathrm{FO}}(S^1 \times \Sigma) = \lambda(\Sigma)$. We wish to extend the above calculation of λ_{SW} and verify the conjecture in the next simplest case, that of mapping tori with finite order monodromy.

7. Mapping Tori: Free Case

Let Σ be an oriented integral homology sphere and X the mapping torus of an orientation preserving diffeomorphism $\tau : \Sigma \to \Sigma$ of order n. Suppose that $\tau : \Sigma \to \Sigma$ has no fixed points then the quotient

$\Sigma' = \Sigma/\tau$ is a homology lens space, and X can be viewed as the total space of the circle bundle $\pi : X \to \Sigma'$ whose Euler class generates $H^2(\Sigma';\mathbb{Z}) = \mathbb{Z}/n$. Let $i\eta$ be the connection form of this bundle and g' a metric on Σ'. Endow X with the metric $g = \eta^2 + \pi^* g'$. Furthermore, given a perturbation 1-form β' on Σ', lift it to the perturbation 1-form $\beta = \pi^*\beta'$ on X.

Note that the unique spin^c-structure on X is pulled back from Σ'. Therefore, according to [5, Theorem B], the pull-back map

$$\pi^* : \mathcal{M}^*(\Sigma', g', \beta') \to \mathcal{M}^*(X, g, \beta)$$

provides a bijective correspondence between the *irreducible* portions of the Seiberg–Witten moduli spaces on Σ' and on X. Moreover, for a generic choice of (g', β'), there are no reducibles on Σ' and the above correspondence is an orientation preserving diffeomorphism between compact oriented 0-dimensional manifolds. The full moduli space $\mathcal{M}(X, g, \beta)$ may in principle contain reducibles because perturbation forms $\beta = \pi^*\beta'$ as above are not dense in the space of all perturbations. That $\mathcal{M}(X, g, \beta)$ is actually free of reducibles can be verified by a Fourier analysis calculation using the observation that the infinite cyclic cover of X is isometric to a product.

The same observation tells us that $\eta(X, g) = \eta_{\text{Dir}}(\Sigma)$ hence we conclude that

$$\lambda_{\text{SW}}(X) = \sum_{\sigma'} \#\mathcal{M}(\Sigma', \sigma') + \frac{1}{2}\eta_{\text{Dir}}(\Sigma) + \frac{1}{8}\eta_{\text{Sign}}(\Sigma),$$

where we broke the moduli space $\mathcal{M}(\Sigma', g', \beta')$ into a disjoint union of the moduli spaces $\mathcal{M}(\Sigma', \sigma')$ corresponding to the n distinct spin^c-structures on Σ', and suppressed metrics and perturbations in our notations. On the other hand, consider the rational number

$$\mathbf{sw}^0(\Sigma', \sigma') = \#\mathcal{M}(\Sigma', \sigma') + \frac{1}{2}\eta_{\text{Dir}}(\Sigma', \sigma') + \frac{1}{8}\eta_{\text{Sign}}(\Sigma'),$$

where $\eta_{\text{Dir}}(\Sigma', \sigma')$ stands for the η-invariant of the spin^c Dirac operator corresponding to the spin^c structure σ'. Lim [13] showed that $\mathbf{sw}^0(\Sigma', \sigma')$ is a topological invariant, and Marcolli and Wang

[15, Theorem 1.1] later proved that[a]

$$\sum_{\sigma'} \mathbf{sw}^0(\Sigma', \sigma') = -\lambda_W(\Sigma'), \tag{3}$$

where $\lambda_W(\Sigma')$ is the Casson–Walker invariant normalized as in Lescop [12], meaning that $\lambda_W(\Sigma')$ equals $n/2$ times the Casson–Walker invariant defined in Walker [26]. Combining the last three formulas, we obtain

$$\lambda_{SW}(X) = -\lambda_W(\Sigma') + \frac{1}{8}(\eta_{\text{Sign}}(\Sigma) - n \cdot \eta_{\text{Sign}}(\Sigma'))$$
$$+ \frac{1}{2}\left(\eta_{\text{Dir}}(\Sigma) - \sum_{\sigma'} \eta_{\text{Dir}}(\Sigma', \sigma')\right).$$

We will next identify the last two terms on the right. The last term actually vanishes: since $\eta_{\text{Dir}}(\Sigma', \sigma')$ are just the η-invariants of the spin Dirac operator twisted by representations $\alpha : \pi_1\Sigma' \to U(1)$, their sum over all α clearly equals the η-invariant of the Dirac operator on Σ. Using the ρ-invariants of [4] we can write

$$\eta_{\text{Sign}}(\Sigma) - n \cdot \eta_{\text{Sign}}(\Sigma') = \sum_{\alpha} \rho_\alpha(\Sigma'),$$

where the summation extends to all the representations $\alpha : \pi_1\Sigma' \to U(1)$. To calculate $\rho_\alpha(\Sigma')$, view Σ' as the result of (n/q)-surgery along a knot k in an integral homology sphere Y, where q is relatively prime to n and $0 < q < n$.

Proposition 7.1. *Let $\alpha : \pi_1\Sigma' \to U(1)$ send the meridian of the knot k to $e^{2\pi i m/n} \in U(1)$. Then*

$$\rho_\alpha(\Sigma') = \rho_\alpha(L(n,q)) - \text{sign}^{m/n} k,$$

where $L(n,q)$ is the lens space obtained by (n/q)-surgery on the unknot in S^3, and $\text{sign}^{m/n} k$ *is the Tristram–Levine equivariant knot signature of k.*

[a] The orientation conventions in Marcolli–Wang [15] differ from ours, which accounts for the extra negative sign in our formula compared to theirs, cf. Nicolaescu [18].

Proof. For knots in S^3, this is essentially the formula of Kirk, Klassen, and Ruberman stated on p. 388 of [10] right before Theorem 4.4. We offer here a slight modification of their argument that extends that formula to knots in arbitrary homology spheres. Start as in the second paragraph on p. 385 of [10] with the manifold $U = ([0,1] \times Y) \cup H$, where H is a 2-handle attached to Y along the knot k with zero framing. Choose a Seifert surface F for k and let $\bar{F}$ be the union of F pushed slightly into $[0,1] \times Y$ and the core of the 2-handle. Remove a tubular neighborhood of $\bar{F}$ from U to obtain a manifold W with boundary. The boundary of W will consist of three components: the 3-manifold obtained from Y by 0-surgery on k, the product $\bar{F} \times S^1$, and the homology sphere Y. Let B be a handlebody of genus equal to the genus of F, and V a simply-connected smooth 4-manifold with boundary Y. Let

$$M_X = V \cup W \cup (B \times S^1).$$

By Novikov additivity, $\operatorname{sign} M_X = \operatorname{sign} V + \operatorname{sign} W$ and $\operatorname{sign}_\alpha M_X = \operatorname{sign} V + \operatorname{sign}_\alpha W$ since α restricts to a trivial representation on $\pi_1 V$. The calculation of $\operatorname{sign} W$ proceeds as before, and the contributions from V cancel in the expression $3 \operatorname{sign} N - \operatorname{sign}_\alpha N$ on p. 388 of [10]. □

Together with the surgery formula for the Casson–Walker invariant, see Walker [26, Proposition 6.2], this gives the following formula:

$$\begin{aligned}\lambda_{\mathrm{SW}}(X) = -n \cdot \lambda(Y) - \frac{1}{8} \sum_{m=0}^{n-1} \operatorname{sign}^{m/n} k - \frac{q}{2}\, \Delta_k''(1) \\ - \lambda_{\mathrm{W}}(L(n,q)) + \frac{1}{8} \sum_{\alpha} \rho_\alpha(L(n,q)),\end{aligned}$$

where $\Delta_k(t)$ is the Alexander polynomial of the knot $k \subset Y$ normalized so that $\Delta_k(1) = 1$ and $\Delta_k(t^{-1}) = \Delta_k(t)$. This should be compared with the formula for $\lambda_{\mathrm{FO}}(X)$ from our paper [22] :

$$\lambda_{\mathrm{FO}}(X) = n \cdot \lambda(Y) + \frac{1}{8} \sum_{m=0}^{n-1} \operatorname{sign}^{m/n} k + \frac{q}{2}\, \Delta_k''(1).$$

We conclude that

$$\lambda_{\mathrm{SW}}(X) + \lambda_{\mathrm{FO}}(X) = -\lambda_{\mathrm{W}}(L(n,q)) + \frac{1}{8}\sum_{\alpha} \rho_\alpha(L(n,q)),$$

where

$$\lambda_{\mathrm{W}}(L(n,q)) = \frac{1}{8}\sum_{\alpha} \rho_\alpha(L(n,q)) = -\frac{1}{8}\sum_{k=0}^{n-1} \cot\left(\frac{\pi q k}{n}\right)\cot\left(\frac{\pi k}{n}\right),$$

see [26, Proposition 6.3] for the Casson–Walker invariant and [4, Proposition 2.12] for the ρ-invariants. This leads to the desired equality $\lambda_{\mathrm{SW}}(X) = -\lambda_{\mathrm{FO}}(X)$.

8. Mapping Tori: Non-Free Case

Assume now that X is the mapping torus of a finite order diffeomorphism $\tau : \Sigma \to \Sigma$ of an oriented integral homology sphere Σ which has fixed points. The quotient space $\Sigma' = \Sigma/\tau$ is then an integral homology sphere, and the projection $\Sigma \to \Sigma'$ is an n-fold branched cover with branch set a knot $k \subset \Sigma'$. We know from [22] that

$$\lambda_{\mathrm{FO}}(X) = n \cdot \lambda(\Sigma') + \frac{1}{8}\sum_{m=0}^{n-1} \operatorname{sign}^{m/n} k, \tag{4}$$

where $\lambda(\Sigma')$ is the Casson invariant of Σ' and $\operatorname{sign}^{m/n} k$ are the Tristram–Levine equivariant knot signatures of k. As a first step towards computing $\lambda_{\mathrm{SW}}(X)$, we could use [5] to express $\mathcal{M}(X)$ in terms of the Seiberg–Witten moduli spaces on the *orbifold* Σ'. While the Seiberg–Witten theory on orbifolds has been actively studied, see for instance Chen [7], one still lacks an orbifold version of the formula (3) which was crucial for the calculation in the fixed point free case.

While the general case is still outstanding, we have been able in [23] to verify our conjecture in some special cases. To be specific, let $\Sigma = \Sigma(a_1, \ldots, a_n)$ be a Seifert fibered homology sphere oriented as a link of a Brieskorn–Hamm complete intersection singularity with real coefficients, and let X be the mapping torus of the involution on Σ induced by complex conjugation. We showed in [16] that, for a natural metric g realizing the Thurston geometry on Σ, the pair $(g, 0)$ is regular and $\mathcal{M}(X)$ is empty. Since the infinite cyclic cover of X is isometric to a product, the invariant $\lambda_{\mathrm{SW}}(X)$ equals

$$\frac{1}{2}\eta_{\mathrm{Dir}}(\Sigma) + \frac{1}{8}\eta_{\mathrm{Sign}}(\Sigma),$$

which in turn equals negative $\bar{\mu}(\Sigma)$, the $\bar{\mu}$-invariant of Neumann and Siebenmann, see [23]. The latter was identified with $\lambda_{\mathrm{FO}}(X)$ in [22] thus leading to the conclusion that $\lambda_{\mathrm{SW}}(X) = -\lambda_{\mathrm{FO}}(X)$ in this case.

Acknowledgments

This paper grew out of a joint project with Tom Mrowka; we truly appreciate his ongoing collaboration. We are also thankful to Liviu Nicolaescu and Weimin Chen for sharing their insight on the material discussed in the last two sections.

The first author was partially supported by NSF Grants 1105234 and 1065827. The second author was partially supported by NSF Grant 1065905.

References

[1] S. Akbulut and J. McCarthy, *Casson's Invariant for Oriented Homology 3-Spheres. An Exposition*, Mathematical Notes, Vol. 36. Princeton Univ. Press, Princeton, 1990.

[2] S. Akbulut, Scharlemann's manifold is standard, *Ann. Math. (2)* **149** (1999) 497–510.

[3] M. Atiyah, V. Patodi and I. Singer, Spectral asymmetry and Riemannian geometry. I, *Math. Proc. Camb. Phil. Soc.* **77** (1975) 43–69.

[4] M. Atiyah, V. Patodi and I. Singer, Spectral asymmetry and Riemannian geometry. II, *Math. Proc. Camb. Phil. Soc.* **78** (1975) 405–432.

[5] S. Baldridge, Seiberg–Witten invariants, orbifolds, and circle actions, *Trans. Amer. Math. Soc.* **355** (2003) 1669–1697.

[6] W. Chen, Casson's invariant and Seiberg–Witten gauge theory, *Turkish J. Math.* **21** (1997) 61–81.

[7] W. Chen, Seiberg–Witten invariants of 3-orbifolds and non-Kähler surfaces, *J. Gökova Geom. Topol.* **6** (2012) 1–27; arXiv:1112.0717v1.

[8] M. Furuta and H. Ohta, Differentiable structures on punctured 4-manifolds, *Topology Appl.* **51** (1993) 291–301.

[9] R. Kirby and L. Taylor, A survey of 4-manifolds through the eyes of surgery, in Surveys on Surgery Theory, Vol. 2, pp. 387–421; S. Cappell, A. Ranicki,

and J. Rosenberg, eds., Ann. of Math. Stud, Vol. 149, Princeton Univ. Press, Princeton, 2001.
[10] P. Kirk, E. Klasses and D. Ruberman, Splitting the spectral flow and the Alexander matrix, *Comment. Math. Helv.* **69** (1994) 375–416.
[11] P. Kronheimer and T. Mrowka, *Monopoles and Three-Manifolds*, Cambridge Univ. Press, Cambridge, 2007.
[12] C. Lescop, Global Surgery Formula for the Casson–Walker Invariant. Princeton Univ. Press, Princeton, 1996.
[13] Y. Lim, Seiberg–Witten invariants for 3-manifolds in the case $b_1 = 0$ or 1, *Pacific J. Math.* **195** (2000) 179–204.
[14] Y. Lim, The equivalence of Seiberg–Witten and Casson invariants for homology 3-spheres, *Math. Res. Lett.* **6** (1999) 631–643.
[15] M. Marcolli and Bai-Ling Wang, Seiberg–Witten and Casson–Walker invariants for rational homology 3-spheres, *Geom. Dedicata* **91** (2002) 45–58.
[16] T. Mrowka, D. Ruberman and N. Saveliev, Seiberg–Witten equations, end-periodic Dirac operators, and a lift of Rohlin's invariant, *J. Differential Geom.* **88** (2011) 333–377.
[17] T. Mrowka, D. Ruberman and N. Saveliev, An index theorem for end-periodic Dirac operators, arXiv:1105.0260.
[18] L. Nicolaescu, Seiberg–Witten invariants of rational homology 3-spheres, *Comm. Contemp. Math.* **6** (2004) 833–866.
[19] C. Okonek and A. Teleman, Seiberg–Witten invariants for 4-manifolds with $b_+ = 0$, in *Complex Analysis and Algebraic Geometry*, pp. 347–357, de Gruyter, Berlin, 2000.
[20] D. Ruberman and N. Saveliev, Casson-type invariants in dimension four, in *Geometry and Topology of Manifolds*, pp. 281–306, Fields Institute Communications, Vol. 47, Amer. Math. Soc., Providence, 2005.
[21] D. Ruberman and N. Saveliev, Dirac operators on manifolds with periodic ends, *J. Gökova Geom. Topol.* **1** (2007) 33–50.
[22] D. Ruberman and N. Saveliev, Rohlin's invariant and gauge theory. II. Mapping tori, *Geom. Topol.* **8** (2004) 35–76 (electronic).
[23] D. Ruberman and N. Saveliev, The $\bar{\mu}$-invariant of Seifert fibered homology spheres and the Dirac operator, *Geom. Dedicata* **154** (2011) 93–101.
[24] M. Scharlemann, Constructing strange manifolds with the dodecahedral space, *Duke Math. J.* **43** (1976) 33–40.
[25] C. Taubes, Gauge theory on asymptotically periodic 4-manifolds, *J. Differential Geom.* **25** (1987) 363–430.
[26] K. Walker, *An Extension of Casson's Invariant.* Princeton Univ. Press, 1992.
[27] C. T. C. Wall, *Surgery on Compact Manifolds*, Academic Press, New York, 1970.
[28] E. Witten, Monopoles and four-manifolds, *Math. Res. Lett.* **1** (1994) 769–796.

SERIES ON KNOTS AND EVERYTHING

Editor-in-charge: Louis H. Kauffman *(Univ. of Illinois, Chicago)*

The Series on Knots and Everything: is a book series polarized around the theory of knots. Volume 1 in the series is Louis H Kauffman's Knots and Physics.

One purpose of this series is to continue the exploration of many of the themes indicated in Volume 1. These themes reach out beyond knot theory into physics, mathematics, logic, linguistics, philosophy, biology and practical experience. All of these outreaches have relations with knot theory when knot theory is regarded as a pivot or meeting place for apparently separate ideas. Knots act as such a pivotal place. We do not fully understand why this is so. The series represents stages in the exploration of this nexus.

Details of the titles in this series to date give a picture of the enterprise.

Published:*

Vol. 1: Knots and Physics (3rd Edition)
by L. H. Kauffman

Vol. 2: How Surfaces Intersect in Space — An Introduction to Topology (2nd Edition)
by J. S. Carter

Vol. 3: Quantum Topology
edited by L. H. Kauffman & R. A. Baadhio

Vol. 4: Gauge Fields, Knots and Gravity
by J. Baez & J. P. Muniain

Vol. 5: Gems, Computers and Attractors for 3-Manifolds
by S. Lins

Vol. 6: Knots and Applications
edited by L. H. Kauffman

Vol. 7: Random Knotting and Linking
edited by K. C. Millett & D. W. Sumners

Vol. 8: Symmetric Bends: How to Join Two Lengths of Cord
by R. E. Miles

Vol. 9: Combinatorial Physics
by T. Bastin & C. W. Kilmister

Vol. 10: Nonstandard Logics and Nonstandard Metrics in Physics
by W. M. Honig

Vol. 11: History and Science of Knots
edited by J. C. Turner & P. van de Griend

*The complete list of the published volumes in the series can also be found at
http://www.worldscibooks.com/series/skae_series.shtml

Vol. 12: Relativistic Reality: A Modern View
edited by J. D. Edmonds, Jr.

Vol. 13: Entropic Spacetime Theory
by J. Armel

Vol. 14: Diamond — A Paradox Logic
by N. S. Hellerstein

Vol. 15: Lectures at KNOTS '96
by S. Suzuki

Vol. 16: Delta — A Paradox Logic
by N. S. Hellerstein

Vol. 17: Hypercomplex Iterations — Distance Estimation and Higher Dimensional Fractals
by Y. Dang, L. H. Kauffman & D. Sandin

Vol. 18: The Self-Evolving Cosmos: A Phenomenological Approach to Nature's Unity-in-Diversity
by S. M. Rosen

Vol. 19: Ideal Knots
by A. Stasiak, V. Katritch & L. H. Kauffman

Vol. 20: The Mystery of Knots — Computer Programming for Knot Tabulation
by C. N. Aneziris

Vol. 21: LINKNOT: Knot Theory by Computer
by S. Jablan & R. Sazdanovic

Vol. 22: The Mathematics of Harmony — From Euclid to Contemporary Mathematics and Computer Science
by A. Stakhov (assisted by S. Olsen)

Vol. 23: Diamond: A Paradox Logic (2nd Edition)
by N. S. Hellerstein

Vol. 24: Knots in HELLAS '98 — Proceedings of the International Conference on Knot Theory and Its Ramifications
edited by C. McA Gordon, V. F. R. Jones, L. Kauffman, S. Lambropoulou & J. H. Przytycki

Vol. 25: Connections — The Geometric Bridge between Art and Science (2nd Edition)
by J. Kappraff

Vol. 26: Functorial Knot Theory — Categories of Tangles, Coherence, Categorical Deformations, and Topological Invariants
by David N. Yetter

Vol. 27: Bit-String Physics: A Finite and Discrete Approach to Natural Philosophy
by H. Pierre Noyes; edited by J. C. van den Berg

Vol. 28: Beyond Measure: A Guided Tour Through Nature, Myth, and Number
by J. Kappraff

Vol. 29: Quantum Invariants — A Study of Knots, 3-Manifolds, and Their Sets
by T. Ohtsuki

Vol. 30: Symmetry, Ornament and Modularity
by S. V. Jablan

Vol. 31: Mindsteps to the Cosmos
by G. S. Hawkins

Vol. 32: Algebraic Invariants of Links
by J. A. Hillman

Vol. 33: Energy of Knots and Conformal Geometry
by J. O'Hara

Vol. 34: Woods Hole Mathematics — Perspectives in Mathematics and Physics
edited by N. Tongring & R. C. Penner

Vol. 35: BIOS — A Study of Creation
by H. Sabelli

Vol. 36: Physical and Numerical Models in Knot Theory
edited by J. A. Calvo et al.

Vol. 37: Geometry, Language, and Strategy
by G. H. Thomas

Vol. 38: Current Developments in Mathematical Biology
edited by K. Mahdavi, R. Culshaw & J. Boucher

Vol. 39: Topological Library
Part 1: Cobordisms and Their Applications
edited by S. P. Novikov & I. A. Taimanov

Vol. 40: Intelligence of Low Dimensional Topology 2006
edited by J. Scott Carter et al.

Vol. 41: Zero to Infinity: The Fountations of Physics
by P. Rowlands

Vol. 42: The Origin of Discrete Particles
by T. Bastin & C. Kilmister

Vol. 43: The Holographic Anthropic Multiverse
by R. L. Amoroso & E. A. Ranscher

Vol. 44: Topological Library
Part 2: Characteristic Classes and Smooth Structures on Manifolds
edited by S. P. Novikov & I. A. Taimanov

Vol. 45: Orbiting the Moons of Pluto
Complex Solutions to the Einstein, Maxwell, Schrödinger and Dirac Equations
by E. A. Rauscher & R. L. Amoroso

Vol. 46: Introductory Lectures on Knot Theory
edited by L. H. Kauffman, S. Lambropoulou, S. Jablan & J. H. Przytycki

Vol. 47: Introduction to the Anisotropic Geometrodynamics
by S. Siparov

Vol. 48: An Excursion in Diagrammatic Algebra: Turning a Sphere from Red to Blue
by J. S. Carter

Vol. 49: Hopf Algebras
by D. E. Radford

Vol. 50: Topological Library
Part 3: Spectral Sequences in Topology
edited by S. P. Novikov & I. A. Taimanov

Vol. 51: Virtual Knots: The State of the Art
by V. O. Manturov & D. P. Ilyutko

Vol. 52: Algebraic Invariants of Links (2nd Edition)
by J. Hillman

Vol. 53: Knots and Physics (4th Edition)
by L. H. Kauffman

Vol. 54: Scientific Essays in Honor of H Pierre Noyes on the Occasion of His 90th Birthday
edited by J. C. Amson & L. H. Kauffman

Vol. 56: New Ideas in Low Dimensional Topology
edited by L. H. Kauffman & V. O. Manturov